MICROSCOPIC APPROACHES TO QUANTUM LIQUIDS IN CONFINED GEOMETRIES

Series on Advances in Quantum Many-Body Theory

Edited by R. F. Bishop, C. E. Campell, J. W. Clark and S. Fantoni
(International Advisory Committee for the Series of International Conferences on Recent Progress in Many-Body Theories)

Published

Vol. 1: Proceedings of the Ninth International Conference on Recent Progress in Many-Body Theories
Edited by D. Neilson and R. F. Bishop

Vol. 3: Proceedings of the Tenth International Conference on Recent Progress in Many-Body Theories
Edited by R. F. Bishop, K. A. Gernoth, N. R. Walet and Y. Xian

Vol. 5: 150 Years of Quantum Many-Body Theory
A Festschrift in Honour of the 65th Birthdays of John W Clark, Alpo J Kallio, Manfred L Ristig and Sergio Rosati
Edited by R. F. Bishop, K. A. Gernoth and N. R. Walet

Vol. 7: Introduction to Modern Methods of Quantum Many-Body Theory and Their Applications
Edited by A. Fabrocini, S. Fantoni and E. Krotscheck

Forthcoming

Vol. 2: Microscopic Approaches to the Structure of Light Nuclei
Edited by R. F. Bishop and N. R. Walet

Vol. 6: Proceedings of the Eleventh International Conference on Recent Progress in Many-Body Theories
Edited by R. F. Bishop, T. Brandes, K. A. Gernoth, N. R. Walet and Y. Xian

Series on Advances in Quantum Many-Body Theory – Vol. 4

MICROSCOPIC APPROACHES TO QUANTUM LIQUIDS IN CONFINED GEOMETRIES

Editors

Eckhard Krotscheck
Johannes-Kepler University, Austria

Jesus Navarro
CSIC–Univeritat Valencia, Spain

World Scientific
New Jersey • London • Singapore • Hong Kong

Published by

World Scientific Publishing Co. Pte. Ltd.
P O Box 128, Farrer Road, Singapore 912805
USA office: Suite 1B, 1060 Main Street, River Edge, NJ 07661
UK office: 57 Shelton Street, Covent Garden, London WC2H 9HE

British Library Cataloguing-in-Publication Data
A catalogue record for this book is available from the British Library.

MICROSCOPIC APPROACHES TO QUANTUM LIQUIDS IN CONFINED GEOMETRIES

ISBN 981-02-4640-4

Printed in Singapore by Uto-Print

Contents

Preface

The quantum mechanics of large interacting systems and the structure and properties of materials in reduced dimensionality are emerging as important issues in condensed matter physics. Quantum liquids and solids provide a particularly well defined and controllable set of such systems. Also theory and experiments can be brought together to mutual advantage. The confinement causes a large variety of new and interesting phenomena. For example the internal structure of the liquid becomes more ordered by contact with an external matrix. This has consequential effects on the behavior of thin liquid films. Another example is free quantum-liquid droplets which provide a laboratory for the study of the interaction of atoms and molecules without the complications that arise from interactions with container walls. This is opening up a new field of the chemistry of nanosolvents at very low temperatures.

This volume presents an assembly of review articles describing the *many–body* aspects of these systems. Modern quantum many-body theory has grown to be one of the most fundamental and exciting areas of contemporary theoretical physics. Its aims are to understand and predict the emergent properties of macroscopic matter that have their origins in the underlying interactions between the elementary constituents. The variety of current approaches to the microscopic many-body problem (including density functional theory, the hypernetted chain formalism, the coupled cluster method, the correlated basis function method and quantum Monte Carlo simulation techniques) present a striking diversity of forms and formalisms, but nevertheless all have essentially the same physical content. Their differences lie in the implementation and not in substance.

Confined quantum liquids are ideal systems for the application of modern theoretical tools, because they are, on the one hand, described by a Hamiltonian that is sufficiently simple and well enough understood such that quantitative theoretical methods can be fruitfully applied; on the other hand these systems are —unlike "model systems" — actually occurring in nature and are, therefore, experimentally accessible. Thus, research is in the very satisfactory situation because its objective are systems where fruitful interactions of fundamental theory and high–precision experiments are possible. Several accurate approaches are available that let us describe these systems in a quantitative manner without modeling uncertainty and uncontrolled assumptions. Among others, dynamic situations of direct experimental relevance can be modeled with high accuracy.

The scope of this volume is to assemble a number of review articles that describe the status of front-line research in this field in a manner that makes the material accessible to the educated, but non-specialist, reader. The articles specifically focus on the *many–body* aspects of the theory of quantum liquids in confined geometry. The theoretical approaches to be discussed in their application to the subject matter are simulation methods, those semi-analytic many–body techniques that have been proven successful in the field, and phenomenological density–functional theories. Each of these methods has strength and weaknesses, and we hope that a collection of comprehensive review articles in one volume will provide sufficient material to the reader to intelligently assess the theoretical problems, and the physical predictions of the individual theories.

The collection starts with a general introduction by C. E. Campbell on the basics of helium physics, taking the "view from the top" on various theoretical methods and experimental prospects. We also include two experimental review articles that highlight front-line research on specific experimental questions such as neutron scattering off confined quantum liquids, and superfluidity in small ^{4}He and para-H_2 clusters.

The emphasis of this collection of articles is, however, quantum many–body theory, and in this sense it fits well into the present series of books which embraces proceedings of conferences and workshops, collections of lecture notes of specialized schools, as well as monographs and collection of review volumes. Our collection was specifically motivated by the striking absence of pedagogical material and even monographs describing the most important developments of the past two decades in a readable manner. We hope that it will fullfill this purpose.

Let us end this preface with a personal touch. On behalf of all contributing authors, the editors wish to express their appreciation for one of our colleagues whose insights and ideas have had a particularly profound influence on the theory of quantum fluids over the past three decades. During the completion of this book, we have realized that Chuck Campbell will celebrate his 60th birthday at the end of the year. We would like this book to serve as a tribute to his meritorious contributions to the field of quantum fluids.

E. Krotscheck
J. Navarro

CHAPTER 1

HELIUM LIQUIDS IN CONFINED GEOMETRIES

C. E. Campbell

School of Physics and Astronomy, University of Minnesota,
Minneapolis, Minnesota 55455, USA
E-mail: chuck@physics.umn.edu

1. Introduction

The objective of this Chapter is to present a brief overview of the contents of this book together with some background which might make the book of more use to those less familiar with the subject of the theories of helium fluids . It should be noted, however, that each Chapter has its own introductory section, generally with sufficient references to the literature to set the context of the work and provide more details. We also include here some definitions of quantities which are used in several chapters but not always defined there. Important examples are the reduced distribution functions and structure functions such as the dynamical structure function that are essential to the measurement and analysis of quantum fluids.

Most of this book is focused on the helium liquids in confined geometries. This includes helium adsorbed to surfaces and wedges, helium clusters and droplets , impurities in droplets, helium in tortuous geometries such as aerogel and porous solids, helium in quasi-one dimensional geometries, and cavitation phenomena in helium.

There are several reviews of inhomogeneous liquid helium cast at the introductory level that are recommended as valuable introductory material. [1–6] Similarly, there are several monographs on the helium liquids, which should serve as accessible resources and historical background for this subject. [7–13]

A variety of theoretical, simulational and experimental methods are used to study these systems, several of which are reported upon in the remaining chapters. The latter include experimental techniques such as inelastic neutron scattering and spectroscopic studies of atoms and molecules in helium droplets.

Several different theoretical and simulational methods are used to study these systems. We will first give a brief overview of the these methods, followed by some observations about the general subject of confined and inhomogeneous liquid helium.

We then elaborate further on droplets and films , briefly survey other systems of interest, and close with some more specifics about theories followed by a brief conclusion.

The theoretical methods used herein fall into two different categories: microscopic theories and phenomenological theories . The work by Saarela, Apaja and Halinen in Chapter 4, and by Apaja and Krotscheck in Chapter 5, and the simulations used by Boronat in Chapter 2 and by Huang, Kwon and Whaley in Chapter 3 are microscopic in the sense that they begin with the best representation of the many-body helium Hamiltonian available, of the form

$$H = \sum_{i=1}^{N} \left[\frac{\hbar^2}{2m} \nabla_i^2 + V_1(\mathbf{r}_i) \right] + \sum_{1 \le i < j \le N} V_2(r_{ij}) \,, \tag{1.1}$$

where N is the number of particles. The one-body potential $V_1(\mathbf{r}_i)$ is the external interaction, *e.g.* the interaction between the helium atoms and an adsorbing wall, or between the helium atoms and an impurity. $V_2(r_{ij})$ is the interaction between the helium atoms. The important physical features of $V_2(r_{ij})$ and in some cases $V_1(\mathbf{r}_i)$ are the strong, short-ranged repulsion and the slightly longer-ranged, relatively weak attraction. The repulsion is due to the fact that helium is a closed shell atom and thus empty shell electronic energy states must be virtually excited when closed shells on neighboring atoms begin to overlap. Consequently the repulsive energy scale is of order an electron volt when the distance between two atoms is comparable to twice the Bohr radius. Beyond this distance, the attraction is due to the van der Waals dipole-induced-dipole correlation. The magnitude of the attraction between helium atoms is of order 10 K, which occurs just outside the repulsive core at about 2.5 Å separation, while the other characteristic constant that sets the scale of zero point motion in ^{4}He is $\frac{\hbar^2}{2m} = 6.06$ K Å^{-2}. Because of the short-ranged repulsion, helium liquids are the most strongly correlated quantum many-body systems known. An indicator of correlation as used here is the density of particles in the neighborhood of a typical particle; thus we are referring to short-range correlations . This is best represented quantitatively by the pair density $\rho_2(\mathbf{r}, \mathbf{r}')$ or the corresponding pair distribution function $g(\mathbf{r}, \mathbf{r}')$ which are defined by:

$$\rho_2(\mathbf{r}, \mathbf{r}') = \left\langle \sum_{i \neq j}^{N} \delta(\mathbf{r} - \mathbf{r}_i)\delta(\mathbf{r}' - \mathbf{r}_j) \right\rangle \tag{1.2}$$

and

$$\rho_2(\mathbf{r}, \mathbf{r}') = \rho_1(\mathbf{r})\rho_1(\mathbf{r}')g(\mathbf{r}, \mathbf{r}') \,, \tag{1.3}$$

where $\rho_1(\mathbf{r})$ is the particle number density:

$$\rho_1(\mathbf{r}) = \left\langle \sum_{i}^{N} \delta(\mathbf{r} - \mathbf{r}_i) \right\rangle \,. \tag{1.4}$$

Note that, for a uniform density system, $g(\mathbf{r}, \mathbf{r}')$ depends only on the distance $|\mathbf{r} - \mathbf{r}'|$, and is called the radial distribution function. Moreover, the square root of $\rho_2(\mathbf{r}, \mathbf{r}')$

can be viewed essentially as the wavefunction of two helium atoms in the background of the remaining atoms. Information about these functions is obtained experimentally from X-ray scattering or frequency integrated neutron scattering . The strong, short-range repulsion between two helium atoms results in a very strong correlation hole around a given atom, which is seen in a graph of $g(\mathbf{r}, \mathbf{r}')$ (see Fig. 5 of chapter 2, or Fig. 2 of chapter 4). In this correlation hole, the density of other helium atoms vanishes within a radius of approximately 2.5 Å. With increasing distance, there is a positive correlation in the form of a local maximum corresponding to the location of helium atom nearest neighbors. This short-range structure is comparable to that which would be found in an electron gas at an r_s value greater than 30, or the correlations between protons which is expected in liquid metallic hydrogen at pressures greater than two megabars. Surprisingly, it is far stronger than the correlation between nucleons in nuclear matter.

Dealing with this strong short-range correlation structure is the challenge which confounded the application of traditional many-body theory for so many years. The problem lies simply in the fact that the Fourier transform of the repulsive part of the interaction potential energy is so very large; indeed its simpler representations (*e.g.,* the Lennard-Jones 6-12 potential) are not Fourier transformable. Although simple classical fluids, such as liquid argon, are even more strongly correlated than the helium liquids, this does not show up in the kinetic energy of the classical fluids, which is just $\frac{3}{2}Nk_BT$ at an absolute temperature T. In contrast, the kinetic energy of the helium liquids is comparable to the potential energy even at $T = 0$; the biggest contribution to the kinetic energy is proportional to the integral of the curvature of the square root of $g(r)$. This is strictly a quantum effect, due to the low mass of the helium atoms and to the small space in which a typical helium atom is "confined" by its neighboring atoms, due to the strong repulsion between the atoms. But it is this zero point energy which keeps the helium liquids from solidifying all the way down to absolute zero unless a significant external pressure is applied.

Much of the successful theory and all simulations of the helium liquids has been formulated in coordinate space where the strong short-range correlations may be dealt with directly. The theories have been wave function theories at zero temperature, and density matrix theories in coordinate representation at finite temperatures. A general discussion of the theory at zero temperature is provided in chapters 4 and 5 below and in Ref. 14.

Phenomenological theories have also been employed extensively in research on the helium liquids, beginning with the theories by Landau of the low lying excitations of liquid ^{4}He and the resulting understanding of the critical velocity of the superfluid , and the Landau-Fermi liquid theory [15,16] of ^{3}He. Pines and collaborators further developed this phenomenological approach for the bulk fluids. [17,18] More recently, the density functional method has been employed to develop phenomenological theories of inhomogeneous helium liquids. In this approach, one takes as input the experimental properties of the bulk, homogeneous systems, such as the liquid structure function , and then derives an effective, state dependent interaction potential which is then used to extrapolate from the uniform system to the inhomogeneous system. The method is discussed and used by Hernández and Navarro in Chapter 6 to explore especially inhomogeneous ^{3}He in constrained and confined geometry, a subject which is much more difficult for the microscopic theories and simulations to deal with because of the statistics and the concomitant sign problem. The application of this approach to the also very challenging problem of cavitation in quantum fluids is presented by Barranco *et al.* in Chapter 7.

The excitations and dynamics of these systems have been explored both theoretically and experimentally by absorption, emission and scattering probes. Thus the nature and energetics of excited states is the subject of theoretical work reported herein, especially as they are found in ^{4}He droplets and films . Excitations of molecules absorbed into nanodroplets are used as a tool to study the superfluidity of ^{4}He droplets. Similarly, these ^{4}He nanodroplets are found to provide a new, relatively inert matrix for studying the molecules themselves. This is discussed at length in Chapters 3, 6, and 9.

The primary probe of the elementary excitations of quantum fluids has been inelastic neutron scattering . *E.g.,* Fåk and Glyde report in Chapter 8 on inelastic neutron scattering from liquid ^{4}He in inhomogeneous and constrained liquid ^{4}He systems, including the tortuous, fractal-like geometry of aerogel and porous media, and comparing their results to bulk ^{4}He. Other dynamical probes such as Brillouin and Raman scattering that have been used to study the dynamics of bulk helium liquids are not reported here.

In inelastic neutron scattering , the neutron cross section is proportional to the dynamical structure function of the system, which may be defined by

$$S(\mathbf{q},\omega) = (\mathbf{1/N})\langle\rho_{\mathbf{q}}\delta(\mathbf{H}-\mathbf{E_0}-\hbar\omega)\rho_{-\mathbf{q}}\rangle\,, \tag{1.5}$$

where

$$\rho_{\mathbf{q}} = \sum_{j}^{N} e^{i\mathbf{q}\cdot\mathbf{r_j}}\,. \tag{1.6}$$

$\hbar\omega$ is the energy lost by the scattered neutron to the fluid, and $\hbar\mathbf{q}$ is the momentum exchanged between the neutron and the fluid. At zero temperature this dynamic structure function may be re-expressed in terms of the excited states by an appropriate insertion of a complete set of states, giving

$$S(\mathbf{q},\omega) = (1/N)\sum_{n} |\langle\Psi_n\,|\rho_{\mathbf{q}}|\,\Psi_0\rangle|^2\,\delta(E_n - E_0 - \hbar\omega)\,. \tag{1.7}$$

This is non-zero for values of $\hbar\omega$ that are excitation energies of the fluid whose wavefunction overlaps with $\rho_{\mathbf{q}}\Psi_0$, where Ψ_0 is the fluid ground state and E_0 is the ground state energy. In bulk ^{4}He, where momentum is a good quantum number, $S(q,\omega)$ has a sharp peak at $\hbar\omega = \epsilon_q$, the phonon roton elementary excitation spectrum , which is infinitely sharp at long wavelengths where $\rho_{\mathbf{q}}\Psi_0$ is an exact eigenfunction. This is also the case for ^{4}He films if they are translationally invariant and the momentum $\hbar\mathbf{q}$ is parallel to the plane of the film. However for inhomogeneous systems, including the films, there are many more excitation modes, including surface modes (such as ripplons) and layer interface modes . If the corresponding excited states have sufficient overlap with $\rho_\mathbf{q}\Psi_0$, then they will contribute to the neutron scattering . While this can produce a much more complex cross section than the bulk fluid, careful data analysis and theoretical modeling has extracted very useful information about many of these modes. [19–23]

2. General Observations on Confined and Inhomogeneous Liquid Helium

In the first instance, confinement and inhomogeneities are annoyances to be avoided when trying to understand the magnificent properties of the bulk helium liquids. The primary way to dispense with them is to do theories and simulations for infinite, uniform density systems. This is accomplished most easily by using periodic boundary conditions, which assures that even a finite size fluid has a constant density. Simulations are only applicable to a small number of particles compared to Avogadro's number, but a systematic approach where the volume and number of particles is increased while the number density (number of particles per unit volume) remains uniform and fixed reveals that most of the bulk properties are already obtained using periodic boundary conditions with a few hundred or a few thousand particles, easily within the capabilities of computers for many years now. Theoretical approaches can take this thermodynamic limit to infinity without difficulty.

However there is already an interesting problem when this is implemented for liquids (as opposed to gases) having to do with the fact that a liquid is a self-bound system . Consequently, if one could do an exact simulation or theory of a liquid, one should find that there is a minimum density below which the *uniform density* liquid cannot exist. Certainly the liquid is thermodynamically unstable below the density where its pressure is zero at temperature absolute zero. At finite temperatures, the liquid coexists with its own vapor at saturated vapor pressure. To be more precise, this density is the thermodynamic equilibrium density where the liquid is in equilibrium with its vapor. However, theoretically and, with great care, experimentally one can actually go to a slightly lower density, where the liquid has negative pressure but is mechanically stable, until one reaches a density where it is mechanically unstable; this is the so-called spinodal density , which is discussed further in Chapter 7. The corresponding spinodal pressure near absolute zero is -9 atm for ^{4}He and -3 atm for ^{3}He. This region between the equilibrium density and the spinodal density is where the phenomena of cavitation may occur [24] which is the subject of Chapter 7. At very low temperatures, cavitation in helium fluids is dominated by quantum tunneling. At the spinodal density, the compressibility diverges, and the sound velocity goes to zero. This would be the signal of a soft mode (continuous) phase transition if it were found within the thermodynamically stable phase. [25]

More generally, from the point of view of the three-dimensional world, the bulk fluids are inhomogeneous systems in the sense that the density is not translationally invariant because the fluid must have free surfaces and/or interfaces with a containing vessel. Of course any finite liquid must have surfaces, and thus cannot be translationally invariant. However a sufficiently large system can often times be described by the dominant "bulk" properties and the quantitatively less important surface and edge properties. *E.g.*, the free energies of the system may be usefully written as a sum of terms, the first of which is proportional to the volume of the system, the second proportional to the surface area, the third proportional to edges if edge is a meaningful concept:

$$F = f_0 V + f_1 S + f_2 L \,, \tag{2.1}$$

where F is, *e.g.*, the Helmholtz free energy at finite temperature or the ground state energy at zero temperature, V is the volume of the system, S is its surface area, and L would be the edge length. The latter would have more meaning if the system were actually confined to two dimensions, where there would be no volume term. In that case, the "surface" of a finite system would actually be an edge. Likewise an edge would describe the interface between two surfaces of different curvature in a three dimensional system. Of similar importance would be the rate of change of the curvature of a surface. Presumably edge curvature would be of even less significance in a three-dimensional system.

Under further study and experimentation, inhomogeneities associated with surfaces, interfaces, and other non-translationally symmetric features of real systems are the source of rich physics, some of which is described in this volume.

3. Droplets

One can idealize the bulk situation without the artifice of periodic boundary conditions by imagining that a large number of helium atoms are brought together in a vacuum in outer space. In the absence of the gravitational or other external forces, the lowest energy state would certainly be a liquid sphere. As more atoms are brought together, the volume of the sphere increases, as does its surface area. Thus one would expect that, as the volume increases, the factors f_0 and f_1 would be independent of S and V. Then f_0 would be the bulk free energy per unit volume and f_1 would be the surface energy per unit area, or the surface tension. The idealized situation that we described would be realized if f_0 is negative, so that the atoms are bound together, and f_1 is positive, so that the shape of minimum energy is a sphere in the absence of other forces, as is the case in the helium liquids.

We have ignored one important point: at finite temperature, some atoms will escape because their thermal distribution includes high kinetic energy states. A theorist can deal with this in several ways: either confine the system to a "box" with hard walls, in which case there will develop a vapor pressure for the confined system if the volume is larger than the equilibrium sphere (though the spherical shape will be influenced by the shape of the containing vessel, and the center of mass will not be instantaneously localizable), or work at absolute zero, where the vapor pressure is zero and thus the atoms do not leak away into the vacuum. In fact, helium nanodroplets produced in gas expansions appear to reach a quasi-equilibrium state in the absence of a vapor pressure at an internal temperature of 0.37 K for ^{4}He and 0.15 K for ^{3}He (see Chapter 9).

These idealizations need to be made more realistic in order to deal with experiments. First, in the case of confinement, real walls of the confinement vessel will not be infinitely hard. In fact, all walls have at least a van der Waals attraction for the helium atoms, and in almost every case, the lowest energy state will have a film of helium on the walls. The notable experimental exception to the latter is a wall coated with cesium, because (as discussed in Chapter 5) a Cs wall is the only known material which helium does not "wet" at low temperatures. In fact, most other wall materials will attract enough helium to make one or more layers of helium as a film on the wall. These films have been the subject of considerable study throughout the history of research on helium, including research reported later in this volume. This is discussed further below.

The notion that one may do useful theory at zero temperature is more realistic than might first appear, since it is possible to perform experiments at very low temperatures. In fact, the helium liquids do not liquefy until near absolute zero (the critical point of bulk ^{4}He being 5.2 K , and ^{3}He being 3.3 K, and neither solidify until 25-30 atm of pressure), and there is apparently no bulk phase transition at a critical temperature of absolute zero in the bulk system (the more complicated situation with 2D and quasi 2D systems will be discussed below). Consequently, the properties of helium at finite temperatures near absolute zero are, to first approximation, described well by the ground state, except of course their quantitative temperature dependence. Moreover, the temperature dependence of most properties at low temperatures, particularly in the case of liquid ^{4}He, can be understood in terms of the low energy excited states through a non-interacting quasi-particle model, as pointed out by Landau in the earliest successful theory of superfluid liquid ^{4}He. [26]

The idealization that one works in outer space to avoid gravitational effects (or in the microgravity environment provided by orbiting space shuttles and space stations) is really only necessary when the gravitational effects are significant. But bulk properties such as the behavior of the fluid near a phase transition are sensitive to the uniformity of the system. In particular, the theory of critical points and other critical behavior is sufficiently advanced that fluid experiments, including the superfluid transition in liquid ^{4}He, have been carried out in space shuttle microgravity environments and other experiments have been scheduled in order to test these theories. However, the gravitational effect on very small droplets is not important because the gravitationally induced pressure gradient is very small. Thus one would expect that clusters of less than 10^6 helium atoms would not be significantly affected by gravity except for their center of mass motion. *I.e.*, such clusters would to an excellent approximation be spherical in shape. Various estimates of the thickness of the surface, defined roughly by the distance that it takes to go from 90% to 10% of the internal (bulk) density, as 5 to 7 Å, and thus clusters with radii significantly larger than that should have an interior region which is bulk-like.

It should be noted that clusters of ^{4}He atoms of size two and three atoms have been made recently, [27–29] but they have a very tiny binding energy, as predicted. [32] In contrast, calculations predict that small ^{3}He clusters of less than about 30 atoms are unstable. [30,31] In Chapter 9, Toennies reviews experiments that have been done on clusters of ^{4}He atoms and clusters of ^{3}He atoms in the range $10^3 - 10^4$ atoms, which are large enough to be called nanodroplets . Microscopic theoretical studies of ^{4}He clusters of a few dozen to $O(10^3)$ atoms are reported in Chapters 3 and 6, and the density functional theory is applied in Chapter 5, with emphasis on ^{3}He. Many interesting questions can be addressed experimentally in clusters of these sizes. Clearly, surface properties should play a very important role in smaller clusters, while a more or less continuous transition to bulk properties should be seen with increasing number of particles. This is particularly true for helium droplets , which seem to remain liquid-like down to clusters of tens of atoms. By way of contrast, it should be noted that other noble gas clusters such as argon show quite a different behavior. [33,34] At clusters of less than a few hundred atoms, the low temperature structures have icosahedral symmetry, which is incompatible with an infinite crystalline system (although it is an important structure in the more complex systems which form bulk quasi-crystals.) However, there is a transition from icosahedral structure to close-packed structure when the cluster has a larger number of atoms, consistent with the structure of bulk noble gas crystals at low temperatures. Thus these are more appropriately referred to as solid clusters instead of droplets.

One would expect helium clusters to exhibit liquid behavior down to quite small sizes, though it is an interesting question of what one means by a liquid at the smallest sizes. Qualitatively the question may addressed in terms of the number of nearest neighbors around a given atom, the sharpness of the distribution of atoms on neighboring sites, and the range of crystalline-like order, as seen in X-ray scattering. The same question arises in solid clusters like Ar. X-ray structure measurements of the low temperature argon clusters show very clear evidence of solid structure and the transition between icosahedral and close-packed structure with increasing number of particles in the cluster. Dynamical measurements may also provide qualitative answers. *E.g.*, a solid cluster should have bulk-like collective modes that include transverse modes resembling those found in a bulk solid. These features should be missing in liquid droplets, but the droplets should support bulk-like longitudinal modes similar to phonons, and surface modes seen in liquids, such as surface waves.

In the case of ^{4}He, even more interesting is the question of whether the droplets exhibit superfluid properties, which raises the question of what it really means to have superfluidity in a droplet. This is closely related to the question of whether ^{4}He droplets are Bose-Einstein condensed , though superfluidity is a dynamic property which is properly addressed in terms of flow and viscosity. The lambda transition of liquid ^{4}He is to a superfluid state which also exhibits other anomalous thermodynamic and dynamic properties, most notably the lambda signature in the specific heat, anomalous thermal conductivity, and the existence of second sound modes. It is generally acknowledged that this is a transition to a low temperature phase which possesses off-diagonal long range order , which is a property shared with the non-interacting Bose gas in its Bose condensed phase. However, the Bose gas has a zero critical velocity and does not have quantized vorticity ; superfluidity and quantized vorticity both require the particles to have at least a repulsive interaction, though the repulsion can be incredibly weak as long as an attraction is insufficient to solidify the system.

Each of these questions can be addressed qualitatively and quantitatively. *E.g.*, formally speaking one knows that one has sharp phase transitions only for infinite systems. Thus one can theoretically discuss infinite three-dimensional systems, and infinite two-dimensional and infinite one-dimensional systems, and in principle precisely describe the phase diagrams of these systems with sharp boundaries separating the different phases in a phase diagram of the thermodynamic variables. On the other hand, one expects these sharp boundaries to at least be smeared for finite sized systems. But no real system is infinite, though large systems show clearly defined phase transitions, with the measurable effects of finite size only appearing for relatively small systems. How small can a system be and still clearly have some property such as superfluidity or Bose-Einstein condensation ?

The relationship between superfluidity and Bose fluids was explained most simply by Landau [26] and Feynman [35,36] in terms of the nature of the low energy excited states of a Bose fluid. Landau postulated and Feynman demonstrated that the low temperature properties of liquid ^{4}He and other Bose fluids can be explained by the fact that the lowest energy excited states are long wavelength phonons with a linear dispersion relation: $\epsilon_k = \hbar k c$ as $k \to 0$, where k is the wavenumber of the excitation ($2\pi/\lambda$ where λ is the wavelength) and c is the velocity of (zero) sound. Superfluidity occurs because the fluid will not lose energy to a surface or impurity that it is flowing by until it reaches a critical flow velocity where it is energetically favorable to create an excitation, leading to a transfer of momentum to the walls. A condition for this critical velocity to be non-zero is that the slope of the excitation energy as a function of wavenumber be non-zero, or better yet, as in the case of fermion superfluids , that there be a gap in the excitation energy at infinite wavelength. If there are lower lying excitations , as in the case of the non-interacting Fermi gas and Fermi liquids with their particle-hole continuum, superfluidity cannot exist. In the non-interacting Bose gas , the phonon disappears, replaced by the free particle spectrum $\frac{\hbar^2}{2m}k^2$, and thus the critical velocity is zero; the ideal Bose gas is Bose condensed, but not a superfluid . However the weakest of repulsive interactions is sufficient to change the dispersion relation to linear at the origin, and thus the critical velocity will be finite (though it must be said that it is possible that some kind of higher energy mode with a small ratio of energy to wavenumber actually determines the critical velocity, as appears to be the case in liquid ^{4}He at low pressures).

Qualitatively one can argue that the same considerations can be used to answer the question of superfluidity in a finite system such as a ^{4}He droplet . Of course the excited states of such finite systems are discrete, and thus one does not have the convenience of the dependence of energy on the continuum wavenumber to aid the discussion. But the real physics has to do with whether one can set up a condition where the fluid flows by a wall or impurity without loss of energy. This has been realized beautifully in the case of doped nanodroplets , as discussed in Chapters 3, 6 and 9. There it is seen that a molecule added to the droplet will reside in most cases at the center of the droplet. Then one may excite rotational modes of the molecule in the droplet and observe their spectrum. If there are many low lying modes within the helium that couple to the molecular rotational modes, this shows up as a broadening of the rotational spectrum, corresponding to a damping of the rotation by these low lying modes. This is clearly seen in doped ^{3}He nanodroplets. On the other hand, the rotational modes in the case of doped ^{4}He nanodroplets are very sharp, which is consistent with the notion of the absence of viscosity between the liquid ^{4}He and the molecule.

One would expect to see a rapid onset of damping (*i.e.,* a spectral broadening) when sufficiently high rotational levels are excited, corresponding to a critical velocity , perhaps better described as a critical angular momentum of the rotational state. This puts one in mind of another critical rotation rate in superfluid ^{4}He, namely the rotation rate of the fluid at which a macroscopic vortex appears in the superfluid. This occurs when the circulation is quantized in units of h/m, which for a cylinder of ^{4}He rotating about its axis corresponds to quantization of the angular momentum of the fluid in units of $N\hbar$ where N is the number of helium atoms in the system. At lower total angular momentum the fluid does not recognize the rotation of the walls of the "bucket" in which it resides. There are arguments against a stable vortex state in a droplet , but the presence of an anisotropic impurity in the center of the droplet may serve to stabilize a vortex state. We are unaware of whether it is feasible or whether there have been any attempts to produce such high rotational states in the nanodroplets .

One may ask what the relationship is between superfluidity and Bose condensation in such small geometries. This is closely related to the theoretical question of the determination of the superfluid density in systems of finite dimensions such as nanodroplets.

It is generally, though not unanimously, accepted that superfluidity and Bose condensation are inseparable features of liquid ^{4}He. We first emphasize an important quantitative difference. At absolute zero, the superfluid density is identical to the total density of a bulk system, while the Bose condensate density is of the order of 10% of the fluid density. To understand this, one should recall that, in a bulk system of uniform density, the condensate fraction n_0, which is the ratio of the condensate density to the fluid density ρ/ρ_0, is the fraction of particles which reside in the zero momentum single particle state:

$$n_0 = \langle a_0^\dagger a_0 \rangle / N\,, \tag{3.1}$$

where $a_0^\dagger$ and a_0 are the creation and annihilation operators for the zero momentum single-body state. Because the helium atoms interact so strongly, only a small fraction of the particles reside in this zero momentum state. However, the system is Bose condensed when this fraction is finite, *i.e.*, when $N_0 = \langle a_0^\dagger a_0 \rangle$ is macroscopic. This notion must be refined when one is dealing with an inhomogeneous system. If the inhomogeneity of the system is due to an external potential, one may expect to replace the zero momentum state by the one-particle ground state of the external potential. However, in the case of a droplet , there is no external potential. The formulation appropriate for inhomogeneous systems, including self-bound systems, was implicit in the paper by Oliver Penrose which first introduced the concept of off-diagonal-long-range-order (ODLRO) which further elucidates the relationship between Bose condensation and superfluidity . [37–39] ODLRO in Bose systems is seen in the long-range behavior of the one-body density matrix $\Gamma_1(\mathbf{r}, \mathbf{r}')$:

$$\Gamma_1(\mathbf{r}, \mathbf{r}') = \langle \psi^\dagger(\mathbf{r}) \psi(\mathbf{r}') \rangle\,, \tag{3.2}$$

where $\psi^\dagger(\mathbf{r})$ is the particle creation operator at position $\mathbf{r}$ and $\psi(\mathbf{r}')$ is the particle annihilation operator at position $\mathbf{r}'$. Note that when $\mathbf{r} = \mathbf{r}'$, then Γ_1 is just the density $\rho_1(\mathbf{r})$. But it is the behavior of Γ_1 in the opposite limit, *i.e.*, the *off – diagonal* limit defined by $|\mathbf{r} - \mathbf{r}'| \to \infty$, that one finds the defining relation of the order parameter or macroscopic wavefunction for the superfluid :

$$\lim_{|\mathbf{r}-\mathbf{r}'| \to \infty} \Gamma_1(\mathbf{r}, \mathbf{r}') = N_0 \Phi_0(\mathbf{r}') \Phi_0^*(\mathbf{r}')\,, \tag{3.3}$$

where the macroscopic wave function Φ_0 is normalized. If there is a macroscopic Bose condensate , then N_0 is macroscopic and is the correct generalization of the condensate number discussed above, and the macroscopic order parameter $N_0^{1/2}\Phi_0(\mathbf{r})$ is of order unity. If the phase of this order parameter is position dependent, then there is a superfluid flow with velocity proportional to the gradient of the phase.

An alternative approach that is well-suited to path integral Monte Carlo simulations is clearly discussed in Chapter 3. It is straightforward to see that the non-zero off diagonal limit of the one-body density matrix is sufficient to satisfy the projected area criterion used there.

When this analysis is applied to finite systems, there appears the usual problem of the meaning of an infinite limit in a finite system. This is solved approximately by noting that infinite may be taken to mean at distances large compared to the length over which ODLRO sets in, which in the case of ^{4}He is less than four angstroms. Consequently one can find unambiguous indications of ODLRO in droplets , as first shown by Lewart *et al.* using variational Monte Carlo simulations [40] and more recently by Siu Chin. [41] This was shown earlier by Krotscheck in the case of thin ^{4}He films , [42] where it was shown how to define a local condensate fraction , and observed that the condensate fraction approaches 100% as the density approaches zero in the surface of the film, a feature that is also seen in the droplets.

4. Films

One faces a similar set of questions and challenges with adsorbed films , which we discuss briefly here. Considerable attention is given to this subject by Apaja and Krotscheck in Chapter 5.

Historically, the first observation of helium films was by Onnes, who was the first to liquefy helium four and found that, at the lowest accessible temperatures, the liquid was accompanied by a film on the walls of the container extending above the surface level of the fluid, and that these films actually move up the walls and out of the containing vessel and, if it is open, moves to the outside walls and down until the level of bulk fluid on the outside of the container reaches the same level as the fluid inside, or the container is emptied. This remarkable effect was noticed even before the superfluid property–flow without dissipation–was discovered.

There has been extensive research on these saturated films. Saturation means that they are in quasi-equilibrium with the saturated vapor at the saturated vapor pressure.

Subsequent research focused on unsaturated films . This was encouraged by the interesting question of how physics might depend on the dimensionality of the space. Thus finding a surface that is very smooth to a helium atom but holds the adsorbed particles tightly to the surface was an important step in producing a Flatland environment, *i.e.* an approximately two-dimensional world. In that case one might expect that a very low coverage of atoms on the adsorbing plane would create a situation where the atoms are in a single layer. These are the so-called submonolayer films, *i.e.,* single-layer films at a density lower than the maximum capacity of the first layer. Thus the submonolayer regime is where the average two-dimensional density is less than the density where the second layer begins to form. This was first seen for quantum fluids, namely ^{4}He, when Bretz and Dash adsorbed ^{4}He onto the basal plane of graphite. [43] In that and subsequent work, [44] a phase diagram was mapped out thermodynamically in this submonolayer regime, with a quasi-2D quantum liquid phase at lowest densities, in equilibrium with a quasi-2D (and 3D) vapor, and a higher density quasi-2D solid phase. However the structure of the substrate showed through in a spectacular fashion when the 2D density of the system matched well to a small-index superlattice of the substrate. Particularly noticeable and useful was the regime close to a ^{4}He number density equal to the one third of the possible adsorption sites on the basal plane structure (centers of the carbon hexagons), corresponding to a triangular superlattice of the substrate. The strong singularity in the heat capacity, appearing at first like a 2D Ising model, was later shown to be in excellent agreement with a three-state Potts model, where the three states correspond to the occupation of sites on the three equivalent superlattices corresponding to this number density.

As to the rest of the phase diagram in this submonolayer region, there are quantitative differences between purely two-dimensional systems and a real system in which there is some extension of the wavefunction into the third dimension. In particular, this somewhat reduces the effect of the repulsion, which results in a larger binding energy per particle within the submonolayer, and increases the equilibrium density somewhat. The biggest effect, however, is that it results in a finite capacity of the first layer. When the two-dimensional density gets high enough, it becomes energetically more favorable to add the next atom to the second layer. The first layer then becomes a quasi-two dimensional solid. Depending upon how strong the adsorbing potential is, this process may repeat itself for a number of layers before one sees a transition to a three-dimensional system. As pointed out in Chapter 5, these layering transitions are well-established in both the microscopic theory [45,46] and in experiments. [44]

The first few layers on most substrates are at relatively high density and are well-approximated as solid layers with no significant exchange between the helium atoms in succeeding liquid layers of the unsaturated films .

In addition to the structure and phase diagram, questions such as the nature of the excited states , results from neutron scattering , the existence and/or nature of superfluidity are important in these systems just as they are in the droplets . Remarkably, there are excellent neutron scattering results for unsaturated helium films, as discussed in Chapter 5 and Refs. 19, 20. The measured $S(\mathbf{q},\omega)$ is very complex, but most of the features can be understood in terms of the various collective modes that have been determined by theory, including surface and interface modes.

As to the question of superfluidity, it is well known that, strictly speak, the Bogoliubov inequality precludes Bose-Einstein condensation in two dimensional systems. Nevertheless superfluidity has been measured in ^{4}He films. The well known solution to this dilemma is the existence of quasi off-diagonal-long-range-order in two-dimensional quantum fluids at finite temperature, as first pointed out by Kosterlitz and Thouless. [47] The difference with three-dimensional systems is that the one-body density matrix goes to zero algebraically in the off diagonal limit in the quasi-two dimensional systems as opposed to going to a finite value in the Bose condensed three-dimensional fluid. This algebraic decay actually leads to a Bose condensate that is sub-macroscopic but supra-microscopic, by which is meant that N_0 is proportional to N^γ where γ is a temperature dependent exponent which lies between zero and one, going to one as the temperature goes to zero. This is sufficient to support the macroscopic wave function $\Phi_0(\mathbf{r})$ which satisfies a Ginzburg-Landau wave equation that is necessary to understand superfluidity . It was shown by Reatto and Chester that the thermal population of phonons in a two-dimensional system is sufficient to produce this algebraic decay of the one-body density matrix. [48]

5. Other Systems of Recent Interest

As we have mentioned, there are other systems where ^{4}He is studied in confined and reduced geometries. Closely related to adsorbed films is adsorption (or absorption) in hectorite , which provides wedge shaped regions in which very interesting capillary condensation effects are reported in by Apaja and Krotscheck in Chapter 5.

There has been considerable work done on the helium liquids in aerogel of various different sizes, discussed in Chapter 8. Also discussed in that chapter is work on porous media like vycor, which has perhaps been studied for a longer period than any other such medium. Of considerable recent interest are quasi-one dimensional systems, which can be realized in carbon nanotubes .

6. Theories

The most successful microscopic theory of the ground state, low excited states , and dynamics of liquid ^{4}He is presented in some detail by Apaja, Saarela and Halinen in Chapter 4 and employed by Apaja and Krotscheck in Chapter 5 in the form adapted for application to inhomogeneous and confined systems. This approach began as a wavefunction theory formulated to deal with the strong short-range correlations, and then was advanced to a more general theory by using functional variational methods to improve upon early version of the theory and extend it to dynamics.

Inhomogeneities were first addressed in many-body theory by using a mean field, which can be formulated easily in coordinate space and using variational theory. The mean field theory of the boson ground state (whether or not formulated in coordinate space) is equivalent to putting every particle into the same single-body state, which may then be chosen variationally:

$$\Psi_0 = \prod f(\mathbf{r}_i) = \prod e^{\frac{1}{2}u_1(\mathbf{r}_i)}\,, \tag{6.1}$$

where the second equality makes use of the fact that the ground state wave function has a constant phase, which can be chosen to be real and positive, and thus can be written in this form with u_1 being a real function. Then f (or u_1) is chosen by minimizing the expectation value of the Hamiltonian (maintaining the normalization constraint) with respect to f or, equivalently, u_1:

$$\frac{\delta}{\delta u_1(\mathbf{r})}\frac{\langle\Psi_0|H|\Psi_0\rangle}{\langle\Psi_0|\Psi_0\rangle} = 0\,. \tag{6.2}$$

This Euler-Lagrange equation for the mean field wave function is the Gross-Pitaevskii equation , first applied to study the single vortex ground state of a rotating nearly ideal Bose fluid (in which case $u_1(\mathbf{r}_i)$ must contain a term in the azimuthal angle, ϕ_i). The potential energy expectation value in this mean field ground state already illustrates the difficulties of dealing with the strong repulsion in liquid ^{4}He:

$$\sum_{i<j}\langle\Psi_0|V_2(r_{ij})|\Psi_0\rangle = \frac{N(N-1)}{2J}\int e^{u_1(\mathbf{r}_1)+u_1(\mathbf{r}_2)}V_2(r_{12})d\mathbf{r}_1 d\mathbf{r}_2\,, \tag{6.3}$$

where J is a normalization integral. (This should have just the real part of u_1 if u_1 is complex, as in the vortex case.) This integral will be of the order of lowest atomic excitation energy of a helium atom, — in temperature units, $O(10^4\ \mathrm{K})$ — compared to the experimental energy scale of $O(10\ \mathrm{K})$. Thus this matrix element is effectively infinite, as will be all matrix elements between non-interacting states if they don't vanish by momentum conservation. The problem is, of course, that the mean field ground state and all non-interacting eigenstates contain no correlations that would keep the atoms apart to avoid this large repulsion. The method which was developed in many-body field theory to solve this problem, *i.e.*, a massive resummation in the form of a t-matrix, worked well for nucleon matter, but has failed when applied to liquid ^{4}He and has worked only poorly for liquid ^{3}He.

It was pointed out by Robert Jastrow [49] that the simplest way to deal with this problem is to introduce two-body correlations via a two-body function for each pair of particles which can be chosen to be very small when the potential energy is very large, *i.e.*, when the two particles are within 2.5 Å of one another in the case of helium atoms. In the case of the boson ground state, this two-body factor can also be written as an exponentiated real function. This then gives a trial ground state wave function of the form

$$\Psi_0(\mathbf{r}_1,\ldots,\mathbf{r}_N) = \prod_{i}^{N} e^{\frac{1}{2}u_1(\mathbf{r}_i)} \prod_{i<j}^{N} e^{\frac{1}{2}u_2(\mathbf{r}_i,\mathbf{r}_j)}\,. \tag{6.4}$$

If this wavefunction is used to describe the uniform density Bose fluid, u_1 vanishes and u_2 depends only on the distance between its two arguments, r_{ij}. For an inhomogeneous fluid it turns out to be important that u_1 does not vanish, and u_2 depends on the particle positions separately. Indeed, it is actually easier and in a certain sense self-consistent to keep this full dependence on the coordinates.

The obvious way to proceed is to choose the function u_2 to minimize the expectation of the Hamiltonian in Ψ_0. Parameterized forms for u_2 were first employed in the 1960's. However, the short-range structure of $u_2(r)$ describes the two-body wavefunction for a pair of helium atoms upon close approach, and thus can be well approximated by the WKB form of the two-body wavefunction for r less than approximately 2.5 Å. However it was known from early on that this choice of u_2 does not correctly describe the long-wavelength structure of the wave-function. Instead, the large r behavior of u_2 can be shown to be A/r^2, where A depends only upon the sound velocity of the fluid; this structure exactly accounts for the zero-point motion of the long wavelength phonons which are the low-lying excitations of the system. It contributes very little to the energy, but it does significantly affect the long-range structure and the long-wavelength properties of the elementary excitation spectrum, as we shall see below. It is, then, only necessary to interpolate sensibly between these two limits to obtain physically realistic results for most ground-state properties of the fluid, particularly the ground-state energy, X-ray structure function (liquid structure function), and condensate fraction .

Alternatively, one may determine this function u_2 by following the lead of mean field theory to derive and solve its Euler-Lagrange equation if it is practical, as it turns out to be. [50] In the case of inhomogeneous systems, this two-body Euler-Lagrange equation

$$\frac{\delta \langle H \rangle}{\delta u_2(\mathbf{r}, \mathbf{r}')} = 0\,. \tag{6.5}$$

At the same time one must satisfy the one-body Euler-Lagrange equation:

$$\frac{\delta \langle H \rangle}{\delta u_1(\mathbf{r})} = 0\,. \tag{6.6}$$

These two equations may be rewritten in several forms, each exhibiting a different aspect of the physics. Obviously the one-body equation leads to an effective Hartree, or Gross-Pitaevskii equation, with an effective one-body interaction that includes the effects of correlations induced by u_2. Similarly, these two equations may be manipulated into a form that appears to be a two-body wave equation for the relative effective wavefunction of the particles, subject to the constraint that the two-body correlations die away at large distances, as is expected for a liquid. The solution produces the correct long and short range structure mentioned above. These equations are discussed by Apaja and Krotscheck in Chapter 5.

The original motivation for this approach in the uniform system was actually the structure of low-lying excited states , where it was found that there are hybridizations between Feynman phonons which may be eliminated by requiring the ground state wave function to satisfy the Euler-Lagrange equation for u_2. As one studies more complicated systems, such as inhomogeneous systems, there is a further motivation to remove any bias which must appear when u_2 is parameterized instead of being the solution of its Euler-Lagrange equation. A rewarding consequence of this approach is that there are no solutions to the Euler-Lagrange equation at spinodal instability points (spinodal *lines* at finite temperature); examples include the spinodal limit for metastable expansion of a liquid to the negative pressure regime, and the phase-separation metastability line in mixtures. This coincides with the softening of the relevant excitation spectrum at those lines, and the divergence of the relevant susceptibility (as in critical opalescence). This feature is an example of the fact that the optimized Jastrow function is self-consistent with the Feynman theory of the low-lying excited states, as we shall discuss further below.

The relation between these Euler-Lagrange equations and the dynamics is even more useful in the case of inhomogeneous systems, as Krotscheck, Saarela and their collaborators have amply shown, and is discussed in Chapters 4 and 5 of this book. In the most useful formulation, discussed in Chapter 4, time dependence is put into u_1 and u_2, and the action is minimized in the presence of a time dependent external potential. This permits the formulation of linear response theory for the dynamical structure of the system. It is noteworthy that there is a self-consistency condition that emerges which requires the ground-state two-body function u_2 to satisfy its Euler-Lagrange equation to obtain the correct behavior of collective modes at the

Feynman level. Similarly, improvements in the collective modes (*e.g.*, the phonon-roton spectrum) are obtained by including the time dependence in u_2 in the linear response theory, but again self-consistency requires that an optimized higher order correlation function, $u_3(\mathbf{r}_i, \mathbf{r}_j, \mathbf{r}_k)$ in the log of the ground-state wavefunction.

Of course one may expect further improvement in the ground state and dynamics by including higher order correlation functions in the ground state, but little is gained qualitatively or quantitatively by taking this path. Other methods for going beyond the Jastrow level in the theory of the ground state and the related theory of excitations has been tried, most notably the state dependent Jastrow approach of Pandharipande, and the shadow wavefunction approach by Reatto *et al*, with a quantitatively similar level of improvement over the Jastrow level results.

As noted above, simulations have played an important role in the study of the helium liquids. While we distinguish between simulations and the microscopic theory described above, they have been developed in parallel with this microscopic theory since the seminal work of Kalos. [51] Moreover, these are microscopic in the same sense as the theories above since they also begin with the best microscopic description of the many-body hamiltonian of the systems studied. For Bose liquids, these simulations provide a method for sampling the "exact" ground state or equilibrium finite temperature density matrix for systems of a finite number of particles enabling one to calculate many of the properties of these systems. The quotes on *exact* are to call attention to the fact that these are stochastic methods, and thus are limited in accuracy by a combination of the number of particles in the simulated system and the number of Monte Carlo steps employed.

An early application of simulations in quantum fluids was the variational Monte Carlo method by MacMillan, which adapted the Metropolis algorithm to obtain the radial distribution function $g(r)$ for a Jastrow function with a given $u_2(r)$. This method may also be used in inhomogeneous systems, as can be seen from the work by Lewart *et al.* [40] and by Chin [52] on droplets . However this variational Monte Carlo approach is "exact" only in finding the correct $g(r)$ for a given wavefunction, defined by u_2, u_1 and possibly u_3.

The first "exact" Bose ground state simulations were the Green Function Monte Carlo simulations (GFMC) developed by M. H. Kalos and applied to a 32 particle ^{4}He liquid with periodic boundary conditions. [53]

This method and others discussed below take advantage of the fact that the Bose ground state wavefunction (and the many-body density matrix) are real and positive semi-definite. Application of these methods to fermion systems such as liquid ^{3}He are hampered by the fact that the ground state wavefunction, while still real, must take on both positive and negative signs, changing sign upon the exchange of particle coordinates. Methods for improving upon this situation for fermions, specifically ^{3}He, are discussed by Boronat in Chapter 2. However it should be noted that progress toward an "exact" fermion method has been reported recently by Kalos and Pederiva. [54]

Excellent introductions and descriptions of applications of two other "exact" simulation methods are given by Boronat in Chapter 2 and by Huang, Kwon, and Whaley in Chapter 3. Boronat provides a review and recent results of the application of diffusion Monte Carlo simulations to helium fluids, including the phonon-roton spectrum of bulk ^{4}He and the ground state of ^{3}He. In Chapter 3, Huang, Kwon, and Whaley review the path integral Monte Carlo simulation method (PIMC), and describe its application to ^{4}He nanodroplets at finite temperatures, including the determination of superfluidity in these droplets .

The well-known Hohenberg-Kohn theorem insures that the energy per particle of a many body system can be expressed in terms of a unique functional of the density, but the theorem does not provide any clue about the specific form of such a functional, which has to be written down resorting to phenomenological approaches . As compared with microscopic methods, density functional methods are computationally quite straightforward. Therefore, if used conscientiously they can serve a number of useful purposes to explore areas of potentially interesting physics before more advanced methods are used, or to provide a guide for choosing the quantities to be determined with optimal accuracy.

Excellent accounts of density functional theory as applied to helium liquids are given by Hernández and Navarro in Chapter 6 and by Barranco *et al.* in Chapter 7. This is particularly useful because it is applied to situations which are still major challenges for the simulations and the microscopic theories , namely the fermion systems and the cavitation problem. The flexibility of a non-local density functional permits its application to non-homogeneous helium systems such as the liquid free surface, drops, adsorbed films , adsorbed gases and liquids in aerogels and nanotubes, or the study of nucleation either in form of drops or bubbles. A comparison of such a non-local density functional approach with the microscopic wavefunction approach described earlier in this section is given by Apaja and Krotscheck in Chapter 5.

7. Conclusions

We can hardly do justice to the variety of systems and the different methodologies reported on in the remainder of this book. Moreover, the subfield of inhomogeneous quantum fluids in confined geometries is burgeoning with new measurements and new theoretical and simulational results, only a portion of which can be contained in a single volume such as this. We can only hope that the patient reader who has read this far will still be encouraged to delve further into the remaining chapters.

References

1. R. B. Hallock, *Physics Today* **51**, 30 (1998).
2. R. B. Hallock, in *Progress in Low Temperature Physics*, edited by W. P. Halperin (North-Holland, Amsterdam, 1995), Vol. XIV, Chap. 5, pp. 321–443.
3. J. P. Toennies, A. F. Vilesov, and K. B. Whaley, *Physics Today* **54**, 31 (2001).
4. J. G. Dash and M. Schick, in *The Physics of Liquid and Solid Helium*, edited by K. H. Bennemann and J. B. Ketterson (Wiley, New York, 1978), Vol. 2.
5. A. J. Dahm and W. F. Vinen, *Physics Today* **43**, 43 (1987).
6. D. F. Brewer, in *The Physics of Liquid and Solid Helium, Part II*, edited by K. H. Bennemann and J. B. Ketterson (Wiley, New York, 1978), Chap. 6, pp. 573–673.
7. D. S. Greywall, *Physica* **B 197**, 1 (1994).
8. K. Mendelssohn, *The Quest for Absolute Zero* (McGraw-Hill, New York, 1966).
9. J. Wilks, *The Properties of Liquid and Solid Helium* (Clarendon, New York, Oxford, 1967).
10. K. R. Atkins, *Liquid Helium* (Cambridge University Press, Cambridge, U. K., 1959).
11. W. E. Keller, *Helium-3 and Helium-4* (Plenum, New York, 1969).
12. H. Glyde, *Excitations in liquid and solid helium* (Oxford University Press, Oxford, 1994).
13. J. Wilks and D. S. Betts, *An introduction to liquid helium* (Clarendon, New York, Oxford, 1987).
14. E. Krotscheck and M. Saarela, *Physics Reports* **232**, 1 (1993).
15. L. D. Landau, *Sov. Phys. JETP* **3**, 920 (1957).
16. L. D. Landau, *Sov. Phys. JETP* **5**, 101 (1957).
17. C. H. Aldrich and D. Pines, *J. Low Temp. Phys.* **25**, 677 (1976).
18. W. Hsu and D. Pines, *J. Stat. Phys.* **38**, 273 (1985).
19. H. J. Lauter, H. Godfrin, and P. Leiderer, *J. Low Temp. Phys.* **87**, 425 (1992).
20. H. J. Lauter, H. Godfrin, V. L. P. Frank, and P. Leiderer, *Phys. Rev. Lett.* **68**, 2484 (1992).
21. B. E. Clements, E. Krotscheck, and C. J. Tymczak, *Phys. Rev.* **B 53**,12253 (1996).
22. C. E. Campbell, E. Krotscheck, and M. Saarela, *Phys. Rev. Lett.* **80**, 2169 (1998).
23. E. Krotscheck and R. Zillich, *J. Chem. Phys.* **22**, 10161 (2001).
24. H. Maris and S. Balibar, *Physics Today* **53**, 29 (2000).
25. C. E. Campbell, R. Folk, and E. Krotscheck, *J. Low Temp. Phys.* **105**, 13 (1996).
26. L. Landau, *J. Phys. U.S.S.R.* **5**, 71 (1941).
27. F. Luo *et al.*, *J. Chem. Phys.* **98**, 3564 (1993).
28. F. Luo, C. F. Giese, and W. R. Gentry, *J. Chem. Phys.* **104**, 1151 (1996).
29. W. Schöllkopf and J. P. Toennies, *J. Chem. Phys.* **104**, 1155 (1996).
30. M. Barranco, J. Navarro and A. Poves, *Phys. Rev. Lett.* **78**, 4729 (1997).
31. R. Guardiola and J. Navarro, *Phys. Rev. Lett.* **84**, 1144 (2000).
32. L. W. Bruch and I. J. McGee, *J. Chem. Phys.* **46**, 2959 (1961).
33. J. Xie, J. A. Northby, D. L. Freeman, and J. D. Doll, *J. Chem. Phys.* **91**, 612 (1989).
34. J. A. Northby, D. L. Freeman, and J. D. Doll, *Zeitschrift fur Physik* **D 12**, 69 (1989).
35. R. P. Feynman, *Phys. Rev.* **94**, 262 (1954).
36. R. P. Feynman and M. Cohen, *Phys. Rev.* **102**, 1189 (1956).
37. O. Penrose, *Phil. Mag.* **42**, 1373 (1951).
38. G. V. Chester, in *Quantum Fluids and Nuclear Matter*, Vol. XI B of *Lectures in Theoretical Physics*, edited by K. T. Mahanthappa and W. E. Britten (Gordon and Breach, New York, 1969), pp. 253–296.
39. C. E. Campbell, *J. Low Temp. Phys.* **93**, 907 (1993).
40. D. S. Lewart, V. R. Pandharipande, and S. C. Pieper, *Phys. Rev.* **B 37**, 4950 (1988).

41. S. A. Chin, *J. Low Temp. Phys.* **93**, 921 (1993).
42. E. Krotscheck, *Phys. Rev.* **B 32**, 5713 (1985).
43. M. Bretz *et al.*, *Phys. Rev.* **A 8**, 1589 (1973).
44. D. S. Greywall and P. A. Busch, *Phys. Rev. Lett.* **67**, 3535 (1991).
45. B. E. Clements, E. Krotscheck, and H. J. Lauter, *Phys. Rev. Lett.* **70**, 1287 (1993).
46. C. E. Campbell, B. E. Clements, E. Krotscheck, and M. Saarela, *Phys. Rev.* **B 55**, 3769 (1997).
47. J. M. Kosterlitz and D. J. Thouless, *J. Phys.* **C 6**, 1181 (1973).
48. L. Reatto and G. V. Chester, *Phys. Rev.* **155**, 88 (1967).
49. R. Jastrow, *Phys. Rev.* **98**, 1479 (1955).
50. C. E. Campbell and E. Feenberg, *Phys. Rev.* **188**, 396 (1969).
51. M. H. Kalos, *Phys. Rev.* **A 2**, 250 (1970).
52. S. A. Chin and E. Krotscheck, *Phys. Rev.* **B 45**, 852 (1992).
53. D. Ceperly and M. Kalos, in *Monte Carlo Methods in Statistical Mechanics*, edited by K. Binder (Springer, Berlin, Heidelberg, and New York, 1979).
54. M. H. Kalos and F. Pederiva, *Phys. Rev. Lett.* **85**, 3547 (2000).

CHAPTER 2

MONTE CARLO SIMULATIONS AT ZERO TEMPERATURE: HELIUM IN ONE, TWO, AND THREE DIMENSIONS

J. Boronat
Departament de Física i Enginyeria Nuclear, Campus Nord B4-B5, Universitat Politècnica de Catalunya, E-08034 Barcelona, Spain
E-mail: jordi.boronat@upc.es

Quantum Monte Carlo methods which work at zero temperature are reviewed and discussed. These methods are nowadays standard tools in condensed matter physics, a field where they have proven their reliability and accuracy. The present review is mainly concerned with recent results obtained by using quantum Monte Carlo for studying the energetics and structural properties of homogeneous liquid helium in different geometries.

1. Monte Carlo methods and condensed helium

The history of quantum Monte Carlo methods is closely related to the physics of the condensed phases of helium. [1] Apart from the intrinsic interest which superfluid helium offers to theoretical physicists, [2,3] the study of its fascinating properties is a very appealing issue for Monte Carlo applications. In fact, the closed-shell structure of the He atom makes it very plausible to consider them as spherical entities. This feature is even more stressed by looking at the energy scales involved: in liquid helium the interaction energy is in the range of tens of Kelvin, five orders of magnitude smaller than typical electronic energies inside the atom. [4] Therefore, the interatomic potential is radial, pairwise, and spin-independent. It is nowadays known with high accuracy. Considering the interatomic potential as the only real input in the *exact* Monte Carlo algorithms (Green's function Monte Carlo (GFMC) [1,5] and diffusion Monte Carlo (DMC) [5,6]), it has been possible to reproduce many experimental data with unprecedented accuracy. It is difficult to find another field in condensed-matter physics where microscopic approaches have arrived to similar precision, especially in systems in which quantum theory is absolutely necessary as in superfluid helium.

The hard core of the interatomic potential, which in the first calculations was considered of the Lennard-Jones type, makes the application of a standard perturbative schemes unreliable. This problem is not present in a variational approach if the trial wave function is properly chosen. Focusing the discussion to bosonic liquid ^{4}He at zero temperature, the trial wave function must vanish when two particles overlap. That constraint is satisfied by defining the model wave function as a product of two-body correlation factors $f_2(r_{ij})$,

$$\psi = \prod_{i<j} f_2(r_{ij}) \ , \tag{1.1}$$

with $f_2(r_{ij}) = 0$ for $r_{ij} < \sigma$, with σ the "diameter" of the atom. The trial function (1.1) was introduced by Bijl [7] and then elaborated by Dingle [8] and Jastrow, [9]; it is nowadays known as Jastrow trial wave function. McMillan [10] was the first to observe the connection between the energy expectation value with the Jastrow function (1.1) and the method for generating random variables, from a general probability distribution function, proposed by Metropolis *et al.* [11] some years before. The seminal work of McMillan introduced the variational Monte Carlo (VMC) method to study quantum fluids, [5,12] a method that has been extremely fruitful in the field for many years.

In a VMC approach, the particular form of the two-body correlation factor influences directly the quality of the upper bound obtained. McMillan [10] introduced a simple form

$$f(r) = \exp\left[-\frac{1}{2}\left(\frac{b\sigma}{r}\right)^m\right] \ , \tag{1.2}$$

that for $m = 5$ and $b = 1.134$ satisfies, at leading order, the condition of the local energy being constant in a two-body problem at small interparticle distance and when the interaction is of the Lennard-Jones type. The parameter $\sigma = 2.556$ Å corresponds to the Lennard-Jones "diameter" and is used to introduce a reduced radial unit in condensed helium physics. In spite of the simplicity of that correlation factor, the variational energy obtained is ~ -6.0 K, to be compared with the experimental value -7.17 K. [13] The results obtained by McMillan were later on corroborated by Schiff and Verlet, [14] and Murphy and Watts. [15] The McMillan correlation factor (1.2) is manifestly not the optimal one. This can be obtained by solving the optimal Euler-Lagrange equation [16]

$$\frac{\delta \langle H \rangle}{\delta f(r)} = 0 \ . \tag{1.3}$$

Eq. (1.2) does not contain, for example, the right behavior $1/r^2$ when $r \to \infty$ which is necessary to reproduce the linear behavior of the static structure factor $S(k)$ when $k \to 0$. The introduction of that long-range behavior, and other terms to improve the middle-range behavior of $f(r)$ does not improve, in a significant way, the energy per particle (a gain of approximately 0.2 K is obtained). [17] A more significant effect in the binding energy is achieved by introducing three-body correlations in the trial wave function

$$\psi = \prod_{i<j} f_2(r_{ij}) \prod_{i<j<k} f_3(r_{ij}, r_{ik}, r_{jk}) \ . \tag{1.4}$$

The emergence of three-body correlations to properly describe the ground state of liquid ^{4}He was put forward by Chang and Campbell [18] using correlated basis function (CBF) theory. Further variational calculations using both VMC [19] and hypernetted-chain (HNC) theory [20] confirmed the significant size of the energy correction by reducing the difference between the Jastrow energy and the experimental data by 65 percent.

A new family of trial wave functions, known as shadow wave functions, was introduced in 1988 by Vitiello *et al.* [21] In this model, a subsidiary particle (shadow) is associated to each real particle introducing shadow-particle and shadow-shadow correlation functions,

$$\psi = \prod_{i<j} f_{\mathrm{pp}}(r_{ij}) \left[\int d\xi_1 \ldots d\xi_N \left(\prod_{i<j} f_{\mathrm{ss}}(\xi_{ij}) \right) \left(\prod_k f_{\mathrm{ps}}(|\mathbf{r}_k - \boldsymbol{\xi}_k|) \right) \right] . \qquad (1.5)$$

The shadow variables $\{\xi_i\}$ introduce extra correlations between the particles $\{\mathbf{r}_i\}$ and thus it goes beyond the Jastrow approximation. The translationally invariant character of this wave function has been especially useful in the study of the liquid-solid phase transition in liquid ^{4}He. [22]

The VMC method has proved its high usefulness in the variational theory applied to quantum liquids. The successive improvements in the trial wave functions have reduced the differences between the upper bounds and the experimental data. Nevertheless, significant differences at the quantitative level remain. A more ambitious and powerful method was developed by Kalos in the sixties and seventies. [1] This method, known as Green's function Monte Carlo (GFMC), allows for an exact solution of the Schrödinger equation in many-boson problems. The Schrödinger equation, written in imaginary time, is solved stochastically by approximating the time-independent Green function by means of a Born series in a certain domain. After some initial applications to few-body problems, [23–25] Kalos introduced importance sampling in the method by means of an auxiliary trial wave function. The introduction of importance sampling was crucial in order to improve the efficiency of the method. GFMC was first applied to a homogeneous hard-sphere gas [26,27] and then to study the properties of liquid and solid ^{4}He. [17] That opened a new era in the application of the Monte Carlo method to condensed matter theory with a scientific work that has been growing since then.

Anderson [28] proposed in 1975 an alternative method to GFMC but with the same scope: the ground-state solution of the many-body Schrödinger equation. This new approach, known as diffusion Monte Carlo (DMC) method, is based on short-time approximations to the time-dependent Green function. Anderson applied the DMC method to some atoms and molecules, [29,30] and some years later on Ceperley and Alder [31] extended the formalism to homogeneous systems. Since those pioneering calculations, the DMC method [5,6,32] has been extensively applied to the study of atomic, molecular, and quantum-liquid areas. In spite of the time-step dependence, which is present in DMC, but not in the original GFMC (domain GFMC), the greater simplicity of DMC compared with GFMC has increased its popularity. The original DMC method results in a linear dependence in the time step used in the simulation; it is then necessary to remove that bias by extrapolating to time step zero. A significant improvement in this technical problem was made by Chin [33] who proved the reliability of new DMC algorithms, which are second-order in the time step. More recently, Forbert and Chin [34] have extended satisfactorily the DMC method up to fourth order thus eliminating, from the practitioner's point of view, the time-step dependence.

Our group has been working for the last years in the study of quantum liquids using a second-order DMC algorithm. [35] We have studied ground-state and excited-state properties of liquid ^{4}He at zero temperature in three, [35,36] two, [37,38] and one dimensions. [39] Other groups have also obtained relevant results on the momentum distribution, [40] clusters, [41–43] slabs, [44] and films [45] of liquid ^{4}He.

The application of quantum Monte Carlo methods to a Fermi system as liquid ^{3}He is more involved due to the antisymmetry of its wave function. In principle, ^{3}He is an easier system than for example nuclei or nuclear matter since the interatomic potential does not include spin-isospin terms. However, its density is comparatively higher and dynamical correlations in ^{3}He play a fundamental role. Nevertheless, in all these systems GFMC and DMC methods have to tackle the well-known sign problem. The sign problem arises in Monte Carlo due to the non positive-definite character of the wave function for a Fermi system. This feature is not a problem in a VMC calculation since the probability distribution function is the squared wave function. The antisymmetry of the variational wave function implies the use of Slater determinants, whose calculation and updating require a larger computational effort, but VMC does not introduce any bias. The sign problem emerges in DMC or GFMC in which the walkers evolve according the product of the trial wave function, used for importance sampling, and the ground-state wave function. That product is not positive, a fact that hinders its probabilistic interpretation. [5] A simple and very useful approach to the problem is the fixed-node (FN) approximation [28,29,32] in which the nodal surface of the wave function, solution of the Schrödinger equation, is assumed to be the same as that of the trial wave function ψ. This restriction, which implies the solution of the Schrödinger equation with boundary conditions imposed by ψ, generates an upper bound to the exact energy. [32]

The elimination of the fixed-node constraint is made possible by introducing a releasing mechanism that allows for crossings through the model nodal surface. The implementation of that idea is the released-node (RN) method [31] that is an asymptotic estimation in the released time which guarantees that there is no bias in the ground-state energy. However, the approach to the asymptotic regime is overwhelmed by increasing statistical noise due to the growth of the bosonic component. The success of the RN method depends therefore on the system studied, and for a given system, on the *quality* of the nodal surface of the trial wave function. Recent calculations [46] of liquid ^{3}He have shown that RN-DMC is not able to generate the ground-state energy starting from a simple trial wave function. Nevertheless, these calculations have shown the usefulness of the RN method served as a check or a measure of the quality of the upper bound obtained in a FN calculation.

Kalos and collaborators have pursued for years the development of a stable and unbiased Monte Carlo method to efficiently tackle the sign problem. The final goal is to devise a method to cancel efficiently positive and negative walkers. At present, the most promising strategy has been the introduction of correlated walkers inside a DMC method. That methodology has been recently applied to a small ensemble of ^{3}He atoms with promising results. [47] Nevertheless, the generalization of the method to a larger and more realistic number of particles seems still nowadays a cumbersome work.

2. Monte Carlo methods at zero temperature

In the present review we limit our analysis to Monte Carlo methods that work at zero temperature. The extension at finite temperature is presented in Chapter 3 of this volume[48]. The goal of this chapter is to give a brief account of the VMC and DMC methods with some specific comments on the specialized algorithms for fermions. The GFMC is not discussed but it shares with DMC the final objective and some terminology and approximations. A detailed description of the GFMC method is given in Refs. 1 and 49. The list of references that include a more or less technical description of QMC methods is quite large, but from a beginner's point of view Refs. 5 and 6 are the most appropriate ones.

2.1. *Variational Monte Carlo*

Let us consider a homogeneous system of identical particles which interact via a pairwise purely radial interatomic potential $V(r_{ij})$. The Hamiltonian is then

$$H = -\frac{\hbar^2}{2m} \sum_{i=1}^{N} \boldsymbol{\nabla}_i^2 + \sum_{i<j} V(r_{ij}) \ , \tag{2.1}$$

and a fruitful guess for the variational wave function is the Jastrow-Feenberg [2] ansatz

$$\psi = F\,\Phi \ . \tag{2.2}$$

In this model, the correlation factor F incorporates the dynamical correlations driven by the interatomic potential

$$F = \prod_{i<j} f_2(r_{ij}) \prod_{i<j<k} f_3(r_{ij}, r_{ik}, r_{jk}) \ \ldots \ . \tag{2.3}$$

Functions f_2, f_3 are the two- and three-body correlation factors which satisfy the general properties of becoming zero when two particles "overlap" and approaching the unity when one particle is moved far away from the rest. Wave function Φ models the ground state of the system in the non-interacting case: it is the unity for bosons and a Slater determinant for fermions. The variational principle states that the expectation value of H is an upper bound to the ground-state energy E_0,

$$E = \frac{\langle\psi|H|\psi\rangle}{\langle\psi|\psi\rangle} \geq E_0 \ . \tag{2.4}$$

The variational energy E is a function of the correlation factors f_i for a fixed model Φ. The optimum functions f_i are the solutions of the coupled Euler-Lagrange differential equations [16]

$$\frac{\delta E}{\delta f_i} = 0 \ . \tag{2.5}$$

These equations are currently solved in an approximate way in the framework of hypernetted-chain (HNC) theory. A not so ambitious project for the functional optimization consists in choosing reasonable analytic functions with some variational parameters α_j to be adjusted through

$$\frac{\partial E}{\partial \alpha_j} = 0 \ . \tag{2.6}$$

Actually, this second procedure leads to upper bounds for the energy that are of the same quality than the ones which result from the full optimization (2.5). However, the stability and existence conditions contained in the Euler-Lagrange equations are lost in the simplified approach (2.6).

The calculation of the energy for a given trial wave function is not an easy task since a multidimensional integral has to be calculated,

$$E = \frac{\int d^3r_1 \ldots d^3r_N \ \psi^\star(\mathbf{r}_1, \ldots, \mathbf{r}_N) H \psi(\mathbf{r}_1, \ldots, \mathbf{r}_N)}{\int d^3r_1 \ldots d^3r_N \ |\psi(\mathbf{r}_1, \ldots, \mathbf{r}_N)|^2} \ . \tag{2.7}$$

HNC summation techniques allow for a quite accurate, but still approximate, evaluation of the energy. [16,50] It is at this stage of the theoretical problem where the Monte Carlo method is extremely useful. As it is well known, multidimensional integration is relatively easy by standard MC and does not increase significantly in algorithmic complexity when the dimensionality grows. Variational Monte Carlo is able to calculate the energy in an exact way with the only prize of a statistical noise that can be well estimated. [6,12]

If one defines the multivariate probability density function (pdf)

$$f(\mathbf{R}) = \frac{|\psi(\mathbf{R})|^2}{\int d\mathbf{R} \ |\psi(\mathbf{R})|^2} \ , \tag{2.8}$$

which is always positive and normalized, and the local energy $E_\mathrm{L}(\mathbf{R})$ as

$$E_\mathrm{L}(\mathbf{R}) = \frac{1}{\psi(\mathbf{R})} H \psi(\mathbf{R}) \ , \tag{2.9}$$

the expectation value of the Hamiltonian H turns to

$$\langle H \rangle_\psi = \int d\mathbf{R} \ E_\mathrm{L}(\mathbf{R}) f(\mathbf{R}) \ . \tag{2.10}$$

By generating multidimensional points (*walkers*) $\mathbf{R} = \{\mathbf{r}_1, \ldots, \mathbf{r}_N\}$ according the pdf $f(\mathbf{R})$, the energy is obtained as the mean value of $E_\mathrm{L}(\mathbf{R})$,

$$\langle H \rangle_\psi = \frac{1}{n_s} \sum_{i=1}^{n_s} E_\mathrm{L}(\mathbf{R}_i) \ , \tag{2.11}$$

n_s being the number of sampling points. Also stochastically, and in an unbiased way, the variance of the measure, [5]

$$\sigma^2 = \frac{1}{n_I - 1} \left[\frac{1}{n_I} \sum_{i=1}^{n_I} E_{\rm L}^2(\mathbf{R}_i) - \left(\frac{1}{n_I} \sum_{i=1}^{n_I} E_{\rm L}(\mathbf{R}_i) \right)^2 \right] , \tag{2.12}$$

can be determined (n_I is the number of statistically independent measures). Obviously, a crucial point to ensure the success of the method is to have a means of generating a random walk that follows $f(\mathbf{R})$ at one's disposal. A simple (not unique) solution to this inverse problem was put forward by Metropolis *et al.* [11] in a seminal paper appeared in 1953.

Let us consider a continuous system and $\mathbf{r}_i^0$ the initial position of particle i. The stochastic matrix $T(\mathbf{r}_i^0, \mathbf{r}_i^{\rm f})$ gives the transition probability to a final position $\mathbf{r}_i^{\rm f}$, following the pdf $f(\mathbf{R})$. If one defines an auxiliary stochastic and symmetric matrix $S(\mathbf{r}_i^0, \mathbf{r}_i^{\rm f})$, the Metropolis solution for $T(\mathbf{r}_i^0, \mathbf{r}_i^{\rm f})$ is

$$\begin{aligned} T(\mathbf{r}_i^0, \mathbf{r}_i^{\rm f}) &= S(\mathbf{r}_i^0, \mathbf{r}_i^{\rm f}) && \text{if } f(.., \mathbf{r}_i^{\rm f}, ..) \geq f(.., \mathbf{r}_i^0, ..) \\ T(\mathbf{r}_i^0, \mathbf{r}_i^{\rm f}) &= S(\mathbf{r}_i^0, \mathbf{r}_i^{\rm f}) \frac{f(.., \mathbf{r}_i^{\rm f}, ..)}{f(.., \mathbf{r}_i^0, ..)} && \text{otherwise} \end{aligned} \tag{2.13}$$

A simple and frequently used model for $S(\mathbf{r}_i^0, \mathbf{r}_i^{\rm f})$ is a displacement drawn from a uniform distribution around the initial position $\mathbf{r}_i^{\rm f} \in U(\mathbf{r}_i^0 - \mathbf{\Delta}, \mathbf{r}_i^0 + \mathbf{\Delta})$. The parameter $\mathbf{\Delta}$ is the maximum allowed displacement and has to be empirically adjusted in each simulation. An additional parameter which helps in the control of $\mathbf{\Delta}$ is the acceptance ratio, defined as the quotient between the accepted and the proposed moves. A reasonable choice for $\mathbf{\Delta}$ is the one that generates an acceptance ratio in the range $40 - 70$ %.

A schematic algorithm for a Metropolis move in a continuous system is the following:

→ `Initial state 0:` $\mathbf{R}_i^0$ $(i = 1, \ldots, N)$

→ `Proposed movement:` $\mathbf{R}_i^{\rm f} = \mathbf{R}_i^0 + (2. * \text{ran}() - 1.) * \mathbf{\Delta}$

→ `Metropolis: if` $(|\psi(\mathbf{R}_i^{\rm f})|^2 / |\psi(\mathbf{R}_i^0)|^2) > \text{ran}()$ `then` $\mathbf{R}_i^0 = \mathbf{R}_i^{\rm f}$

→ `Sampling properties: energy, distribution functions, etc.`

→ `Next step`

In the algorithm, one decides if the proposed movement concerns only one particle, a subset of the walker or the full walker. Consequently, the value for Δ depends on this choice for a preassigned value of the acceptance ratio.

Apart from statistical noise, the VMC method calculates in an exact way the expected value of the Hamiltonian for a given trial wave function ψ. Once an analytical model for ψ has been chosen, the variational energy is a function of the values given to the set of parameters $\{\beta_i\}$ entering ψ. The optimization of $\{\beta_i\}$ is therefore crucial to improve the quality of the upper-bound to the ground-state energy of the system. The search of the optimal set $\{\beta_i\}$ is, in general, not easy and its complexity increases severely with the number of parameters. An auxiliary Monte Carlo technique that helps in the optimization is the reweighting configuration method, [5] based on the more general ideas of correlated sampling. [51] In this method, the random walk is generated by a pdf with a determined set of parameters $\{\beta_i\}$ but energies with different values $\{\beta_i^I\}$ ($I = 1, \ldots, n_s$, n_s number of different sets) are also estimated

$$E(\{\beta_i^I\}) = \frac{\int d\mathbf{R}\; E_{\rm L}(\{\beta_i^I\})\, w(\{\beta_i^I\})\, |\psi(\{\beta_i\})|^2}{\int d\mathbf{R}\; w(\{\beta_i^I\})\, |\psi(\{\beta_i\})|^2} \;. \tag{2.14}$$

$E_{\rm L}(\{\beta_i^I\}) = \psi(\{\beta_i^I\})^{-1} H \psi(\{\beta_i^I\})$ is the local energy corresponding to the set $\{\beta_i^I\}$, and

$$w(\{\beta_i^I\}) = \left| \frac{\psi(\{\beta_i^I\})}{\psi(\{\beta_i\})} \right|^2 \tag{2.15}$$

is the weight that "corrects" the pdf. This method produces positive correlations between the different energies $E(\{\beta_i\})$ that reduce significantly the statistical noise of the energy differences with respect to independent calculations. Nevertheless, the reweighting procedure works well only when the relative differences between the different parameters are fairly small.

The optimum variational set is also the one which minimizes the variance. It is worth noticing that in the case $\psi = \Psi_0$, with Ψ_0 the ground-state eigenfunction, the variance of the energy is strictly zero. Therefore, the measure of the variance for a given choice of $\{\beta_i\}$ is also a direct measure of the quality of the trial wave function. Usually, in optimization processes both the minimum of the energy and the variance are pursued in separate estimators or in a combined one. If the dimension of the parameter space is large enough it is more convenient to optimize the trial wave function by calculating the energies on a fixed number of representative walkers previously calculated. On top of the later proposal one can also use general purpose optimization methods as the simplex algorithm. [52]

2.2. *Diffusion Monte Carlo*

The Green's function Monte Carlo methods go a step further than variational Monte Carlo. Their goal is to solve directly the N-body Schrödinger equation exploiting the resemblance between this, when it is written in imaginary time, and an ordinary diffusion equation. In the present review, we are mainly concerned with the DMC method but some of the concepts and techniques here discussed are common with domain GFMC.

The starting point in the DMC method is the Schrödinger equation, written in imaginary time

$$-\frac{\partial \Psi(\mathbf{R},t)}{\partial t} = (H - E)\,\Psi(\mathbf{R},t)\ , \tag{2.16}$$

where $\mathbf{R} \equiv (\mathbf{r}_1, \ldots, \mathbf{r}_N)$ is a $3N$-dimensional vector (*walker*) and t is the imaginary time measured in units of $\hbar$. As it is usual in quantum mechanics, the time-dependent wave function of the system $\Psi(\mathbf{R},t)$ can be expanded in terms of a complete set of eigenfunctions $\phi_i(\mathbf{R})$ of the Hamiltonian,

$$\Psi(\mathbf{R},t) = \sum_n c_n \exp\left[-(E_i - E)t\right]\,\phi_i(\mathbf{R})\ , \tag{2.17}$$

where E_i is the eigenvalue associated to $\phi_i(\mathbf{R})$. The asymptotic solution of Eq. (2.16), for any value E close to the energy of the ground state and for long times ($t \to \infty$), gives $\phi_0(\mathbf{R})$, provided that there is a nonzero overlap between $\Psi(\mathbf{R}, t = 0)$ and the ground-state wave function $\phi_0(\mathbf{R})$.

A direct Monte Carlo implementation of Eq. (2.16) is hardly able to work efficiently, especially when the interatomic potential contains a hard core. Actually, this is the most common case in condensed matter. A fundamental progress, in the line of making the method operative, was put forward by Kalos by introducing importance sampling in the solution of the diffusion equation. [1] In fact, the importance sampling technique is a general concept in Monte Carlo [51] and is one of the best methods to reduce the variance of any MC calculation. The importance sampling method, applied to Eq. (2.16), consists in rewriting the Schrödinger equation in terms of the wave function

$$f(\mathbf{R},t) \equiv \psi(\mathbf{R})\,\Psi(\mathbf{R},t)\ , \tag{2.18}$$

where $\psi(\mathbf{R})$ is a time-independent trial wave function that describes approximately the ground state of the system at the variational level. Considering a Hamiltonian of the form

$$H = -\frac{\hbar^2}{2m}\,\nabla^2_{\mathbf{R}} + V(\mathbf{R})\ , \tag{2.19}$$

Eq. (2.16) turns out to be

$$-\frac{\partial f(\mathbf{R},t)}{\partial t} = -D\,\nabla^2_{\mathbf{R}} f(\mathbf{R},t) + D\,\nabla_{\mathbf{R}}\left(\mathbf{F}(\mathbf{R})\,f(\mathbf{R},t)\right) + \left(E_L(\mathbf{R}) - E\right)\,f(\mathbf{R},t)\ , \tag{2.20}$$

where $D = \hbar^2/(2m)$, $E_L(\mathbf{R}) = \psi(\mathbf{R})^{-1} H \psi(\mathbf{R})$ is the local energy, and

$$\mathbf{F}(\mathbf{R}) = 2\,\psi(\mathbf{R})^{-1}\nabla_{\mathbf{R}}\psi(\mathbf{R}) \tag{2.21}$$

is called drift or quantum force. $\mathbf{F}(\mathbf{R})$ acts as an external force which guides the diffusion process, involved by the first term in Eq. (2.20), to regions where $\psi(\mathbf{R})$ is large.

The r.h.s. of Eq. (2.20) may be written as the action of three operators A_i acting on the wave function $f(\mathbf{R},t)$,

$$-\frac{\partial f(\mathbf{R},t)}{\partial t} = (A_1 + A_2 + A_3)\,f(\mathbf{R},t) \equiv A\,f(\mathbf{R},t) \tag{2.22}$$

The three terms A_i may be interpreted by similarity with classical differential equations. The first one, A_1, corresponds to a free diffusion with a a diffusion coefficient D; A_2 acts as a driving force due to an external potential, and finally A_3 looks like a birth/death term. In Monte Carlo, the Schrödinger equation (2.22) is best suited when it is written in a integral form by introducing the Green function $G(\mathbf{R}', \mathbf{R}, t)$, which gives the transition probability from an initial state $\mathbf{R}$ to a final one $\mathbf{R}'$ during a time t,

$$f(\mathbf{R}', t + \Delta t) = \int G(\mathbf{R}', \mathbf{R}, \Delta t)\, f(\mathbf{R}, t)\, d\mathbf{R} \,. \tag{2.23}$$

More explicitly, the Green function is given in terms of the operator A by

$$G(\mathbf{R}', \mathbf{R}, \Delta t) = \langle\, \mathbf{R}' \,|\, \exp(-A\Delta t) \,|\, \mathbf{R} \,\rangle \,. \tag{2.24}$$

The domain GFMC method works with the whole Green function which is sampled by means of a Born series expansion. Instead, DMC algorithms rely on reasonable approximations of $G(\mathbf{R}', \mathbf{R}, \Delta t)$ for small values of the time-step Δt. Considering such a short-time approximation, Eq. (2.24) is then iterated repeatedly until to reach the asymptotic regime $f(\mathbf{R}, t \to \infty)$, a limit in which one is effectively sampling the ground state.

In the first applications of the DMC method, the simplest version of the short-time approximation was used,

$$\exp\left(-A\Delta t\right) = \exp\left(-A_3\Delta t\right) \exp\left(-A_2\Delta t\right) \exp\left(-A_1\Delta t\right) + \mathcal{O}\left((\Delta t)^2\right) \,. \tag{2.25}$$

This expansion generates a linear time-step dependence of the energy. A significantly better behavior is obtained by expanding the exponential of the operator A to higher orders in Δt. A good compromise between algorithmic complexity and efficiency is obtained by using a second-order expansion (quadratic DMC). In this case, the Green function $G(\mathbf{R}', \mathbf{R}, \Delta t)$ is approximated by

$$\begin{aligned} \exp\left(-A\Delta t\right) = {} & \exp\left(-A_3 \frac{\Delta t}{2}\right) \exp\left(-A_2 \frac{\Delta t}{2}\right) \exp\left(-A_1 \Delta t\right) \\ & \times \exp\left(-A_2 \frac{\Delta t}{2}\right) \exp\left(-A_3 \frac{\Delta t}{2}\right) + \mathcal{O}\left((\Delta t)^3\right) \,. \end{aligned} \tag{2.26}$$

This decomposition, which is the one we have used in our calculations, is not unique as pointed out by Rothstein *et al.* [53] and Chin. [33] Introducing the above expansion (2.26) in Eq. (2.23) the Schrödinger equation, written in integral form, becomes

$$\begin{aligned} f(\mathbf{R}', t + \Delta t) = \int \Bigg[& G_3\left(\mathbf{R}', \mathbf{R}_1, \frac{\Delta t}{2}\right) G_2\left(\mathbf{R}_1, \mathbf{R}_2, \frac{\Delta t}{2}\right) G_1\left(\mathbf{R}_2, \mathbf{R}_3, \Delta t\right) \\ & \times G_2\left(\mathbf{R}_3, \mathbf{R}_4, \frac{\Delta t}{2}\right) G_3\left(\mathbf{R}_4, \mathbf{R}, \frac{\Delta t}{2}\right)\Bigg] f(\mathbf{R}, t)\, d\mathbf{R}_1 \ldots d\mathbf{R}_4 d\mathbf{R} \,. \end{aligned} \tag{2.27}$$

In Eq. (2.27), the total Green function G is split into the product of individual Green functions G_i, each one associated to the single operator A_i. G_1 is the Green function corresponding to the free diffusion term (A_1), and thus it is the well-known solution for a noninteracting system,

$$G_1(\mathbf{R}', \mathbf{R}, t) = (4\pi D t)^{-\frac{3N}{2}} \exp\left[-\frac{(\mathbf{R}' - \mathbf{R})^2}{4Dt}\right] . \tag{2.28}$$

In the MC simulation, the evolution given by G_1 corresponds to an isotropic gaussian movement of size proportional to $\sqrt{Dt}$. The Green function G_2 describes the movement due to the drift force appearing in A_2; its form is given by

$$G_2(\mathbf{R}', \mathbf{R}, t) = \delta\left(\mathbf{R}' - \mathbf{R}(t)\right), \qquad \text{where} \begin{cases} \mathbf{R}(0) = \mathbf{R} \\ \dfrac{d\mathbf{R}(t)}{dt} = D\,\mathbf{F}(\mathbf{R}(t)), \end{cases} . \tag{2.29}$$

Under the action of G_2, the walkers evolve in a deterministic way according to the drift force $\mathbf{F}(\mathbf{R}(t))$. In order to preserve the second-order accuracy in the time step, the differential equation (2.29) must also be solved with a second-order integration method. Finally, the third individual Green function G_3 has an exponential form, with an argument that depends on the difference between the local energy of a given walker and the prefixed value E,

$$G_3(\mathbf{R}', \mathbf{R}, t) = \exp\left[-(E_L(\mathbf{R}) - E)\, t\right] \delta(\mathbf{R}' - \mathbf{R}). \tag{2.30}$$

This third term, which is called the branching factor, assigns a weight to each walker according to its local energy. Depending on the value of this weight the walker is replicated or eliminated in the population list. The branching mechanism is a key ingredient in DMC; it is easy to prove that, if this weight is externally fixed to be always equal to one, then the asymptotic solution is $|\psi|^2$, *i.e.*, a VMC estimation.

The stochastic characterization of the wave function $f(\mathbf{R}, t)$ is achieved by representing it by a set of n_w walkers $\mathbf{R}_i$. This set of walkers evolves in imaginary time according to the three mechanisms G_i given above; after a sufficiently long time, the walkers follow the pdf $\psi\Psi$ from which ground-state properties may be sampled. A second-order algorithm, [35] following the decomposition (2.27), is symbolically shown in the following scheme which corresponds to the evolution of a given walker during a time step Δt.

→ `Select a walker` $\mathbf{R}$ `of the list: its index, ipop; its local energy,` E_L^0

→ `Gaussian displacement:` $\mathbf{R}_1 = \mathbf{R} + \chi$; χ `randomly drawn from the` $3N$ `gaussian distribution` $\exp(-\chi^2/(4D\Delta t))$

→ `Calculation of the drift force` $\mathbf{F}_1(\mathbf{R}_1)$

→ `Auxiliary drift movement:` $\mathbf{R}_2 = \mathbf{R}_1 + 0.5 * D * \Delta t * \mathbf{F}_1$

→ `Calculation of the drift force` $\mathbf{F}_2(\mathbf{R}_2)$

→ `Drift movement to the middle point:` $\mathbf{R}_2 = \mathbf{R}_1 + 0.25 * D * \Delta t * (\mathbf{F}_1 + \mathbf{F}_2)$

→ `Calculation of the local energy` E_L, `drift force` $\mathbf{F}(\mathbf{R}_2)$, `and other properties`

→ `Final drift movement:` $\mathbf{R} = \mathbf{R}_1 + D * \Delta t * \mathbf{F}$

→ `Branching weight:` $w = \exp\left(-\Delta t * (0.5 * (E_L + E_L^0) - E)\right)$

→ `Randomly replicate each walker` n_r `times, with` $n_r = \mathrm{int}(w + ran())$

→ `Next walker of the initial list: ipop = ipop + 1`

In the above algorithm, the replication or elimination of a walker is performed at each step of the simulation. An alternative, which produces a slightly lower fluctuation in n_w, consists in carrying on a multiplicative weight w associated to each walker. In this second method, the replication or death is produced from time to time when the weights are large or small enough, respectively. More specific details of the algorithm may be drawn from Ref. 35.

When the asymptotic limit ($t \to \infty$) is reached, the sampling of an operator A is carried out according to the *mixed* distribution $\psi\Psi$, with Ψ the ground-state wave function. Thus, the natural output in DMC corresponds to the so called *mixed* estimators. The mixed estimator of an operator $A(\mathbf{R})$ is, in general, biased by the trial wave function ψ used for importance sampling. Only when $A(\mathbf{R})$ is the Hamiltonian of the system, or commutes with it, the mixed estimation is the exact one. A simple method that has been widely used to remove the bias present in the mixed estimations is the extrapolated estimator, [1]

$$\langle A(\mathbf{R})\rangle_e = 2\,\langle A(\mathbf{R})\rangle_m - \langle A(\mathbf{R})\rangle_v \ , \tag{2.31}$$

from the knowledge of the mixed estimator $\langle A(\mathbf{R})\rangle_m$ and the variational one

$$\langle A(\mathbf{R})\rangle_v = \frac{\langle \psi(\mathbf{R}) \,|\, A(\mathbf{R}) \,|\, \psi(\mathbf{R})\rangle}{\langle \psi(\mathbf{R}) \,|\, \psi(\mathbf{R})\rangle} \ . \tag{2.32}$$

The expectation values obtained through the extrapolation method (2.31) are not completely independent of the trial wave function ψ used for importance sampling. In spite of using good trial wave functions, the extrapolated estimator is biased and therefore introduces a systematic error that is difficult to assess *a priori*. In order to overcome that important restriction, one can go a step further and calculate *pure* (exact) expectation values,

$$\langle A(\mathbf{R})\rangle_p = \frac{\langle \Psi(\mathbf{R}) \,|\, A(\mathbf{R}) \,|\, \Psi(\mathbf{R})\rangle}{\langle \Psi(\mathbf{R}) \,|\, \Psi(\mathbf{R})\rangle} \ . \tag{2.33}$$

Having in mind that walkers evolve according to the mixed distribution $\psi\Psi$, the pure estimator is more conveniently written as

$$\langle A(\mathbf{R})\rangle_p = \left\langle \Psi(\mathbf{R}) \left| A(\mathbf{R}) \frac{\Psi(\mathbf{R})}{\psi(\mathbf{R})} \right| \psi(\mathbf{R}) \right\rangle \Big/ \left\langle \Psi(\mathbf{R}) \left| \frac{\Psi(\mathbf{R})}{\psi(\mathbf{R})} \right| \psi(\mathbf{R}) \right\rangle \ . \tag{2.34}$$

Some time ago, Liu *et al.*, [54] proved that $\Psi(\mathbf{R})/\psi(\mathbf{R})$ can be obtained from the asymptotic offspring of the $\mathbf{R}$ walker. Assigning to each walker $\mathbf{R}_i$ a weight $W(\mathbf{R}_i)$ proportional to its number of future descendants

$$W(\mathbf{R}) = n(\mathbf{R}, t \to \infty) , \tag{2.35}$$

Eq. (2.34) becomes

$$\langle A(\mathbf{R})\rangle_p = \frac{\sum_i A(\mathbf{R}_i)\, W(\mathbf{R}_i)}{\sum_i W(\mathbf{R}_i)} , \tag{2.36}$$

where the sum $\sum_i$ runs over all walkers and all times in the asymptotic regime. The difficulty of the method, known as forward walking, lies in the estimation of the weight $W(\mathbf{R})$ (2.35). The weight of a walker existing at time t is not known until a future time $t' \geq t+T$, T being a time interval long enough so that Eq. (2.35) could be replaced by $W(\mathbf{R}(t)) = n(\mathbf{R}(t'))$.

The evaluation of Eq. (2.36) has traditionally required the implementation of a tagging algorithm. [55,56] The purpose of this algorithm is to know, at any time during the simulation, which walker of any preceeding configuration originated an actual walker. In this way, one can determine the number of descendants of the former $\mathbf{R}_i$, and accumulate its contribution to Eq. (2.36). We have proposed [57] an alternative method that is much easier to implement in a DMC program. In this second algorithm, we work only with the present values of $A(\mathbf{R}_i)$ in such a way that a weight proportional to its future progeny is automatically introduced.

The schedule of the algorithm, which we have extensively used in our calculations, is the following: The set of walkers at a given time $\{\mathbf{R}_i\}$, and the values that the operator A takes on them $\{A_i\}$, evolve after a time step to

$$\{\mathbf{R}_i\} \to \{\mathbf{R}'_i\} \tag{2.37}$$

$$\{A_i\} \to \{A'_i\} . \tag{2.38}$$

In the same time interval, the number of walkers N changes to N'. In order to sample the pure estimator of A, we introduce an auxiliary variable $\{P_i\}$, associated to each walker, with an evolution law given by

$$\{P_i\} \to \{P'_i\} = \{A'_i\} + \{P^t_i\} , \tag{2.39}$$

where $\{P^t_i\}$ is the old set $\{P_i\}$ *transported* to the new one, in the sense that each element P_i is replicated as many times as the $\mathbf{R}_i$ walker, without any other changes. $\{P_i\}$ is initialized to zero when the run starts.

With this procedure, and after M addition steps (2.39) one ends up with a set of N_f values $\{P_i\}$. The pure estimator of A is given by

$$\langle A(\mathbf{R})\rangle_p = \sum_{i=1}^{N_f} \{P_i\} \,/\, (M \times N_f) . \tag{2.40}$$

The contributions to the $\{P_i\}$, entering Eq. (2.40), can be determined following the evolution of the series. The values $A(\mathbf{R}_i)$, existing at a given time t, carry a weight one (although they contribute together with other values corresponding to previous times which have already been weighted). Then, if any of the descendants of $\mathbf{R}(t)$ disappears or replicates, the former contribution does the same. As a result, $A(\mathbf{R}(t))$ appears in as many rows of $\{P\}$ as descendants of $\mathbf{R}(t)$ exist, and therefore its contribution to Eq. (2.40) is proportional to the weight $W(\mathbf{R}(t))$ (2.35).

In order to ensure the asymptotic condition (2.35), the series are continued for a predefined time only with the reweighting law

$$\{P_i\} \rightarrow \{P_i'\} = \{P_i^t\} \ . \tag{2.41}$$

Since a calculation is usually divided into blocks, one can collect data during a block and allow for a further reweighting in the next one. In this second block, new information can be accumulated to be reweighted in the next block. Thus, after a first initialization block, each new block gives a number for the pure expectation value of A.

An alternative to the simple branching algorithm, implicitly assumed in the above method, is the use of weights $p(\mathbf{R}_i)$ related to the branching factor. In that case, the evolution laws (2.39,2.41) become

$$\{P_i\} \rightarrow \{P_i'\} = \{p(\mathbf{R}_i') \times A_i'\} + \left\{ \frac{p(\mathbf{R}_i')}{p(\mathbf{R}_i)} \times P_i^t \right\} \tag{2.42}$$

$$\{P_i\} \rightarrow \{P_i'\} = \left\{ \frac{p(\mathbf{R}_i')}{p(\mathbf{R}_i)} \times P_i^t \right\} \ , \tag{2.43}$$

whereas the expression of the pure expectation value (2.40) is only modified by a normalization factor.

The large fluctuations observed in the asymptotic offsprings, and therefore in the corresponding weights (Ψ/ψ), have precluded the consideration of the forward walking as a stable and reliable method. In contrast with these considerations, we have found [57] that those statistical fluctuations (of unphysical origin) show a highly depressed effect over integrated quantities, and that in order to accurately sample pure estimators stable regions can be reached.

3. Diffusion Monte Carlo in Fermi systems

DMC relies on the fact that the only contribution which survives, when the imaginary time goes to infinity, is the ground-state one. This is true if there is a nonzero overlap between the trial wave function used for importance sampling $\psi(\mathbf{R})$ and the ground-state wave function $\Psi(\mathbf{R})$. If, on the contrary, the overlap between both wave functions is zero, *i.e.*, both are orthogonal, the imaginary-time wave function $\Psi(\mathbf{R}, t)$ will converge to the next-lowest excited state $\Psi_e(\mathbf{R})$. This approach is used in the calculation of the ground-state energy of a Fermi system, which may be considered as the first fully antisymmetric excited state of the Hamiltonian. In this case, $\psi(\mathbf{R})$ usually incorporates a Slater determinant to satisfy antisymmetry, and therefore the orthogonality to the Bose ground state. The fact that in Fermi systems the wave function is not positive-definite introduces the well-known sign problem: the Monte Carlo simulation needs to work with a positive density of walkers and simultaneously enforce the existence of nodal surfaces delimiting the positive and negative regions of the real wave function.

Despite some recent success in the simulation of a small number of fermions, [47] using correlated sampling, there is presently no method as robust as the DMC algorithm for bosons introduced in the previous section. Nevertheless, several alternatives exist that have proved their reliability in a number of systems. In the rest of this section, we will describe the three methods that we have used to study both Fermi systems [46] and excitations of Bose liquids: [58] fixed node, released node, and an analytical method to improve the Fermi/excited trial wave function.

3.1. *Fixed node*

The Monte Carlo interpretation of the imaginary-time Schrödinger equation requires that $f(\mathbf{R},t) = \psi(\mathbf{R})\Psi(\mathbf{R},t)$ be a density, *i.e.*, $\psi\Psi \geq 0$ in all the domain. This boundary condition can be satisfied if ψ and Ψ change sign together and thus share the same nodes. This approximation, known as fixed-node (FN) method, [28,29,32] has been extensively used in the ground-state calculations of fermions [59,60] and also in the calculations of excited-state properties of small molecules [56] and quantum liquids. [36,38]

In the asymptotic regime

$$f(\mathbf{R}, t \to \infty) = \psi(\mathbf{R})\Psi^{\rm FN}(\mathbf{R}) \, , \tag{3.1}$$

where $\Psi^{\rm FN}(\mathbf{R})$ is an approximation to the exact eigenfunction $\Psi(\mathbf{R})$ of the Schrödinger equation. The nodes are imposed by the trial wave function, and not modified along the calculation. It can be proved that, due to that nodal constraint, the fixed node energies are variational upper bounds to the exact eigenvalues for a given symmetry. [32] In particular, if the nodes of ψ were exact then $\Psi^{\rm FN}$ would also be exact. Therefore, the FN results depend significantly on the *quality* of the trial wave function that can be analytically improved following the procedure proposed in Sec. 3.3.

The requirement of common nodes between ψ and Ψ is naturally fulfilled by the infinite drift force that walkers feel when they come close to the nodal surface of $\psi(\mathbf{R})$. The use of a finite time step Δt originates, however, some spurious crossings through the nodal surface which have to be controlled to reduce its influence in the final energy.

3.2. *Released node*

In order to go beyond the variational bound provided by the FN approximation, it is necessary to generalize the problem by considering the fermionic wave function.

This can be written in a generic form,

$$\Psi(\mathbf{R}) = \Phi^+(\mathbf{R}) - \Phi^-(\mathbf{R}) \, , \tag{3.2}$$

where $\Phi^+(\mathbf{R})$ and $\Phi^-(\mathbf{R})$ are positive and then can be considered as densities. With this wave function $\Psi(\mathbf{R})$ the Schrödinger equation can be split in two separate equations, one for Φ^+ and another for Φ^-, which may be translated into the Monte Carlo language due to their fixed sign. From their definition, it is clear that Φ^+ and Φ^- have a finite overlap with both the bosonic and fermionic ground states. A simple realization of $\Phi^\pm$ at the beginning of the walk can be [5]

$$\Phi^\pm(t=0) = \frac{1}{2} \left(|\Psi| \pm \Psi \right) \, , \tag{3.3}$$

which are positive everywhere and trivially satisfy (3.2). Once the higher-energy excited states have decayed, only the antisymmetric fermionic state and the bosonic components contribute to $\Phi^\pm$,

$$\Phi^\pm(t \to \infty) = \pm c_{\mathrm{f}} \Psi_{\mathrm{f}} + c_{\mathrm{b}} \Psi_{\mathrm{b}} e^{(E_0^{\mathrm{f}} - E_0^{\mathrm{b}})t} \, . \tag{3.4}$$

The difference $\Phi^+ - \Phi^-$ is then proportional to the fermionic wave function Ψ_{f}, but the boson component in Φ^+ and Φ^- is not cancelled, and increases exponentially with time. The attainment of a regime in which the signal of Ψ_{f} is not masked by the bosonic ground-state component depends crucially on the difference $E_0^{\mathrm{f}} - E_0^{\mathrm{b}}$ which is always positive. As a general trend, the existence of this regime is related to the degree of fulfillment of the inequality $E_1^{\mathrm{f}} - E_0^{\mathrm{f}} > E_0^{\mathrm{f}} - E_0^{\mathrm{b}}$, E_1^{f} being the next antisymmetric excited state above E_0^{f}.

In the MC calculation of the energy, the bosonic component does not contribute due to the orthogonality between Ψ_{b} and ψ:

$$E_{\mathrm{MC}} = \frac{\int d\Omega \, \Psi(t \to \infty) \, H \, \psi}{\int d\Omega \, \Psi(t \to \infty) \, \psi} \tag{3.5}$$

$$= \frac{\int d\Omega \, \Phi^+ \, H \, \psi}{\int d\Omega \, (\Phi^+ - \Phi^-) \, \psi} - \frac{\int d\Omega \, \Phi^- \, H \, \psi}{\int d\Omega \, (\Phi^+ - \Phi^-) \, \psi} = E_0^{\mathrm{f}} \, . \tag{3.6}$$

However, the bosonic component in the estimate of the variance of the energy, which requires the calculation of $(\Psi H)^2$, is not cancelled and introduces a noise proportional to $\exp[(E_0^{\mathrm{f}} - E_0^{\mathrm{b}})t]$. The key issue is thus to achieve a rapid decay to the antisymmetric state before the bosonic noise overwhelms the fermionic signal. Obviously, the success of the procedure also depends on the proximity between the nodes of the trial fermionic wave function ψ and the exact ones.

A Monte Carlo realization of the above formalism is the released-node (RN) method. [31] In this approach, the walkers are allowed to cross the nodal surface and survive for a finite lifetime t_{r}. A positive or negative sign is attached to each walker if it has crossed an even or an odd number of times, respectively. The unequal flux of walkers through the model nodal surface is the origin of the small displacement of the nodes towards their exact location.

In order to facilitate a flux of walkers through the nodal surface, they are guided by a positive trial wave function $\psi_g(\mathbf{R})$. For a given trial wave function $\psi(\mathbf{R})$, the guiding function $\psi_g(\mathbf{R})$ has to approach $|\psi(\mathbf{R})|$ away from the nodal surface and must be nonzero in the nodes to make the flux of walkers through it possible. The model for ψ_g that we have used in He simulations is [36,46,61]

$$\psi_g(\mathbf{R}) = \left(\psi(\mathbf{R})^2 + a^2\right)^{1/2} , \tag{3.7}$$

which satisfies both requirements for a proper choice of the value of the parameter a. In fact, the value given to a, which has to be of the same order of magnitude than the mean value of $|\psi(\mathbf{R})|$, governs the flux of walkers through the nodes. However, the RN energies are independent of the explicit value of a, and also of the explicit analytic form of $\psi_g(\mathbf{R})$, the influence of both being only on the variance of the Monte Carlo estimation.

The released-node energy is obtained by projecting out the excited state modelled by $\psi(\mathbf{R})$. This projection is carried out assigning a weight $W(\mathbf{R})$ to each walker, given by

$$W(\mathbf{R}) = \sigma(\mathbf{R}) \, \frac{|\psi(\mathbf{R})|}{\psi_g(\mathbf{R})} , \tag{3.8}$$

$\sigma(\mathbf{R})$ being $+1$ (-1) for an even (odd) number of crossings. The released-node energy is thus determined through

$$E_{\rm RN}(t_r) = \frac{\sum_{t\le t_r} W(\mathbf{R}) \, E_{\rm L}(\mathbf{R})}{\sum_{t\le t_r} W(\mathbf{R})} , \tag{3.9}$$

where the sums are extended to all the surviving walkers with a lifetime less than t_r, and $E_{\rm L}(\mathbf{R}) = \psi(\mathbf{R})^{-1} H \psi(\mathbf{R})$.

The evolution towards an asymptotic regime is usually analyzed by plotting a sequence of energies $E_{\rm RN}(t_{r,1})$, $E_{\rm RN}(t_{r,2})$,...,$E_{\rm RN}(t_{r,\max})$ with $t_{r,1} < t_{r,2} < \ldots < t_{r,\max}$. In this form, one can determine the size of the systematic error of the released-node energy and the time evolution of the stochastic noise due to the growth of the population of negative walkers.

3.3. *Analytic improvement of the trial wave function*

The *quality* of the variational upper bound generated by the FN-DMC method, and the success of the RN method in discriminating the fermionic signal above the statistical noise, are a function of ψ for a given system. It is therefore useful, and complementary to the numerical optimization of those methods, to have a mechanism available for analytically improving the trial wave function.

Let us write the imaginary-time dependent wave function as the product of a phase and a modulus

$$\Psi(\mathbf{R}, t) = e^{i\Omega(\mathbf{R},t)} \, \Phi(\mathbf{R}, t) . \tag{3.10}$$

With this decomposition, the Schrödinger equation for $\Psi(\mathbf{R},t)$ splits in two coupled equations, one for the modulus and one for the phase,

$$-\frac{\partial\Phi}{\partial t} = D(\boldsymbol{\nabla}_{\mathbf{R}}\Omega)^2\,\Phi - D(\boldsymbol{\nabla}^2_{\mathbf{R}}\Phi) + (V(\mathbf{R}) - E)\Phi \tag{3.11}$$

$$\frac{\partial\Omega}{\partial t} = D\left[(\boldsymbol{\nabla}^2_{\mathbf{R}}\Omega) + 2(\boldsymbol{\nabla}_{\mathbf{R}}\Omega)\cdot\frac{(\boldsymbol{\nabla}_{\mathbf{R}}\Phi)}{\Phi}\right] \,. \tag{3.12}$$

If an initial guess for Φ and Ω is given (Φ_0, Ω_0), the above equations provide the first order correction for both functions. In the case of the modulus, and writing $\Phi(\mathbf{R},t) = \exp(U(\mathbf{R},t))$, the first correction to an initial U_0 is

$$U = U_0 - t\left[D\left((\boldsymbol{\nabla}_{\mathbf{R}}\Omega_0)^2 - (\boldsymbol{\nabla}^2_{\mathbf{R}}U_0) - (\boldsymbol{\nabla}_{\mathbf{R}}U_0)^2\right) + (V(\mathbf{R}) - E)\right] \,. \tag{3.13}$$

Let us consider, for example, a boson system in which $\Omega = 0$. If a pair potential is assumed $V(\mathbf{R}) = \sum_{i<j} v(r_{ij})$, and we start with U_0 =const., the Jastrow form for the boson wave function is recovered, $U = \sum_{i<j} u(r_{ij})$ with $u(r) \sim -v(r)$. In a next iteration, considering U_0 as the Jastrow wave function one may identify in (3.13) a three-body term with an analytic form extensively used in variational calculations of quantum liquids, [19]

$$(\boldsymbol{\nabla}_{\mathbf{R}}U_0)^2 \sim \sum_k \mathbf{G}_k\cdot\mathbf{G}_k \tag{3.14}$$

with

$$\mathbf{G}_k = \sum_{l\neq k}\frac{u'(r_{kl})}{r_{kl}}\,\mathbf{r}_{kl} \,. \tag{3.15}$$

More important in the excited-state and fermionic problems is the improvement of the phase. The first-order correction to an initial choice Ω_0 is given by [36,62]

$$\Omega = \Omega_0 + Dt\left[(\boldsymbol{\nabla}^2_{\mathbf{R}}\Omega_0) + 2(\boldsymbol{\nabla}_{\mathbf{R}}\Omega_0)\cdot(\boldsymbol{\nabla}_{\mathbf{R}}U_0)\right] \,. \tag{3.16}$$

A straightforward application of this approximation is the case of the spectrum of a ^{3}He impurity dissolved in liquid ^{4}He. Considering a Jastrow approximation for the modulus, and the Feynman model for the excitation $\Omega_0 = \mathbf{q}\cdot\mathbf{r}_{\mathrm{I}}$, Eq. (3.16) generates the first correction, which contains the backflow correlations introduced by Feynman and Cohen [63]

$$\Omega = \mathbf{q}\cdot\left(\mathbf{r}_{\mathrm{I}} + 2Dt\sum_j\frac{u'(r_{\mathrm{I}j})}{r_{\mathrm{I}j}}\,\mathbf{r}_{\mathrm{I}j}\right) \,. \tag{3.17}$$

Applying Eq. (3.16), one also arrives to the Feynman-Cohen wave function for the study of the phonon-roton excitation branch in liquid ^{4}He,

$$\psi^{\mathrm{BF}}(\mathbf{R}) = \sum_{i=1}^{N} e^{i\mathbf{q}\cdot\bar{\mathbf{r}}_i}\,\psi^0(\mathbf{R}) \,, \tag{3.18}$$

where

$$\tilde{\mathbf{r}}_i = \mathbf{r}_i + \sum_{j \neq i} \eta(r_{ij})\, \mathbf{r}_{ij} \ . \tag{3.19}$$

If, in a next step, the following correction is looked for, one arrives to an improved excited-state wave function [36]

$$\psi(\mathbf{R}) = \sum_{i=1}^{N} c^{iq_a \cdot \tilde{r}_i^a}\, \psi^0(\mathbf{R}) \ , \tag{3.20}$$

with

$$\tilde{r}_i^a = r_i^a + AZ_i^a + BT_i^a + C((W_i)_b^a Z_i^b - Y_i^a) \ , \tag{3.21}$$

A, B, C being constants, and

$$Z_i^a = \sum_{j \neq i} \eta(r_{ij})(r_{ij})^a \tag{3.22}$$

$$T_i^a = \sum_{j \neq i} (5\beta(r_{ij})(r_{ij})^a + \beta'(r_{ij})r_{ij}(r_{ij})^a) \tag{3.23}$$

$$(W_i)_b^a = \sum_{j \neq i} (\beta(r_{ij})(r_{ij})^a (r_{ij})_b + \eta(r_{ij})\delta_b^a) \tag{3.24}$$

$$Y_i^a = \sum_{\substack{j \neq i \\ k \neq j}} (\beta(r_{ij})\eta(r_{jk})(r_{ij})^a (r_{ij})_b (r_{jk})^b + \eta(r_{ij})\eta(r_{jk})(r_{jk})^a) \ . \tag{3.25}$$

The function $\eta(r)$ is the one used in $\psi^{\rm BF}(\mathbf{R})$ (3.18,3.19), and $\beta(r) \equiv \eta'(r)/r$. In Eq. (3.21), the first two terms on the r.h.s. correspond to $\psi^{\rm BF}(\mathbf{R})$ and the other three are new contributions. It is worth noticing the appearance of explicit three-body correlations $((W_i)_b^a Z_i^b - Y_i^a)$ in the excitation operator.

The splitting of the Schrödinger equation has also been used to study, from a Monte Carlo viewpoint, states which require a complex wave function. The method, which has been implemented presently, is the fixed-phase (FP) approximation. In this method, one takes a model for the phase and solve by DMC the Schrödinger equation for the modulus. [62] The result is variational and depends on the phase, that enters in the Schrödinger equation for the modulus as an external potential $(D(\boldsymbol{\nabla}_{\mathbf{R}}\Omega)^2)$.

An alternative way [46] of deriving analytic improvements in the trial wave function ψ emerges from the analysis of the local energy coming from a Jastrow-Slater wave function

$$\psi = \psi_{\rm J}\, \psi_{\rm A} \ , \tag{3.26}$$

with $\psi_{\rm A} = D_\uparrow D_\downarrow$, $D_\uparrow$ ($D_\downarrow$) being the Slater determinant of the spin-up (spin-down) atoms with single-particle orbitals $\varphi_{\alpha_i}(\mathbf{r}_j) = \exp(i\, \mathbf{k}_{\alpha_i} \cdot \mathbf{r}_j)$. In this variational description (3.26), the dynamical correlations induced by the interatomic potential are well modelled by the Jastrow factor. On the other hand, the statistical correlations implied by the antisymmetry are introduced with a Slater determinant of plane waves, which is the exact wave function of the free Fermi sea. The two factors account well for the dynamical correlations and the Fermi statistics when these effects are independently considered. However, their product is only a relatively poor approximation for a strongly correlated Fermi liquid. This feature appears reflected in the local energy $E_{\rm L}(\mathbf{R})$, where the crossed kinetic term

$$-D \sum_{i=1}^{N} \frac{\boldsymbol{\nabla}_i \psi_{\rm A}}{\psi_{\rm A}} \cdot \frac{\boldsymbol{\nabla}_i \psi_{\rm J}}{\psi_{\rm J}} \tag{3.27}$$

diverges when approaching the nodal surface $\psi_{\rm A} = 0$. The nodal surface has to change according to the dynamical correlations in order to reduce, or even eliminate, that divergence. One is therefore looking for a trial wave function able to move the nodes of the noninteracting system, *i.e.*, a new ψ with adaptable nodes.

The way to the solution is contained in the time-dependent Schrödinger equation. Let us consider a time-dependent wave function $\phi(\mathbf{R},t)$, with $\phi(\mathbf{R},t=0) = \psi_{\rm A}(\mathbf{R})\psi_{\rm J}(\mathbf{R})$ the initial guess for the trial wave function, satisfying

$$-\frac{\partial \phi(\mathbf{R},t)}{\partial t} = H\phi(\mathbf{R},t) \; . \tag{3.28}$$

A natural choice for a more accurate trial wave function is obtained by solving Eq. (3.28) to first order in t. Near the nodes, which is the relevant region to our purposes, one readily captures the main correction to the original ψ in the form

$$\phi(\mathbf{R},t) = \psi_{\rm A}(\mathbf{R})\,\psi_{\rm J}(\mathbf{R}) + 2tD\sum_{i=1}^{N} \boldsymbol{\nabla}_i\psi_{\rm A}(\mathbf{R}) \cdot \boldsymbol{\nabla}_i\psi_{\rm J}(\mathbf{R}) \; . \tag{3.29}$$

The second term in the r.h.s. of Eq. (3.29) is the first dominant correction to the initial guess $\psi_{\rm A}\psi_{\rm J}$. Such a correction can be *analytically* introduced in a new trial wave function with arguments slightly shifted,

$$\psi_{\rm A}(\tilde{\mathbf{R}}) \equiv \psi_{\rm A}(\mathbf{R}+\delta\mathbf{R}) = \psi_{\rm A}(\mathbf{R}) + \sum_{i=1}^{N} \boldsymbol{\nabla}_i\psi_{\rm A}(\mathbf{R}) \cdot \delta\mathbf{r}_i \; . \tag{3.30}$$

If the equality between $\psi_{\rm J}(\mathbf{R})\psi_{\rm A}(\tilde{\mathbf{R}})$ (3.30) and $\phi(\mathbf{R},t)$ (3.29) is imposed, one identifies the spatial displacement introduced in Eq. (3.30),

$$\delta\mathbf{r}_i = 2Dt\,\frac{\boldsymbol{\nabla}_i\psi_{\rm J}(\mathbf{R})}{\psi_{\rm J}(\mathbf{R})} = Dt\mathbf{F}_{\rm J}(\mathbf{R}) \; . \tag{3.31}$$

The above result shows how the single-particle orbitals of the Slater determinant have to be modified in order to include dynamical effects: the radial coordinates are displaced according to the Jastrow drift force $\mathbf{F}_{\rm J}(\mathbf{R})$. In a general form,

$$\varphi_{\alpha_i}(\mathbf{r}_j) = \exp(i\,\mathbf{k}_{\alpha_i} \cdot \tilde{\mathbf{r}}_j) \; , \tag{3.32}$$

with

$$\tilde{\mathbf{r}}_j = \mathbf{r}_j + \lambda_B \sum_{k \neq j} \eta(r_{jk})\,\mathbf{r}_{jk} \tag{3.33}$$

The function $\eta(r)$ is related to the two-body Jastrow factor and λ_B is a parameter to be determined variationally. As the first method presented in this subsection, the latter approach concludes with the necessary introduction of backflow correlations in the model wave function. If that second method is iterated, starting on the backflow wave function, the next-order wave function shows the appearance of explicit three-body correlations in $\tilde{\mathbf{R}}$, in agreement with the results obtained by the first method.

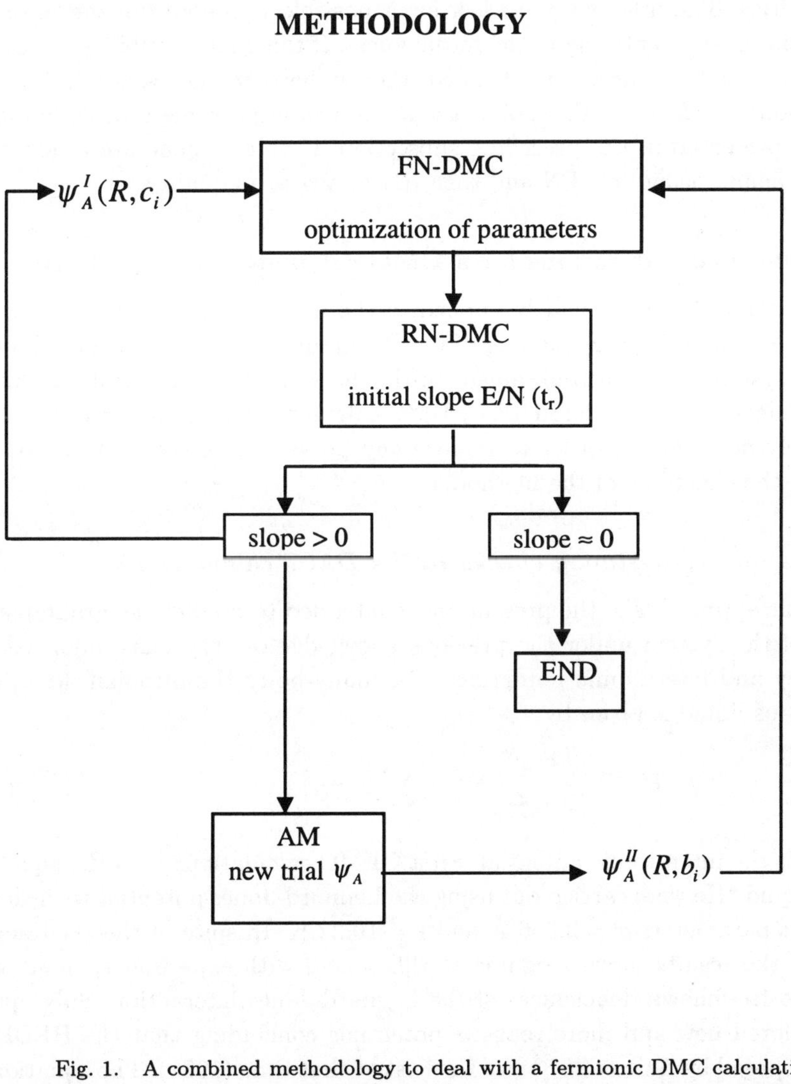

Fig. 1. A combined methodology to deal with a fermionic DMC calculation.

3.4. *A combined strategy*

An efficient calculation of the properties of a fermionic system, like liquid ^{3}He, requires a combined use of the three methods presented in the above subsections. The strategy we have currently used is schematically illustrated in Fig. 1. One of the key points is to have available a measure of the quality of the upper bounds provided by the FN method. Information on the difference between the FN energy and the eigenvalue can be drawn from the RN method. The slope of the released energy versus t_r, at small t_r values, provides a direct measure of the quality of the input nodal surface (the true antisymmetric ground-state wave function would generate a zero slope), and constitutes a means of comparing different trial wave functions. If an initial model for the trial wave function ψ shows a nonzero slope there appear two possibilities. The first one is to ask for a possible better optimization of the variational parameters entering in the initial guess. If this first possibility is exhaustively explored and no success is attained, then it becomes necessary to improve the model nodal surface. At this point, the analytical improvement of ψ, following the method presented in the preceding subsection, is able to generate a new ψ to be explored again, using first FN and then RN to stress its quality.

4. Preliminary considerations for a DMC calculation of liquid ^{4}He

The DMC method has been widely applied to the microscopic study of superfluid ^{4}He. [34,35,37] Nowadays, a large number of results, mainly concerned with the ground state of the system, have been obtained. In this Section, we review some technical aspects that have to be considered in any DMC calculation. The tests discussed here are absolutely necessary in order to remove any possible bias that can negatively influence on the exactness of the method.

4.1. *Inputs and consistency checks in the DMC calculations*

A microscopic approach like the present one is intended to predict the ground-state properties of the system under the previous knowledge of very basic information: mass, density and interatomic potentials. The many-body Hamiltonian describing a homogeneous liquid is given by

$$H = -\frac{\hbar^2}{2m} \sum_{i=1}^{N} \nabla_i^2 + \sum_{i<j} V(r_{ij}) , \tag{4.1}$$

where $V(r)$ is the interatomic potential. First GFMC calculations [17] of the equation of state of liquid ^{4}He were carried out using the Lennard-Jones potential with the de Boer-Michels parameters $\sigma = 2.556$ Å and $\varepsilon = 10.22$ K. In spite of the accuracy of the method, the results showed significant differences with experiment, these were attributed to the known deficiencies of the Lennard-Jones interaction. Subsequent work [27] explored new and more realistic potentials concluding that the HFDHE2 potential proposed by Aziz *et al.* [64] was the best at that time (1981). The equation of state of liquid ^{4}He improved significantly with that Aziz potential, which become the standard one for a decade or more. In 1987, Aziz *et al.* [65] suggested a new potential (HFD-B(HE)), hereafter referred to Aziz II. This new potential was brought about as a consequence of several new theoretical and experimental results which appeared between the publication of the two potentials. First, Ceperley and Partridge [67] pointed out, by means of a quantum Monte Carlo calculation of the interaction energy of two He atoms with internuclear separations less than 1.8 Å, that the Aziz potential is too repulsive below this distance. On the other hand, new experimental measurements of the second virial coefficients and transport properties for ^{3}He and ^{4}He showed evidence of some small inconsistencies of the Aziz potential. Apart from a softer core, the Aziz II potential has its minimum at $\varepsilon = 10.95$ K, $r_m = 2.963$ Å, while Aziz potential has its minimum at $\varepsilon = 10.80$ K, $r_m = 2.967$ Å. Therefore, the new potential is only slightly deeper with the minimum localized at a smaller interatomic separation.

DMC calculations [35] using the Aziz II potential have proved to be able of reproducing, with excellent accuracy, the density dependence of the pressure and the speed of sound. Later models, also proposed by Aziz and collaborators, [68] have shown no additional improvements in the equation of state. Recently, accurate *ab initio* interatomic potentials have been calculated, [69–71] but a complete analysis of the equation of state generated by them is still lacking. Nevertheless, in many condensed-matter simulations the empirical or semiempirical potentials, like the Aziz ones, are more accurate than the *ab initio* models since include to some extent the influence of the medium on the pair interaction.

One of the technical aspects that has to be addressed before a DMC calculation is the choice of the trial wave function ψ used for importance sampling. That function guides the diffusion process to regions where the maximum of the exact wave function is reasonably expected and, what is perhaps more relevant, excludes regions where the interatomic potential becomes extremely large or even singular. For example, in homogeneous liquid ^{4}He, whose potential presents a hard core for distances $r \leq \sigma$ ($\sigma = 2.556$ Å), the introduction of a Jastrow wave function with a proper two-body factor makes the local energy nearly constant even at short interparticle distances. Within reasonable assumptions for ψ, the DMC energy does not depend on ψ but the variance is usually reduced when ψ is improved. Once a specific model for ψ is chosen, a VMC optimization is carried out in order to obtain the best set of variational parameters entering in ψ. In the case of liquid ^{4}He we have mainly worked with three different trial wave functions. The first one is the McMillan two-body factor (1.2) with *optimal* parameter $b = 1.20\ \sigma$ at the experimental equilibrium density $\rho_0^{\text{expt}} = 0.365\ \sigma^{-3}$.

The McMillan two-body factor is a simple and good solution at short distances in order to deal with the hard-core of $V(r)$. However, it is not optimal at intermediate and long distances. An Euler-Lagrange variational optimization [16] shows that there is some structure at intermediate distance, and that there is a long-range behavior $f(r) \simeq 1 - A/r^2$ when $r \to \infty$. The intermediate range can be well approximated

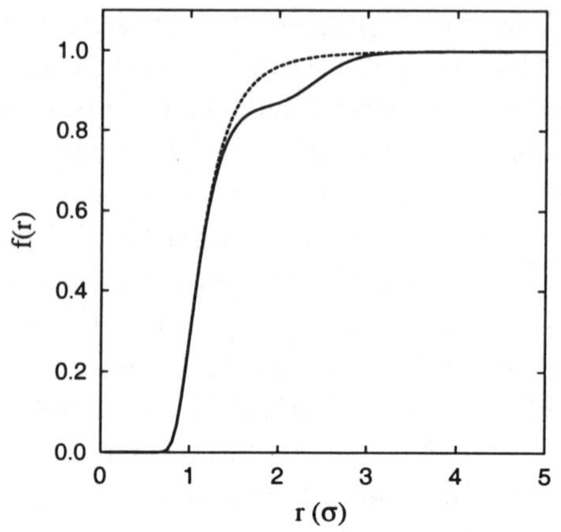

Fig. 2. Two-body correlation factor in liquid ^{4}He at the experimental equilibrium density $\rho_0^{\rm expt}$. Solid line, Reatto (4.2); dashed line, McMillan (1.2).

by the two-body correlation function

$$f(r) = \exp\left[-\frac{1}{2}\left(\frac{b}{r}\right)^5 - \frac{L}{2}\exp\left[-\left(\frac{r-\lambda}{\Lambda}\right)^2\right]\right] , \tag{4.2}$$

suggested by Reatto, [72] with additional parameters L, λ, and Λ. Optimal values of these parameters at $\rho_0^{\rm expt}$ are $L = 0.2$, $\lambda = 2.0\ \sigma$, and $\Lambda = 0.6\ \sigma$. The two-body correlation factors (1.2) and (4.2) are compared in Fig. 2. In MC calculations the long-range tail of $f(r)$ is not included since the use of periodic boundary conditions requires finite-range functions. Nevertheless, the contribution to the energy of that long-range (phonon) contribution has been estimated perturbatively to be less than $1 \cdot 10^{-3}$ K. [17]

Even when the optimal two-body correlation factor is used, the variational binding energy is manifestly smaller than the experimental value. The high density of liquid helium makes that also three-body correlations have to be considered in order to substantially improve the variational bound to the experimental energy. [18–20] The emergence of three-body correlations in the wave function may be drawn from different methods. One of them, which is adequate to the MC approach, is to directly use the imaginary-time Schrödinger equation (see Sec. 3.3). The most used trial wave function is the one suggested by Schmidt *et al.*, [19]

$$\psi_{\rm JT} = \psi_{\rm J} \exp\left[-\frac{1}{4}\lambda\sum_k \mathbf{G}_k \cdot \mathbf{G}_k + \frac{1}{2}\lambda\sum_{i<j}\xi^2(r_{ij})r_{ij}^2\right] , \tag{4.3}$$

where

$$\mathbf{G}_k = \sum_{l \neq k} \xi(r_{kl})\, \mathbf{r}_{kl} \; , \tag{4.4}$$

and

$$\xi(r) = \exp\left[- \left[\frac{r - r_t}{r_\omega} \right]^2 \right] \; . \tag{4.5}$$

The values for the triplet parameters, roughly optimal at the equilibrium density, are: $\lambda = -1.08\ \sigma^{-2}$, $r_t = 0.82\ \sigma$, and $r_\omega = 0.50\ \sigma$.

The introduction of three-body correlations improves the VMC energy appreciably, reducing the difference between the DMC and the Jastrow values by more than a half. Using the HFDHE2 Aziz potential, one obtains at $\rho_0^{\rm expt}$ -5.683 ± 0.014 K and -5.881 ± 0.005 K at the Jastrow level and considering the McMillan (1.2) and Reatto (4.2) two-body factors, respectively. The two- plus three-body wave function (4.3) lowers the energy down to -6.617 ± 0.007 K, much closer to the DMC result -7.120 ± 0.024 K and to the experimental value -7.17 K. In what concerns the choice of ψ as importance sampling wave function we have verified, as expected, that the DMC energy is the same for the three models here presented. [35] The influence of the trial reduces only to improve the variance of the estimation. A good compromise between efficiency and complexity is achieved by using the Reatto two-body factor (4.2).

In a DMC calculation, there are several technical features that have to be considered in order to reduce any systematic bias down to a predefined value: the mean number of walkers, the time-step to be used and, if a homogeneous system is simulated, the dependence on the number of particles. Concerning the first one, it is necessary to ensure that the mean population of walkers $n_{\rm w}$ is large enough to effectively sample the probability distribution $\psi\,\Psi$. That number $n_{\rm w}$ slightly depends on the trial wave function ψ used for importance sampling: $n_{\rm w}$ decreases when the overlap between ψ and Ψ increases. In liquid ^{4}He, and using a Jastrow wave function $\psi_{\rm J}$, a mean number $n_{\rm w} = 400$ has proved to be enough to remove the possible bias associated to $n_{\rm w}$.

The DMC algorithm requires the use of a finite time-step value Δt whose dependence has also to be analyzed. In Fig. 3, a characteristic result of the total energy as a function of the time-step Δt is shown. It corresponds to a quadratic DMC calculation of the liquid ^{4}He energy at the experimental equilibrium density $\rho_0^{\rm expt}$. One can see in the figure that there is a clear departure from the linear time dependence supplied by linear DMC algorithms. If a second order polynomial fit $E/N = (E/N)_0 + A(\Delta t)^2$ (solid line in the Fig. 3) to the QDMC results is performed, one obtains an extrapolated value $(E/N)_0 = -7.124 \pm 0.003$ K, which is indistinguishable from the values obtained working with the two smallest Δt. Therefore, it is plausible to calculate the properties of the system accurately using a single value for Δt, lying in the stated range, without the necessity of a complete analysis in time to extrapolate the correct results.

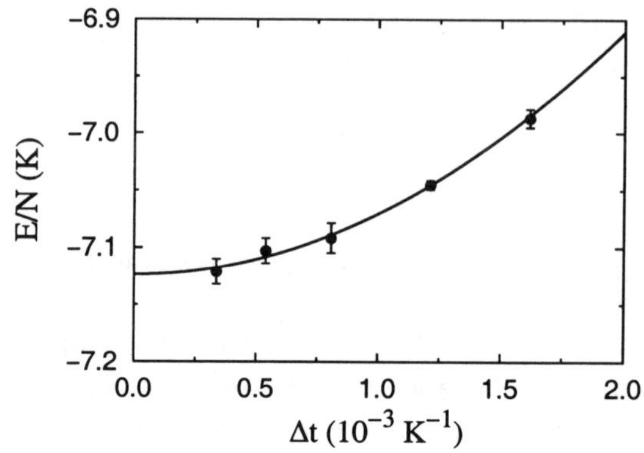

Fig. 3. Time-step dependence of the binding energy of liquid ^{4}He at the experimental equilibrium density $\rho_0^{\rm expt}$. Points with error bars, DMC results; line, second-order polynomial fit.

The size dependence of the simulation is the third aspect that has to be analyzed if the calculation corresponds to a homogeneous system. At a given density ρ, which is the physical input, the calculation is performed with a prefixed number of particles N. That determines the maximum length over which one can extract *real* information, $L/2$ with $L = (N/\rho)^{1/3}$. The contribution to the energy coming from distances larger than $L/2$ have to be estimated in some way. The usual approach to take these corrections into account (tails) consists of considering a uniform background or, in other words, to assume that the two-body radial distribution function $g(r) = 1$ for $r > L/2$. With this *uniform* approximation, the tail correction to the potential energy is

$$(V/N)_{\rm t}(\rho) = 2\pi\rho \int_{L/2}^{\infty} dr\; r^2 V(r) \; . \tag{4.6}$$

At large interparticle distances, the three-body terms of the kinetic energy become much smaller than the two-body term. Thus, the tail correction to the kinetic energy is well approximated by

$$(T/N)_{\rm t}(\rho) = -\frac{\hbar^2}{m}\, 2\pi\rho \int_{L/2}^{\infty} dr\; r^2 \left(\frac{2}{r}\, u'(r) + u''(r) \right) \; , \tag{4.7}$$

$u(r)$ being the pseudopotential appearing in the two-body correlation factor $f(r) = \exp(u(r))$.

For a given number of particles N, the energy of a homogeneous system is defined as the sum of the mixed energy, derived from the DMC calculation, and the tail corrections (4.6, 4.7). At a fixed density ρ, a study on the N dependence of the total energy has to be carried out in order to determine the lowest value of N from

Table 1. VMC energy per particle of liquid ^{4}He as a function of the number of particles N used in the simulation. The results (in K) correspond to a VMC calculation using the Reatto trial wave function (4.2) at density $\rho_0^{\rm expt} = 0.365\ \sigma^{-3}$. The interatomic potential is the HFD-B(HE) Aziz model. [64] The figures in parenthesis are the statistical errors.

N	E/N	T/N	V/N	$(T/N)_{\rm t}$	$(V/N)_{\rm t}$
64	-6.202(17)	15.053(24)	-21.255(20)	2.946	-1.309
128	-6.069(15)	15.227(24)	-21.297(22)	0.241	-0.647
190	-6.084(11)	15.271(20)	-21.355(18)	0.102	-0.433
256	-6.059(11)	15.253(20)	-21.312(18)	0.068	-0.321

on the energy is nearly constant (within the statistical errors). Table 1 shows the behavior of the energy, with increasing N, from a VMC calculation of liquid ^{4}He at $\rho_0^{\rm expt}$. For this particular choice of the trial wave function, the Reatto model (4.2), one can see that the energy becomes constant for $N \geq 128$.

The size corrections can also be included in a rather different way. In this second method, it is required that both the pseudopotential $u(r)$ and the interatomic potential $V(r)$ become zero at $r = L/2$. Therefore, the kinetic tail disappears and only the tail contribution to the potential energy remains. However, for the same number of particles the total tail correction is larger than in the first method, which is the one used in our work. Moreover, this second method implies the use of trial wave functions that explicitly depend on the size of the simulation box, a not very pleasant feature. Obviously, if a sufficiently large N is used both methods converge to a common value.

4.2. *Unbiased estimators*

The implementation of unbiased (pure) estimators based on the forward walking strategy was discussed in subsection 2.2. In the present subsection, we explore the efficiency of the method by calculating pure estimators of radial operators. The analysis has been performed in a calculation of liquid ^{4}He using the HFD-B(HE) Aziz potential and three different trial wave functions for importance sampling. They are a McMillan form ψ_{J1} (1.2) with $b = 1.20\sigma$, a Reatto's factor ψ_{J2} (4.2) with parameters $b = 1.20\sigma$, $\lambda = 2.0\sigma$, $\Lambda = 0.6\sigma$, and finally, a third one which includes three-body correlations ψ_{JT} (4.3,4.4,4.5), with $\lambda = -1.08\sigma^{-2}$, $r_t = 0.80\sigma$, and $r_\omega = 0.41\sigma$. In the three cases, the values of the parameters are those which optimize the variational energy at the experimental equilibrium density $\rho_0^{\rm expt} = 0.365\ \sigma^{-3}$.

Results for the potential energy per particle using ψ_{J1}, ψ_{J2}, and ψ_{JT} are reported in Table 2. A small but significant difference between the extrapolated results $\langle V/N\rangle_e$ appears, pointing to a bias related to the quality of the trial wave function. The bias is completely removed when the pure estimator is calculated, as one can see in the last row of Table 2. The three values for $\langle V/N\rangle_p$ are indistinguishable

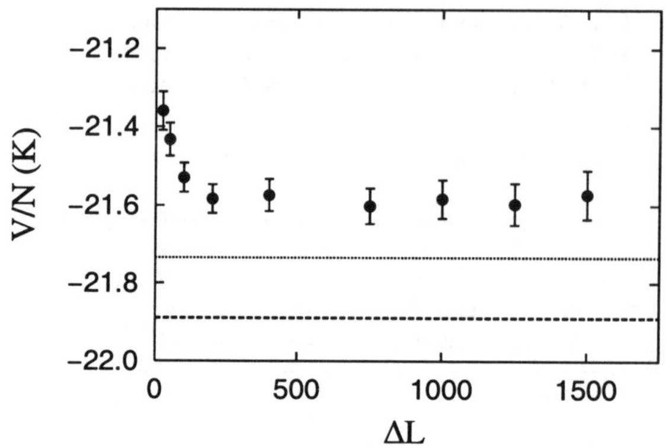

Fig. 4. Pure expectation value of V/N for liquid ^{4}He at $\rho_0^{\rm expt}$ as a function of the block length ΔL. The long-dashed and short-dashed lines stand for the extrapolated estimations using ψ_{J2} and ψ_{JT}, respectively.

and, what is more important, they evidence a systematic error of the extrapolation approximation. In fact, none of the extrapolated values is statistically compatible with the common pure value, the closest estimation being the one obtained with ψ_{JT} which actually is the best variational choice. Subtracting the pure potential energy from the total energy $E/N = -7.267 \pm 0.013$ K, [35] un unbiased value for the kinetic energy is obtained, $T/N = 14.32 \pm 0.05$ K. Experimental determinations from analysis of deep inelastic scattering data predicts a slightly lower value $(T/N)^{\rm expt} = 13.3 \pm 1.3$ K, [73] the difference being due mainly to the significant errors in the experimental measurement of the tail of the response function.

To assess the stability of the method, the dependence of the pure expectation value of V/N on the length of the forward walking is plotted in Fig. 4. After a transition regime, and already for relatively small block lengths ΔL ($\Delta L \geq 250$), an asymptotic limit is reached where the systematic error is practically negligible. Notice that, according the algorithm we have presented in Subsection 2.2, a forward

Table 2. Variational, mixed, extrapolated and pure expectation values of V/N (in K) for liquid ^{4}He at $\rho_0^{\rm expt}$ using different trial wave functions ψ.

	ψ_{J1}	ψ_{J2}	ψ_{JT}
$\langle V/N \rangle_v$	-21.054(26)	-21.311(18)	-21.348(20)
$\langle V/N \rangle_m$	-21.459(8)	-21.600(8)	-21.541(8)
$\langle V/N \rangle_e$	-21.864(30)	-21.889(24)	-21.734(25)
$\langle V/N \rangle_p$	-21.56(5)	-21.59(5)	-21.58(5)

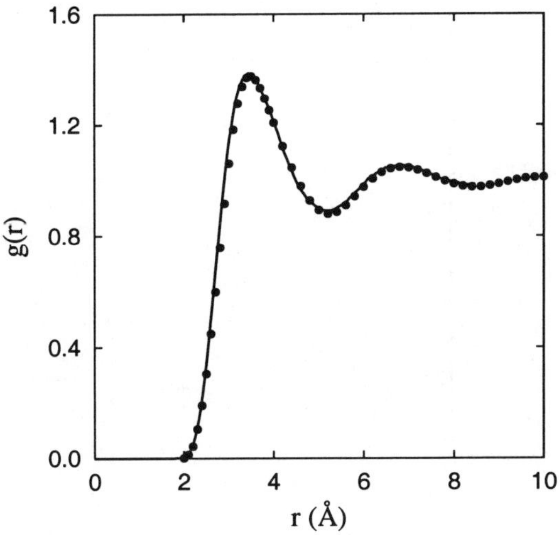

Fig. 5. Pure expectation result of the two-body radial distribution function (solid line) for liquid ^{4}He at $\rho_0^{\rm expt}$, in comparison with experimental data (points) from Ref. 74.

walking of length ΔL is constructed from data ranging from L to $2L$, and hence the length of the forward walking is not the same for all the walkers. This effect is not relevant provided that a region of stability exists. On the other hand, one can determine the asymptotic value within a single run collecting data for different block lengths. The statistical errors associated with each individual ΔL–value can be lowered in the same way as mixed estimators, *i.e.*, by continuing the evolution of the series. The biases associated to the extrapolated expectation values are also shown in Fig. 4, where $\langle V/N \rangle_e$ using ψ_{J2} and ψ_{JT} are represented by a long-dashed and a short-dashed line, respectively.

Other important quantities in the study of quantum liquids can also be calculated with the pure algorithm. In particular, the two-body radial distribution function

$$g(r_{12}) = \frac{N(N-1)}{\rho^2} \frac{\int |\Phi_0(\mathbf{r}_1,\ldots,\mathbf{r}_N)|^2 \, d\mathbf{r}_3 \ldots d\mathbf{r}_N}{\int |\Phi_0(\mathbf{r}_1,\ldots,\mathbf{r}_N)|^2 \, d\mathbf{r}_1 \ldots d\mathbf{r}_N} , \tag{4.8}$$

and the static structure function

$$S(q) = \frac{1}{N} \frac{\langle \Phi_0 | \rho_{\mathbf{q}} \, \rho_{-\mathbf{q}} | \Phi_0 \rangle}{\langle \Phi_0 | \Phi_0 \rangle} , \tag{4.9}$$

with

$$\rho_{\mathbf{q}} = \sum_{i=1}^{N} e^{i\mathbf{q}\cdot\mathbf{r}_i} . \tag{4.10}$$

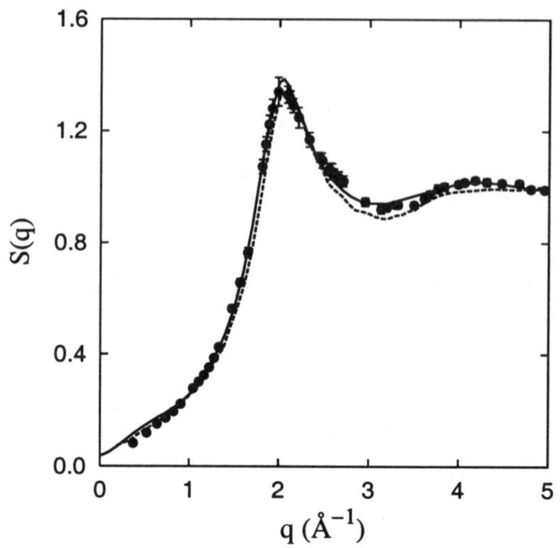

Fig. 6. Pure estimation of the static structure function (points) for liquid ^{4}He at $\rho_0^{\rm expt}$, in comparison with experimental determinations from Refs. 74 (solid line) and 75 (dashed line). The error bars of the theoretical points are only shown where larger than the size of the symbols.

The result obtained for $g(r)$ is shown in Fig. 5 in comparison with the experimental data of Ref. 74. As one can see, the pure expectation value of $g(r)$ is in a good agreement with the experimental $g(r)$ for all the calculated r values. In Fig. 6, the pure structure function $S(q)$ is plotted together with the experimental measures of Refs. 74 and 75. An overall agreement between the theoretical and experimental $S(q)$ is obtained, our result lying well between the two experimental determinations. The extrapolated estimations of $g(r)$ and $S(q)$ are not significantly different of the pure result. Nevertheless, the use of pure estimations is always more secure because it removes possible biases present in the extrapolated estimations.

5. Bulk liquid ^{4}He: ground-state and excitations

A great deal of microscopic information is nowadays available from GFMC and DMC calculations. The accuracy of the interatomic potentials and the efficiency of the QMC methods have allowed for a nearly complete description of liquid ^{4}He properties like the equation of state, the spatial structure or the momentum distribution. The extension of the QMC methods to the calculation of excited-state energies is more recent. Actually, only the phonon-roton spectrum and the vortex excitation energies have been calculated with an accuracy similar to the one achieved in the ground-state. On the other hand, the full dynamics of the system is still far from being solved by stochastic methods. In this Section, we present some of the results obtained in the last years concerning the ground- and excited-state properties.

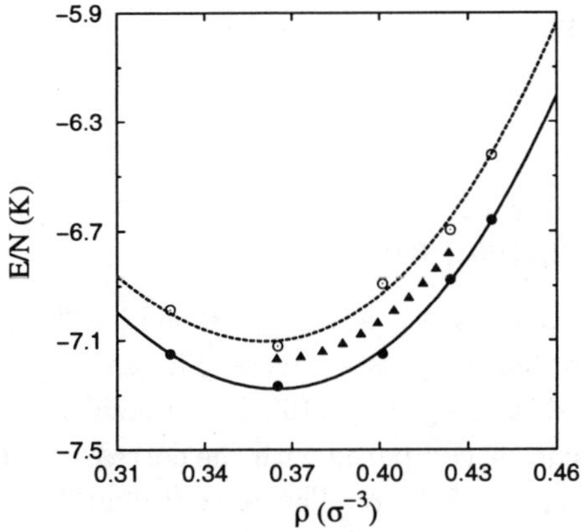

Fig. 7. Equation of state of liquid ^{4}He. The open circles are the QDMC results with the Aziz potential and the dashed line is a fit to the calculated energies. The full circles correspond to the QDMC energies with the Aziz II potential; the solid line is a fit to these energies. The experimental values, represented by solid triangles, are taken from Ref. 13. The error bars of the QDMC results are smaller than the size of the symbols.

5.1. *Equation of state and other ground-state properties*

A complete analysis of the ground-state properties of liquid ^{4}He using quadratic DMC and the Aziz II potential is nowadays available. [35,66] Table 3 contains results for the total and partial energies in the vicinity of the experimental equilibrium density $\rho_0^{\rm expt} = 0.3648\sigma^{-3}$. The potential energies are calculated by means of the pure estimator and the kinetic energies as the difference between the total and pure potential energies. The comparison with experimental data, also contained in Table 3, shows a slight overbinding in all the density range considered. This feature is better illustrated in Fig. 7, where the experimental data [13] are compared with both Aziz and Aziz II results for the total energy. The lines in Fig. 7 correspond to a numerical fit to the DMC data using the analytical form,

$$e(\rho) = b\rho + c\rho^{1+\gamma} , \tag{5.1}$$

Table 3. Total and partial energies of liquid ^{4}He using the Aziz II potential. Experimental data from Ref. 13. All energies are in K.

$\rho(\sigma^{-3})$	E/N	V/N	T/N	$(E/N)^{\rm expt}$
0.328	-7.150(10)	-19.14(6)	11.99(8)	
0.365	-7.267(13)	-21.59(5)	14.32(5)	-7.17
0.401	-7.150(16)	-23.88(9)	16.73(9)	-7.03
0.424	-6.877(22)	-25.45(8)	18.57(8)	-6.77

proposed by Stringari and Treiner [76] in the density functional framework ($e \equiv E/N$). The best set of parameters, for the two Aziz potentials, are

$$\begin{array}{ll} \text{AzizII} & \text{Aziz} \\ b = (-27.258 \pm 0.017)\ K\sigma^3 & b = (-26.947 \pm 0.016)\ K\sigma^3 \\ c = (114.95 \pm 0.22)\ K\sigma^{3(1+\gamma)} & c = (115.72 \pm 0.21)\ K\sigma^{3(1+\gamma)} \\ \gamma = 2.7324 \pm 0.0020 & \gamma = 2.7160 \pm 0.0020 \end{array} \tag{5.2}$$

According to the above fits, the equilibrium densities are $\rho_0 = 0.3647\ \sigma^{-3}$ and $\rho_0 = 0.3606\ \sigma^{-3}$ for the Aziz II and Aziz potentials, respectively. The energies at the equilibrium densities are $e_0 = -7.277$ K and $e_0 = -7.103$ K for Aziz II and Aziz, respectively. These theoretical results can be compared with the corresponding experimental data, $\rho_0^{\rm expt} = 0.3648\ \sigma^{-3}$ and $e_0 = -7.17$ K. The Aziz II equilibrium density matches exactly $\rho_0^{\rm expt}$ but the binding energy is slightly larger. It is worth noticing that in recent temperature scales (EPT76) [13] the experimental energy is larger ($e_0 = -7.23$ K) and approaches the Aziz II result.

Once the equation of state $e(\rho)$ is known, it is straightforward to calculate the isothermal compressibility,

$$\kappa(\rho) = \frac{1}{\rho} \left(\frac{\partial \rho}{\partial P} \right)_T , \tag{5.3}$$

where $P(\rho) = \rho^2\ (\partial e \,/\, \partial \rho)$ is the pressure, and the speed of sound

$$c(\rho) = \left(\frac{1}{m\, \kappa\, \rho} \right)^{1/2} . \tag{5.4}$$

In Table 4, results of the pressure, the compressibility and the speed of sound using the two Aziz potentials are compared with the corresponding experimental values at $\rho_0^{\rm expt}$. It is remarkable the accuracy provided by the Aziz II potential, which generates results for these quantities in a complete agreement with experimental data. The equation of state corresponding to the Aziz potential leads to results which are slightly worse.

Recent experiments [77–79] on cavitation in liquid ^{4}He at low temperatures have motivated the theoretical study [66,80–86] of liquid helium properties at low temperature and negative pressure (see Chapter 7 in this volume [87]). Some interesting

Table 4. DMC results for the pressure P, the compressibility κ and the speed of sound c at the experimental equilibrium density using the Aziz and Aziz II potentials. The last row contains experimental values from Ref. 13.

	P(atm)	κ(atm^{-1})	c(m/sec)
Aziz	0.878±0.073	0.01199±0.00004	241.53±0.44
Aziz II	-0.019±0.075	0.01241±0.00004	237.40±0.46
Expt.	0.	0.0124	237.2

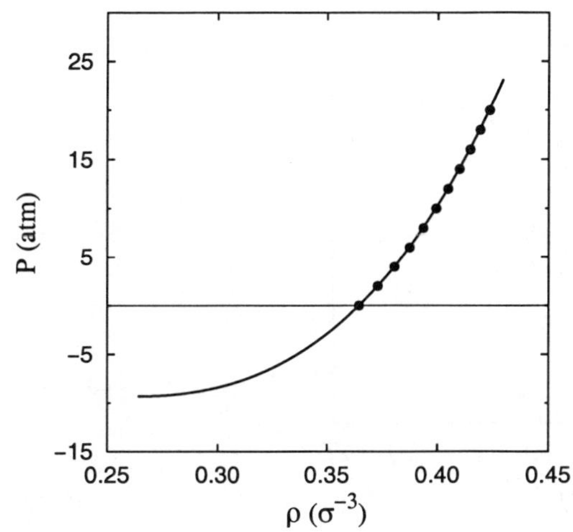

Fig. 8. Pressure of liquid ^{4}He as a function of density. Points: experimental results, [13]; solid line: QDMC result.

questions have been raised, *e.g.*, the determination of the tensile strength (*i.e.*, the magnitude of the negative pressure at which cavitation becomes likely), and the spinodal pressure (*i.e.*, the pressure at which liquid helium becomes macroscopically unstable against density fluctuations). In Ref. 80 the spinodal pressure was estimated by fitting to the measured speed of sound [88] c as a function of pressure P several polynomial and Padé forms, and then extrapolating into the negative pressure region to determine the zero of $c(P)$. From a different point of view, the spinodal pressure was calculated in Ref. 82 using two different phenomenological models that reproduce the equation of state in the measured positive pressure region. Although an overall agreement between the phenomenological calculations and the empirical results was obtained, some questions arise, as for instance to what extent the extrapolated results depend on the form used in the fit, or on the density functional used in the calculations.

We have extended our QDMC calculations to lower densities to explore the negative pressure regime. [66] Quantities derived from the energy such as the pressure or the speed of sound have been obtained through a third and fourth degree interpolation, with unnoticeable changes when higher orders were introduced in the interpolation method. The QDMC prediction of $P(\rho)$ is shown in Fig. 8 (solid line) for the whole range of densities, in comparison with experimental data for positive pressures. [13] The agreement between the Aziz II results and the experimental data is quite impressive.

The speed of sound c as function of pressure P is displayed in Fig. 9. The

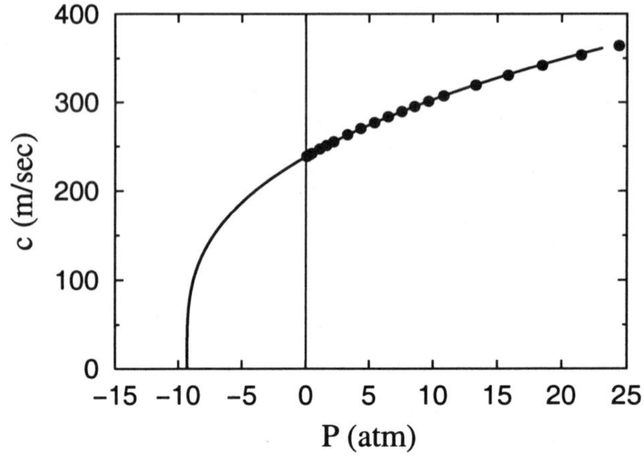

Fig. 9. Speed of sound in liquid ^{4}He as a function of pressure. The experimental points are taken from Ref. 88, and the solid line corresponds to the QDMC results.

points are the experimental values of Ref. 88, and the solid line corresponds to the QDMC results. The accuracy provided by the Aziz II potential is again remarkable, giving results for the speed of sound in close agreement with the experiment. It can be seen that c drops to zero very rapidly when approaching the spinodal point. According to our results, the spinodal point of liquid ^{4}He is located at a density $\rho_s = 0.264 \pm 0.002\ \sigma^{-3}$ and pressure $P_s = -9.30 \pm 0.15$ atm.

The DMC calculations yield other relevant quantities of the system. In particular, information on the spatial structure of the liquid can be drawn from the two-body radial distribution function $g(r)$ and also from its Fourier transform, the static structure function $S(q)$. As commented in Sec. 4.2, the calculation of radial operators others than the Hamiltonian is more involved if an unbiased estimation is desired. In Figs. 5 and 6, respectively, pure estimations of $g(r)$ and $S(q)$ are compared with experimental data at the experimental equilibrium density $\rho_0^{\rm expt}$. The agreement between theoretical results and experimental data is excellent.

The influence of the density on $g(r)$ is shown in Fig. 10. When the density increases, the degree of localization increases and the mean next-neighbor distance becomes smaller. The corresponding results for $S(q)$ are reported in Fig. 11. The results reported in Fig. 11 are Fourier transforms of the $g(r)$ functions shown in Fig. 10 and, therefore, the low-q behavior is not accessible. In order to get more accurate $S(q)$ values when $q \to 0$ it is necessary to carry out an explicit calculation; that is the case, for example, of the $S(q)$ reported in Fig. 6.

The Bose-Einstein statistics of ^{4}He atoms is dramatically manifested in the off-

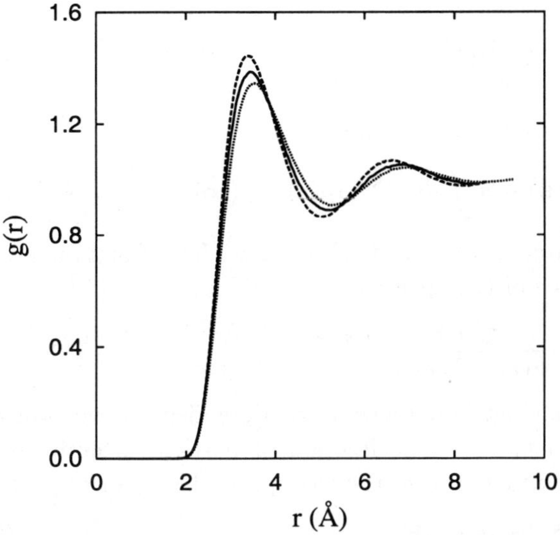

Fig. 10. Two-body radial distribution function of liquid ^{4}He at several densities. Dotted, solid, and dashed lines correspond to densities 0.365 σ^{-3}, 0.401 σ^{-3}, and 0.424 σ^{-3}, respectively.

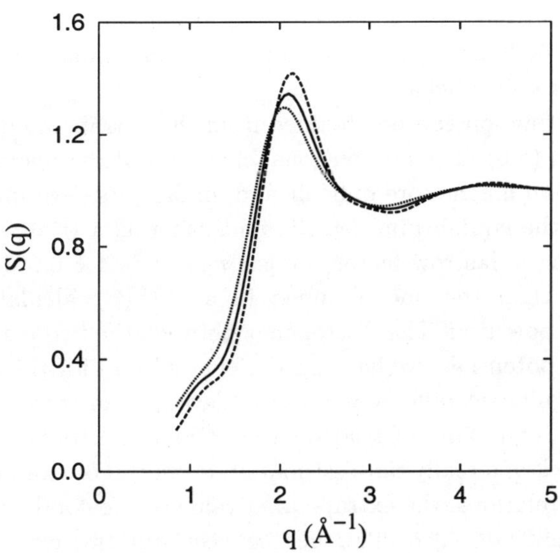

Fig. 11. Static structure function of liquid ^{4}He at several densities. Same notation than in Fig. 10.

diagonal long range order (ODLRO) present in the one-body density matrix $\rho(r)$ defined as

$$\rho(r_{11'}) = N \, \frac{\int \Psi(\mathbf{r}'_1,\ldots,\mathbf{r}_N)\,\Psi(\mathbf{r}_1,\ldots,\mathbf{r}_N)\; d\mathbf{r}_2\ldots d\mathbf{r}_N}{\int |\Psi(\mathbf{r}_1,\ldots,\mathbf{r}_N)|^2 \;\; d\mathbf{r}_1\ldots d\mathbf{r}_N} \tag{5.5}$$

and its Fourier transform, the momentum distribution

$$n(k) = (2\pi)^3\,\rho\, n_0\, \delta(\mathbf{k}) + \rho \int d\mathbf{r}\; e^{i\mathbf{k}\cdot\mathbf{r}}\; (\rho(r) - \rho(\infty)) \; . \tag{5.6}$$

Both quantities can be calculated using the QDMC code. The function $\rho(r)$ is obtained as the expectation value of the operator

$$\left\langle \frac{\Psi(\mathbf{r}_1,\ldots,\mathbf{r}_i+\mathbf{r},\ldots,\mathbf{r}_N)}{\Psi(\mathbf{r}_1,\ldots,\mathbf{r}_N)} \right\rangle \tag{5.7}$$

evaluated in configuration space, considering a set of random displacements of particle i. The condensate fraction n_0, *i.e.*, the fraction of particles occupying the zero-momentum state, may be extracted from the asymptotic behavior of $\rho(r)$

$$n_0 = \lim_{r\to\infty} \rho(r) \; . \tag{5.8}$$

In Fig. 12 the ^{4}He momentum distribution is plotted as $k\,n(k)$, for three values of the density. The correlations between particles cause that the population of states with high momenta increases with density. The shoulder observed at $k \simeq 2$ Å^{-1} for the three curves, which has been observed in other theoretical calculations of $n(k)$, [89,90] has been attributed in the past to the zero-point motion of the rotons. [91] On the other hand, it has been proven that, if the condensate fraction is non-zero, $n(k)$ diverges as $1\,/\,k$ when $k \to 0$. [92] However, the finite size of the simulation cell precludes the observation of this $k \to 0$ behavior.

The value of the condensate fraction can be obtained from the extrapolated estimation of $\rho(r)$ and the relation (5.8). The non-diagonal character of the operator $\rho(r)$ makes its unbiased estimation much more difficult and, in fact, its determination is not yet satisfactory. At the equilibrium density, and using as a trial wave function for importance sampling a Jastrow factor, we get $n_0 = 0.084 \pm 0.001$. [35] This result is slightly smaller than the one obtained in a GFMC calculation (0.092 ± 0.001) [90] using the Aziz potential. The discrepancy between the two results is not due to the use of different potentials; we have also calculated $\rho(r)$ for the two Aziz potentials and found no significant differences. More relevant than the choice between several Aziz potentials is the kind of trial wave function ψ introduced in the DMC calculation, and the lack of a really pure estimation for $\rho(r)$. For example, if ψ incorporates three-body correlations, the extrapolated value of the condensate fraction decreases, $n_0 = 0.078 \pm 0.001$. The difference between our two extrapolated values is larger than the statistical errors and, consequently, an uncontrolled bias remains. From an experimental view, n_0 is an elusive quantity that can be extracted from deep inelastic neutron scattering. [93] The most recent data [94] point to an extrapolated zero-temperature value $n_0 = 0.072 \pm 0.007$ which is closer to the

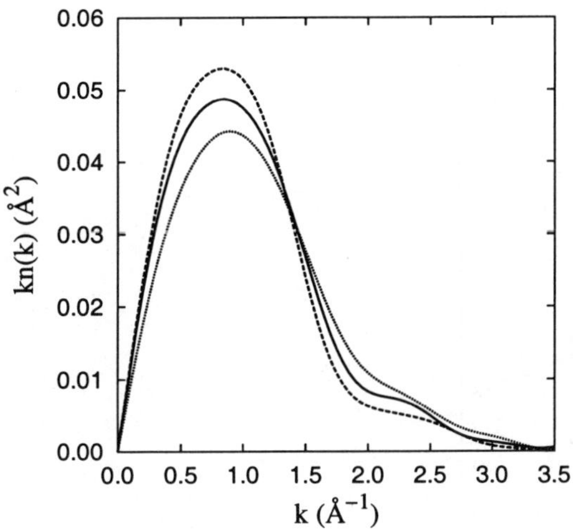

Fig. 12. ^{4}He momentum distribution plotted as $kn(k)$. The dashed, solid, and dotted lines stand for values at densities 0.328 σ^{-3}, 0.365 σ^{-3}, and 0.401 σ^{-3}, respectively.

three-body result than to the Jastrow one. Nevertheless, the experimental estimation is also influenced by theoretical models which are necessary to extract n_0 and, therefore, a clean measure is not yet available.

A final point of interest is the density dependence of the condensate fraction. In Fig. 13, the change in the value of n_0 is shown for a wide range of densities. The condensate fraction decreases with the density, following a nearly quadratic law with ρ. In the figure, a quadratic fit to the results is shown to guide the eye.

5.2. *Excited states: phonon-roton spectrum*

The physical nature of excitations in superfluid ^{4}He at low momenta is nowadays still not completely understood, [93,95] in contrast to the vast knowledge about its static properties. The collective excitation energy follows a characteristic dispersion curve which starts with a phonon excitation, reaches a first maximum (maxon), lowers to a local minimum (roton), and then increases again up to a plateau with an energy that is twice the roton energy. From general ideas on the nature of the excitations in an interacting Bose fluid, Feynman [96] proposed the first microscopic approach to the problem. The Feynman trial wave function provides a qualitative description of the excitation spectrum but fails in reproducing the roton energy by a factor two. Later on, Feynman and Cohen [63] included backflow correlations in the trial wave function, reducing the differences between the experimental data and the original Feynman prediction by one half . Following Feynman's language, the phonon-roton branch corresponds to collective density-like excitations where the

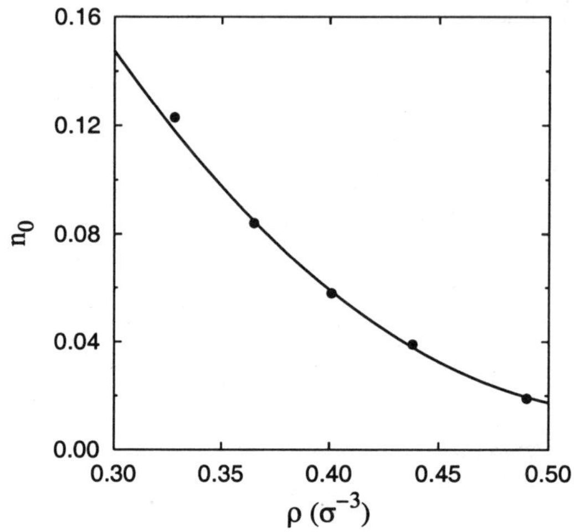

Fig. 13. Condensate fraction in liquid ^{4}He as a function of the density. The solid line is a second-order polynomial fit to the calculated values. The error bars are smaller than the size of the symbols.

condensate fraction does not enter in an explicit way. On the other hand, it has been argued by Glyde and Griffin [97] that the continuous spectrum results from the superposition of density excitations, dominating in the phonon region, and single-particle excitations, important in the roton minimum. This theory has emerged after the experimental determination of the temperature dependence of the excitation spectrum [98,99] which shows the phonon peak in the dynamic structure function $S(q,\omega)$ at both sides of the λ-transition whereas the roton practically disappears in the normal phase.

In the last years a considerable effort has been made towards quantitative improvement of the microscopic predictions for the phonon-roton excitation spectrum $\varepsilon(q)$. Manousakis and Pandharipande [100] calculated $\varepsilon(q)$ by means of the correlated basis function (CBF) method using a basis of Feynman-Cohen states. The variational Monte Carlo (VMC) method using shadow wave functions has also proved to be quantitatively quite efficient in the calculation of $\varepsilon(q)$ [101] in spite of its approximate description of the ground state. The application of *ab initio* Monte Carlo methods to this problem has been, however, severely hindered by the sign problem associated to the excited wave function. Boninsegni and Ceperley [102] have calculated $\varepsilon(q)$ by means of a path integral Monte Carlo (PIMC) calculation of $S(q,\omega)$ from a Laplace inversion of the imaginary-time correlation factor $S(q,t)$. However, for noisy data this inversion is an ill-posed problem that prevents a model-independent determination.

The DMC method has also been applied to the study of the phonon-roton spectrum of liquid ^{4}He. [36] The imaginary-time Schrödinger equation is solved for the function $f(\mathbf{R},t) = \psi(\mathbf{R})\Phi(\mathbf{R},t)$, $\psi(\mathbf{R})$ being a trial wave function eigenstate of the momentum operator and, therefore, orthogonal to the ground state. Within these conditions, the simplest model for ψ was proposed by Feynman, [96]

$$\psi^{\rm F}(\mathbf{R}) = \sum_{i=1}^{N} e^{i\mathbf{q}\cdot\mathbf{r}_i}\, \psi^0(\mathbf{R}) \,, \tag{5.9}$$

$\psi^0(\mathbf{R})$ being the ground-state wave function. The first correction to $\psi^{\rm F}(\mathbf{R})$, originally proposed by Feynman and Cohen, [63] includes backflow correlations. In this case, the excited wave function is given by

$$\psi^{\rm BF}(\mathbf{R}) = \sum_{i=1}^{N} e^{i\mathbf{q}\cdot\tilde{\mathbf{r}}_i}\, \psi^0(\mathbf{R}) \,, \tag{5.10}$$

where

$$\tilde{\mathbf{r}}_i = \mathbf{r}_i + \sum_{j\neq i} \eta(r_{ij})\, \mathbf{r}_{ij} \,. \tag{5.11}$$

The inclusion of backflow correlations improves the variational results for $\varepsilon(q)$ appreciably but the quantitative agreement with experiment is still poor, especially near the roton minimum.

In the DMC implementation, the excited problem is mapped onto a fermion-like calculation in which, instead of a complex wave function, a real probability distribution function $f(\mathbf{R},t)$ is used. With the aim of achieving that, we choose as importance sampling wave function the superposition of two excitations of momenta $\mathbf{q}$ and $-\mathbf{q}$ which are degenerate in energy,

$$\psi^{\rm F}(\mathbf{R}) = \sum_{i=1}^{N} \cos(\mathbf{q}\cdot\mathbf{r}_i)\, \psi^0(\mathbf{R}) \tag{5.12}$$

at Feynman's level, and

$$\psi^{\rm BF}(\mathbf{R}) = \sum_{i=1}^{N} \cos(\mathbf{q}\cdot\tilde{\mathbf{r}}_i)\, \psi^0(\mathbf{R}) \tag{5.13}$$

when backflow correlations are included.

In a first step, the excitation energy

$$\varepsilon(q) = \frac{\langle\psi(q)|H|\Phi(q)\rangle}{\langle\psi(q)|\Phi(q)\rangle} - \frac{\langle\psi^0|H|\Phi^0\rangle}{\langle\psi^0|\Phi^0\rangle} \tag{5.14}$$

was calculated using the FN method and the two models for the excited wave functions (5.12, 5.13). These calculations provide upper bounds to the exact energies that can, in principle, be improved down to the exact energies using the released node method. In the RN method, it is necessary to introduce a positive-defined

Table 5. Excitation energies at $\rho_0^{\rm expt}$ in comparison with experimental data. The FN-$\psi^{\rm F}$ and FN-$\psi^{\rm BF}$ columns are the fixed-node energies using $\psi^{\rm F}$ and $\psi^{\rm BF}$, respectively. The RN column corresponds to the released-node estimation. Experimental data is taken from Ref. 103.

q (Å^{-1})	FN-$\psi^{\rm F}$ (K)	FN-$\psi^{\rm BF}$ (K)	RN (K)	Expt. (K)
0.369	7.56 ± 0.49	7.24 ± 0.38	7.02 ± 0.49	7.0
1.106	18.47± 0.49	16.52 ± 0.43	13.82 ± 0.43	13.8
1.844	13.82± 0.54	10.37 ± 0.59	9.18 ± 0.59	8.9

guiding wave function $\psi_{\rm g}(\mathbf{R})$ that has to approach $|\psi(\mathbf{R})|$ away from the nodal surface and must be non-zero in the nodes to make possible the flux of walkers through them. We have used $\psi_{\rm g}(\mathbf{R}) = \left(\psi(\mathbf{R})^2 + a^2\right)^{1/2}$, which satisfies both requirements for a proper choice of the value of the parameter a.

In Table 5, fixed-node values using $\psi^{\rm F}$ and $\psi^{\rm BF}$, and the released-node estimation are compared with experimental data [103] at the equilibrium density $\rho_0^{\rm expt}$. The FN results with backflow correlations improve the Feynman ones for the three values of q in a magnitude which depends on q. Thus, the inclusion of backflow correlations seems slightly more relevant in the roton than in the maxon. On the other hand, the RN excitation energies agree with the experimental data for the three values of q within the statistical errors.

The asymptotic regime of the RN energy has been studied empirically by fitting the function

$$E(t_r) = E_\infty + C\, e^{-t_r/\tau} \tag{5.15}$$

to the largest t_r values. The guideline has been the acceptance of the RN values only when there are no discrepancies between the largest t_r data and the asymptotic limit E_∞. The fit (5.15) has been used to decide whether to trust or not the MC values but not to provide the asymptotic limit.

We have verified that, at $\rho_0^{\rm expt}$ and for values $q < 2.5$ Å^{-1}, the RN energies using the Feynman wave function do reach the expected constant regime, the difference between the largest t_r calculated energy and the value of E_∞ predicted by the χ^2-fit (5.15) being less than the statistical error. This is not the case for higher densities. For example, near the freezing density, at $\rho_P = 0.438\ \sigma^{-3}$, that agreement only subsists for the lowest q and for the value of q nearest to the roton. For the other values we have had to include backflow correlations to reach the asymptotic limit. In Fig. 14, the RN excitation energy per particle is plotted as a function of t_r for $q = 1.11$ Å^{-1} and $q = 1.84$ Å^{-1}. Near the roton, $q = 1.84$ Å^{-1}, both the Feynman and backflow results show a common asymptotic value without a significant slope. At $q = 1.11$ Å^{-1}, near the maxon energy, the situation is clearly different: the backflow results have reached a constant behavior whereas the Feynman ones show a slow approach to the correct value, which has not yet been achieved for the maximum value of t_r. The latter behaviour is also observed for the highest value

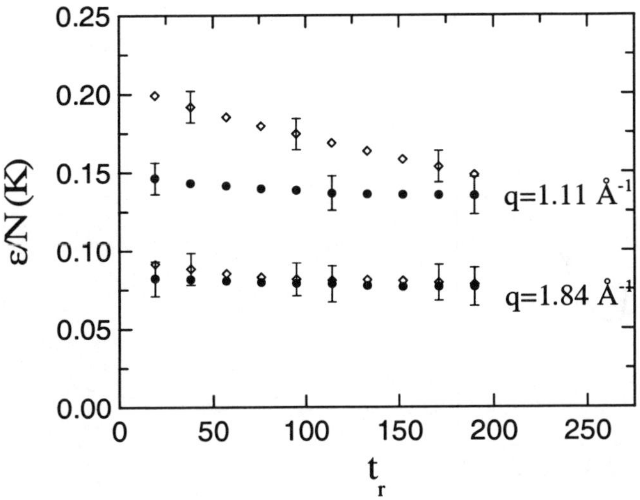

Fig. 14. Excitation energies per particle as a function of the lifetime t_r at ρ_P. The full circles are obtained using $\psi^{\rm BF}$ and the diamonds using $\psi^{\rm F}$.

of q ($q \simeq 2.6$ Å^{-1}) both at $\rho_0^{\rm expt}$ and ρ_P. In this case, the inclusion of backflow correlations in the wave function is not enough to eliminate the bias and a significant difference exists between the largest t_r energy and the asymptotic value predicted by the numerical fit (5.15).

In Fig. 15, the RN excitation energies are compared with the experimental spectrum [103] at $\rho_0^{\rm expt}$. The RN results correspond, for each q, to the last point in the release process, the error bars being only the statistical errors. As commented before, the systematic errors are less than the statistical ones except for the highest q result ($q = 2.58$ Å^{-1}). For this latter value of q, we also report an estimate coming from the extrapolation supplied by the fit (5.15). Apart from this point, where the RN method shows the shortcomings of the backflow wave function at a such high value of q, the agreement between the RN results and the experiment is excellent. As a matter of comparison, the FN results using $\psi^{\rm BF}$ are also plotted. It is worth noticing the difference between the FN energies in the maxon and in the roton regimes; the roton is reproduced quite accurately whereas in the maxon the backflow correlations overestimate appreciably the excitation energy. At the highest value of q, where the spectrum bends down, the FN energy is quite far from the experimental data. Recent and more accurate experimental data, [104,105] beyond the roton excitation, have shown that the collective spectrum terminates at a value of q that is twice the roton momentum and with an energy that asymptotically approaches twice the roton energy. According to the theory proposed by Pitaevskii [106] the flattening of the spectrum is due to a double-roton excitation. In fact, a VMC calculation with shadow wave functions [101] has provided evidence that the excita-

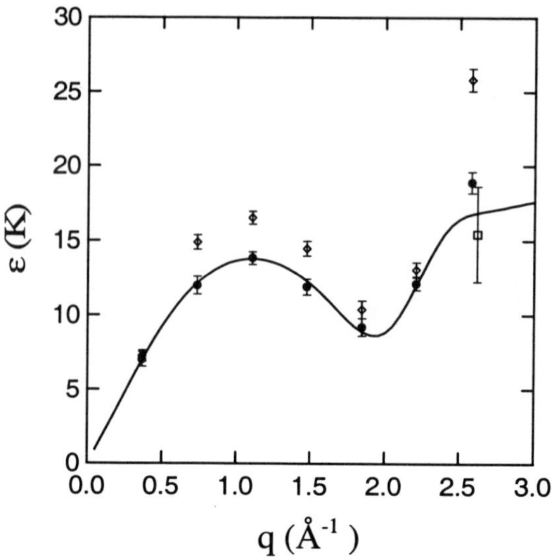

Fig. 15. Phonon-roton spectrum at the equilibrium density ρ_0^{expt}. The full circles are the RN results and the diamonds correspond to a FN calculation with ψ^{BF}. The open square, which has been slightly shifted to the right for clarity, is the result of the extrapolation with the fit (5.15). The solid line is the experimental data from Ref. 103.

tion energy in this region is effectively improved if a double excitation wave function is used. Nevertheless, more *ab initio* work is still necessary to reproduce that final part of the phonon-roton branch.

Is well known from neutron scattering data that the location and depth of the roton minimum depends on density. In Fig. 16, the dependence of the phonon, maxon, and roton energies on ρ is shown. The calculations were extended to the negative pressure region in which no experimental data is available. Specifically, the figure contains results at a density $\rho = 0.285\ \sigma^{-3}$, close to the spinodal density [66] ($\rho_s = 0.264\,\sigma^{-3}$). At the lowest density, the roton minimum has not disappeared but the difference between its energy and the maxon one is quite small. Both the maxon and phonon energies increase with ρ, the former in a smooth form and the latter with a more pronounced slope which is practically constant in accordance with the dominant linear increase of the speed of sound. Fig. 16 contains also experimental data at ρ_0^{expt} [104] and ρ_P [107] which agrees satisfactorily with the DMC results.

6. Two-dimensional liquid ^{4}He

In recent years a great deal of interest has been devoted to the study of quantum boson liquids in restricted geometries. [45,108–115] From a theoretical point of view, thin films of liquid ^{4}He adsorbed on different solid substrates have been studied using variational techniques, based on the hypernetted-chain/Euler-Lagrange (HNC–EL)

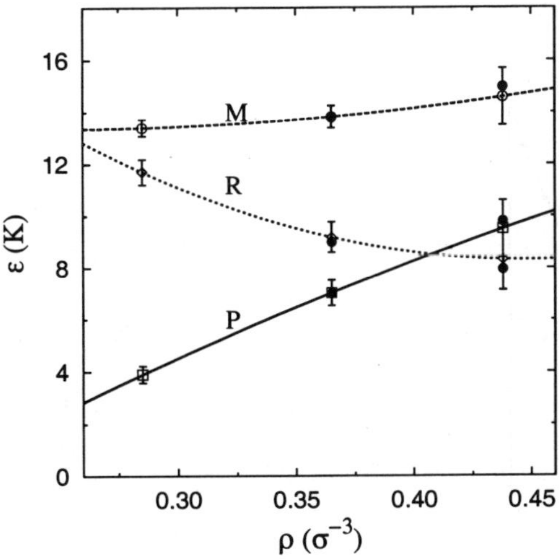

Fig. 16. Phonon (P), roton (R) and maxon (M) energies as a function of density; the lines are "guides to the eye". The solid circles are experimental data.

theory, [108–110] Density Functional theories (DFT), [111,112] and also Monte Carlo techniques. [45,113–115] According to these investigations, liquid ^{4}He films display different behaviors depending on the strength of the substrate potential.

The role of a reduced dimensionality appears mainly when the substrate-helium interaction contains deep and narrow potential wells. In these cases, the degree of freedom perpendicular to the surface is practically frozen out and low-coverage films are stabilized at the surface forming nearly two-dimensional (2D) systems. To this category belongs graphite, solid H_2 on glass. and some alkali metal substrates like Li and Mg. On these substrates the growth of the first liquid layers is predicted to proceed via layering transitions: [108,110,111] by increasing the surface coverage single atomic monolayers develop and become stable one on the top of the other. This prediction has been confirmed experimentally for helium on graphite by heat capacity [116] and third sound measurements. [117] On the other hand, variational calculations of ground-state and dynamic properties of single ^{4}He monolayers at zero temperature reveal a striking 2D behaviour over a wide range of coverages. [108–110] Such calculations seem to indicate that single monolayers of liquid helium adsorbed on strong-binding substrates represent physical realizations of 2D homogeneous quantum liquids.

From the point of view of Monte Carlo simulations, ^{4}He in confined geometries has already been the subject of several studies. Variational Monte Carlo techniques have been applied to the study of both inhomogeneous films on substrates, [118] and more extensively to homogeneous 2D ^{4}He, [119,120] where the simulation is easier.

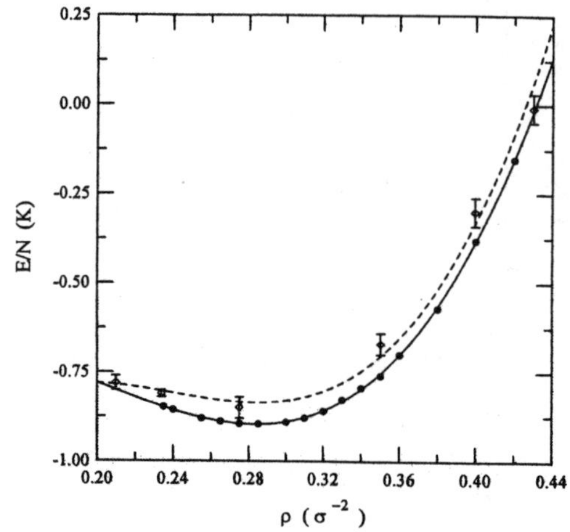

Fig. 17. Equation of state for 2D liquid ^{4}He. The solid circles correspond to the DMC energies obtained with the Aziz II potential (the statistical error bars are smaller than the size of the symbols); the solid line is the polynomial fit (6.1) to the calculated energies. The open diamonds are the GFMC results of Ref. 113 with the Aziz potential and the dashed line is the reported fit to these values.

Strictly two-dimensional ^{4}He has also been the subject of more exact Monte Carlo calculations. At $T = 0$ Whitlock *et al.* [113] calculated, using Green's function Monte Carlo (GFMC), the equation of state and other ground-state properties of liquid and solid ^{4}He in two dimensions, giving an estimate of the freezing and melting densities. Some years later, and using DMC, the equation of state of 2D liquid ^{4}He was updated [37] using the Aziz II potential. At finite temperature, PIMC techniques have been employed by Ceperley and Pollock [114] to investigate the superfluid transition, which in 2D belongs to the Kosterlitz-Thouless universality class. Recently, Gordillo and Ceperley [121] have calculated the phase diagram of 2D liquid ^{4}He at low temperatures using PIMC.

6.1. *Ground-state properties*

As it has been shown in the previous section, the Aziz II potential is slightly more attractive than the original Aziz potential. Therefore, the binding energies at the different densities are somewhat larger than the ones obtained in Ref. 113 which used the Aziz potential. For example, at $\rho = 0.275\ \sigma^{-2}$ (near to the equilibrium density) the energies per particle are $E/N = -0.8519 \pm 0.0044$ K and $E/N = -0.8950 \pm 0.0019$ K for the Aziz and Aziz II potentials, respectively. In Fig. 17, the DMC equation of state [37] calculated with the Aziz II potential is compared with the Aziz data from Ref. 113.

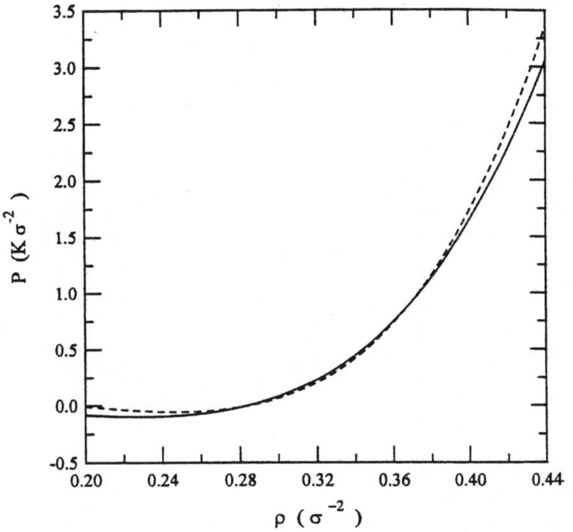

Fig. 18. Density dependence of the pressure in 2D liquid ^{4}He. The solid line corresponds to the Aziz II data and the dashed line to the Aziz data from Ref. 113.

The equation of state of 2D liquid ^{4}He is well parametrized by a third-degree polynomial

$$e = e_0 + B\left(\frac{\rho - \rho_0}{\rho_0}\right)^2 + C\left(\frac{\rho - \rho_0}{\rho_0}\right)^3 , \tag{6.1}$$

where $e = E/N$ and ρ_0 is the equilibrium density. In Table 6, the values of the parameters, which fit the DMC results best, are compared with the values reported in Ref. 113. The results for the equilibrium density are very close, whereas the B and C parameters are appreciably different. These differences affect the predictions for the density dependence of the pressure and compressibility, as well as the estimate of the spinodal density. The two fits are shown in Fig. 17 together with the Monte Carlo data. The cubic polynomial fit (6.1) fits the data rather well and no significant improvement in the χ^2 quality of the fit is found by increasing the order of the polynomial function.

Once the functional dependence of the energy on the density is established, the pressure and speed of sound are straightforwardly derived. The pressure obtained from the Aziz II equation of state and the Aziz result from Ref. 113 are compared in Fig. 18. An appreciable difference is found in the regimes of low and high densities. A larger difference is found by comparing the predictions of the two fits for the speed of sound, shown in Fig. 19. The speed of sound at the 2D equilibrium density is $c(\rho_0) = 92.8 \pm 0.6$ m/sec, which is nearly 3 times smaller than the speed of sound at the saturation density of 3D bulk liquid ^{4}He, $c^{3D}(\rho_0) = 238.3$ m/sec.

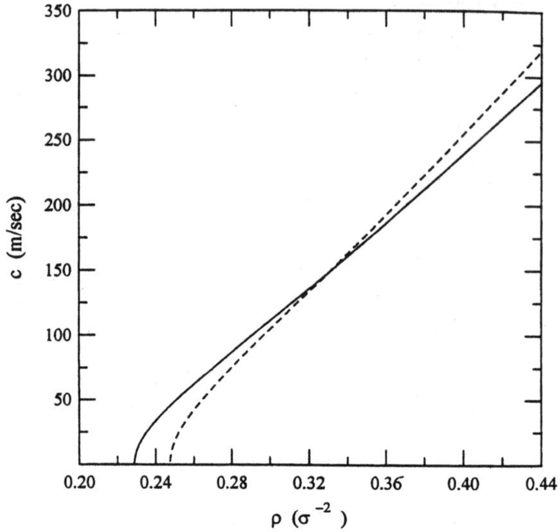

Fig. 19. Density dependence of the speed of sound in 2D liquid ^{4}He. The solid line corresponds to the Aziz II data and the dashed line to the Aziz data from Ref. 113.

The speed of sound is a linear function of the density over a wide range. Only when one approaches the spinodal density, where the system becomes unstable against infinitesimal density fluctuations, the speed of sound drops suddenly to zero and the compressibility diverges. The estimate of the spinodal density from our fit gives $\rho_s = 0.228 \pm 0.002\ \sigma^{-2}$, which is significantly smaller than the value $\rho_s = 0.247\ \sigma^{-2}$ obtained from the data of Ref. 113.

Important information on the structure of the ground state is obtained from the two-body radial distribution function and from its Fourier transform, the static structure factor. Both these quantities can be calculated using the method for pure estimators described in Sec 4.2. In Fig. 20, the DMC results obtained for the radial distribution function at three different densities are reported. As the density increases, more peaks at large interparticle distances appear. At the highest density shown $\rho = 0.420\ \sigma^{-2}$, which is just before freezing, four peaks are clearly visible

Table 6. Parameters of the equation of state (6.1) of 2D liquid ^{4}He.

$\rho_0(\sigma^{-2})$	0.28380 ± 0.00015	0.28458
$e_0(K)$	-0.89706 ± 0.00061	-0.8357
$B(K)$	2.065 ± 0.014	1.659
$C(K)$	2.430 ± 0.035	3.493
χ^2/ν	0.99	1.45

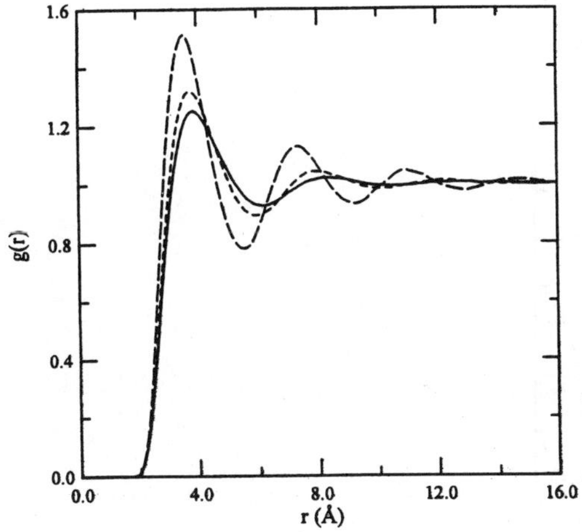

Fig. 20. Radial distribution function of 2D liquid ^{4}He for three densities: $\rho = 0.275\ \sigma^{-2}$ (solid line), $\rho = 0.320\ \sigma^{-2}$ (short-dashed line), $\rho = 0.420\ \sigma^{-2}$ (long-dashed line).

in the distribution function. This is a signal of the proximity to solidification. The height of the first peak in $g(r)$ increases with the density and it shifts towards smaller interparticle distances. It is interesting to compare the height of the first peak in the radial distribution function for the 2D and 3D systems: in 2D at the equilibrium density $g(r_m) \simeq 1.25$ whereas the corresponding value in 3D is 1.38. This is a clear indication that the 2D system is more dilute and possesses less correlations at equilibrium than its 3D counterpart. Close to the freezing density the heights of the first peak in $g(r)$ of the 2D and 3D systems become comparable.

In Fig. 21, the static structure factor for three values of the surface density is plotted. As the density increases, the peak in $S(k)$ increases and the values at the lowest momenta accessible in our calculation increase. Due to phonon excitations, the structure factor $S(k)$ is expected to go to zero in the long wavelength limit as $S(k) \sim k/2mc$. As the density decreases, and the spinodal density is approached, the speed of sound c drops to zero and consequently the slope in $S(k)$ diverges. This behavior, which has been observed in the variational calculations of Refs. 108 and 109, agrees qualitatively with the DMC results.

Finally, Fig. 22 contains DMC results for the condensate fraction obtained at different densities, ranging from the spinodal point up to the freezing density. The data have been fitted with the quadratic polynomial

$$n_0(\rho) = n_0(\rho_0) + a\left(\frac{\rho - \rho_0}{\rho_0}\right) + b\left(\frac{\rho - \rho_0}{\rho_0}\right)^2 , \tag{6.2}$$

where ρ_0 is the equilibrium density, $\rho_0 = 0.284\ \sigma^{-2}$. The values of the parameters

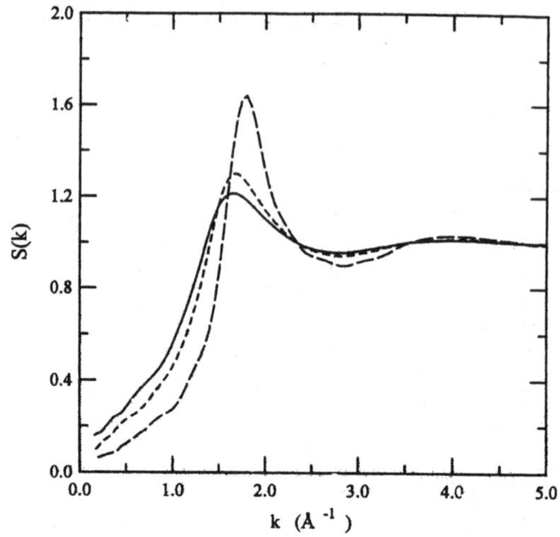

Fig. 21. Static structure function of 2D liquid ^{4}He for three densities: $\rho = 0.275\ \sigma^{-2}$ (solid line), $\rho = 0.320\ \sigma^{-2}$ (short-dashed line), $\rho = 0.420\ \sigma^{-2}$ (long-dashed line).

giving the best fit are the following

$$\begin{aligned} n_0(\rho_0) &= 0.233 \pm 0.001 \\ a &= -0.583 \pm 0.006 \\ b &= 0.44 \pm 0.02\ . \end{aligned} \tag{6.3}$$

The value of the condensate fraction at equilibrium density $n_0(\rho_0)$ is consistent with the estimate reported in Ref. 114 $n_0(\rho_0) \simeq 0.22$, obtained by extrapolating the PIMC results for the algebraic decay of the one-body density matrix to zero temperature. The Aziz II results for the condensate fraction are somewhat smaller than the ones reported in Ref. 113; for example at low density, $\rho = 0.275\ \sigma^{-2}$, we find $n_0 = 0.251 \pm 0.005$ whereas the GFMC calculation of Ref. 113 gives $n_0^{\rm GFMC} = 0.36 \pm 0.05$. This discrepancy between DMC and GFMC results persists over the whole density range and only at very high density, $\rho = 0.400\ \sigma^{-2}$, $n_0^{\rm DMC}$ and $n_0^{\rm GFMC}$ become consistent. The reason for this is unclear, but certainly it is not due to the revised version of the Aziz potential used in our DMC simulation. We have repeated our DMC calculation of the condensate fraction using the Aziz potential for the two densities, $\rho = 0.275\ \sigma^{-2}$ and $\rho = 0.400\ \sigma^{-2}$ and no difference was found with the results obtained with Aziz II. It is interesting to notice that near the freezing density, where the mean interparticle distance in the 2D and 3D systems are comparable, also the condensate fraction is nearly the same ($n_0 \simeq 4\%$).

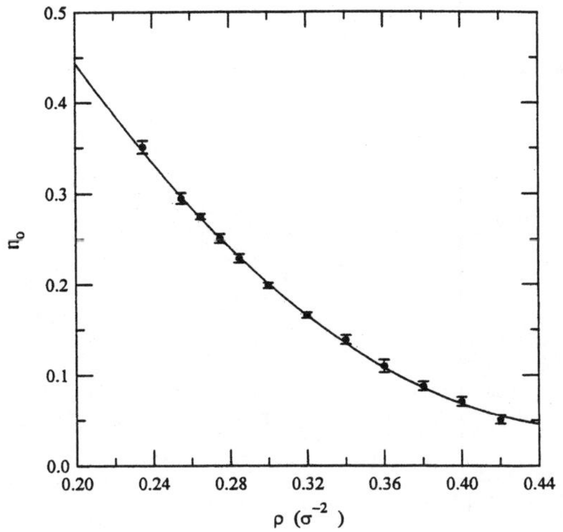

Fig. 22. Density dependence of the condensate fraction in 2D liquid ^{4}He. Solid circles with error bars: results of DMC calculations; solid line: fit from Eq. 6.2.

6.2. *Vortex excitation*

A vortex excitation is an eigenstate of the N-particle Hamiltonian H and of the z-component of the angular momentum L_z with eigenvalue $\hbar N\ell$, corresponding to an integer number ℓ of quanta of circulation. [122] The simplest microscopic wave function to describe a vortex state was introduced by Feynman: [123]

$$\psi^{\mathrm{F}}(\mathbf{R}) = e^{i\ell\varphi_{\mathrm{F}}} \prod_{i=1}^{N} f(r_i)\Phi_0(\mathbf{R}) , \tag{6.4}$$

where $\varphi_{\mathrm{F}} = \sum_{i=1,N} \theta_i$ is the Feynman phase with θ_i the azimuthal angle of the i-th particle, $\Phi_0(\mathbf{R})$ describes the ground state of the system and $f(r_i)$ is a function of the radial distance of each particle from the vortex axis, which models the density near the core. Only vortices with one quantum of circulation, *i.e.*, $\ell = \pm 1$, are considered. Starting from the Feynman phase φ_{F} as zeroth order ansatz, the first correction to the phase can be obtained following the method described in Sec. 3.3. The new phase, which includes backflow correlations, is of the form [62]

$$\varphi_{\mathrm{BF}} = \varphi_{\mathrm{F}} - 2Dt \sum_{j \neq i} \frac{r_j}{r_i} \frac{u'(r_{ij})}{r_{ij}} \sin(\theta_i - \theta_j) . \tag{6.5}$$

The wave function constructed with the phase φ_{BF} is the vortex analogue of the Feynman-Cohen [63] backflow wave function for the phonon-roton excitation branch. In order to deal with a real walker probability distribution function $f(\mathbf{R}, t)$, one

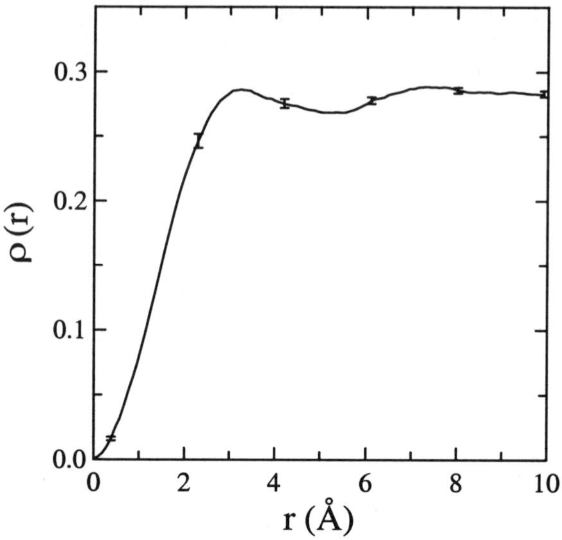

Fig. 23. Pure density profile of a ^{4}He vortex at the 2D equilibrium density.[38] Some of the points calculated are shown with errorbars.

considers as a trial function the superposition of two vortex states, one with positive and one with negative circulation, which are degenerate in energy.

The DMC calculations [38] have been carried out with two models for the radial function $f(r)$: $f_1(r) = 1 - e^{(-r/a)}$, and $f_2(r) = 1$. The first function gives, in the trial function, a density which decreases to zero at the vortex axis over a distance of order a, for which we take the value $a = 1$ Å, whereas the second one does not contain any parameter associated with the vortex core. For the backflow function entering Eq. (6.5) we have used the same functional form and the same values for the parameters as in Ref. 62. The calculation was performed using three different trial wave functions: $\psi^{\rm F1}$ and $\psi^{\rm F2}$ which correspond to the Feynman phase $\varphi_{\rm F}$ with the radial terms f_1 and f_2 respectively, and $\psi^{\rm BF}$ corresponding to the backflow phase $\varphi_{\rm BF}$ with f_2.

The common pure-estimated profile for the particle density $\rho(r)$ is presented in Fig. 23, where it clearly appears that a zero particle density is reached on the vortex axis. This result contradicted the prediction of a significant non-zero particle density on the axis obtained with the backflow phase in Ref. 62. It is worth noting that the result of Ref. 62 was probably biased due to the extrapolation technique used by the authors to improve the mixed estimator result. Variational calculations based on shadow wave functions have shown a non-zero density in the axis.[124] At the variational level, this result is also obtained using the backflow model commented before.[38] However, when the calculation is improved up to the exact level that finite density in the core unambiguously disappears.

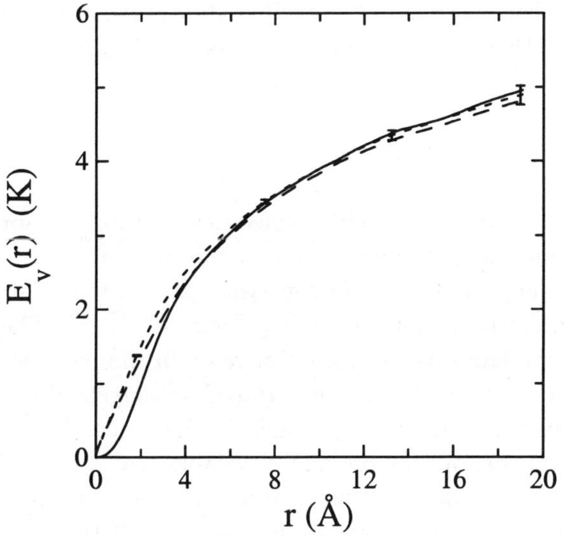

Fig. 24. Radial dependence of the vortex excitation energy. Solid line, with $\psi^{\rm F1}$; short-dashed line, with $\psi^{\rm F2}$, and long-dashed line with $\psi^{\rm BF}$.

The excitation energy $E_{\rm v}(r)$ of a vortex inside a disk of radius r is obtained as the difference between the total energy of the disk with and without the vortex $E_{\rm v}(r) = E(r) - E_0(r)$. For large distances from the vortex axis, $E_{\rm v}(r)$ is usually decomposed in an hydrodynamic tail, which depends logarithmically on r, and a core energy $E_{\rm c}$: $E_{\rm v}(r) = \pi\hbar^2\rho_0/m \ \log(r/\xi) + E_{\rm c}$, where ρ_0 is the homogeneous density far from the axis and ξ is the vortex core radius. In Fig. 24, the vortex excitation energy $E_{\rm v}(r)$ is shown for the different choices of the trial wave function. For distances $r > 6$ Å $E_{\rm v}(r)$ shows the expected hydrodynamic behavior with a small negative shift of the backflow energy with respect to the Feynman one. For small values of r, the estimate of $E_{\rm v}(r)$ is not exact and exhibits the influence of the trial wave function used.

The core radius ξ can be estimated as the position of the maximum in the azimuthal circulating current $J_\theta(r)$. The radial dependence of $J_\theta(r)$ has been estimated from the pure density profile using the expression for the current at the Feynman level, $J_\theta(r) = \rho(r)/r$. The value obtained is $\xi = 2.10 \pm 0.20$ Å,in agreement with the result reported in Ref. 62. By calculating the hydrodynamic contribution to the total energy for the square simulation box, the core energy results $E_{\rm c}^{\rm F1} = 1.23 \pm 0.25$ K, $E_{\rm c}^{\rm F2} = 1.18 \pm 0.26$ K and $E_{\rm c}^{\rm BF} = 1.00 \pm 0.26$ K for the Feynman and backflow phases, respectively. These values coincide with the results obtained by a fit to $E_{\rm v}(r)$ for $r > 6$ Å. Our results for $E_{\rm c}$ are significantly smaller from the ones obtained in Ref. 62 and the values of $E_{\rm c}^{\rm F}$ are close to the variational results of Ref. 125 based on the Feynman phase. On the other hand, the results on the core energies point to a very small influence of the backflow correlations in the excited-state wave function and therefore the use of the RN method in this problem would not significantly change the FN estimations.

7. One-dimensional liquid ^{4}He

The discovery of carbon nanotubes by Ijima [126] in 1991 opened new possibilities for physical realizations of quasi-one-dimensional systems. One of the most attractive features of nanotubes is, certainly, the possibility of filling with different materials both their inner cavities and the interstitial channels among them. [39,127–132] The interest in this field is twofold. On one hand, the expected increase in the particle-substrate potential energy with respect to a flat carbon surface has suggested the use of nanotubes as storage devices for molecular hydrogen in fuel cells. [129,133] On the other hand, more theoretical, nanotubes provide a reliable realization of one-dimensional systems in the same way that a substance adsorbed on graphite manifests trends that are characteristic of a two-dimensional medium. If the nanotubes are filled with light atoms (He) or molecules (H_2) and the temperature is low enough, one is dealing with quasi-one-dimensional quantum fluids. Such an experimental realization has been carried out for the first time by in a honeycomb of FSM-16. [134,135] This is a mesoporous substrate with tubes approximately 18 Å in diameter. Using a torsional oscillator, this group proved the existence of superfluidity of the ^{4}He atoms adsorbed in the pores below a critical temperature of $\sim$ 0.7 K. More recently, Teizer *et al.* [130] have studied experimentally the desorption of ^{4}He previously adsorbed in the interstitial sites of carbon nanotube bundles. In this case, the data points unambiguously to the one-dimensional nature of helium inside the nanotubes.

From a theoretical point of view, it has been recently established using both the hypernetted chain (HNC) variational approach [136] and the DMC method [39,137] that strictly one dimensional (1D) ^{4}He is a self-bound liquid at zero temperature. The equation of state of that ideal system is shown in Fig. 25. It corresponds to a DMC calculation using the Aziz II potential, like in the study of bulk and 2D liquid ^{4}He. The solid line in the figure corresponds to a polynomial fit (6.1) around the equilibrium density λ_0. The optimal values for the parameters B, C, λ_0, and e_0 are reported in Table 7. For densities larger than the ones reported in Fig. 25, the energy increases very fast and such a simple analytic function can not reproduce the DMC data.

In agreement with the DMC calculation of Boninsegni and Moroni [137] and the variational one of Krotscheck and Miller, [136] our results show that ^{4}He is self–bound in a 1D array. However, the binding energy (-0.0036±0.0002 K) is much smaller than that in 2D (-0.897±0.002 K) and 3D (-7.267±0.013 K). It is worth noting that such a small total energy results from a big cancellation between the potential and kinetic energies. At λ_0, $T/N = 0.2706 \pm 0.0004$ K and $V/N = -0.2742 \pm 0.0004$ K. In fact, the influence of the ^{4}He interatomic potential in this system is very large. A

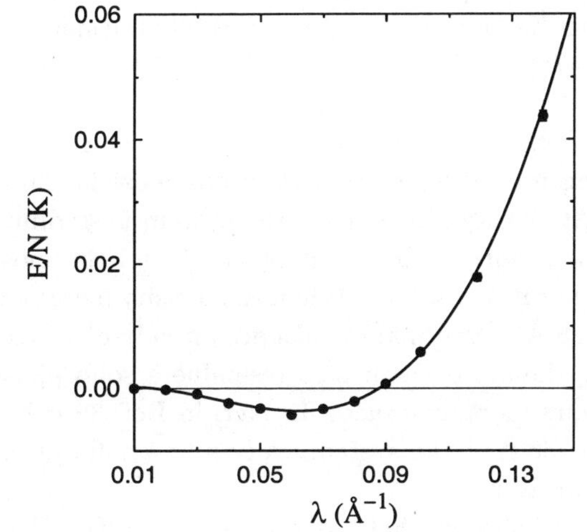

Fig. 25. Equation of state of 1D liquid ^{4}He using the Aziz II potential. The line corresponds to a polynomial fit (6.1) around the equilibrium linear density λ_0.

Table 7. Parameters of the equation of state (6.1) of 1D liquid ^{4}He in the vicinity of the equilibrium density λ_0.

λ_0 (Å^{-1})	0.062 $\pm$ 0.001
e_0 (K)	-0.0036 $\pm$ 0.0002
B (K)	0.0156 $\pm$ 0.0009
C (K)	0.0121 $\pm$ 0.0008
χ^2/ν	2.2

calculation at the equilibrium density λ_0 for the 1D system using the HFDHE2 Aziz potential [64] indicates that ^{4}He is still a liquid, but the total energy is two times smaller (-0.0018 ± 0.0003 K), with a potential energy -0.2724 ± 0.0004 K and the same kinetic energy. These sizeable differences partially explain the discrepancies of our DMC calculation [39] with both the results of Boninsegni and Moroni [137] and Krotscheck and Miller [136] who used the HFDHE2 Aziz potential.

From the values of the energy, one can obtain the linear system pressure and the speed of sound. Fig. 26 displays both observables as a function of the ^{4}He density. Around the equilibrium point the behavior is quite similar to the one observed in 2D and 3D. However, with increasing density, both the pressure and the speed of sound increases very fast. This is a consequence of the strictly 1D character and the reduction of the mean distance (essentially $1/\lambda$) to values where the repulsive core of the potential starts to emerge. The speed of sound at the equilibrium density is

$c = 7.98(7)$ m/sec, a tiny fraction of the corresponding 2D ($c = 92.8$ m/sec) and 3D (238.3 m/sec) ^{4}He liquids. On the other hand, according to our results the spinodal point is located at a density $\lambda_s = 0.047(1)$ Å^{-1} and pressure $P_s = -1.93(5) \cdot 10^{-4}$ KÅ^{-1}.

Results for the function $g(r)$ for 1D ^{4}He as a function of the density are shown in Fig. 27. The wide range of linear densities at which the liquid phase exists allows for very different spatial structures. At the equilibrium λ_0, the system is so dilute that $g(r)$ is quite a monotonic function: only an incipient peak around the most probable interatomic distance can be (hardly) observed. When the density increases, a clearer structure emerges: at $\lambda = 0.15$ Å^{-1} two main peaks are already observed. If the density is increased even more, the system starts to resemble a solid phase with a structure that survives up to quite long distances. In fact, in Ref. 39 it has been observed that at densities $\lambda > 0.358$ Å^{-1} the system evolves to a solid phase following a quasi-continuous phase transition.

Fig. 28 contains the static structure functions $S(k)$ at the same densities than Fig. 27. Here, the emergence of a local order is still more evident in the high main peak at the highest density. The behavior of $S(k)$ when $k \to 0$ is also very different for the three cases reported. As in 2D and bulk, in this limit $S(k)$ is linear with k with a slope inversely proportional to the speed of sound. Therefore at small densities, where the sound velocities are very small, $S(k)$ increases quickly. Instead, at high density the slope is very small due to the large sound speed.

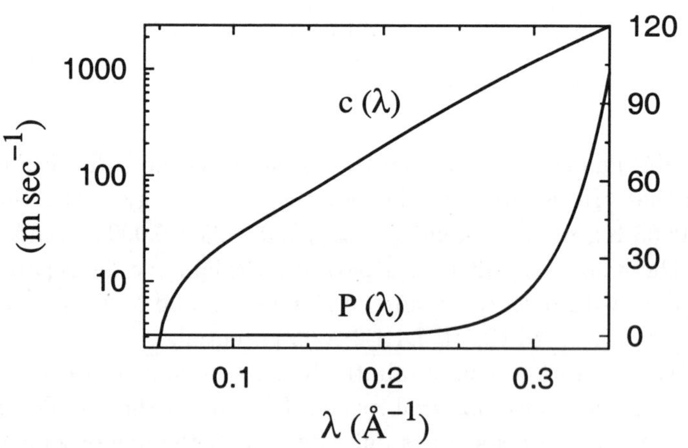

Fig. 26. Pressure and speed of sound of 1D liquid ^{4}He as a function of the linear density.

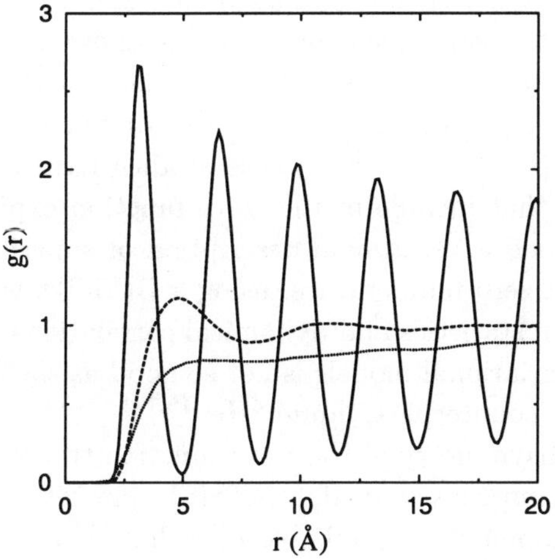

Fig. 27. Two-body distribution function in 1D liquid ^{4}He as a function of the linear density λ. The dotted, dashed, and solid lines stand for densities $\lambda = 0.062$, 0.150, and 0.300 Å^{-1}, respectively.

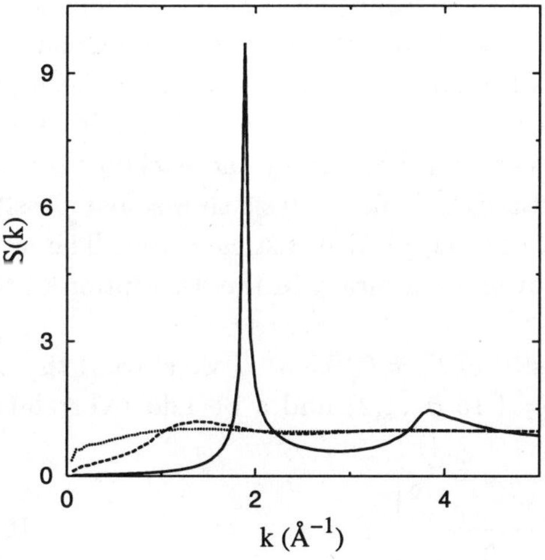

Fig. 28. Static structure function in 1D liquid ^{4}He as a function of the linear density λ. The dotted, dashed, and solid lines stand for densities $\lambda = 0.062$, 0.150, and 0.300 Å^{-1}, respectively.

8. Bulk liquid ^{3}He

Liquid ^{3}He has been, for many years, one of the benchmarks of quantum many-body theory. Among the different theoretical approaches, the variational method has been the most fruitful one. Starting from a Jastrow-Slater trial wave function, and incorporating also three-body correlations in the dynamical part and backflow correlations in the Slater determinant, the variational method has allowed for a reasonable description of the ground state of liquid ^{3}He. That has been achieved by solving the Fermi-hypernetted-chain equations (FHNC) [138,139] or via a multidimensional Monte Carlo integration (VMC). [140–142] These studies have been complemented with FHNC calculations that include in the wave function explicit spin correlations, [143] and other that intend a variational description of superfluid ^{3}He, which is actually the real system at zero temperature, using a BCS-like wave function. [144] The interplay between Fermi statistics and dynamical correlations has made, however, that the quality of the variational models is not so good as the one attained in the description of its bosonic counterpart, liquid ^{4}He. [19,141]

The necessary requirement of the antisymmetry of the wave function translates into the well-known sign problem that impedes a straightforward application of DMC or GFMC. Waiting for the development of a reliable cancellation method that can work in a many-body system, [47] the approximate fixed-node method (FN) [28,29,32] has become the standard tool. Normal liquid ^{3}He at zero temperature has been studied with the FN methodology including up to backflow correlations in the trial wave function. [59,60] The results obtained showed a significant improvement with respect to VMC calculations using the same trial wave functions but the agreement with the experimental equation of state was not completely satisfactory. All these calculations failed in the exigent test of reproducing the density dependence of the pressure, a feature that pointed to possible inaccuracies of the nodal surface.

The combined strategy to deal with a fermion calculation, which has been presented in Sect. 3, has been applied recently [46] to the study of normal liquid ^{3}He. In that work the high quality of the nodal surface generated by the backflow correlations has been verified using the RN method. [31] This analysis shows that possible residual systematic errors are at the level of the typical statistical errors. The interatomic potential is Aziz II, which has proven its accuracy in the description of ^{4}He experimental data.

At the experimental equilibrium density $\rho_0^{\text{expt}} = 0.273\ \sigma^{-3}$, we started the calculation with a McMillan parameter $b = 1.15\ \sigma$ (1.2) and a backflow correlation (5.10, 5.11) with

$$\eta(r) = \exp\left[-\left(\frac{r - r_{\rm B}}{\omega_{\rm B}}\right)^2\right] , \tag{8.1}$$

and optimal parameters [59] $\lambda_{\rm B} = 0.14$, $r_{\rm B} = 0.74\ \sigma$, and $\omega_{\rm B} = 0.54\ \sigma$. With this initial set of parameters, the results obtained were clearly biased by the trial wave function. We found that these backflow parameters correspond to a local minimum of the FN energy, and that a narrower but deeper minimum exists with $\lambda_{\rm B} = 0.35$ and $r_{\rm B}$ and $\omega_{\rm B}$ unchanged. The relation of initial slopes of the corresponding RN energies, $1 : 0.27 : 0.016$ for $\lambda_{\rm B} = 0,\ 0.14,\ 0.35$, provides information on the accuracy of $\psi(\mathbf{R})$. In the optimal case, $\lambda_{\rm B} = 0.35$, the slope is practically zero and the energy

Table 8. DMC total and kinetic energies for liquid ^{3}He at $\rho_0^{\rm expt}$ as a function of the backflow parameter $\lambda_{\rm B}$. The experimental values for the total and kinetic energies are taken form Ref. 13 and Ref. 145, respectively.

	E/N (K)	T/N (K)
$\lambda_{\rm B}=0$	-2.128 ± 0.015	12.603 ± 0.031
$\lambda_{\rm B}=0.14$	-2.330 ± 0.014	12.395 ± 0.035
$\lambda_{\rm B}=0.35$	-2.477 ± 0.014	12.239 ± 0.030
Expt	-2.473	8.1 ± 1.7

correction would be < 0.01 K if the asymptotic regime could be reached. In Table 8, results for the total and kinetic energies are shown as a function of $\lambda_{\rm B}$. As one can see, the value with the optimal $\lambda_{\rm B}$ reproduces the experimental value.

In order to get additional evidence on the size of a possible correction beyond the $\lambda_{\rm B}=0.35$ result, we included for the first time corrections to the backflow trial wave function using the analytical method previously described in Sect. 3. The new terms incorporate explicit three-body correlations in $\varphi_{\alpha_i}(\tilde{\mathbf{r}}_j)$ of the form

$$\tilde{\mathbf{r}}_j^{\rm BFT} = \tilde{\mathbf{r}}_j^{\rm BF} + \lambda_{\rm BT} \sum_{k \neq j} \eta(r_{jk})\,(\mathcal{F}_j - \mathcal{F}_k)\;, \tag{8.2}$$

with $\mathcal{F}_i = \sum_{l \neq i} \eta(r_{il})\,\mathbf{r}_{il}$. A FN-DMC calculation with this new trial wave function at $\rho_0^{\rm expt}$ was carried out and the result for the energy correction was found < 0.01 K. Both this analytical check and the numerical findings provided by the RN method underscore the excellent description of the nodal surface in liquid ^{3}He provided by backflow correlations.

More critical than a single value is the entire equation of state. The DMC calculation was extended to a wide range of densities ranging from the spinodal point up to a maximum value $\rho = 0.403\ \sigma^{-3}$, located near to the experimental freezing density $\rho_{\rm f}^{\rm expt} = 0.394\ \sigma^{-3}$. $N = 114$ atoms were used only at the highest density and, below that, $N = 66$ proved to be accurate enough. In all cases the tail contributions to the energy were estimated according to the procedure described in Sect. 4 in order to practically eliminate size corrections. Among the three variational parameters entering the backflow wave function (5.10,5.11) only $\lambda_{\rm B}$ showed a density dependence which was nearly linear in the range studied ($\lambda_{\rm B} = 0.42$ at

Table 9. Parameters of the equation of state (6.1) of liquid ^{3}He.

ρ_0 (σ^{-3})	0.274 ± 0.001
e_0 (K)	-2.464 ± 0.007
B (K)	6.21 ± 0.11
C (K)	3.49 ± 0.50
χ^2/ν	1.2

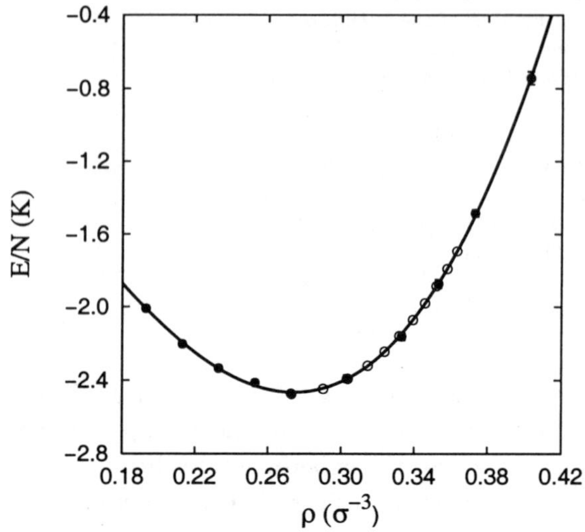

Fig. 29. Energy per particle of normal liquid ^{3}He as a function of the density. The full circles are the DMC results (the error bars are depicted only when larger than the size of the symbol), and the open circles are experimental data from Ref. 13. The line is a polynomial fit to the MC data.

$\rho = 0.403\ \sigma^{-3}$). The results are displayed in Fig. 29 and compared with the experimental data of Ref. 13. The solid line in the same figure is a third-degree polynomial fit (6.1) with optimal values reported in Table 9.

In Fig. 30, the behavior of the pressure and the speed of sound with the density is shown in comparison with experimental data from Refs. 13, 146, and 147. The theoretical prediction for both quantities, derived form the polynomial fit to $E/N(\rho)$ (Fig. 29), shows again an excellent agreement with the experimental data from equilibrium up to freezing. The speed of sound is $c = 182.2(6)$ m/ sec at $\rho_0^{\rm expt}$ in close agreement with the experimental value $c^{\rm expt} = 182.9$ m/ sec. [146] It goes to zero at the spinodal point. The location of this point has been obtained both by extrapolation of experimental data at positive pressures [148,149] and from density-functional theories. [150] The present microscopic calculation allows for an accurate calculation, free from extrapolation uncertainties, that locates the spinodal point at a density $\rho_{\rm s} = 0.202(2)\ \sigma^{-3}$ corresponding to a negative pressure $P_{\rm s} = -3.09(20)$ atm, much closer to the equilibrium than in liquid ^{4}He where $P_{\rm s} = -9.30(15)$ atm. [66]

The structure properties of normal liquid ^{3}He have also been studied. [151] In Fig. 31 the two-body radial distribution function $g(r)$ at $\rho_0^{\rm expt}$ is shown in comparison with available experimental data. The comparison with the experimental data of Achter and Meyer [152] is not completely satisfactory but these data, which are quite old, suffers from some normalization problems and spurious behavior at small distances.

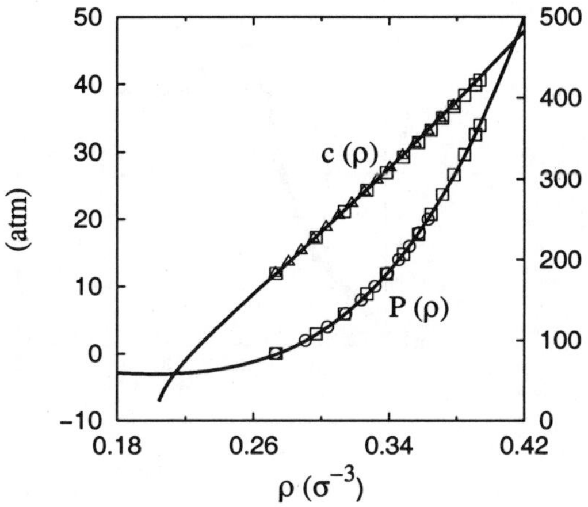

Fig. 30. Pressure and speed of sound of liquid ^{3}He as a function of the density. The lines are the FN-DMC results and the circles, triangles and squares are experimental data from Refs. 13, 146, and 147, respectively.

A function more closely related to experiment is the static structure function $S(k)$. We have compared our result for $S(k)$ at $\rho_0^{\rm expt}$ with two different sets of experimental data [152,153] in Fig. 31. There are significant differences between both experimental determinations, the DMC result being closer to the most recent one. However, the latter is restricted to the range $k < 2$ Å^{-1}. New measures of $S(k)$ that could enlarge its knowledge, especially beyond the first maximum, will be welcome in order to test more accurately present and future theoretical data.

The momentum distribution of normal liquid ^{3}He has been calculated both using GFMC [90] and DMC. [40]. The form of $n(k)$ resembles that of the step function of the free Fermi gas, with significant modifications induced by the strong interatomic correlations. The most prominent one is the population of states with momenta $k > k_{\rm F}$, with $k_{\rm F}$ the Fermi momentum. The depletion is large, the value of $n(k=0)$ being reduced to approximately one half of the gas value $n(k=0) = 1$. Consequently, the size of the discontinuity Z at $k = k_{\rm F}$ is also considerably reduced (from 1 to about 0.2 at the experimental equilibrium density). The long tails of the momentum distribution explain, at least partially, the disagreement between the theoretical value of the kinetic energy and the experimental measures extracted from deep inelastic neutron scattering experiments (see Table 8). Recent DINS results [154] of ^{3}He kinetic energies of the solid phase and of the liquid at high density are in much better agreement with theory.

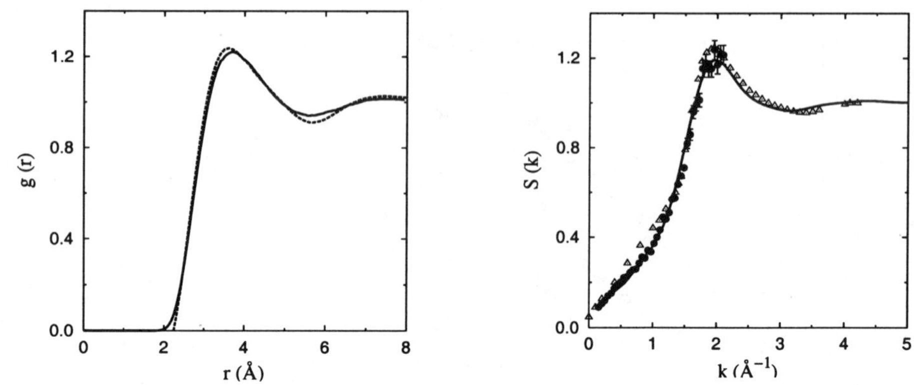

Fig. 31. Left: Two-body radial distribution function of liquid ^{3}He at $\rho_0^{\rm expt}$ (solid line) in comparison with experimental data [152] (dashed line). Right: Static structure function at $\rho_0^{\rm expt}$ (line) in comparison with experimental data of Ref. 152 (triangles) and Ref. 153 (circles).

9. Two-dimensional ^{3}He

^{3}He adsorbed on strongly interacting substrates like graphite or on top of bulk ^{4}He of ^{4}He films are experimental realizations of quasi-two dimensional Fermi systems. [155] In the last decades there has been a continued experimental effort to understand the fascinating properties of such a nearly perfect two-dimensional Fermi liquid. Among these unique features it is of particular relevance the observed possibility of increasing continuously the areal density from an almost ideal gas behavior up to a strongly correlated regime. In other words, there are clear experimental findings that indicate the non-existence of a self-bound ^{3}He system. This point has been discussed for a long time from both the experimental [156–163] and theoretical side. [164–170]

The number of theoretical calculations on the properties of a 2D ^{3}He system or a ^{3}He film is considerably smaller than the corresponding one for the boson isotope ^{4}He. Usually, the Fermi statistics of ^{3}He has presented an additional problem to the one of dealing with a strongly correlated system like helium. In one of the pioneering works on the field, Novaco and Campbell [164] calculated the equation of state of ^{3}He adsorbed on graphite. Using lowest-order Fermi corrections, they concluded that the ^{3}He film is in a gas state, contrary to ^{4}He which exhibits a well-established self-bound character. A comparative study of bosons and fermions in 2D was performed by Miller and Nosanow [165] using the variational method. According to their variational approach, and using a Wu-Feenberg expansion [169] at lowest order, ^{3}He cannot condense in 2D. More recently, Brami *et al.* [166] calculated

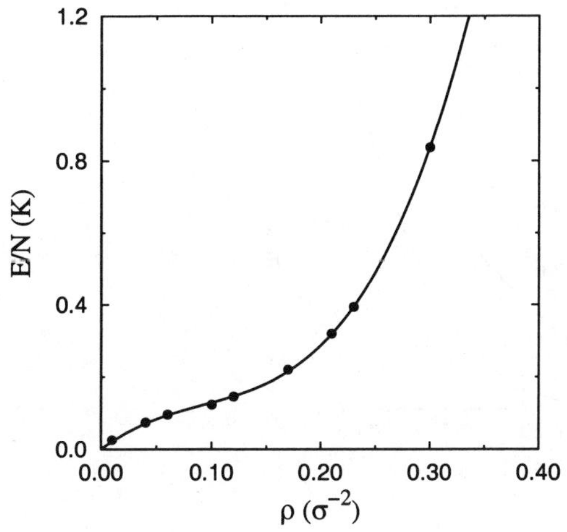

Fig. 32. DMC energies per particle of 2D ^{3}He as a function of density. The solid line corresponds to a polynomial fit to the data.

the properties of a 2D ^{3}He film using VMC. They showed that the presence of a transverse degree of freedom, not present in 2D, allows the system to gain enough additional binding energy to guarantee a liquid phase with a very small energy ($\sim$ 200 mK). This result seems to be in contradiction with experimental data from several groups.[157,159] Moreover, a recent GFMC calculation of a ^{4}He film adsorbed on graphite by Whitlock *et al.*[45] has shown that the energy gain with respect to the ideal 2D system is much smaller than the one estimated in Ref. 166. Extrapolating the ^{4}He results[45] to the ^{3}He density regime one can see that such a liquid phase can hardly exist.

Using the same methodology as in bulk ^{3}He, which has led to a good agreement between theoretical and experimental results, our group has calculated recently the properties of 2D ^{3}He.[171] As in the bulk calculation, the trial wave function used for importance sampling incorporates backflow correlations in the Slater determinant. With optimal parameters ($\lambda_{\rm B} = 0.40$, $r_{\rm B} = 0.75\ \sigma$, and $\omega_{\rm B} = 0.54\ \sigma$), the behavior of the released-node energies with the released time becomes flat with a slope compatible with zero. The gain in energy due to the inclusion of backflow correlations is in 2D smaller than in 3D, with a contribution that increases nearly quadratically with density. The interatomic potential is the Aziz II model used previously in the bulk phases of ^{4}He and ^{3}He. The DMC energies of the 2D fluid are shown in Fig. 32.

Our results confirm the non-existence of a self-bound system. The DMC energies are smaller than previous VMC results in all the density regime, the differences being appreciably larger at the highest densities. The line in Fig. 32 corresponds to

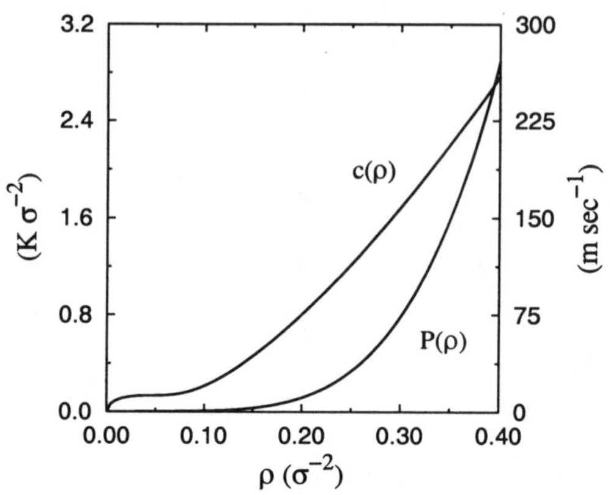

Fig. 33. DMC results for the pressure and speed of sound of 2D ^{3}He as a function of density. Both quantities are derived from the analytic fit (9.1) to the energies.

a third-degree polynomial fit

$$\frac{E}{N} = A\rho + B\rho^2 + C\rho^3 \ , \tag{9.1}$$

with optimal parameters $A = 2.376(74)$ Kσ^2, $B = -16.87(82)$ Kσ^4, and $C = 6.08(21) \cdot 10$ Kσ^6. At very small densities ($\rho < 0.05\ \sigma^{-2}$), the energy increases linearly since the dominant contribution comes from the free Fermi energy, which in 2D is linear with ρ ($E_{\rm F} \propto k_{\rm F}^2$ and $k_{\rm F} = \sqrt{2\pi\rho}$). At medium densities, the slope decreases but it is still positive. Probably, with a slightly greater mass one would observe in that medium region the emergence of a minimum corresponding to a liquid phase. When the density is increased even farther, the energy increases quite rapidly due to the strong interatomic correlations, mainly of dynamical origin.

It is worth mentioning that the fictitious 2D boson ^{3}He system is a self-bound system with an equilibrium point located at density $\rho_0 = 0.13\ \sigma^{-2}$ and energy $E/N(\rho_0) = -0.117(2)K$. In the first variational estimates of the properties of a 2D ^{3}He system, the energy was approximated by the sum of the boson energy and the Fermi energy at each density. [164] Playing that game, one can see that this oversimplified approach holds only at very small densities ($\rho < 0.05\ \sigma^{-2}$), and beyond that regime this model overestimates the exact energy by an amount that grows dramatically with ρ.

Results for the pressure and speed of sound are shown in Fig. 33. They have been obtained from the analytic fit (9.1) to the DMC energies. The pressure increases with the density but with a very small slope up to densities $\rho \sim 0.15\ \sigma^{-2}$ and much faster from then on. More singular is the behavior of the speed of sound at small densities. After a first increase from zero, the speed of sound reaches a plateau up to $\rho \sim 0.10\ \sigma^{-2}$, and then increases with density. Therefore, the system has a small

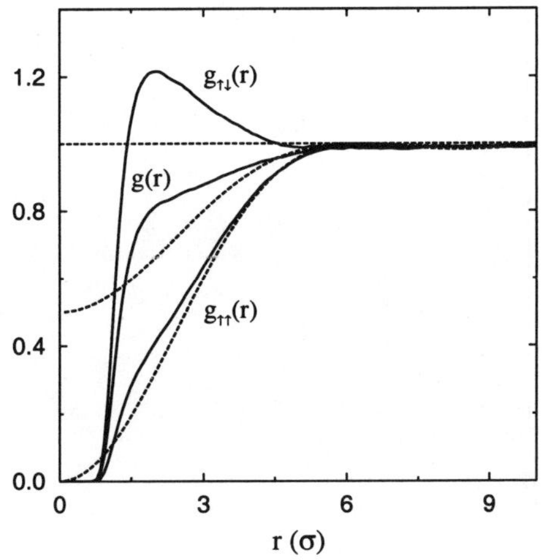

Fig. 34. Spin-dependent and total two-body radial distribution functions of 2D ^{3}He at density $\rho = 0.06\ \sigma^{-2}$. The dashed lines are the Fermi gas functions (9.3) at the same density.

density regime where the speed of sound remains nearly constant.

The fact that the system is thermodynamically accessible from zero to high densities allows for the study of 2D ^{3}He from nearly a free Fermi gas to a strongly correlated regime. That evolution is shown explicitly in the atomic spatial structure. At very low densities, one expects the system approaches a Fermi gas with two-body radial distribution functions

$$g_{\uparrow\downarrow}^{\rm FG}(r) = 1 \tag{9.2}$$
$$g_{\uparrow\uparrow}^{\rm FG}(r) = 1 - \frac{2}{\pi\rho}\,\frac{J_1(k_{\rm F}r)^2}{r^2}$$

with normalization

$$g(r) = \frac{1}{2}\,\left(g_{\uparrow\downarrow}(r) + g_{\uparrow\uparrow}(r)\right)\ . \tag{9.3}$$

In Fig. 34, results for the spin-dependent and total $g(r)$ at $\rho = 0.06\ \sigma^{-2}$ are shown in comparison with the same functions for the Fermi gas (9.3). The agreement between the functions of the free system and those corresponding to ^{3}He is only satisfactory for the $g_{\uparrow\uparrow}(r)$ case, in which the statistical correlations dominate (except logically the hard-core hole at very small r). In order to reach such an agreement for the total $g(r)$ it is necessary to decrease the density even more.

The evolution of $g(r)$ with increasing density is shown in Fig. 35. From nearly a free Fermi gas at the lowest density ($\rho = 0.01\,\sigma^{-2}$), the two-body distribution function evolves progressively up to a more localized structure. At the highest density reported ($\rho = 0.30\,\sigma^{-2}$) the gas shows already a pronounced structure that is a

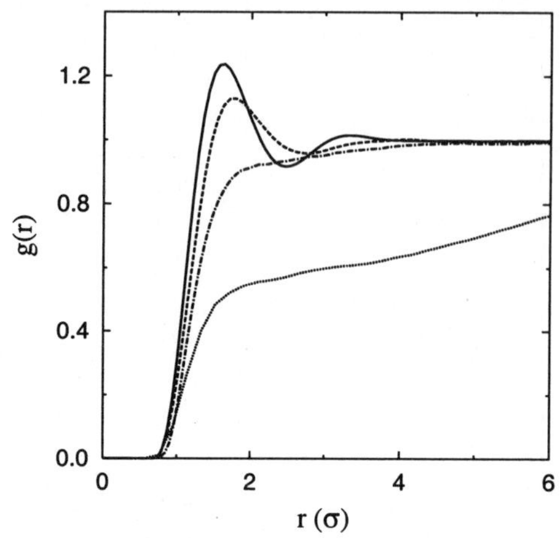

Fig. 35. Two-body radial distribution function of 2D ^{3}He at several densities. Dotted, dashed-dotted, dashed, and solid lines correspond to $\rho = 0.01$, 0.10, 0.21, and 0.30 σ^{-2}, respectively.

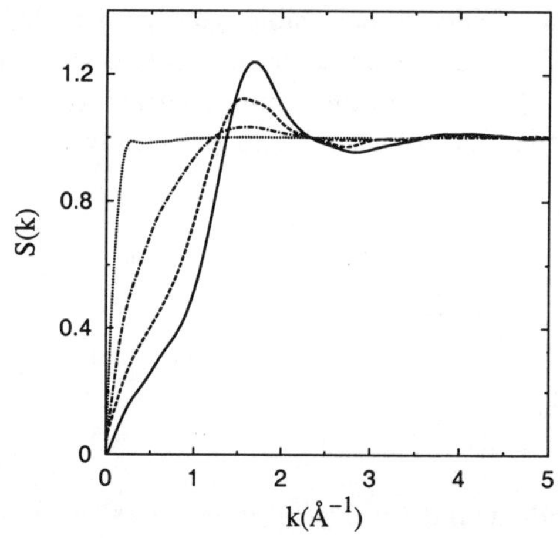

Fig. 36. Static structure function of 2D ^{3}He at several densities. Dotted, dashed-dotted, dashed, and solid lines correspond to $\rho = 0.01$, 0.10, 0.21, and 0.30 σ^{-2}, respectively.

precursor of the solidification of the system. That solidification is expected to be at a density $\rho \sim 0.4\ \sigma^{-2}$ according to experimental measures on ^{3}He films adsorbed on graphite.[162] An accurate theoretical estimate of that phase transition is not yet

Table 10. Summary of some relevant DMC results for ^{4}He and ^{3}He in different dimensions D. $\bar{r}$ is defined as $(3/(4\pi\rho_0))^{1/3}$, $(1/(4\pi\rho_0))^{1/2}$, and $1/\rho_0$ for 3D, 2D, and 1D, respectively. The subscript *mp* stands for main peak.

	^{4}He			^{3}He	
	3D	2D	1D	3D	2D†
ρ_0 (σ^{-D})	0.365	0.284	0.158	0.274	0.100
$E/N(\rho_0)$ (K)	-7.267	-0.897	-0.0036	-2.464	0.130
$T/N(\rho_0)$ (K)	14.32	4.00	0.27	12.24	1.20
$c(\rho_0)$ (m/sec)	237.4	92.8	8.0	182.2	20.1
ρ_s (σ^{-D})	0.264	0.228	0.120	0.202	–
P_s $(K\sigma^{-D})$	-1.141	-0.097	-0.00049	-0.379	–
$\bar{r}$ (σ)	0.87	0.53	6.33	0.95	0.89
$r_{\rm mp}$ (σ)	1.34	1.56	–	1.46	–
$g(r_{\rm mp})$	1.39	1.27	–	1.22	–
$k_{\rm mp}$ (σ^{-1})	5.08	4.20	–	5.06	–
$S(k_{\rm mp})$	1.34	1.22	–	1.19	–

† 2D ^{3}He is a gas and therefore ρ_0 is meaningless. A value $\rho = 0.10\ \sigma^{-2}$ has been chosen to make a comparison with the rest of the data possible.

performed.

Finally, results for the static structure function are reported in Fig. 36, at the same densities as in Fig. 35. At the lowest density, $S(k)$ is very similar to the free case which is equal to one for $k \geq 2k_{\rm F}$. When the density increases, the main peak emerges progressively at wave vectors that increase with k.

10. Concluding remarks

The application of *ab initio* quantum Monte Carlo methods to the study of quantum liquids has allowed for a wide and accurate set of results. Some of these results are contained in Table 10. Going through the different dimensionalities, both ^{4}He and ^{3}He show different behaviors and different characteristic length and energy scales. The quantum nature of all these fluids, combined with their different dimensionality, provide a unique opportunity in condensed matter theory. Moreover, the ideal 2D and 1D geometries have a physical connection with the real world through the adsorption of helium in surfaces and nanotubes, respectively. The substrate effects have also been studied using MC but more effort is necessary to arrive to a full description. This is one of the future lines in the field. Other topics that still need additional MC work are mixed ^{4}He-^{3}He fluids in several geometries. The interplay between Bose and Fermi statistics introduces appealing features from the theoretical viewpoint that nowadays, with the available methodology to deal with fermions, can be accurately studied. More ambitious is the study of the dynamics of these systems. That is perhaps the most difficult problem MC has still to solve.

Acknowledgments

I want to acknowledge with special emphasis the fruitful and friendly collaboration of Joaquim Casulleras throughout the past decade. The work reported here has also benefited from the collaboration of Stefano Giorgini, M. Carmen Gordillo, and Víctor Grau. Exchange of ideas and discussions with my colleagues have encouraged our work along the time. Among them I want to name M. Barranco, C. E. Campbell, S. A. Chin, H. R. Glyde, R. Guardiola, E. Krotscheck, F. Mazzanti, J. Navarro, A. Polls, and M. Saarela.

References

1. D. M. Ceperley and M. H. Kalos, in *Monte Carlo Methods in Statistical Physics*, ed. K. Binder (Springer, Berlin, 1979).
2. E. Feenberg, *Theory of Quantum Fluids* (Academic Press, New York, 1969).
3. P. Nozières and D. Pines, *The Theory of Quantum Liquids*, Vol. I and II (Addison-Wesley, New York, 1990).
4. J. Wilks and D. S. Betts, *An Introduction to Liquid Helium* (Clarendon Press, Oxford, 1987).
5. B. L. Hammond, W. A. Lester Jr., and P. J. Reynolds, *Monte Carlo Methods in Ab Initio Quantum Chemistry* (World Scientific, Singapore, 1994).
6. R. Guardiola, in *Microscopic Quantum Many-Body Theories and Their Applications*, ed. by J. Navarro and A. Polls (Springer, Berlin, 1998).
7. A. Bijl, *Physica* **7**, 869 (1940).
8. R. B. Dingle, *Phil. Mag.* **40**, 573 (1949).
9. R. Jastrow, *Phys. Rev.* **98**, 1479 (1955).
10. W. L. McMillan, *Phys. Rev.* **138**, 442 (1965).
11. N. Metropolis, A. W. Rosenbluth, M. N. Rosenbluth, A. M. Teller, and E. Teller, *J. Chem. Phys.* **21**, 1087 (1953).
12. *Monte Carlo Methods in Theoretical Physics*, ed. by S. Caracciolo and A. Fabrocini (ETS Editrice, Pisa, 1991).
13. R. De Bruyn Ouboter and C. N. Yang, *Physica* B **44**, 127 (1987).
14. D. Schiff and L. Verlet, *Phys. Rev.* **160**, 208 (1967).
15. R. D. Murphy and R. O. Watts, *J. Low Temp. Phys.* **2**, 507 (1970).
16. E. Krotscheck, in *Microscopic Quantum Many-Body Theories and Their Applications*, ed. by J. Navarro and A. Polls (Springer, Berlin, 1998).
17. P. A. Whitlock, D. M. Ceperley, G. V. Chester, and M. H. Kalos, *Phys. Rev.* B **19**, 5598 (1979).
18. C. C. Chang and C. E. Campbell, *Phys. Rev.* B **15**, 4238 (1977).
19. K. Schmidt, M. H. Kalos, M. A. Lee, and G. V. Chester, *Phys. Rev. Lett.* **45**, 573 (1980).
20. Q. N. Usmani, S. Fantoni, and V. R. Pandharipande, *Phys. Rev.* B **26**, 6123 (1982).
21. S. Vitiello, K. Runge, and M. H. Kalos, *Phys. Rev. Lett.* **60**, 1970 (1988).
22. F. Pederiva, A. Ferrante, and L. Reatto, *Phys. Rev.* B **52**, 7564 (1995).
23. M. H. Kalos, *Phys. Rev.* **128**, 1791 (1962).
24. M. H. Kalos, *Nucl. Phys.* A **126**, 609 (1969).
25. M. H. Kalos, *J. Comp. Phys.* **1**, 127 (1966).
26. M. H. Kalos, D. Levesque, and L. Verlet, *Phys. Rev.* A **9**, 2178 (1974).
27. M. H. Kalos, M. A. Lee, P.A. Whitlock, and G. V. Chester, *Phys. Rev.* B **24**, 115 (1981).

28. J. B. Anderson, *J. Chem. Phys.* **63**, 1499 (1975).
29. J. B. Anderson, *J. Chem. Phys.* **65**, 4121 (1976).
30. J. B. Anderson, *J. Chem. Phys.* **73**, 3897 (1980).
31. D. M. Ceperley and B. J. Alder, *Phys. Rev. Lett.* **45**, 566 (1980).
32. P. J. Reynolds, D. M. Ceperley, B. J. Alder, and W. A. Lester Jr., *J. Chem. Phys.* **77**, 5593 (1982).
33. S. A. Chin, *Phys. Rev.* B **42**, 6991 (1990).
34. H. A. Forbert and S. A. Chin, *Phys. Rev.* B **63**, 144518 (2001).
35. J. Boronat and J. Casulleras, *Phys. Rev.* B **49**, 8920 (1994).
36. J. Boronat and J. Casulleras, *Europhys. Lett.* **38**, 291 (1997).
37. S. Giorgini, J. Boronat, and J. Casulleras, *Phys. Rev.* B **54**, 6099 (1996).
38. S. Giorgini, J. Boronat, and J. Casulleras, *Phys. Rev. Lett.* **77**, 2754 (1996).
39. M. C. Gordillo, J. Boronat, and J. Casulleras, *Phys. Rev.* B **61**, R878 (2000).
40. S. Moroni, G. Senatore, and S. Fantoni, *Phys. Rev.* B **55**, 1040 (1997).
41. S. A. Chin and E. Krotscheck, *Phys. Rev.* B **45**, 852 (1992).
42. M. Lewerenz, *J. Chem. Phys.* **106**, 4596 (1997).
43. R. N. Barnett and K. B. Whaley, *Phys. Rev.* A **47**, 4082 (1993).
44. J. L. Vallés and K. E. Schmidt, *Phys. Rev.* B **38**, 2879 (1988).
45. P. A. Whitlock, G. V. Chester, and B. Krishnamachari, *Phys. Rev.* B **58**, 8704 (1998).
46. J. Casulleras and J. Boronat, *Phys. Rev. Lett.* **84**, 3121 (2000).
47. M. H. Kalos and F. Pederiva, *Phys. Rev. Lett.* **85**, 3547 (2000).
48. P. Huang, Y. Kwon and K.B. Whaley, The finite-temperature path integral Monte Carlo Method and its application to superfluid helium clusters, 2001, this volume.
49. R. Guardiola, in *First International Course on Condensed Matter*, ed. by D. Prosperi *et al.* (World Scientific, Singapore, 1988).
50. S. Rosati, in *From Nuclei to Particles*, Proceedings of the International School of Physics Enrico Fermi, Course LXXIX, Varenna, ed. by A. Molinari (North Holland, Amsterdam, 1981).
51. M. H. Kalos and P. Whitlock, *Monte Carlo Methods*, Vol. I (John Wiley&Sons, New York, 1986).
52. W. A. Press, S. A. Teukolsky, W. T. Vettering, and B. P. Flannery, *Numerical Recipes in Fortran*, 2nd. ed. (Cambridge University Press, Cambridge, 1992).
53. S. M. Rothstein, J. Vrbik, and N. Patil, *J. Comput. Chem.* **8**, 412 (1987).
54. K. S. Liu, M. H. Kalos, and G. V. Chester, *Phys. Rev.* A **10**, 303 (1974).
55. R. N. Barnett, P. J. Reynolds, and W. A. Lester Jr., *J. Comput. Phys.* **96**, 258 (1991).
56. P. J. Reynolds, R. N. Barnett, B. L. Hammond, and W. A. Lester Jr., *J. Stat. Phys.* **43**, 1017 (1986).
57. J. Casulleras and J. Boronat, *Phys. Rev.* B **52**, 3654 (1995).
58. J. Boronat, in *Microscopic Quantum Many-Body Theories and Their Applications*, ed. by J. Navarro and A. Polls (Springer, Berlin, 1998).
59. R. M. Panoff and J. Carlson, *Phys. Rev. Lett.* **62**, 1130 (1989).
60. S. Moroni, S. Fantoni, and G. Senatore, *Phys. Rev.* B **52**, 13547 (1995).
61. J. B. Anderson, C. A. Traynor, and B. M. Boghosian, *J. Chem. Phys.* **95**, 7418 (1991).
62. G. Ortiz and D. M. Ceperley, *Phys. Rev. Lett.* **75**, 4642 (1995).
63. R. P. Feynman and M. Cohen, *Phys. Rev.* **102**, 1189 (1956).
64. R. A. Aziz, V. P. S. Nain, J. S. Carley, W. L. Taylor, and G. T. McConville, *J. Chem. Phys.* **70**, 4330 (1979).
65. R. A. Aziz, F. R. W. McCourt, and C. C. K. Wong, *Mol. Phys.* **61**, 1487 (1987).
66. J. Boronat, J. Casulleras, and J. Navarro, *Phys. Rev.* B **50**, 3427 (1994).

67. D. M. Ceperley and H. J. Partridge, *J. Chem. Phys.* **84**, 820 (1986).
68. R. A. Aziz, M. J. Slaman, A. Koide, A. R. Allnatt, and W. J. Meath, *Mol. Phys.* **77**, 321 (1992).
69. A. R. Janzen and R. A. Aziz, *J. Chem. Phys.* **107**, 914 (1997).
70. T. Korona, H. L. Williams, R. Bukowskj, B. Jeziorski, and K. Szalewicz, *J. Chem. Phys.* **106**, 5109 (1997).
71. K. T. Tang, J. P. Toennies, and C. L. Yiu, *Phys. Rev. Lett.* **74**, 1546 (1995).
72. L. Reatto, *Nucl. Phys.* A **328**, 253 (1979).
73. T. R. Sosnick, W. H. Snow, and P. E. Sokol, *Phys. Rev.* B **41**, 11185 (1990).
74. E. C. Svensson, V. F. Sears, A. D. B. Woods, and P. Martel, *Phys. Rev.* B **21**, 3638 (1980).
75. F. H. Wirth and R. B. Hallock, *Phys. Rev.* B **35**, 89 (1987).
76. S. Stringari and J. Treiner, *Phys. Rev.* B **36**, 8369 (1987).
77. J. A. Nissen, E. Bodegom, L. C. Brodie, and J. S. Semura, *Phys. Rev.* B **40**, 6617 (1989).
78. Q. Xiong and H. J. Maris, *J. Low Temp. Phys.* **82**, 105 (1991).
79. M. S. Pettersen, S. Balibar, and H. J. Maris, *Phys. Rev.* B **49**, 12062 (1994).
80. H. J. Maris and Q. Xiong, *Phys. Rev. Lett.* **63**, 1078 (1989).
81. H. J. Maris, *Phys. Rev. Lett.* **66**, 45 (1991).
82. M. A Solís and J. Navarro, *Phys. Rev.* B **45**, 13080 (1992).
83. M. Guilleumas, M. Pi, M. Barranco, J. Navarro, and M. A. Solís, *Phys. Rev.* B **47**, 9116 (1993).
84. D. M. Jezek, M. Guilleumas, M. Pi, M. Barranco, and J. Navarro, *Phys. Rev.* B **48**, 16582 (1993).
85. G. H. Bauer, D. M. Ceperley, and N. Goldenfeld, *Phys. Rev.* B **61**, 9055 (2000).
86. C. E. Campbell, R. Folk, and E. Krotscheck, *J. Low Temp. Phys.* **105**, 13 (1996).
87. M. Barranco, M. Guilleumas, M. Pi, and D.M. Jezek, Cavitation in liquid helium, 2001, this volume.
88. B. M. Abraham, Y. Eckstein, J. B. Ketterson, M. Kuchnir, and P. R. Roach, *Phys. Rev.* A **1**, 250 (1970).
89. E. Manousakis, V. R. Pandharipande, and Q. N. Usmani, *Phys. Rev.* B **31**, 7022 (1985); **43**, 13587 (1991).
90. P. Whitlock and R. M. Panoff, *Can. J. Phys.* **65**, 1409 (1987).
91. H. A. Mook, *Phys. Rev. Lett.* **51**, 1454 (1983).
92. J. Gavoret and P. Nozières, *Ann. Phys. (N.Y.)* **28**, 349 (1964).
93. H. R. Glyde, *Excitations in Liquid and Solid Helium* (Clarendon Press, Oxford, 1994).
94. H. R. Glyde, R. T. Azuah, and W. G. Stirling, *Phys. Rev.* B **62**, 14337 (2000).
95. A. Griffin, *Excitations in a Bose-Condensed Liquid* (University Press, Cambridge, 1993).
96. R. P. Feynman, *Phys. Rev.* **94**, 262 (1954).
97. H. R. Glyde and A. Griffin, *Phys. Rev. Lett.* **65**, 1454 (1990).
98. E. F. Talbot, H. R. Glyde, W. G. Stirling, and E. C. Svensson, *Phys. Rev.* B **38**, 11229 (1988).
99. W. G. Stirling and H. R. Glyde, *Phys. Rev.* B **41**, 4224 (1990).
100. E. Manousakis and V. R. Pandharipande, *Phys. Rev.* B **30**, 5062 (1984).
101. D. E. Galli, L. Reatto, and S. Vitiello, *J. Low Temp. Phys.* **101**, 755 (1995).
102. M. Boninsegni and D. M. Ceperley, *J. Low Temp. Phys.* **104**, 339 (1996).
103. R. J. Donnelly, J. A. Donnelly, and R. N. Hills, *J. Low Temp. Phys.* **44**, 471 (1981).
104. H. R. Glyde, M. R. Gibbs, W. G. Stirling, and M. A. Adams, *Europhys. Lett.* **43**, 422 (1998).

105. B. Fåk and J. Bossy, *J. Low Temp. Phys.* **112**, 1 (1998).
106. L. P. Pitaevskii, *Zh. Eksp. Teor. Fiz.* **36**, 1168 (1959) (*Sov. Phys. JETP* **9**, 830 (1959)).
107. A. D. B. Woods and R. A. Cowley, *Rep. Prog. Phys.* **36**, 1135 (1973).
108. B. E. Clements, J. L. Epstein, E. Krotscheck, and M. Saarela, *Phys. Rev.* B **48**, 7450 (1993).
109. B. E. Clements, H. Forbert, E. Krotscheck, H. J. Lauter, M. Saarela, and C. J. Tymczak, *Phys. Rev.* B **50**, 6958 (1994).
110. B. E. Clements, H. Forbert, E. Krotscheck, and M. Saarela, *J. Low Temp. Phys.* **95**, 849 (1994).
111. E. Cheng, M. W. Cole, W. F. Saam, and J. Treiner, *Phys. Rev.* B **46**, 13967 (1992).
112. F. Dalfovo, A. Lastri, L. Pricaupenko, S. Stringari, and J. Treiner, *Phys. Rev.* B **52**, 1193 (1995).
113. P. A. Whitlock, G. V. Chester, and M. H. Kalos, *Phys. Rev.* B **38**, 2418 (1988).
114. D. M. Ceperley and E. L. Pollock, *Phys. Rev.* B **39**, 2084 (1989).
115. M. E. Pierce and E. Manousakis, *Phys. Rev. Lett.* **83**, 5314 (1999).
116. D. S. Greywall and P. A. Bush, *Phys. Rev. Lett.* **67**, 3535 (1991).
117. G. Zimmerli, G. Mistura, and M. H. W. Chan, *Phys. Rev. Lett.* **68**, 60 (1992).
118. C. Carraro and M. W. Cole, *Phys. Rev.* B **46**, 10947 (1992).
119. K. S. Liu, M. H. Kalos, and G. V. Chester, *Phys. Rev.* B **38**, 2418 (1988).
120. A. Belic and S. Fantoni, *Physica* B **194-196**, 517 (1994).
121. M. C. Gordillo and D. M. Ceperley, *Phys. Rev.* B **58**, 6447 (1998).
122. R. J. Donnelly, *Quantized Vortices in Helium II* (University Press, Cambridge, 1991).
123. R. P. Feynman, in *Progress in Low Temperature Physics I*, ed. by C. J. Gorter (North-Holland, Amsterdam, 1955).
124. S. A. Vitiello, L. Reatto, G. V. Chester, and M. H. Kalos, *Phys. Rev.* B **54**, 1205 (1996).
125. M. Saarela and F. V. Kusmartsev, *Phys. Lett.* A **202**, 317 (1995).
126. S. Ijima, *Nature (London)* **354**, 56 (1991).
127. M. R. Pederson and J. Q. Broughton, *Phys. Rev. Lett.* **69**, 2689 (1992).
128. P. M. Ajayan and S. Ijima, *Nature (London)* **361**, 333 (1993).
129. C. Dillon, K. M. Jones, T. A. Bekkedahl, C. H. Kiang, D. S. Bethune, and M. J. Heben, *Nature (London)* **386**, 377 (1997).
130. W. Teizer, R. B. Hallock, E. Dujardin, and T. W. Ebbesen, *Phys. Rev. Lett.* **82**, 5305 (1999); **84**, 1844(E) (2000).
131. S. Talapatra, A. Z. Zambano, S. E. Weber, and A. D. Migone, *Phys. Rev. Lett.* **85**, 138 (2000).
132. G. Stan, M. J. Bojan, S. Curtarolo, S. M. Gatica, and M. W. Cole, *Phys. Rev.* B **62**, 2173 (2000).
133. F. Darkrim and D. Levesque, *J. Chem. Phys.* **109**, 4981 (1998).
134. H. Yano, S. Yoshizaki, S. Inagaki, Y. Fukushima, and N. Wada, *J. Low Temp. Phys.* **110**, 573 (1998).
135. N. Wada, J. Taniguchi, H. Ikagami, S. Inagaki, and Y. Fukushima, *Phys. Rev. Lett.* **86**, 4322 (2001).
136. E. Krotscheck and M. D. Miller, *Phys. Rev.* B **60**, 13038 (1999).
137. M. Boninsegni and S. Moroni, *J. Low Temp. Phys.* **118**, 1 (2000).
138. E. Manousakis, S. Fantoni, V. R. Pandharipande, and Q. N. Usmani, *Phys. Rev.* B **28**, 3770 (1983).
139. E. Krotscheck, *J. Low Temp. Phys.* **119**, 103 (2000).
140. K. E. Schmidt, M. A. Lee, M. H. Kalos, and G. V. Chester, *Phys. Rev. Lett.* **47**, 807

(1981).
141. S. Moroni, S. Fantoni, and G. Senatore, *Europhys. Lett.* **30**, 93 (1995).
142. F. Pederiva, S. A. Vitiello, K. Gernoth, S. Fantoni, and L. Reatto, *Phys. Rev.* B **53**, 15129 (1996).
143. M. Viviani, E. Buendía, S. Fantoni, and S. Rosati, *Phys. Rev.* B **38**, 4523 (1988).
144. J. P. Bouchard and C. Lhuillier, *Z. Phys.* B **75**, 283 (1989).
145. P. E. Sokol, K. Sköld, D. L. Price, and R. Klebb, *Phys. Rev. Lett.* **54**, 909 (1985).
146. R. A. Aziz and K. R. Pathria, *Phys. Rev.* A **7**, 809 (1973).
147. J. C. Wheatley, *Rev. Mod. Phys.* **47**, 415 (1975).
148. H. J. Maris, *J. Low Temp. Phys.* **98**, 403 (1995).
149. F. Caupin, P. Roche, S. Marchand, and S. Balibar, *J. Low Temp. Phys.* **113**, 473 (1998).
150. D. M. Jezek, M. Pi, and M. Barranco, *Phys. Rev.* B **60**, 3048 (1999).
151. J. Boronat and J. Casulleras, in *Proceedings of the Tenth International Conference on Recent Progress in Many-Body Theories*, ed. by R. F. Bishop *et al.* (World Scientific, Singapore, 2000).
152. E. K. Achter and L. Meyer, *Phys. Rev.* **188**, 291 (1969).
153. R. B. Hallock, *J. Low Temp. Phys.* **26**, 109 (1972).
154. R. Senesi, C. Andreani, D. Colognesi, A. Cunsolo, and M. Nardone, *Phys. Rev. Lett.* **86**, 4584 (2001).
155. R. B. Hallock, *Physics Today*, June 1998, p. 30.
156. D. O. Edwards, S. Y. Shen, J. R. Eckardt, P. P. Fatouros, and F. M. Gasparini, *Phys. Rev.* B **12**, 892 (1975).
157. J. M. Valles Jr., R. H. Higley, R. B. Johnson, and R. B. Hallock, *Phys. Rev. Lett.* **60**, 428 (1988).
158. K. D. Morhard, J. Bossy, and H. Godfrin, *Phys. Rev.* B **51**, 446 (1995).
159. K. D. Morhard, C. Bäuerle, J. Bossy, Y. Bunkov, S. N. Fisher, and H. Godfrin, *Phys. Rev.* B **53**, 2658 (1996).
160. P. A. Sheldon and R. B. Hallock, *Phys. Rev. Lett.* **77**, 2973 (1996).
161. A. Casey, M. Patel, J Nyéki, B. P. Cowan, and J. Saunders, *J. Low Temp. Phys.* **110**, 293 (1998).
162. C. Bäuerle, Y. Bunkov, A. S. Chen, S. N. Fisher, and H. Godfrin, *J. Low Temp. Phys.* **110**, 333 (1998).
163. P. C. Ho and R. B. Hallock, *Phys. Rev. Lett.* **87**, 135301 (2001).
164. A. D. Novaco and C. E. Campbell, *Phys. Rev.* B **11**, 2525 (1975).
165. M. D. Miller and L. H. Nosanow, *J. Low Temp. Phys.* **32**, 145 (1978).
166. B. Brami, F. Joly, and C. Lhuillier, *J. Low Temp. Phys.* **94**, 63 (1994).
167. E. Krotscheck, *Phys. Rev.* B **32**, 5713 (1985).
168. B. E. Clements, E. Krotscheck, and M. Saarela, *Phys. Rev.* B **55**, 5959 (1997).
169. F. Y. Wu and E. Feenberg, *Phys. Rev.* **128**, 943 (1962).
170. B. Krishnamachari and G. V. Chester, *Phys. Rev.* B **59**, 8852 (1999).
171. V. Grau, J. Casulleras, and J. Boronat, to be published.

CHAPTER 3

THE FINITE-TEMPERATURE PATH INTEGRAL MONTE CARLO METHOD AND ITS APPLICATION TO SUPERFLUID HELIUM CLUSTERS

P. Huang[†], Y. Kwon[‡], and K. B. Whaley[†]

[†] *Department of Chemistry and Kenneth S. Pitzer Center for Theoretical Chemistry, University of California, Berkeley, CA 94720-1460, USA*

[‡] *Department of Physics, Konkuk University, Seoul 143-701, Korea*

E-mail: whaley@socrates.berkeley.edu

We review the use of the path integral Monte Carlo (PIMC) methodology to the study of finite-size quantum clusters, with particular emphasis on recent applications to pure and impurity-doped ^{4}He clusters. We describe the principles of PIMC, the use of the multilevel Metropolis method for sampling particle permutations, and the methods used to accurately incorporate anisotropic molecule-helium interactions into the path integral scheme. Applications to spectroscopic studies of embedded atoms and molecules are summarized, with discussion of the new concepts of local and nanoscale superfluidity that have been generated by recent PIMC studies of the impurity-doped ^{4}He clusters.

1. Introduction

Over the past 15 years, the path integral Monte Carlo (PIMC) method has evolved into a uniquely powerful computational tool for the study of bulk and finite quantum systems. In PIMC, one is interested in computing the thermal average of a quantum observable $\hat{O}$ at a given temperature T, which can be expressed with respect to the thermal density matrix $\rho(R, R'; \beta) = \langle R'|e^{-\beta \hat{H}}|R\rangle$:

$$\langle \hat{O} \rangle = Z^{-1} \int dRdR' \, \rho(R, R'; \beta) \langle R|\hat{O}|R'\rangle, \tag{1.1}$$

where $R \equiv (\mathbf{r}_1, \mathbf{r}_2, \ldots, \mathbf{r}_N)$ is a point in the $3N-$dimensional configuration space of an N-particle system, $\hat{H}$ is the Hamiltonian, and $\beta = 1/k_BT$. Here $Z = \int dR \, \rho(R, R; \beta)$ is the partition function. The multidimensional integral of Eq. (1.1) can in principle be evaluated by standard Monte Carlo integration schemes, *i.e.* by taking an average of $\langle R|\hat{O}|R'\rangle$ over the configurations $\{R, R'\}$ sampled from the probability distribution $Z^{-1}\rho(R, R'; \beta)$. However, the full density matrix of an interacting N-particle quantum system is generally not known at low temperatures. Therefore one needs to resort to the discrete representation of the Feynman path integral formula for a low-temperature density matrix, which will be discussed in detail in Section 2.

The finite-temperature nature of PIMC makes this a complementary approach to zero-temperature Monte Carlo methods, such as variational Monte Carlo, or Green's function-based methods such as diffusion Monte Carlo, which are reviewed in Chapter 1 of this volume. [1] PIMC is currently the only numerical method capable of directly addressing finite-temperature superfluidity and the superfluid transition in helium. In addition, unlike the zero-temperature methods, PIMC does not require the use of a trial function, requiring as input only the particle masses, numbers, and interaction potentials, in addition to the temperature and volume. Consequently, it is a numerically exact technique, and is independent of the trial function bias problems that zero-temperature methods may suffer from. However, because PIMC provides thermodynamic averages, state-specific information is generally not available. Although in principle the density matrix contains information on the full eigenspectrum of the Hamiltonian $\hat{H}$, extracting this requires the numerical inversion of a Laplace transform. [2–4] Such inversions are known to be notoriously difficult in the presence of Monte Carlo noise. Thus, to date, path integral Monte Carlo has provided only very limited dynamical information of a direct nature. Nevertheless, it has provided critical microscopic input into dynamical models for physical systems in helium droplets, and in conjunction with zero-temperature, state-specific calculations for these finite helium systems, PIMC has proven to be a powerful means of investigating the dynamic consequences of atomic scale structure of a superfluid. [5]

Feynman first applied the path integral approach to liquid helium in 1953, and provided a consistent clarification of the role of Bose permutation symmetry in the lambda transition of liquid helium. [6] In Feynman's original treatment, the multidimensional integrals of Eq. (1.1) were approximated analytically. In the late 1980's, Ceperley and Pollock subsequently devised a Monte Carlo scheme for the exact numerical evaluation of these multidimensional integrals, in which the combined configuration and permutation spaces were efficiently sampled using a multilevel Metropolis method. [7] This allowed direct quantitative application of Feynman's path integral approach to the superfluid state of helium for the first time.

Since then, the PIMC method has been applied to provide a quantitative description of numerous bulk and finite bosonic systems. In addition to extensive studies of bulk helium, PIMC has now been employed in the study of ^{4}He/^{3}He mixtures, [8] of helium and molecular hydrogen droplets, [5,9] and of helium and hydrogen films on various surfaces. [10,11] In this work, we focus on the application of PIMC to quantum simulations of finite helium clusters, ^4He_N. Since we are primarily concerned with the bosonic isotope of helium, for the remainder of the chapter, we will denote ^{4}He as simply He, unless explicitly stated. The field of helium cluster research has grown very rapidly over recent years due to new possibilities of inserting molecular probes and studying their properties. [12] An overview of the experimental work in this area is provided in Chapter 9 of this volume. Accompanying this rise in experimental studies, there has been a correspondingly increased demand for complementary theoretical study of these finite quantum systems.

The earliest PIMC simulation of pure He_N droplets, made in 1989, demonstrated superfluid behavior for sizes N as small as 64. [13] This result, together with zero-temperature calculations of the size scaling for pure cluster excitation spectra made at that time, [15] was taken to be strong theoretical evidence that these finite-sized clusters were indeed superfluid. This was later confirmed experimentally, through a series of elegant experiments with impurity-doped helium clusters. [16,17] Doped clusters present additional technical challenges for PIMC beyond the requirements posed by a finite cluster of pure He_N. Early PIMC work with doped clusters addressed the widely studied He_NSF_6 system. [18,19] These studies showed that the global superfluid fraction appeared not to be significantly modified by introduction of an impurity. However, it soon became apparent that interesting new local features due to Bose exchange symmetry were present in the immediate vicinity of an impurity. This led to the recognition that a local non-superfluid density could be induced by the molecular interaction with helium. [5] PIMC simulations of doped clusters have now been made with a variety of impurities, including OCS, HCN, benzene, H_2, neutral and ionic alkali atoms, and some complexes of these molecules. As will be outlined here, these studies have revealed a broad range of properties of the dopant as well as insight into the dopant influence on the superfluid properties of the droplets. Key features that have emerged from these studies of doped droplets are the ability to analyze superfluid behavior in nanoscale dimensions, to characterize quantum solvation in a superfluid, and to probe the atomic-scale behavior of a superfluid near a molecular interface.

In this review, we first provide an overview of the Feynman path integral formalism for quantum statistical thermodynamics in Section 2. Following this introduction, we discuss the PIMC implementation of the general Feynman theory, focusing in particular on the multilevel Metropolis sampling method of Ceperley and Pollock. We then review applications of the PIMC method to the study of pure helium droplets, and of doped helium clusters containing atomic or molecular impurities in Section 3. The analysis of superfluidity in finite droplets, and concepts of nanoscale superfluids and local superfluidity are described in Section 3.4. PIMC applications to spectroscopic studies of doped helium clusters are summarized in Section 4. We conclude in Section 5 with a summary of open questions.

2. Theory

2.1. *General formulation*

Here we deal with a cluster of N He atoms doped with a single impurity. In many of the studies made to date, the impurity is assumed to be fixed at the origin without either translational or rotational motion. Neglect of the impurity translational degrees of freedom is not essential, but for impurity particles which are heavy relative to a helium atom, it is reasonable and often convenient to ignore the translational motion of the impurity. However, the neglect of the impurity rotational degrees of freedom should be treated with caution, especially for impurities with small principal moments of inertia. Incorporation of the rotational motion of the impurity is an area of current work. Hence for the present discussion we consider the following system Hamiltonian $\hat{H}$:

$$\hat{H} = \hat{K} + \hat{V} = -\lambda \sum_{i=1}^{N} \nabla_i^2 + \sum_{i<j} V_{\text{He-He}}(r_{ij}) + \sum_{i=1}^{N} V_{\text{He-imp}}(\mathbf{r}_i), \tag{2.1}$$

where $\lambda = \hbar^2/2m_4$, with m_4 being the helium mass. The potential energy $\hat{V}$ includes a sum of He-He pair potentials, $V_{\text{He-He}}$, and He-impurity interactions $V_{\text{He-imp}}$, where the latter are most readily given in the molecular frame. If necessary, *e.g.* for light molecules, the impurity translational degrees of freedom can be incorporated by adding an additional term $-\lambda_I \nabla_I^2$, corresponding to the impurity center-of-mass kinetic energy.

In the path integral approach, one uses the identity $e^{-(\beta_1+\beta_2)\hat{H}} = e^{-\beta_1 \hat{H}} e^{-\beta_2 \hat{H}}$ to express the low-temperature density matrix by an integral over all possible *paths*, $\{R, R_1, R_2, \ldots, R_{M-1}, R'\}$, with the weight for each path given by the product of density matrices at a higher temperature $T' = MT$:

$$\rho(R, R'; \beta) = \int dR_1 dR_2 \ldots dR_{M-1}\, \rho(R, R_1; \tau)\rho(R_1, R_2; \tau) \ldots \rho(R_{M-1}, R'; \tau). \tag{2.2}$$

Here $\tau \equiv \beta/M = (k_B T')^{-1}$ constitutes the imaginary *time step* defining the discrete representation of the path integral. For a sufficiently high temperature T' or, equivalently, for a small enough time step τ, there exist several approximations to the density matrix that are sufficiently accurate for this factorization to be used in numerical work. [20] The simplest of these high-temperature approximations is the *primitive* approximation, which is based upon the Trotter formula: [21]

$$\rho(R_k, R_{k+1}; \tau) \approx \int dR' \, \langle R_k | e^{-\tau \hat{K}} | R' \rangle \langle R' | e^{-\tau \hat{V}} | R_{k+1} \rangle. \tag{2.3}$$

The potential energy operator $\hat{V}$ in Eq. (2.1) is diagonal in the position representation,

$$\langle R' | e^{-\tau \hat{V}} | R_{k+1} \rangle = e^{-\tau V(R_{k+1})} \delta(R' - R_{k+1}), \tag{2.4}$$

while the kinetic term corresponds to the free particle density matrix:

$$\langle R_k | e^{-\tau \hat{K}} | R' \rangle = \rho_0(R_k, R'; \tau) = (4\pi\lambda\tau)^{-3N/2} e^{-(R_k - R')^2/4\lambda\tau}. \tag{2.5}$$

From Eqs. (2.2)–(2.5), the path integral representation for the density matrix in the primitive approximation may be expressed as:

$$\rho(R_0, R_M; \beta) = \tag{2.6}$$
$$(4\pi\lambda\tau)^{-3NM/2} \int dR_1 \cdots dR_{M-1} \exp\left[-\sum_{k=1}^{M} \sum_{i=1}^{N} \frac{(\mathbf{r}_{i,k} - \mathbf{r}_{i,k-1})^2}{4\lambda\tau} - \tau \sum_{k=1}^{M} V(R_k) \right],$$

with

$$V(R_k) = \sum_{i<j}^{N} V_{\text{He-He}}(r_{ij,k}) + \sum_{i=1}^{N} V_{\text{He-imp}}(\mathbf{r}_{i,k}). \tag{2.7}$$

A single R_k is referred to as a *time slice*, and $\mathbf{r}_{i,k}$, the position of the ith particle at the kth time slice, as a *bead*. From here on, i will denote particle index and k will denote time index. Eq. (2.6) may be viewed as a classical configuration integral, with the exponent of its integrand corresponding to an energy function. The first term in the exponent, derived from the kinetic energy, corresponds to a spring potential that connects beads representing the same atom at successive imaginary times, with coupling constant $(2\lambda\tau)^{-1}$. This chain of beads connected by springs is often referred to as a *polymer.* [6,22] The helium-helium interaction is represented by an inter-polymer potential that has non-zero interactions only between beads located on different polymers and indexed by the same imaginary time value. This corresponds to Feynman's original idea of mapping path integrals of a quantum system onto interacting classical polymers, with a special form of polymer interaction potential. [6] In the absence of kinetic contributions from the impurity, the helium-impurity interaction acts as an additional external field for the helium atoms, and hence for these polymers.

The thermal density matrix $\rho(R, R'; \beta) = \langle R|e^{-\beta\hat{H}}|R'\rangle$ can be expanded in terms of the eigenvalues $\{E_n\}$ and the corresponding eigenfunctions $\{\phi_n\}$ of the Hamiltonian $\hat{H}$:

$$\rho(R, R'; \beta) = \sum_n \phi_n(R)\phi_n^*(R')e^{-\beta E_n}. \tag{2.8}$$

This is appropriate for a system of distinguishable particles under Boltzmann statistics. For a Bose system such as He or para-H_2, it should be symmetrized with respect to particle exchanges. This can be done by modifying the sum in Eq. (2.8) to a sum over exchange-symmetrized stationary states $\tilde{\phi}_\alpha$ only:

$$\rho_B(R, R'; \beta) = \sum_\alpha \tilde{\phi}_\alpha(R)\tilde{\phi}_\alpha^*(R')e^{-\beta E_\alpha}. \tag{2.9}$$

A symmetrized eigenfunction $\tilde{\phi}_\alpha(R)$ can be obtained by summing a stationary wavefunction of distinguishable particles, $\phi_n(\mathcal{P}R)$, over all N-particle permutations $\mathcal{P}$: [23]

$$\tilde{\phi}_\alpha(R) = \frac{1}{N!}\sum_{\mathcal{P}} \phi_n(\mathcal{P}R). \tag{2.10}$$

Inserting Eq. (2.10) into Eq. (2.9), we obtain the symmetrized density matrix for a Bose system:

$$\rho_B(R, R'; \beta) = \frac{1}{N!}\sum_{\mathcal{P}} \rho(R, \mathcal{P}R'; \beta). \tag{2.11}$$

2.2. *Density matrix evaluation*

In order to make PIMC calculations more tractable, one wishes to use the smallest possible number of time slices M for a given temperature T. This means that it is essential to find accurate high-temperature density matrices at as *small* a value T' as possible, so that the imaginary time step τ be kept as *large* as possible. It has been found that for the helium-helium interaction, the primitive approximation described in Eqs. (2.3)–(2.5) is accurate enough at temperatures higher than ~1000 K. [24] This implies that a PIMC simulation at $T \sim 0.3 - 0.4$ K, where the spectroscopic measurements for the impurity-doped helium clusters have been performed, [25] would require several thousands of time slices. This computational expense can be avoided by going beyond the primitive approximation to a more sophisticated approximation for the high-temperature density matrices. Based on the Feynman-Kac formula, [24] the high-temperature density matrix $\rho(R, R'; \tau)$ can be approximated by a product of the free particle propagator $\rho_0(R, R'; \tau)$ of Eq. (2.5) and an interaction term $e^{-U(R,R';\tau)}$:

$$\rho(R, R'; \tau) \approx \rho_0(R, R'; \tau)e^{-U(R,R';\tau)}. \qquad (2.12)$$

The interaction term $e^{-U(R,R';\tau)}$ is in turn factored into contributions deriving from the helium-helium and helium-impurity interactions, $\rho_{\mathrm{He-He}}$ and $\rho_{\mathrm{He-imp}}$, respectively. For spherical interactions, one can generate a pair-product form of the exact two-body density matrices using a matrix squaring approach discussed in detail in Ref. 20. We use this for the helium-helium interaction, which is spherically symmetric. Such helium-helium density matrices of the pair-product form have been shown to be accurate for $\tau^{-1}/k_B \geq 40$ K, *i.e.*, $T' \geq 40$ K. [7,24] This same approach can be used for the helium-impurity interaction when this is also isotropic. However, for molecules the helium-impurity interaction $V_{\mathrm{He-imp}}(\mathbf{r})$ is in general not isotropic, and may involve complicated three-dimensional dependencies. This can be dealt with in several ways. One approach is to expand the helium-impurity interaction in spherical terms and then employ pair-product forms as above. We have found it convenient to work within the primitive approximation for the helium-impurity interaction, which allows considerable flexibility when changing impurities. The required time step for the accurate primitive helium-impurity density matrices varies, depending on the impurity molecule involved (*e.g.*, $\tau^{-1}/k_B \geq 80$ K for He-SF_6 and He-OCS, $\tau^{-1}/k_B \geq 160$ K for He-benzene). This must be recalibrated, *e.g.*, by establishing converged helium densities, for every new molecule that is studied. The same re-calibration requirement holds also for spherical expansions.

2.3. *Multilevel Metropolis algorithm*

For a diagonal operator $\hat{O}$ in the position representation, $\langle R|\hat{O}|R'\rangle = O(R)\delta(R-R')$, we need to consider only the diagonal density matrices for evaluation of its thermal average, Eq. (1.1). For the diagonal density matrix, both the sum over permutations in Eq. (2.11) and the multidimensional integration in Eq. (2.2) can be evaluated by a sampling of discrete paths which end on a permutation of their starting positions, *i.e.*, $s = \{R_0, R_1, R_2, \cdots, R_{M-1}, R_M\}$ with $R_M = \mathcal{P}R_0$. This gives rise to an isomorphic mapping onto *ring* polymers. In fact, all physical quantities discussed in this review can be estimated from a set of stochastically sampled ring polymers.

In the sampling process, it will be natural to choose the probability density function as

$$\pi_s = Z^{-1} \prod_{k=0}^{M-1} \rho(R_k, R_{k+1}; \tau). \tag{2.13}$$

The Metropolis algorithm, a widely-used Monte Carlo sampling technique, provides a route to obtain the converged distribution π_s in the limit of many sampled configurations, as long as the detailed balance condition is satisfied for transitions between successive configurations:

$$\pi_s P_{s \to s'} = \pi_{s'} P_{s' \to s}. \tag{2.14}$$

Here $P_{s \to s'}$ is the transition probability from a configuration s to s'. This is factorized into an *a priori* sampling distribution $T_{s \to s'}$ and an acceptance ratio $A_{s \to s'}$:

$$P_{s \to s'} = T_{s \to s'} A_{s \to s'}. \tag{2.15}$$

In order to speed up convergence times in a path integral simulation, in particular one involving permutation moves, it is very important to select an appropriate distribution function $T_{s \to s'}$ for a trial move s' from s. The most efficient way of doing this is the multilevel Metropolis algorithm developed by Pollock and Ceperley. [20,24] Here one first chooses end points of each path by sampling a permutation $\mathcal{P}$. Then the paths are bisected and the configurations at the midpoints sampled. This process of bisection and midpoint sampling is repeated multiple times, resulting in a multilevel scheme that samples whole sections of the paths in a single step. The acceptance ratio at each level of this multilevel Markov process is set so that the combined process of permutation and configuration moves may lead to the probability density function π_s of Eq. (2.13). Detailed procedures are summarized as follows: [24]

(1) Initialize a configuration s. Typically one starts from a *classical* configuration, in which all beads representing each atom are located at the same site. So each polymer corresponds initially to a single point.
(2) Choose a time slice k randomly between 0 and $M-1$ and construct a table for trial permutation transitions between time slices k and $k+n$, where $n = 2^l$ and l is the *level* of this path updating process. For the simulation of a He system, $l = 3$ turns out to be a good choice for the permutation moves. Trial permutations may be restricted to cyclic permutations among 2, 3, or 4 particles. The probability for permutation transitions is proportional to

$$T_{\mathcal{P}} = \exp\left[\frac{-(R_k - \mathcal{P}R_{k+n})^2}{4\lambda n \tau}\right], \tag{2.16}$$

$$C_I = \sum_{\mathcal{P}} T_{\mathcal{P}}. \tag{2.17}$$

Thus the transition probability for permutational moves does not depend on the potential energy. Note that one can explore the entire N-particle permutation space by repeatedly sampling cyclic permutations among a small number of particles.

(3) Select a trial permutation $\mathcal{P}$ involving p atoms such that

$$\sum_{\mathcal{P}'<\mathcal{P}} T_{\mathcal{P}'} < \chi < \sum_{\mathcal{P}'\leq\mathcal{P}} T_{\mathcal{P}'}, \tag{2.18}$$

where χ is a random number on $(0, C_I)$. This selects the permutation with probability $T_{\mathcal{P}}/C_I$. Then compute $\Delta_0 = T_{\mathcal{P}}/T_I$. After this, we will sample the intermediate path coordinates connecting R_k with $\mathcal{P}R_{k+n}$. The coordinates of the $(N-p)$ atoms not on the cycle represented by $\mathcal{P}$ will not change from their old positions. This is level 0 sampling.

(4) Start a bisection algorithm by sampling a new midpoint $R'_{k+\frac{n}{2}}$. For the sampling distribution function $T(R'_{k+\frac{n}{2}}|R_k, \mathcal{P}R_{k+n}; \frac{n}{2}\tau)$, we use a multivariate Gaussian form centered at the mean position $\bar{R} = (R_k + R_{k+n})/2$ (see Eq. (5.16) of Ref. 20). Then compute

$$\Delta_1 = \frac{\rho(R_k, R'_{k+\frac{n}{2}}; \frac{n}{2}\tau)\rho(R'_{k+\frac{n}{2}}, \mathcal{P}R_{k+n}; \frac{n}{2}\tau)}{\rho(R_k, R_{k+\frac{n}{2}}; \frac{n}{2}\tau)\rho(R_{k+\frac{n}{2}}, R_{k+n}; \frac{n}{2}\tau)}. \tag{2.19}$$

Proceed to the next step with probability

$$\frac{\Delta_1 T(R_{k+\frac{n}{2}}|R_k, R_{k+n}; \frac{n}{2}\tau)}{\Delta_0 T(R'_{k+\frac{n}{2}}|R_k, \mathcal{P}R_{k+n}; \frac{n}{2}\tau)}. \tag{2.20}$$

If rejected, go back to step 3 and sample a new trial permutation. This is level 1 sampling.

(5) At the second level, sample $R'_{k+\frac{n}{4}}$ and $R'_{k+\frac{3n}{4}}$ by bisecting the two intervals and continue to the next level with the same procedures as used in step 4. This bisection process is repeated until we get to the final l-th level. At the l-th level, sample $R'_{k+1}, R'_{k+3}, \ldots$, and R'_{k+n-1} with the probability distribution function $T(R'_j|R'_{j-1}, R'_{j+1}; \tau)$. Proceed to the next step with probability

$$\frac{\Delta_l}{\Delta_{l-1}} \prod_{j=k+1,k+3,\cdots}^{k+n-1} \frac{T(R_j|R_{j-1}, R_{j+1}; \tau)}{T(R'_j|R'_{j-1}, R'_{j+1}; \tau)} \tag{2.21}$$

where

$$\Delta_l = \prod_{j=k+1}^{k+n} \frac{\rho(R'_{j-1}, R'_j; \tau)}{\rho(R_{j-1}, R_j; \tau)}. \tag{2.22}$$

If rejected, go back to step 3. Fig. 1 depicts the structure of a multi-level sampling with $l = 3$, in a path integral containing $M = 16$ time slices.

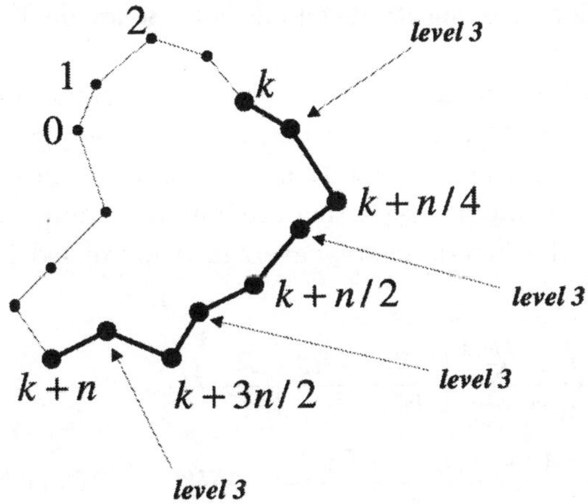

Fig. 1. Schematic of multilevel sampling. The figure shows a ring polymer of configuration beads for a single particle corresponding to $M = 16$ time slices, to be updated with a three-level ($l = 3$) sampling of $2^3 = 8$ time slices simultaneously. The bold connections indicate the section of 8 time slices that is to be updated.

(6) Construct a new permutation table for all 2, 3, or 4 particle exchanges $\mathcal{P}'$ acting on $\mathcal{P}R_{k+n}$. Accept a new path $R_k, R'_{k+1}, \ldots, R'_{k+n-1}, \mathcal{P}R_{k+n}$ with probability $C_I/C_{\mathcal{P}}$, where $C_{\mathcal{P}} = \sum_{\mathcal{P}'} T_{\mathcal{P}'\mathcal{P}}$. If rejected, go again back to step 3.
(7) After replacing old coordinates and permutation table with new ones, repeat steps 3 to 6.
(8) After attempting several hundreds or thousands of permutation moves between times slices k and $k+n$, followed by the bisection procedures to update the midpoints of all l levels, we select a new time slice k and repeat steps 2 to 6.

One can check that the multilevel bisection algorithm described here satisfies the detailed balance condition in Eq. (2.14). Note that $P_{s \to s'}$ is the total transition probability to go through all l levels.

2.4. *Estimators for some physical quantities*

With the generalized Metropolis sampling of the permutation symmetrized density matrix $\rho_B(R, R; \beta)$, the thermal average of an observable $\hat{O}$ diagonal in the position representation can be estimated by taking an arithmetic average of $O(R) = \langle R|\hat{O}|R\rangle$ over the paths sampled. For instance, the helium density distribution around an impurity molecule can be estimated by

$$\rho(\mathbf{r}) = \frac{1}{M} \sum_{k=0}^{M-1} \sum_{i=1}^{N} \langle \delta(\mathbf{r} - \vec{\mathbf{r}}_{i,k}) \rangle. \tag{2.23}$$

Note that all time slices in ring polymers can be considered as equivalent. Unlike importance-sampled diffusion Monte Carlo methods, the PIMC calculation of structural properties such as the density distribution does not involve any trial function bias.

There are many ways to compute the energy in PIMC, discussed in detail in Ref. 20. Most of the PIMC applications discussed here employ the *direct* estimator obtained by directly applying the Hamiltonian operator to the density matrix in the position space. For calculations neglecting the impurity translational and rotational degrees of freedom, the kinetic energy average is expressed in the path integral representation as

$$\langle K \rangle = \left\langle \frac{1}{M} \sum_{k=0}^{M-1} \left[\frac{3N}{2\tau} - \frac{(R_k - R_{k+1})^2}{4\lambda\tau^2} - \frac{(R_k - R_{k+1}) \cdot \nabla_k U^k}{\tau} + \lambda \nabla_k^2 U^k - \lambda (\nabla_k U^k)^2 \right] \right\rangle, \tag{2.24}$$

where $\nabla_k = \partial/\partial R_k$ is the $3N$-dimensional gradient operator, and $U^k \equiv U(R_{k-1}, R_k; \tau)$ is the interaction for link k, *i.e.*, for the spring connecting beads $k-1$ and k (see Eq. (2.12)). Computation of the potential energy is straightforward since this is diagonal in the position representation:

$$\langle V \rangle = \frac{1}{M} \sum_{k=0}^{M-1} \langle V(R_k) \rangle. \tag{2.25}$$

As noted earlier, the inter-polymeric potential acts only between beads defined at the same imaginary time.

One of the most interesting properties of bulk and finite He systems is their superfluid behavior. For bulk systems superfluid estimators are generally derived from linear response theory, *i.e.* by considering the helium response to boundary motion. [26] Pollock and Ceperley showed how to derive momentum density correlation functions that quantify the superfluid response of bulk systems with periodic boundary conditions. [7] Sindzingre *et al.* subsequently developed a global linear response estimator for finite helium clusters with free boundaries. [13] This estimator is based on the response to a rotation of continuous angular frequency, *i.e.* to a classical rotation such as might be appropriate to a macroscopic droplet. Consider the Hamiltonian in a coordinate frame rotating about an axis with frequency ω,

$$\hat{H}_{\text{rot}} = \hat{H}_0 - \hat{\mathbf{L}} \cdot \omega, \tag{2.26}$$

where $\hat{\mathbf{L}}$ is the total angular momentum operator. For a classical fluid, in the limit of an infinitesimally small rotation the entire fluid should rotate rigidly with classical moment of inertia I_{cl}. But in a Bose superfluid, only the normal component responds to the rotation, resulting in an effective moment of inertia

$$I = \left. \frac{\partial \langle \hat{\mathbf{L}} \rangle}{\partial \omega} \right|_{\omega=0}. \tag{2.27}$$

Note that this is to be evaluated in the limit $\omega \to 0$, appropriate to rotation of a macroscopic system. In a homogeneous system the normal fraction can be defined as

$$\frac{\rho_n}{\rho} = \frac{I}{I_{cl}}, \tag{2.28}$$

and the complementary superfluid fraction is then given by

$$\frac{\rho_s}{\rho} = 1 - \frac{\rho_n}{\rho} = \frac{I_{cl} - I}{I_{cl}}. \tag{2.29}$$

For $\omega = \omega\hat{z}$, this linear response estimator can be expressed in the path integral representation as [13]

$$\frac{\rho_s}{\rho} = \frac{4m^2\langle A_z^2\rangle}{\beta\hbar^2 I_{cl}}, \tag{2.30}$$

with the vector quantity $\mathbf{A}$ defined as

$$\mathbf{A} = \frac{1}{2}\sum_{i,k}(\mathbf{r}_{i,k} \times \mathbf{r}_{i,k+1}). \tag{2.31}$$

Here the summation runs over particle i and imaginary-time slice k. The vector quantity $\mathbf{A}$ is the directed total area of closed imaginary-time polymers spanned by all N particles, *e.g.*, A_z is the projection of $\mathbf{A}$ on the $\hat{z}$-direction. The average size of a single polymer is given by its thermal de Broglie wavelength $\Lambda = (\beta\hbar^2/m)^{1/2}$. This becomes negligible in the high-temperature limit, and thus the corresponding superfluid fraction ρ_s/ρ goes to zero (although the projected area can remain finite at the microscopic level). At low temperatures, when the de Broglie wavelength becomes comparable to the inter-polymer spacing, particle exchanges cause polymers to cross-link and form larger ring polymers. The projected area $\mathbf{A}$ increases correspondingly and the helium system attains an appreciable superfluid fraction.

We discuss the application of this finite-system superfluid estimator in detail in Sec. 3, together with analysis of how this can be decomposed into local contributions for an inhomogeneous system. [5,19,27]

3. Superfluidity and quantum solvation of atoms and molecules in bosonic helium clusters

Spectroscopic studies of impurity-doped clusters have allowed experimental investigation of a variety of excitations in helium clusters. [12] The relevant temperature range currently accessible is $T \sim 0.15 - 0.5$ K. [25] Thus, the incorporation of Bose symmetry is essential in simulation of these systems. In this section, we focus on the application of finite-temperature PIMC to bosonic helium clusters. We begin by briefly reviewing studies of pure clusters in Sec. 3.1, and then focus on the more recent work for the clusters doped with various impurities in Secs. 3.2–3.4. Secs. 3.2–3.3 summarize the structural and energetic aspects, while Sec. 3.4 deals with the microscopic analysis of superfluid properties of the doped clusters.

3.1. *Pure clusters*

Most of the previous theoretical studies involving pure clusters are based on zero-temperature methods, and have focused on the cluster elementary excitation spectrum, which qualitatively retains the phonon-roton features characteristic of bulk He II. Current work in this area aims to understand the physical nature of the roton excitations. [28] Zero- and finite-temperature calculations for pure helium clusters have been reviewed previously, [29] and so we shall provide only a brief outline of the finite-temperature results here. The first studies were made by Cleveland *et al.*, using the path integral molecular dynamics approach in which exchange is neglected. [30] This allowed structural analysis, which was used to study the changes in droplet density and diffuseness as a function of size. Permutation exchange symmetry was incorporated by Sindzingre *et al.* in a study of the temperature and size dependence of the global superfluid fraction in finite He_N clusters. [13] These calculations employed the area estimator discussed in Sec. 2.4 and showed that a broad transition to a predominantly superfluid state occurs at a temperature depressed from the bulk superfluid transition temperature, in accordance with expectations from scaling of phase transitions for finite systems. The extent of depression increased as the cluster size decreased. For $N = 64$, the onset of the transition occurs just below $T = 2$ K, and the transition appears complete at $T = 0.5$ K, with about 90% or more of the cluster being superfluid at that temperature. A qualitative examination of the relative contribution of long exchange path lengths to the density revealed that the long exchange path contribution was largest in the interior of the droplets.

3.2. *Atomic impurities*

Neutral and ionic atomic impurities constitute the simplest dopants. For ground electronic states, the helium-impurity interatomic potential can be calculated with fairly high accuracy using standard quantum chemistry methods, and the helium-helium interatomic potential is well-known. Thus, within the two-body approximation, it is possible to construct accurate potential energy surfaces for the ground electronic state. The interactions of excited electronic states with helium are, by comparison, less well-characterized and only a few calculations of electronically excited potential energy surfaces have been even attempted. To date, PIMC calculations have been made for the neutral alkali metal impurities Li, Na, K, [31] and for the ionic impurity Na^+. [32] In general, the solvation characteristics of each impurity are controlled by a balance between different energetic factors. [33,34] These include the helium-impurity interaction strength, the helium-helium interaction strength, the impurity kinetic energy (and thus impurity mass), and the free energy change due to the loss of exchange energy for helium atoms adjacent to the impurity.

The He-Li, He-Na, and He-K ground state potentials typically have well depths of $\sim 1 - 2$ K, [33] smaller than the He-He well depth of ~ 11 K. [35] By considering these potential energy factors alone, one would qualitatively expect that the atomic impurities would reside on the droplet surface in order to minimize the total energy. The PIMC studies, made at $T = 0.5$ K, indicate that the neutral alkali impurity species are indeed surface-attached for cluster sizes of $N \leq 300$. [31] In all cases, the perturbation on the cluster structure due to the presence of the impurity is weak. The neutral impurity atom induces small but distinct modulations in the helium density, starting at the surface and decaying into the interior of the droplet. While no zero-temperature microscopic calculations of these systems have been made to date, it is expected that this behavior would persist to lower temperatures and is therefore also applicable to the experimental studies of these alkali-doped systems that are made at $T = 0.38$ K. [36]

The ionic impurity Na^+ interacts much more strongly with helium and consequently gives a markedly stronger structural perturbation of the local helium density. The He-Na^+ well depth is 407 K, about 40 times larger than the He-He well depth. Here the finite-temperature PIMC studies indicate that the Na^+ ion resides in the center of the cluster, and the strength and range of the He-Na^+ interaction induces a tightly packed helium "snowball" around the ion. [32] In Fig. 2a, the radial helium density profile for the $He_{100}Na^+$ system at $T = 0.625$ K is shown. This is compared with the $N = 64$ radial density profile around the molecular impurity SF_6, at the same temperature using an isotropic He-SF_6 interaction. There is a very strongly modulated layer structure around the Na^+ ion, with a high first coordination shell peak followed by a second peak of lower density. Similar structural features have been seen in variational shadow function calculations for Na^+ and K^+ in bulk He, [38,39] although quantitative differences exist in comparison with those results. In the variational calculations the local angular ordering within the coordination shells was also examined, leading to more conclusive evidence of solid-like structure in the first two shells. These studies indicate that there definitely exists a more strongly layered shell structure in the helium density around an impurity ion than around neutral atomic species, with more solid-like character. This feature can be further explored in the imaginary-time path integral representation by examining the permutation exchanges of helium atoms at specific locations. For the $He_{100}Na^+$ cluster at $T = 0.8$ and 1.25 K, the atoms in the first coordination shell rarely participated in permutations with other particles, and thus are well-localized in the PIMC sense. In the second solvation shell, some atoms are involved in long exchanges at the lower temperature, while in the outer third shell most atoms are involved in long exchanges. From this, it was inferred that the third shell is superfluid, while the second shell has an intermediate, temperature-dependent character. [32] Such an analysis has been made in more quantitative detail for the molecular impurities, which we discuss next.

3.3. *Molecular impurities*

Molecular impurities introduce an additional level of complexity because molecules have internal structure and usually possess an anisotropic interaction with helium.

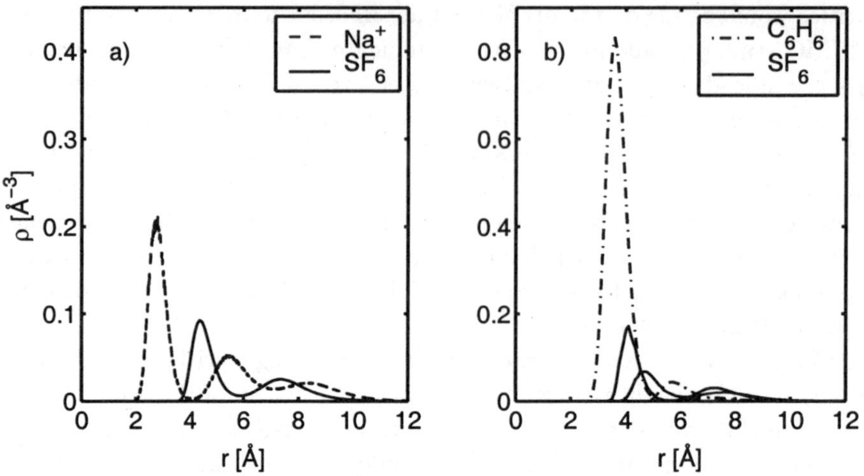

Fig. 2. Helium density profiles relative to the impurity center, at $T = 0.625$ K. The left panel (a) shows the radial density profiles for $He_{100}Na^+$ (dashed lines) and $He_{64}SF_6$ (solid lines), where an isotropic He-SF_6 interaction was used. The right panel (b) shows anisotropic helium densities for SF_6 (solid lines) from a calculation using an anisotropic He-SF_6 interaction, [19] viewed along the molecular C_3 (higher values) and C_4 (lower values) axes. This is compared to anisotropic radial density profiles for He_{39}-benzene (dotted-dashed lines), [37] viewed along the molecular C_6 axis (higher peak) and along the C–H bond (lower peak). The $He_{100}Na^+$ profile is reproduced from Ref. 32.

Especially for the larger molecules, there is a severe lack of accurate two-body molecule-helium interaction potentials. Nevertheless, the study of molecular impurities in helium clusters is currently of great interest, with an increasing number of experiments being performed on a variety of molecules. Even with simple models for the molecule-helium interaction, analysis of these experiments in terms of the perturbation of the helium environment on the molecular internal degrees of freedom has provided much insight into the quantum fluid nature of these clusters. It is important to recall that to date, all PIMC work involving *molecular* impurities in helium have not explicitly incorporated the impurity rotational kinetic energy. This is not an essential restriction, and has been so far made for convenience rather than for any fundamental limitations. Since zero-temperature DMC calculations have recently shown that the helium densities around small molecules may be sensitive to the rotational motion of the molecule, [5,40–42] it would be desirable to incorporate the rotational degrees of freedom in future path integral studies. Only for the heaviest rotors can molecular rotation be justifiably omitted. Several studies have also justified neglecting the translational motion of the molecular impurity, in which case the helium atoms may be regarded as moving in the external potential field of the molecule in the molecular body-fixed frame, given by the Hamiltonian of Eq. (2.1). [5,37] The validity of this assumption can be assessed with the comparative study of molecular delocalization as a function of molecular mass and binding en-

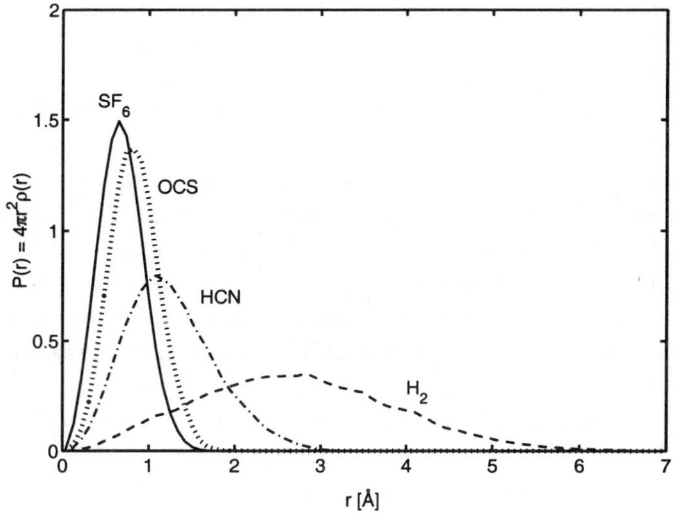

Fig. 3. Radial distribution for several impurity molecules (H_2, HCN, SF_6, and OCS) relative to the cluster center-of-mass, shown as $P(r) = 4\pi r^2 \rho(r)$ such that $\int P(r)\,dr = 1$. All calculations were made from PIMC at $T = 0.312$ K, and include the impurity translational kinetic energy. Isotropic interaction potentials were used, and Bose permutation symmetry was not included. The H_2 and HCN distributions were obtained from a calculation with $N = 128$ He, while the SF_6 and OCS distributions correspond to a cluster of $N = 100$ He. Data courtesy of D. T. Moore.

ergy shown in Fig. 3.[43] There, the probability of finding a molecule at some distance r from the cluster center-of-mass is shown for a series of molecules. As expected, the heaviest dopants such as SF_6 and OCS are well-localized near the center of the cluster, and thus it is a reasonable approximation to neglect their translational motion. On the other hand, H_2 is much more delocalized throughout the interior of the cluster, due to its comparatively smaller mass and weaker helium-impurity interaction, and therefore requires that its translational motion be properly incorporated.

To date, the most extensively studied molecular impurity is the octahedral SF_6 molecule. Early PIMC work on SF_6 in helium clusters employed isotropic molecule-helium interaction potentials,[18] and was later extended to include anisotropic interactions.[19] The helium structure around SF_6 from an isotropic calculation is shown in Figs. 2a and 8. The anisotropic He-SF_6 potential surface has a global minimum of -84 K,[44] considerably deeper than that of the He-He interaction. Thus, SF_6 is expected to reside at the center of the cluster. This has been verified by both zero-temperature DMC[45] and by finite-temperature PIMC calculations. For the $He_{64}SF_6$ cluster in the temperature range of $T = 0.3 - 0.75$ K, there is an anisotropic layering of the helium density around the SF_6. Integration of the helium density over the first solvation shell yields about 23 atoms, independent of whether isotropic or anisotropic interactions are employed.[5] The strength and range of the molecule-helium interaction pins the helium density in the first solvation shell to a total density comparable to that of the more strongly bound He_N-Na^+ system. Detailed analysis of the helium density distribution around the molecule shows that while the angular average of the density in the first solvation shell is independent of temperature below $T = 1.25$ K, there is a small increase in the extent of anisotropy as the temperature is lowered. This is illustrated in Fig. 4, with a comparison of the densities along different molecular symmetry axes for an $N = 64$ cluster at temperatures $T = 0.625$ K and $T = 0.312$ K. As the temperature is increased above $T = 1.25$ K, this trend to an increasingly isotropic distribution is further modified by the onset of evaporation of helium atoms. Evaporation begins with atoms in the second solvation shell, is clearly evident at $T = 2.5$ K, and is essentially complete at $T = 5.0$ K (Fig. 5).

Since the first experiments for doped clusters that employed SF_6 as a probe species, [46] a broad array of molecular impurities have been studied by spectroscopic means. [25] The infrared spectral regime has provided a particularly rich field of study. Vibrational spectra in the infrared at $T \sim 0.4$ K show rotational fine structure in ^{4}He droplets, but not in ^{3}He droplets, providing evidence that quantum statistics play an important role in the spectral properties of the dopant. There is now an increasing collection of experimental data available for the rotational dynamics of molecules possessing varying symmetries and a range of values for the gas phase rotational constant. To date, PIMC has been used to make theoretical studies of the linear rotors OCS [5,47] and HCN, [5] the planar aromatic molecule benzene (C_6H_6), [37] the linear $(HCN)_3$ complex, [27] and the OCS-$(H_2)_M$ complex. [47] From these studies the notion of two different dynamical regimes has emerged, namely that of heavy molecules such as SF_6 that are characterized by gas phase rotational constants $B_0 < 0.5$ cm^{-1}, and a complementary regime of lighter molecules possessing larger gas phase values of B_0. [5] This division into two dynamical regimes based on rotational constants emerges from analysis of the helium solvation density and energetics derived from path integral calculations.

The OCS impurity lies in the regime of relatively heavy molecules, with $B_0 = 0.20$ cm^{-1}. The He-OCS potential has a global minimum of ~ 64 K, [48] which is only about two-thirds that of the He-SF_6 potential. [44] It is important to consider the anisotropy of the intermolecular potential in addition to its strength when assessing the quantum solvation structure. In this respect the linear OCS molecule has lower symmetry than the octahedral SF_6. The minimum angular barrier for rotation of the OCS about an axis perpendicular to the molecular axis (*i.e.*, the angular adiabatic barrier for rotation) is 41.9 K. [48] This barrier is markedly higher than the corresponding value 20.7 K for SF_6, and consequently gives rise to stronger angular modulations in the solvating density. [5] As shown in Fig. 6, PIMC calculations for the He_{64}OCS cluster at $T = 0.312$ K reveal a strongly structured helium density, forming approximately elliptical solvation shells around the OCS impurity. The first shell integrates to ~ 17 atoms. [5] Because of the axial symmetry of the He-OCS potential, the density at the global minimum forms a ring around the OCS molecular

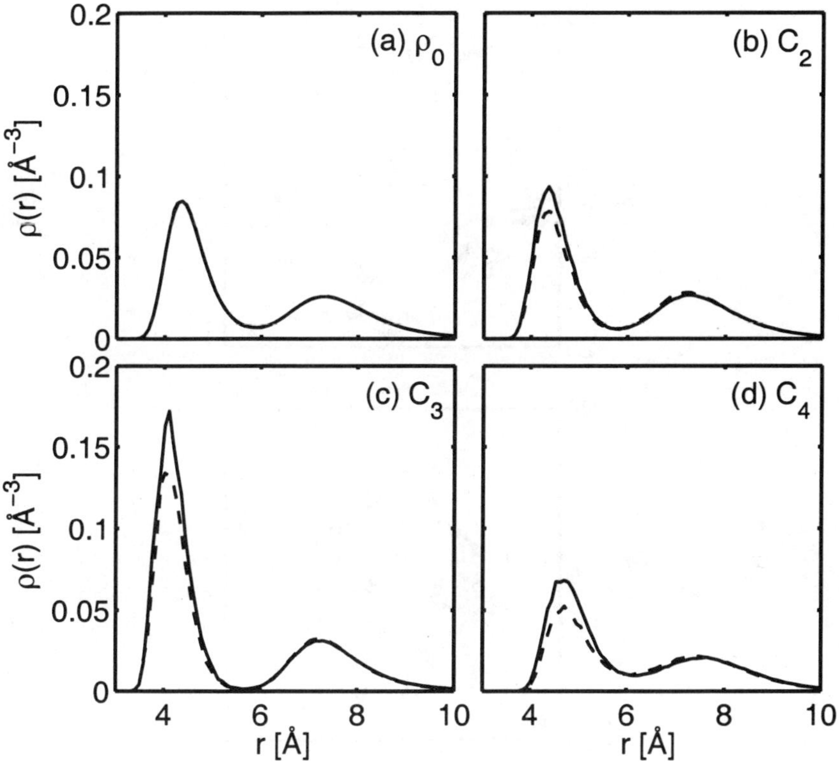

Fig. 4. Comparison of the helium density distribution around the octahedral SF_6 molecule in a $N = 64$ cluster at $T = 0.625$ K and $T = 0.312$ K. Solid lines show the lower temperature densities and dashed lines the higher temperature densities. Panel (a) shows the angular-averaged density ρ_0, for which the profiles at two different temperatures are identical. Panels (b), (c), and (d) show the densities along the three symmetry axes of the molecule C_2, C_3, and C_4, respectively. The higher temperature profiles show consistently smaller peak values in the first solvation shell, indicating a decrease in the anisotropy of the distribution as temperature increases.

axis, consisting of about 6 helium atoms.

The benzene molecule (C_6H_6) also lies in the heavy regime. The benzene π-electron character leads to a highly anisotropic interaction with helium, with two deep, equivalent global potential minimum located on the six-fold axis of symmetry above and below the plane of the molecule. [37,49] A PIMC study of benzene-doped clusters has shown a highly anisotropic helium structure around the impurity molecule that reflects this six-fold symmetry. [37] The sharpest density peak is located along the C_6-axis, at the two equivalent locations of the global potential minima. These two global density maxima are higher than the local density maxima viewed along the in-plane directions by more than a factor of four, reflecting the marked anisotropy of the He-benzene interaction potential. The extreme density anisotropy

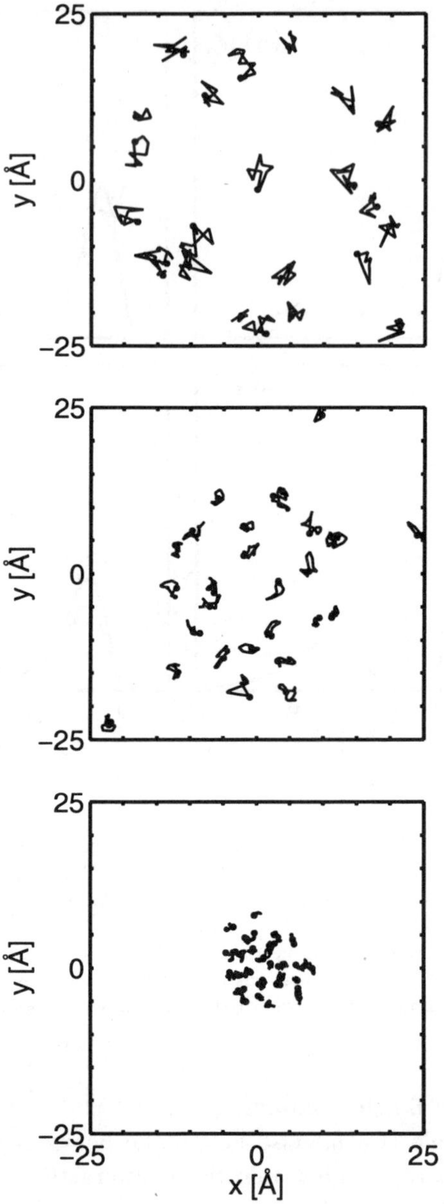

Fig. 5. Helium evaporation for the SF_6He_{39} cluster. The lower panel plots a snapshot of imaginary time paths at $T = 1.25$ K. At this temperature the helium atoms are bound to the cluster. In the middle panel, at $T = 2.5$ K, the cluster begins to dissociate, loosing helium atoms. In the upper panel, at $T = 5.0$ K, the cluster has completely evaporated.

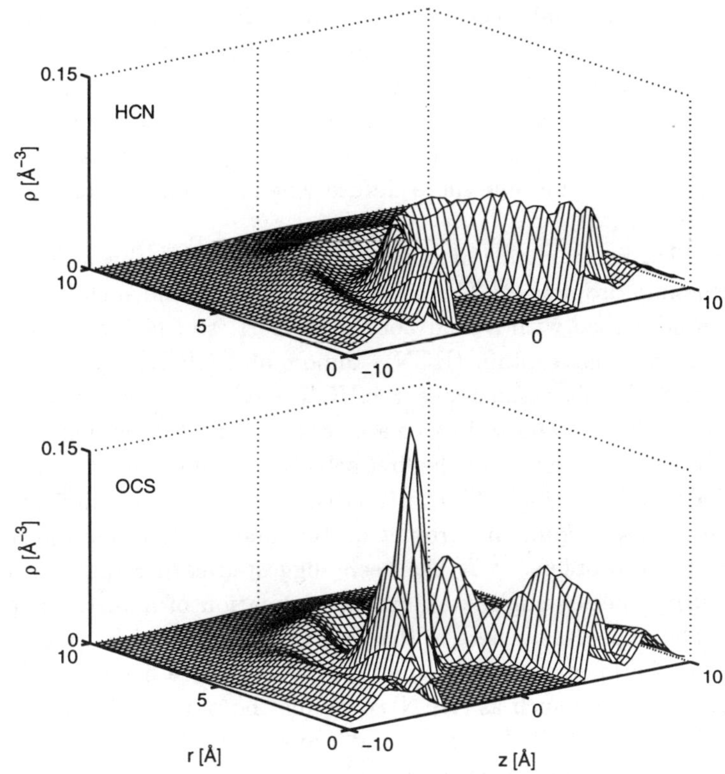

Fig. 6. Total helium density around HCN (top panel) and OCS (bottom panel) for a $N = 64$ cluster at $T = 0.312$ K. [5] The origin is set at the impurity center-of-mass. The OCS is oriented with the oxygen end directed towards the -$\hat{z}$ direction, and the HCN is oriented with the nitrogen end directed towards the -$\hat{z}$ direction.

is summarized in Fig. 2b where the dotted-dashed lines show density profiles along the C_6-axis and along one of the in-plane directions. Integration over any one of the two equivalent global density maxima gives exactly one helium atom. We see an interesting effect of near complete localization of these two helium atoms located at the two global potential minima on either side of the molecular plane. As noted in Ref. 37, this phenomenon can be viewed as a precursor form of helium adsorption onto a molecular nanosubstrate. Extending these studies to larger polyaromatic molecules will allow contact to be made with PIMC studies of helium adsorption on graphite. [10]

In contrast to this highly structured quantum solvation observed around the heavier molecules such as OCS and benzene, the linear HCN molecule falls into the light molecule regime, with a significantly larger gas phase rotational constant. For HCN, $B_0 = 1.48$ cm^{-1}. The He-HCN potential [50] is both weaker (its global minimum is -42 K) and less anisotropic than the He-OCS potential. While there is clearly still an ellipsoidal layering of the helium density around the HCN, within each solvation shell there is now a noticeable lack of angular structure, in contrast to the situation with OCS (Fig. 6). For such a light rotor, neglect of the molecular rotational kinetic energy now becomes a more serious concern. From DMC studies assessing the effect of molecular rotation, [5,42,51] the expectation here is that the helium density will become more diffuse when molecular rotation is explicitly incorporated into PIMC.

Self-assembled linear chains of polymeric $(HCN)_M$ have been detected experimentally in helium droplets. [52] The helium structure around such linear chains has recently been addressed with a study of the properties of helium droplets with up to $N = 500$ atoms that contain $(HCN)_2$ dimers and $(HCN)_3$ trimers. [27] Like the monomeric molecules discussed above, the HCN polymers are found to be located at the center of the droplet and to induce a layering of the helium density. Draeger *et al.* have analyzed the structure of the first solvation shell around the linear polymer in terms of a two-dimensional film, estimated the effective confinement potential for displacement away from the droplet center, and made calculations for vortex formation in these droplets. [27] It has been suggested earlier that the presence of a linear impurity species might stabilize the formation of a vortex line in helium droplets. [53] The expectation here is that a vortex line could be pinned along the molecular axis of a linear molecule such as HCN, or more likely, along the axis of a linear polymeric chain such as $(HCN)_M$. While the physics of vortices constitute an essential feature of bulk He II, [54] and ways of producing and detecting vortices during helium droplet formation have been the subject of much discussion (see Ref. 53 and therein), no experimental evidence has been found so far for existence of vortices in finite helium droplets. Theoretically, vortices have been shown to be unstable in pure droplets, [55] and the situation with regard to doped droplets is still controversial. The energy for formation of a vortex, ΔE_V, is defined as

$$\Delta E_V = E_V - E_0, \tag{3.1}$$

where E_0 is the ground state cluster energy and E_V is the energy of the cluster with a vortex line present. Within the fixed-phase approximation, the PIMC estimate for this vortex formation energy is ~ 30 K for a $He_{500}(HCN)_M$ cluster at $T = 0.38$ K, where $M = 0 - 3$. [27] In this case, the vortex formation energy is found not to be significantly affected by the presence of a linear impurity. In comparison, density functional calculations made for a range of impurities and cluster sizes give values of ΔE_V that are larger than the fixed-phase PIMC estimates by a factor of 3, and that are reduced by $\sim 5 - 10$ K in the presence of an impurity. [56] An exact estimator for the energy of a cluster in an angular momentum state m relative to $m = 0$ has been derived using angular momentum projection methods. [27] Application of this estimator at $T = 2.0$ K indicates that the presence of an impurity actually results in a slight *increase* in the vortex formation energy. More work is required in this direction, in particular the systematic examination of the cluster size and temperature dependence of ΔE_V obtained from the angular momentum projection estimator.

Many other complexes have now been synthesized in helium droplets. [12] Indeed, these droplets are proving to be a remarkably versatile quantum matrix environment for synthesis of unusual or metastable aggregates. Of particular interest from a fundamental point of view are the complexes of OCS with molecular hydrogen, H_2. Recent spectroscopic measurements on $OCS(H_2)_M$ complexes inside He_N clusters have shown an interesting feature that has been interpreted as evidence of nanoscale hydrogen superfluidity. [57] Initial PIMC studies of these systems have been carried out [47] using accurate pair potentials of OCS with He and with H_2. [48,58] Since the H_2-OCS potential surface has a similar angular modulation as that for He-OCS, but a deeper minimum, the OCS molecule is expected to bind preferentially to H_2 over He. Calculations for the $OCS(H_2)_5$ complex in the He_{39} cluster [47] showed that approximately six helium atoms, which would normally occupy the region of the global potential minimum in the absence of H_2, are completely displaced by five H_2 molecules. These H_2 molecules form a complete ring encircling the linear OCS molecule at the region of lowest potential energy. The helium density is pushed either to either the secondary peaks in the first shell, or outwards from first to second shell region.

3.4. *Exchange permutation analysis and impurity-induced non-superfluidity*

In addition to providing structural and energetic information, PIMC is currently the only numerical method capable of providing information on finite-temperature superfluidity in He systems. At high temperatures an N-body system may be described by Boltzmann statistics, *i.e.* in the path integral representation, only the identity permutation is important. At low temperatures however, permutations must be included in the path integral representation for the thermal density matrix. In particular, for liquid helium near the lambda transition, Feynman qualitatively showed that the presence of long exchange cycles gives rise to the sharp increase in the heat capacity, but due to the analytical approximations made in his analysis he was not able to correctly identify the order of the transition. [6] Further refinements in this and numerical PIMC simulations have quantitatively confirmed both the transition temperature and its order. [20]

The area estimator of Eq. (2.30) gives a scalar value for the global superfluid fraction ρ_s/ρ. This provides a complete description for homogeneous helium systems. However, a finite cluster of nanoscale dimensions necessarily contains inhomogeneity deriving from the surface, and atomic and molecular dopants provide additional sources of inhomogeneities. In this situation Eq. (2.30) may be interpreted as providing an estimate of the *global* superfluid fraction averaged over all sources of inhomogeneity. It is notable that the impurity molecule does not significantly perturb this global superfluid fraction. For neutral Na-doped clusters, the area estimator yields a global superfluid fraction of about 95%, [31] consistent with the very weak perturbation of the density noted earlier. For more strongly bound systems such as He_NSF_6 and He_NNa^+, it is found that ρ_s is similarly large, approaching unity for $N > 100$ at $T = 0.625$ K. [5,32] Thus to see a molecular effect on superfluidity, one needs to examine the local solvation structure on microscopic length scales. Here, the density ρ is no longer uniform, particularly in the neighborhood of an inhomogeneity. Thus, the superfluid fraction ρ_s/ρ is expected to be dependent on position. Some indirect indications of this have also been found in analyses of helium films. [59–61]

A simple way to qualitatively estimate the *local* dependence of superfluid character is to examine the probability $\Pi_p(\mathbf{r})$ of a particle at a position $\mathbf{r}$ to participate in an imaginary-time exchange cycle of length p. As discussed previously, Bose superfluidity is associated with the existence of exchange cycles of long p. In a pure cluster, the single source of inhomogeneity is the cluster surface. For a pure cluster, $\Pi_{p\geq 6}(r)$ goes to zero as the radial distance r approaches the surface. In the presence of an impurity, the examples discussed in Secs. 3.2 and 3.3 show that an embedded molecule can significantly modify the total density distribution $\rho(\mathbf{r})$. Consequently one also expects changes in the local superfluid character. Kwon and Whaley have systematically examined $\Pi_p(\mathbf{r})$ for helium clusters doped with a single SF_6 impurity. [19] They define a local superfluid density by

$$\rho_s(\mathbf{r}) = \sum_{p\geq p'}^{N} \Pi_p(\mathbf{r})\rho(\mathbf{r}) \tag{3.2}$$

where $\rho(\mathbf{r})$ is the total density at $\mathbf{r}$, and p' is a cutoff value for the permutation cycle length. This does not account for the tensor nature of the superfluid response, providing a three-dimensional anisotropic representation of a scalar, that may be viewed as an average over the set of tensorial response functions. For a molecule with high symmetry such as a spherical top, this will not be a serious limitation. For linear molecules it will introduce some uncertainty. For clusters of $N > 50$, most of the polymers sampled involve either one or two atoms ($p = 1, 2$) or many atoms (large p). Thus in this size regime a clear cutoff exists. For these sizes, Kwon and Whaley used a value of $p' = 6$. For small clusters ($N < 50$), a clear distinction between short and long exchange cycles cannot be made, which implies that in the small cluster regime a two-fluid interpretation of the density cannot be applied.

For the octahedral SF_6 molecule, the local superfluid estimator of Eq. (3.2) yields an anisotropic superfluid solvation structure around the impurity molecule, whose density modulations are similar to those of the total density $\rho(\mathbf{r})$. The local non-superfluid density $\rho_n(\mathbf{r}) = \rho(\mathbf{r}) - \rho_s(\mathbf{r})$ does depends weakly on temperature, which implies that $\rho_n(\mathbf{r})$ consists of thermal contribution and a molecule-induced component. Fig. 7 shows a three-dimensional representation of the local non-superfluid distribution around the octahedral SF_6 molecule. The red areas of highest non-superfluid density are located at the octahedral sites of strongest binding to the molecule, reflecting the origin of this as a molecule-induced non-superfluid. This is in contrast to the thermal normal density of bulk He II in the Landau two-fluid de-

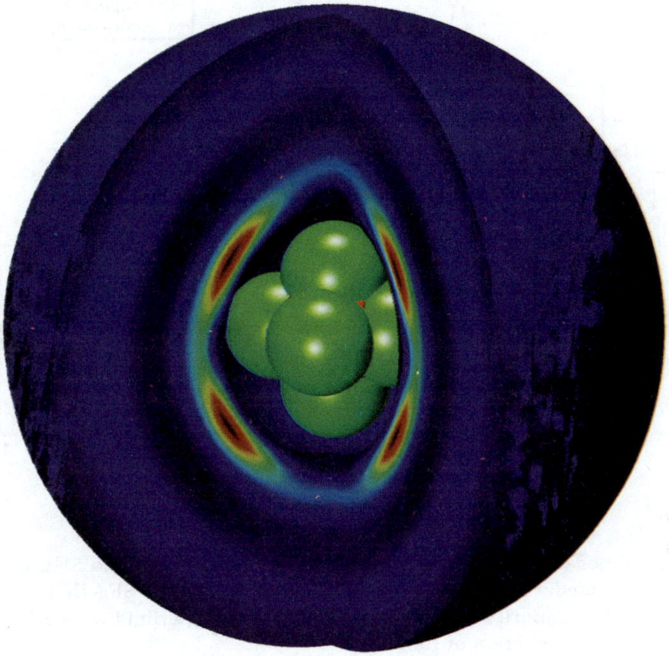

Fig. 7. Local non-superfluid density $\rho_n(\mathbf{r})$ around SF_6 in a $N = 64$ cluster at $T = 0.312$ K, as measured by the exchange path decomposition of the density. [5] The color scale goes from red for highest values of $\rho_n(\mathbf{r})$, to blue for the lowest values of $\rho_n(\mathbf{r})$. The size of the ball corresponds to a distance from the SF_6 molecule of $r = 9.0$ Å. The two cuts display the density in two equivalent planes containing C_3 and C_2 axes. The strong binding to the octahedral sites located along the C_3 axes is evident, with 4 of the 8 octahedral sites visible here.

scription of a homogeneous superfluid. [62] The molecule-induced density component depends on the strength and range of the helium-impurity interaction potential, and is expected to persist at $T = 0$. Detailed analysis shows that it is non-zero only in the first solvation layer around the molecule. [19]

An analysis using the local estimator of Eq. (3.2) has been applied to a number of different molecular impurities in helium clusters, including the linear molecules OCS and HCN, [5,47] and benzene. [37] These systems exhibit a similar layering in both local superfluid density $\rho_s(\mathbf{r})$ and local non-superfluid density $\rho_n(\mathbf{r})$ around the molecule. The non-superfluid density shows slightly stronger modulations, resulting in a weakly anisotropic local superfluid *fraction* in addition to the component densities themselves. [5] In the more strongly bound He-OCS case, the maximum of the non-superfluid component is roughly $\sim 50\%$ of the total density, while for the weakly bound He-HCN, the non-superfluid, or short-exchange path, component

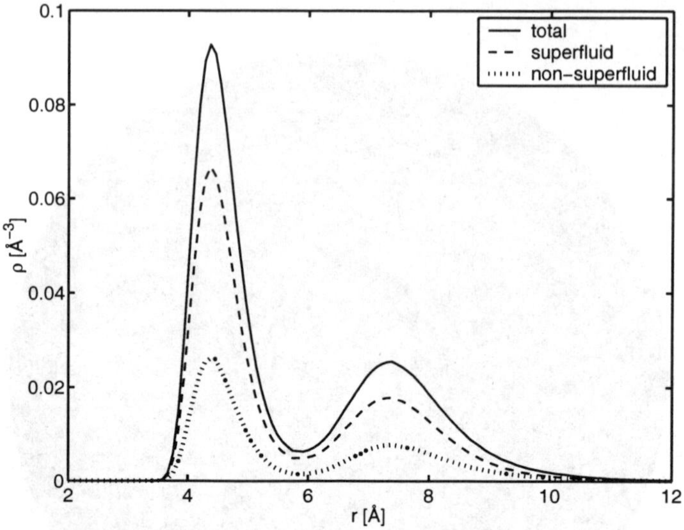

Fig. 8. Total, local non-superfluid, and local superfluid densities around SF_6 in a $N = 64$ cluster at $T = 0.625$ K, calculated with only the isotropic component of the SF_6-He interaction potential. The origin is set at the impurity center-of-mass. The local superfluid density is calculated with the exchange path length criterion of Ref. 19.

is only $\sim 20\%$. We note that the molecule-induced non-superfluid density is also present around an impurity possessing an isotropic interaction with helium, *i.e.*, it is not essential to have an anisotropic interaction. In fact the existence of a molecule-induced non-superfluid density was first seen in calculations of the SF_6 molecule with isotropic interactions potentials, summarized in Fig. 8.

Nakayama and Yamashita have pursued a similar analysis of the local superfluid density for the $He_N Na^+$ cluster, which exhibits a triple-layer structure for $N = 100$. [32] While they did not explicitly compute the local quantities $\rho_s(r)$ or $\rho_n(r)$ in their PIMC study, they observed that the helium atoms in the first solvation shell ($r < 4$ Å) rarely participate in long exchanges. This observation, combined with the pair distribution functions computed with respect to atoms in the first shell, led them to conclude that the first shell is solid-like.

As discussed previously, an even more anisotropic impurity-helium interaction potential is provided by the benzene molecule. For the He_{39}-benzene cluster at $T = 0.625$ K, the two atoms corresponding to the two total density maxima localized at the two global potential minima, undergo less than 2% permutation exchanges with the surrounding helium. This implies that they are effectively removed from the superfluid, *i.e.*, constitute a true "dead" adsorbed pair of atoms. [59–61] This near-complete removal of individual helium atoms in the solvation shell from participation in permutation exchanges of nearby helium atoms has not been seen for other molecules to date. It provides an extreme case of the local non-superfluid density, $\rho_n(\mathbf{r})$, in which there is no longer any partial exchange of helium atoms between the non-superfluid and superfluid densities. These features of the helium solvation around a benzene molecule are expected to appear also in clusters containing larger polyaromatic molecules such as tetracene and naphthalene. A systematic analysis of the effect in planar aromatic molecules of increasing size, making the transition from a molecular to a micron-scale substrate, would be very useful.

Recently another local estimator of superfluidity has been proposed that decomposes the projected area into contributions from each local density bin. [27] This decomposition allows the anisotropy of the response tensor to be evaluated explicitly. Application of this local estimator to the linear HCN trimer embedded in helium droplets has confirmed that the superfluid density is reduced in the first solvation layer, consistent with the presence of a local non-superfluid density induced by the molecule-helium interaction, as first established by Kwon and Whaley. [19] Furthermore, this new estimator shows that there is an asymmetry between the helium response to rotation about the molecular axis, versus rotation about an axis perpendicular to the molecular axis. Draeger *et al.* find that the superfluid response is reduced more for rotation about the perpendicular axes than for rotation about the molecular axis. [27] In both cases it is less than unity, implying that there is a non-superfluid component both when rotation is accompanied by variation in potential energy, and when there is no variation in potential energy. This finding supports the existence of a local non-superfluid induced by an isotropic helium-impurity interaction, using the exchange path analysis of Kwon and Whaley (Fig. 8). Thus the local non-superfluid is not dependent on the presence of anisotropy, but derives primarily from the stronger attraction of helium to the molecule than to itself.

These studies of various molecules embedded in He_N clusters employing different estimators of local superfluidity all point to the existence of a molecule-induced non-superfluid density in the first solvation shell around a molecule. While the details of this non-superfluid density may be somewhat dependent on how it is defined, it is evident from the studies of OCS, benzene, and HCN polymers made to date, that this local non-superfluid component is a general phenomenon to be expected for all heavy molecules. It therefore appears to be one of the defining features of quantum solvation in a superfluid. The extent of exchange between non-superfluid and superfluid densities exhibits a dependence on the strength of the helium interaction with the molecule. Benzene provides an interesting extreme case of negligible exchange between non-superfluid and superfluid density components, while less anisotropic molecules such as SF_6 still possess considerable exchange between local non-superfluid and local superfluid. Thus, both the interaction strength with the molecular impurity and the symmetry of this interaction are important. The benzene example indicates that there are useful analogies with the well-known "dead" or "inert" layer of helium adsorbed into bulk solid surfaces, which will be valuable to pursue in future studies.

4. PIMC and the connection to cluster spectroscopy

4.1. *Electronic spectra in He_N*

Calculations of electronic spectra typically require accurate potential energy surfaces for both ground and excited electronic states. This is particularly challenging for excitations in condensed phases. To date, theoretical work in this area has been limited to relatively to simple systems, where the helium-impurity ground and excited state pair potentials can be computed to good accuracy using standard *ab initio* electronic structure methods. Thermally-averaged electronic absorption spectra for the ${}^2P \leftarrow {}^2S$ transition have been computed for neutral alkali impurities at $T = 0.5$ K, [31] using a modification of the semi-classical Frank-Condon expression for the electronic lineshape, [63,64]

$$I(\omega) \propto |M|^2 \int dR\, \rho(R, R; \beta)\, \delta[V_e(R) - E_g(R) - \hbar\omega], \tag{4.1}$$

where M is the electronic transition dipole moment, and V_e is the potential in the electronic excited state. The quantity E_g is a local ground state energy, which is assumed to take the form

$$E_g(R) = T_{\text{imp}}(R) + \sum_{i=1}^{N} V_{\text{He-imp}}(\mathbf{r}_i) + \sum_{i<j}^{N} V_{\text{He-He}}(r_{ij}), \tag{4.2}$$

and explicitly incorporates the kinetic energy of the impurity atom T_{imp}. The terms $V_{\text{He-imp}}$ and $V_{\text{He-He}}$ correspond to the helium-impurity and helium-helium ground state pair potentials, respectively. The electronic excited state potential V_e is obtained [31] from the diatomics-in-molecules (DIM) model,

$$V_e(R) = V^e_{\text{He-imp}}(R) + \sum_{i<j}^{N} V_{\text{He-He}}(r_{ij}), \tag{4.3}$$

where the first term $V^e_{\text{He-imp}}$ is the adiabatic energies of the alkali atom in the ${}^2P$ manifold interacting with the N helium atoms, and the remaining helium-helium pair potentials $V_{\text{He-He}}$ are taken to be identical to the ground state. Thus, the thermal absorption profile $I(\omega)$ can be computed by sampling this energy difference of Eq. (4.1) from a PIMC simulation. As discussed in Sec. 3.2, the neutral alkali impurities reside on the droplet surface, and the resulting perturbation on droplet properties is weak. The electronic lineshape is therefore most sensitive to the details of the surface structure near the alkali atom. The PIMC calculations for neutral Li, Na, and K on helium clusters of size $N = 100 - 300$ give good qualitative agreement with experiment. The doublet structure (${}^2P$, ${}^2P_{3/2} \leftarrow {}^2S_{1/2}$) observed in the experimental spectra for Na and K on helium droplets [65] can be seen in the PIMC calculations. However, while both experiment and theory show that these transitions are shifted to the blue relative to the experimental gas phase values, the absolute value of these shifts is in general much more difficult to obtain from theory. Due to weak spin-orbit coupling for Li, the doublet splitting is small relative to the linewidth, and thus also difficult to resolve. The PIMC absorption spectra for the He_NLi system exhibits a weak red shift and a long tail towards the blue, both of which are consistent with the experimental spectra. [65]

4.2. *Vibrational shifts in infrared spectroscopy of molecules in He_N*

The first spectroscopic experiment made on a doped helium cluster measured the infrared absorption spectrum of the octahedral SF_6 molecule. [46] This low resolution spectrum, obtained with a pulsed CO_2 laser, revealed that the ν_3 vibrations of SF_6 molecule are red-shifted from the gas phase value by about $1 - 2$ cm^{-1}. They also appeared to show that the three-fold degenerate absorption for these vibrational modes is split into two peaks. The split peaks were interpreted as implying that the SF_6 molecule resides on the cluster surface where the three-fold degeneracy would be expected to be split into parallel and perpendicular modes. However, DMC calculations made at that time showed that the molecule should be located at the cluster center. [66] This was later confirmed by PIMC calculations [18] and verified by subsequent experimental investigations. These include high resolution spectra made with diode lasers [67] which showed a single vibrational absorption, red-shifted by $\Delta\nu = -1.6$ cm^{-1} and having no splitting of the vibrational degeneracy, and analysis of ionization products of SF_6-doped helium clusters. [68] Calculations of the spectral shifts of these triply degenerate intramolecular ν_3 vibrations of SF_6 were made with both PIMC and DMC, [18,66] using the instantaneous dipole-induced dipole (IDID) mechanism originally proposed by Eichenauer and Le Roy [69] to calculate the vibrational spectra of SF_6 inside argon clusters. PIMC allows the calculation of the thermally averaged spectral shift at finite temperatures, while DMC gives the ground state, $T = 0$ K value of the spectral shift. Ref. 18 provides a discussion of the difficulties in calculating spectral line shapes (and hence extracting line widths) from a finite-temperature path integral calculation. The IDID approach is taken because the intramolecular vibrational dependence of the He-SF_6 interaction potential is not known, and it is therefore necessary to approximate this. In the IDID model, the origin of the spectral shift is assumed to be the dipole-dipole interaction between the instantaneous dipole moment of the SF_6 ν_3 vibration and the induced dipole moments of the surrounding helium atoms. The average shift of the ν_3 absorption estimated from the IDID model within PIMC calculations are red-shifted, in agreement with experiment, but the magnitudes of the calculated shift ($\Delta\nu = -0.84$ cm^{-1}) is somewhat smaller than the experimental value of $\Delta\nu = -1.6$ cm^{-1}. Overall, the agreement of the spectral shift value to within a factor of two is quite reasonable, but it is evident that for a proper understanding of the spectral shift of SF_6 inside helium, one needs to also incorporate the contribution from the repulsive part of the He-SF_6 interaction that is neglected in the IDID model.

Recently Gianturco and Paesani have calculated vibrationally adiabatic He-OCS potentials for various internal vibrational states of OCS, which allow for a more fundamental approach to the calculation of vibrational shifts. [70] These vibrationally adiabatic potentials are derived by evaluating the interaction potentials as a function of both external coordinates and an internal vibrational coordinate using *ab initio* electronic structure methods, and then averaging over the internal vibrational wavefunctions. In particular, they have provided vibrationally adiabatic potentials V_{00} and V_{11} that are averaged over the ground state and first excited state of the asymmetric stretching motion of the molecule, respectively. The shift of an intramolecular vibrational mode inside the helium cluster can be estimated within an adiabatic separation of the fast intramolecular vibrational mode from the slow He-He and He-molecule degrees of freedom. [71] In this approach, the spectral shift results from computing the average of the difference between V_{00} and V_{11} over the finite-temperature ensemble sampled in the PIMC simulation. Recent PIMC calculations for He_{39}-OCS at $T = 0.3$ K find a red-shifted asymmetric vibration, with the shift of $\Delta\nu = -0.87(1)\ \mathrm{cm}^{-1}$. [51] The sign of the shift is in agreement with that seen in experimental measurements for OCS made in larger clusters involving more than 1000 helium atoms at $T = 0.38$ K, but its magnitude is somewhat larger than the experimental value of $\Delta\nu = -0.557(1)\ \mathrm{cm}^{-1}$. [72] Detailed analysis indicates that these discrepancies are likely due to small errors in the vibrationally averaged adiabatic potentials. [51]

4.3. *Rotational spectra of molecules embedded in He_N*

The experimental observation of rotational fine structure for infrared spectra of vibrational transitions in the bosonic ^{4}He clusters but not in the corresponding fermionic ^{3}He clusters at the operative temperature of $T = 0.38$ K, [17] led to the conclusion that superfluidity is essential for observation of a free rotor-like spectrum. This has been explained as a result of the weak coupling of molecular rotations to the collective excitations of superfluid He II, compared to the much stronger coupling to particle-hole excitations in the Fermi fluid ^{3}He. [73] Consequently, the rotational lines are considerably broadened in the fermionic clusters, and the fine structure of rovibrational transitions is washed out. This is consistent with the results of direct calculations of rotational energy levels of ^{4}He clusters containing rotationally excited molecules, using zero-temperature DMC-based methods. [5,40,42,74,75] These direct calculations show that the bosonic nature of the ^{4}He is critical in ensuring a free rotor-like spectrum of rotational energy levels of the molecule when embedded in a helium droplet. The corresponding rotational energy levels in fermionic helium droplets have not yet been calculated, and would constitute an interesting theoretical topic for future study. Path integral calculations have not provided any information on the dynamical differences resulting from solvation in fermionic versus bosonic helium droplets so far, since fermionic PIMC simulations have not been made for these systems.

A major feature of the rotational spectra of molecules in ^{4}He droplets is the appearance of free rotor-like spectra with increased effective moments of inertia. In principle, any helium-induced change in the molecular moment of inertia should be directly related to a change in the global superfluid fraction, according to Eqs. (2.29) and (2.30) (assuming that linear response measures are applicable to the quantized rotation of the molecule). Furthermore, a highly anisotropic molecule would be expected to result in some anisotropy in the helium response for rotation around different axes, yielding anisotropy in the tensor of global superfluid response. [37] However, as noted earlier, the global superfluid estimator is relatively insensitive to the presence of an impurity and the statistical errors mask small changes. It is possible that for significantly larger, and more anisotropic molecules than those theoretically studied to date, *e.g.*, for the planar aromatic molecules such as tetracene and phthalocyanine that have already been studied experimentally, [76–78] the global superfluid response may be more affected and yield information. For the relatively small molecules and complexes studied so far however, it has proven necessary to examine the local perturbations of the helium superfluidity in order to develop an understanding of the coupling between this and the molecular rotational dynamics.

The microscopic two-fluid description of the quantum solvation of molecules in He that is provided by path integral calculations has led to a detailed analysis of the effective moments of inertia of molecules solvated in a bosonic superfluid, and hence to a quantitative understanding of the effective rotational constants measured in the infrared and microwave spectroscopy experiments. Since the path integral calculations carried out to date do not explicitly incorporate the molecular rotational degrees of freedom, the connection between the path integral densities and the molecular moments of inertia has to be made within a dynamical model. Kwon and Whaley have proposed a quantum two-fluid model for calculating the effective moment of inertia. [5,19] The main features of this model are summarized below. As will be evident from the assumptions of this quantum two-fluid model for superfluid helium response to molecular rotation, it is applicable only to the regime of heavier molecules, *i.e.*, those possessing gas phase rotational constants less than ~ 0.5 cm^{-1}. Excellent agreement with spectroscopic measurements is obtained for the two instances in which the He-molecule interaction potential is best known, SF_6 and OCS. [5] The theoretical values of rotational constant calculated from the quantum two-fluid model are 0.033 cm^{-1} and 0.067 cm^{-1}, for SF_6 and OCS respectively, compared with the corresponding experimental values 0.034(1) cm^{-1} and 0.073 cm^{-1}. Draeger *et al.* have recently tested this quantum two-fluid model for the linear trimer $(HCN)_3$. [27] They also find excellent agreement between the predictions of the quantum two-fluid model and the experimentally measured rotational constant. While the HCN monomer lies in the regime of light molecules, the trimer is sufficiently massive to fall within the heavy regime, possessing a gas phase rotational constant of $B_0 = 0.015$ cm^{-1}. [79]

The quantum two-fluid model of Kwon and Whaley is a microscopic two-fluid continuum theory for the spectroscopic response of a molecule rotating in superfluid He. It is to be distinguished from the phenomenological two-fluid theory of

Landau for bulk He II. [62] The Kwon and Whaley model examines the helium response to rotation of an embedded molecule that, starting from the local two-fluid decomposition of the molecular solvation density into non-superfluid and superfluid components. Unlike the Landau two-fluid theory for bulk He II, it does not make a two-fluid decomposition of the current densities, but deals only with decomposition of the helium density near an impurity on an atomic length scale, into a non-superfluid component induced by the molecular interaction and the remaining superfluid. This constitutes a significant difference between the well-known phenomenological theory for bulk, homogeneous He II, and the microscopic two-fluid model for molecular rotational dynamics in an inhomogeneous superfluid solvation situation. For the remainder of this section we shall interchangeably use the terms two-fluid model, microscopic two-fluid model, and quantum two-fluid model to refer to the Kwon/Whaley model.

The starting point for the quantum two-fluid model for helium response to molecular rotation is the local two-fluid density decomposition of the molecular solvation density that results from path integral calculations. As described in Sec. 3.4, consistent evidence for the existence of the local non-superfluid density in the first solvation shell around the molecule, induced by a strong molecular interaction with helium, has now been obtained from two different estimators of the local superfluid response. The second feature of the model is the assumption of adiabatic following of some or all of the solvating helium density with the molecular rotation. Adiabatic following means that the helium density follows the molecular rotational motion. Quantitatively, complete adiabatic following of the helium density would imply that when viewed in the rotating molecular frame, the helium density appears stationary. Thus, in the molecular frame it is independent of rotational state. This applies to both classical and quantum descriptions of the molecular rotation. In a classical description the helium density is analyzed as a function of continuous molecular rotation frequency, while for a quantum description it is analyzed as a function of quantum rotational state of the molecule.

The accuracy of the adiabatic following assumption, as well as quantification of the extent of adiabatic following by helium, has been the subject of several studies by Whaley and co-workers. [5,74,80] Within a classical description of molecular rotation, Kwon *et al.* provided a criterion for adiabatic following, namely that the kinetic energy of rotation associated with a particular helium density (total, non-superfluid, or superfluid) be less than the potential energy barrier to rotation around the molecule. [5] This criterion is applicable to densities deriving from any number of helium atoms, and allows simple estimates using either barriers to rigid rotation, or barriers to adiabatic motion between potential minima associated with different molecular orientations. Application of this criterion to the molecules OCS, SF_6 and HCN, for which the molecule-helium interaction potentials are very well characterized, showed that for both OCS and SF_6 it is energetically feasible for the entire helium density to adiabatically follow the molecular rotation. However, for the lighter HCN molecule, it is not energetically feasible for any density component, whether non-superfluid, superfluid, or total, to adiabatically follow the molecular rotation. In a classical sense, the molecular rotation is then too fast for the helium to follow. The consequence of this lack of adiabatic following is that the helium density distribution is more diffuse when viewed in the molecular frame. This cannot be seen directly in the PIMC densities, since molecular rotation is not included in these. However, it can be seen directly in diffusion Monte Carlo calculations of excited rotational states of the molecules in He clusters. In these calculations, made with an importance sampling algorithm for rotational degrees of freedom, [41] the helium density (or wave function) is projected into the rotating molecular frame and compared with the corresponding density (wave function) from a calculation performed without molecular rotation. [5,74] Explicit analysis of the dependence on rotational state can also be performed, although comparison between rotating and non-rotating cases is already very revealing. The original application of this analysis showed that the extent of adiabatic following decreases for lighter molecules, with the helium density in the molecular frame becoming more diffuse as the rotational constant of the molecule increases. [74] Kwon *et al.* showed recently how this comparison may be quantified by evaluation of a quality factor Q that measures changes in the ratio of densities along directions corresponding to strong and weak binding, as a function of molecular rotational state. [5] Complete adiabatic following is measured by $Q = 1$, provided the molecular interaction potential is anisotropic. (For an isotropic interaction with helium, adiabatic following is not applicable, and $Q \equiv 1$ by definition.) Application to the series of molecules, OCS, SF_6, and HCN, shows that $Q \sim 0$ for HCN, and $Q \sim 0.7$ for both OCS and SF_6. [80] This confirms the prediction of the energetic criterion for HCN, *i.e.*, there is negligible adiabatic following around this molecule. The Q-value results for the heavier molecules are quite significant, implying that the extent of adiabatic following is not complete, even for the most strongly bound case of a single He atom attached to SF_6. [5] So only a *fraction* of the helium density can adiabatically follow the molecular rotation, even for a heavy, strongly bound molecule.

The next stage of the quantum two-fluid model is to consider the consequences of adiabatic following for both the local non-superfluid and local superfluid density around a dopant molecule. These two density components show very different response to adiabatic following, deriving essentially from the different spatial extent that results from their corresponding underlying exchange permutation paths. The molecule-induced non-superfluid density is localized close to the molecule, within the first solvation layer, and is composed of very short permutation exchange paths. In order to satisfy adiabatic following, such a localized density must rotate rigidly with the molecule. There is no other obvious way in which a density that is spatially localized within a few angstrom can remain constant in a rotating molecular frame. This results in an increment of moment of inertia from the local, molecule-induced non-superfluid that is given by

$$I_n = m_4 \int_V d\mathbf{r}\, \rho_n(\mathbf{r}) r_\perp^2, \tag{4.4}$$

For the heavy molecules SF_6, OCS, and the linear trimer $(HCN)_3$, the PIMC values for ΔI_n amount to 100%, 90%, and $\sim$ 81% of the corresponding experimentally observed moment of inertia increments, ΔI. It is interesting that for the highly symmetric SF_6 molecule, a very similar result ($\Delta I_n \sim 98\%$ of ΔI) is obtained from calculations with only an isotropic interaction potential. While there is no adiabatic following with an isotropic interaction and hence no mechanism for rigid coupling of the non-superfluid helium density to the molecular rotation, the high symmetry of the octahedral SF_6 molecule nevertheless results in the integrated non-superfluid density in the first shell being very similar in anisotropic and isotropic calculations. In fact, the finding that the isotropic non-superfluid density could account quantitatively for ΔI was obtained prior to calculations of the anisotropic local non-superfluid density. [81] While this result did not have the theoretical justification of rigid coupling as a result of adiabatic following at that time, it was the first indication that a local two-fluid description was dynamically relevant and prompted the application of an microscopic Andronikashvili analysis of experimental rotational spectra for the case of OCS in He. [17]

In contrast to the local non-superfluid density, the superfluid density, while also modulated around the molecule, is not restricted on an angstrom length scale within the quantum solvation structure. By its very definition, consisting of long exchange paths, the superfluid density extends far from the molecule. Thus the equation of continuity can applied to this density over long distances. Kwon *et al.* have shown that for a classical molecular rotation, determined by a continuous frequency ω, the condition of adiabatic following, if satisfied, can be combined with the equation of continuity to eliminate the explicit time dependence of the density and to arrive at an equation for the superfluid velocity: [5]

$$\nabla \cdot [\rho_s(\mathbf{r},t)\mathbf{v}_s(\mathbf{r},t)] = \nabla \rho_s(\mathbf{r},t) \cdot (\omega \times \mathbf{r}). \tag{4.5}$$

The irrotational nature of a superfluid may be used to replace $\mathbf{v}_s$ by $(\hbar/m_4)\nabla u(\mathbf{r},t)$, to arrive at a second-order partial differential equation for the superfluid velocity potential $u(\mathbf{r})$:

$$\nabla \cdot [\rho_s(\mathbf{r})\nabla u(\mathbf{r})] = \left(\frac{m_4}{\hbar}\right) \nabla \rho_s(\mathbf{r}) \cdot (\omega \times \mathbf{r}). \tag{4.6}$$

This equation, discussed in detail in Ref. 5, was first proposed in 1997 before full anisotropic superfluid densities in three dimensions were calculated. [82]

A similar equation was recently presented by Callegari and co-workers, [83] together with the somewhat different assumption that the entire local solvation density is superfluid. Solution of these hydrodynamic equations leads to a hydrodynamic moment of inertia increment ΔI_h that is derived from the excess fluid kinetic energy associated with the flow pattern of $\mathbf{v}_s$. Callegari *et al.* solved the hydrodynamic equations for several linear or rod-like molecules for which the equations become two-dimensional, using total densities derived from density functional calculations. [83] The full solution for a molecule showing true three-dimensional anisotropy was made recently for SF_6 using PIMC densities. [5] Draeger *et al.* have applied the hydrodynamic treatment to the $(HCN)_3$ trimer, using their PIMC densities and also assuming the total density to be superfluid, for the purpose of comparison. It appears that very different results are obtained for different molecules within the hydrodynamic treatment. For octahedral SF_6, the value of ΔI_h is small, irrespective of whether the total density or superfluid density is used as input to the hydrodynamic calculations (6% and 9% of ΔI, respectively). For the linear $(HCN)_3$ trimer, Draeger *et al.* find an upper bound of $\Delta I_h \sim 0.7\Delta I$, when the total density is assumed superfluid ($\Delta I_h = 850$ amu Å^2, compared with an experimental value [79] of $\Delta I = 1240$ amu Å^2). The calculations of Callegari *et al.* for rod-like molecules yielded between $\Delta I_h \sim 67\%$ and 98% of the experimentally measured increments ΔI. These studies differed from those for SF_6 and $(HCN)_3$ in that input densities were obtained from density functional calculations rather than from PIMC, in some cases using simple estimates from pairing rules to construct interaction potentials when no empirical or *ab inito* potentials were available.

The hydrodynamic treatment of the local superfluid density derived from PIMC has a number of questionable aspects. [5] Firstly, the treatment of the molecular rotation as a classical rotation characterized by a continuous frequency ω must be reconciled with the intrinsic quantized nature of spectroscopic transitions between quantum rotational states. The response to classical rotation necessarily gives rise to angular momentum generation in the superfluid, analogous to the rotation of bulk superfluid in a superleak. [84,85] Kwon *et al.* have calculated the angular momentum generation by absorption of a photon within a semiclassical analysis, and shown that significant values of ΔI_h result in large fractions of the photon angular momentum being transferred to the superfluid density component. This contradicts conclusions of a number of zero-temperature DMC-based calculations that indicate there is negligible transfer of angular momentum to the fluid on rotational excitation. [40,41,74] Kwon *et al.* resolved this by adding quantum constraints to the hydrodynamic formulation, and concluding that violation of these indicates invalidity of the hydrodynamic contribution. This in turn may derive from lack of complete adiabatic following, for which considerable evidence now exists, as outlined above, or from the intrinsic lack of applicability of hydrodynamics to the motions of a superfluid on the atomic length scale. An indicator of this breakdown is the fact that the solutions to the hydrodynamic equations with density inputs of atomically modulated helium solvation densities around an embedded molecule, show variations over length scales of 1 to 2 Å. [5,83] Such variations on a distance comparable to or less than the coherence length ξ of helium imply that a hydrodynamic solution is at its limits of validity here, at best, and should be interpreted with great caution.

The overall conclusions of the quantum two-fluid model for the response of helium to rotation of an embedded molecule are thus that the primary contribution to the increased molecular moment of inertia is a rigid coupling to the local non-superfluid density in the first solvation shell. This yields 100%, 90%, and $\sim 81\%$ of ΔI for the heavy molecules SF_6, OCS, and $(HCN)_3$, respectively. The accuracy of these estimates is dependent on the accuracy of the underlying molecule-helium interaction potentials. In contrast, there appears to be negligible contribution from the superfluid, whose response must be restricted by angular momentum constraints. This is consistent with the findings of only partial adiabatic following of the total helium density. It appears reasonable that only the non-superfluid density adiabatically follows the molecular rotation, while the superfluid density, which is defined over much longer length scales, cannot effectively adiabatically follow. For heavy molecules, this two-fluid model provides a complete dynamical picture.

For light molecules such as HCN, the zero-temperature calculations have shown that adiabatic following is questionable even for the non-superfluid density. Consequently, in this situation the two-fluid model cannot be used to estimate effective moments of inertia. At this time, the zero-temperature DMC-based direct calculations of rotational energy levels of doped clusters provide the only route to microscopic theoretical understanding of spectroscopic measurements of rotational transitions for such light molecules in helium droplets. [41] This will hopefully change in the future, when molecular rotational motions are explicitly incorporated into the PIMC.

5. Conclusions and future directions

The path integral approach has provided a powerful theoretical tool for investigating the superfluid properties of finite helium droplets. Path integral Monte Carlo calculations have shown that these systems constitute nanoscale superfluids and offer a unique route to probing the structure and response of a Bose superfluid on a microscopic length scale. They also provide examples of inhomogeneous superfluid density, with the unique feature that this inhomogeneous, nanoscale superfluid density can be probed by molecular and atomic dopants.

The microscopic calculations show that such impurities introduce a local quantum solvation structure into the otherwise smoothly varying helium density. The numerical path integral Monte Carlo method has allowed this quantum solvation structure in a superfluid to be analyzed in terms of the boson permutation exchange properties, and conversely, the effect of the molecular interaction on the superfluid to be quantified. PIMC calculations show that a strongly bound impurity induces a non-superfluid density in the first solvation shell, whose extent is determined by the strength of the molecular interaction. Similar conclusions are derived from analysis of the permutation exchange path lengths into short, strongly localized paths, and long, delocalized paths, and from decomposition of the linear response estimator for global superfluidity. The existence of this local non-superfluid component in the solvation layer around microscopic impurities therefore seems to be a general feature of molecular solvation in superfluid He clusters. Response of this local two-fluid density to the rotation of a molecular impurity gives rise to increments in the molecular moment of inertia, but does not otherwise modify the effective free rotation of the molecule in the superfluid.

These path integral studies of doped helium droplets open the way to study of several intriguing questions. One is the effect of the molecular rotation on the quantum solvation structure in the superfluid local environment. As noted in this article, all PIMC studies to date have not explicitly incorporated the molecular rotational degrees of freedom. We now know from zero-temperature calculations of the quantum rotational excitations that the molecular rotation does result in a smearing out of the angular anisotropy in the quantum solvation structure. [5,42,51] This implies less than perfect adiabatic following of the helium density, even in the rotational ground state. Given the significance of the adiabatic following assumption for models of the helium response and hence for the analysis of rotational spectra of doped molecules, developing a direct route to the solvation density around a rotating molecule is highly desirable. This can be done by incorporating the molecular rotational kinetic energy in the path integral representation. A key question with spectroscopic implications is then how the local two-fluid density decomposition is modified. We noted earlier that the moment of inertia increment of the non-superfluid density around SF_6 is approximately independent of the anisotropy of the interaction potential. This suggests that even if the two-fluid densities are modified with rotation, becoming less anisotropic, the effective moment of inertia of the molecule in ^{4}He will be unchanged. This remains to be verified.

A second direction departing from the analysis of molecular solvation structure in a superfluid is the investigation of localization of helium atoms and their removal from the superfluid state, as a function of the binding to organic molecules of increasing size. In the study of benzene, the key feature responsible for the localization phenomenon was identified as the strong and highly anisotropic interaction of helium with the π-electron system. Systematically varying the extent of the π-electron system by going to larger planar, polyaromatic molecules will allow the transition from a nanosubstrate to a microsubstrate that begins to mimic a bulk solid surface to be investigated. We expect that the "inert" layers familiar from studies of thin films of helium on graphite will evolve from these localized atoms, but the manner in which this happens will depend on the role of lateral confinement and permutation exchanges in the presence of an extended π-electron system.

A third, novel direction is provided by extension of these ideas to nanoscale clusters of molecular hydrogen, H_2. In its rotational ground state, the H_2 molecule is a boson, but bulk superfluidity is preempted by the occurrence of the triple point at $T = 13.6$ K. However, finite-size and reduced dimensionality systems are offer ways of bypassing this solidification of hydrogen by allowing lower densities and thereby moderating the effects of strong interactions. Path integral calculations have already been used in several instances in the search for a superfluid state of molecular hydrogen. Thus, very small finite clusters of $(H_2)_N$ ($N \leq 18$) have been shown with PIMC to be not only liquid-like but also to show a limited extent of superfluidity. [86] Two-dimensional films of hydrogen have been shown to allow a stable superfluid phase at low temperatures provided that an array of alkali atoms is co-adsorbed, providing stabilization of a low density liquid phase. [11] Given these low-dimensional antecedents, it appears possible that a relatively small solvating layer of hydrogen wrapped around a molecule might also show some superfluid behavior. Path integral calculations are now in progress to examine the extent of permutation exchanges in cycles around different axes of a linear molecule wrapped with variable numbers of H_2 molecules. [87] Such studies will provide microscopic theoretical insight into the quantum dynamics underlying recent spectroscopic experiments showing anomalies in the molecular moment of inertia that are consistent with a partial superfluid response of the solvating hydrogen layer. [57]

In summary, the path integral Monte Carlo approach provides a unique tool for analysis of these degenerate quantum systems in finite geometries and with chemically complex impurity dopants. The insights into nanoscale superfluid properties that have resulted, and the interplay between physical and chemical effects afforded by calculations on doped helium droplets offer promise of new opportunities for analysis and manipulation of superfluid at the microscopic level.

Acknowledgments

We acknowledge financial support from the National Science Foundation (KBW, CHE-9616615, CHE-0107541) and the Korea Research Foundation (YK, 2000-015-DP0125). We thank NPACI for a generous allocation of supercomputer time at the San Diego Supercomputer Center, and KORDIC for a generous allocation of its supercomputer time. PH acknowledges the support of an Abramson Fellowship. KBW thanks the Alexander von Humboldt Foundation for a Senior Scientist Award, and Prof. J. P. Toennies for hospitality at the Max-Planck Institut für Strömungsforschung, Göttingen during a sabbatical year 1996-97. We thank D. T. Moore for permission to produce Fig. 3, and A. Nakayama and K. Yamashita for providing some of the data given in Fig. 2a.

References

1. J. Boronat, Monte Carlo simulations at zero temperature: Helium in one, two, and three-dimensions, 2001, this volume.
2. E. L. Pollock and D. M. Ceperley, *Phys. Rev.* **B 30**, 2555 (1984).
3. E. Gallicchio and B. J. Berne, *J. Chem. Phys.* **101**, 9909 (1994).
4. M. Boninsegni and D. M. Ceperley, *J. Low Temp. Phys.* **104**, 339 (1996).
5. Y. Kwon *et al.*, *J. Chem. Phys.* **113**, 6469 (2000).
6. R. P. Feynman, *Phys. Rev.* **91**, 1291 (1953).
7. E. L. Pollock and D. M. Ceperley, *Phys. Rev.* **B 36**, 8343 (1987).
8. M. Boninsegni and D. M. Ceperley, *Phys. Rev. Lett.* **74**, 2288 (1995).
9. D. Scharf, G. J. Martyna, and M. L. Klein, *J. Chem. Phys.* **99**, 8997 (1993).
10. M. E. Pierce and E. Manousakis, *Phys. Rev. Lett.* **83**, 5314 (1999).
11. M. C. Gordillo and D. M. Ceperley, *Phys. Rev. Lett.* **79**, 3010 (1997).
12. J. P. Toennies, A. F. Vilesov, and K. B. Whaley, *Phys. Today* **54**, 31 (2001).
13. P. Sindzingre, M. L. Klein, and D. M. Ceperley, *Phys. Rev. Lett.* **63**, 1601 (1989).
14. J. P. Toennies, Microscopic superfluidity of small ^{4}He and para-H_2 clusters inside Helium droplets, 2001, this volume.
15. M. V. R. Krishna and K. B. Whaley, *J. Chem. Phys.* **93**, 746 (1990).

16. M. Hartmann, F. Mielke, J. P. Toennies, and A. F. Vilesov, *Phys. Rev. Lett.* **76**, 4560 (1996).
17. S. Grebenev, J. P. Toennies, and A. F. Vilesov, *Science* **279**, 2083 (1998).
18. Y. Kwon, D. M. Ceperley, and K. B. Whaley, *J. Chem. Phys.* **104**, 2341 (1996).
19. Y. Kwon and K. B. Whaley, *Phys. Rev. Lett.* **83**, 4108 (1999).
20. D. M. Ceperley, *Rev. Mod. Phys.* **67**, 279 (1995).
21. H. F. Trotter, *Proc. Am. Math. Soc.* **10**, 545 (1959).
22. D. Chandler and P. G. Wolynes, *J. Chem. Phys.* **74**, 4078 (1981).
23. R. P. Feynman, *Statistical Mechanics* (Benjamin/Cummings Publishing Company, Inc., Reading, MA, 1972).
24. D. M. Ceperley and E. L. Pollock, in *Monte Carlo Methods in Theoretical Physics*, edited by S. Caracciolo and A. Fabrocini (ETS Editrice, Pisa, Italy, 1992).
25. J. P. Toennies and A. F. Vilesov, *Annu. Rev. Phys. Chem.* **49**, 1 (1998).
26. G. Baym, in *Mathematical methods in solid state and superfluid theory*, edited by R. C. Clark and G. H. Derrick (Oliver and Boyd, Edinburgh, 1969), p. 121.
27. E. Draeger, Ph.D. thesis, University of Illinois at Urbana-Champaign, 2001.
28. D. E. Galli, E. Cecchetti, and L. Reatto, *Phys. Rev. Lett.* **77**, 5401 (1996).
29. K. B. Whaley, *Int. Rev. Phys. Chem.* **13**, 41 (1994).
30. C. L. Cleveland, U. Landman, and R. N. Barnett, *Phys. Rev.* **B 39**, 117 (1989).
31. A. Nakayama and K. Yamashita, *J. Chem. Phys.* **114**, 780 (2001).
32. A. Nakayama and K. Yamashita, *J. Chem. Phys.* **112**, 10966 (2000).
33. F. Ancilotto, P. B. Lerner, and M. W. Cole, *J. Low Temp. Phys.* **101**, 1123 (1995).
34. K. B. Whaley, in *Advances in Molecular Vibrations and Collision Dynamics*, edited by J. M. Bowman and Z. Bačić (JAI Press Inc., Stamford, CT, 1998), Vol. 3, p. 397.
35. R. A. Aziz, F. R. W. McCourt, and C. C. K. Wong, *Mol. Phys.* **61**, 1487 (1987).
36. C. Callegari, J. Higgins, F. Stienkemeier, and G. Scoles, *J. Phys. Chem.* A **102**, 95 (1998).
37. Y. Kwon and K. B. Whaley, *J. Chem. Phys.* **114**, 3163 (2001).
38. M. Buzzacchi, D. E. Galli, and L. Reatto, *Phys. Rev.* **B 64**, 094512 (2001).
39. C. C. Duminuco, D. E. Galli, and L. Reatto, *Physica* **B284–B288**, 109 (2000).
40. D. Blume, M. Mladenović, M. Lewerenz, and K. B. Whaley, *J. Chem. Phys.* **110**, 5789 (1999).
41. A. Viel, M. V. Patel, P. Niyaz, and K. B. Whaley, *Comp. Phys. Com.* (2001), in press.
42. A. Viel and K. B. Whaley, *J. Chem. Phys.* **115**, 10186 (2001).
43. D. T. Moore and K. B. Whaley, (2001), to be published.
44. R. T. Pack, E. Piper, G. A. Pfeffer, and J. P. Toennies, *J. Chem. Phys.* **80**, 4940 (1984).
45. M. A. McMahon, R. N. Barnett, and K. B. Whaley, *J. Chem. Phys.* **104**, 5080 (1996).
46. S. Goyal, D. L. Schutt, and G. Scoles, *Phys. Rev. Lett.* **69**, 933 (1992).
47. Y. Kwon and K. B. Whaley, *J. Chem. Phys.* **115**, 10146 (2001).
48. K. Higgins and W. Klemperer, *J. Chem. Phys.* **110**, 1383 (1999).
49. P. Hobza, O. Bludský, H. L. Selzle, and E. W. Schlag, *J. Chem. Phys.* **97**, 335 (1992).
50. K. M. Atkins and J. M. Hutson, *J. Chem. Phys.* **105**, 440 (1996).
51. F. Paesani, F. A. Gianturco, and K. B. Whaley, *J. Chem. Phys.* **115**, 10225 (2001).
52. K. Nauta and R. E. Miller, *Science* **283**, 1895 (1999).
53. J. D. Close, F. Federmann, K. Hoffmann, and N. Quaas, *J. Low Temp. Phys.* **111**, 661 (1998).
54. R. J. Donnelly, *Quantized Vorticies in Helium II* (Cambridge University Press, Cambridge, England, 1991).
55. G. H. Bauer, R. J. Donnelly, and W. F. Vinen, *J. Low Temp. Phys.* **98**, 47 (1995).

56. F. Dalfovo, R. Mayol, M. Pí, and M. Barranco, *Phys. Rev. Lett.* **85**, 1028 (2000).
57. S. Grebenev, B. Sartakov, J. P. Toennies, and A. F. Vilesov, *Science* **289**, 1532 (2000).
58. K. Higgins and W. Klemperer, (2001), to be published.
59. P. J. Shirron and J. M. Mochel, *Phys. Rev. Lett.* **67**, 1118 (1991).
60. P. W. Adams and V. Pant, *Phys. Rev. Lett.* **68**, 2350 (1992).
61. G. Zimmerli, G. Mistura, and M. H. W. Chan, *Phys. Rev. Lett.* **68**, 60 (1992).
62. L. D. Landau, *J. Phys.* **5**, 71 (1941).
63. M. Lax, *J. Chem. Phys.* **20**, 1752 (1952).
64. E. Cheng and K. B. Whaley, *J. Chem. Phys.* **104**, 3155 (1996).
65. F. Stienkemeier *et al.*, *Z. Phys.* **D 38**, 253 (1996).
66. R. N. Barnett and K. B. Whaley, *J. Chem. Phys.* **99**, 9730 (1993).
67. M. Hartmann, R. E. Miller, J. P. Toennies, and A. Vilesov, *Phys. Rev. Lett.* **75**, 1566 (1995).
68. A. Scheidemann, B. Schilling, and J. P. Toennies, *J. Phys. Chem.* **97**, 2128 (1993).
69. D. Eichenauer and R. J. L. Roy, *J. Chem. Phys.* **88**, 2898 (1988).
70. F. A. Gianturco and F. Paesani, *J. Chem. Phys.* **113**, 3011 (2000).
71. D. Blume, M. Lewerenz, F. Huisken, and M. Kaloudis, *J. Chem. Phys.* **105**, 8666 (1996).
72. S. Grebenev *et al.*, *J. Chem. Phys.* **112**, 4485 (2000).
73. V. S. Babichenko and Y. Kagan, *Phys. Rev. Lett.* **83**, 3458 (1999).
74. E. Lee, D. Farrelly, and K. B. Whaley, *Phys. Rev. Lett.* **83**, 3812 (1999).
75. F. Paesani, A. Viel, F. A. Gianturco, and K. B. Whaley, (2002), to be published.
76. M. Hartmann, Ph.D. thesis, University of Goettingen, 1997.
77. M. Hartmann, A. Lindinger, J. P. Toennies, and A. F. Vilesov, *Chem. Phys.* **239**, 139 (1998).
78. M. Hartmann, A. Lindinger, J. P. Toennies, and A. F. Vilesov, *J. Chem. Phys.* (2002), submitted.
79. K. W. Jucks and R. E. Miller, *J. Chem. Phys.* **88**, 2196 (1988).
80. M. Patel, Ph.D. thesis, University of California, Berkeley, 2001; M. Patel, (2001), to be published.
81. Y. Kwon and K. B. Whaley (unpublished).
82. B. Sartakov, A. F. Vilesov, and K. B. Whaley, Schloss Ringberg meeting on Quantum Clusters, Fluids, and Solids, 1997.
83. C. Callegari *et al.*, *Phys. Rev. Lett.* **83**, 5058 (1999); C. Callegari *et al.*, *ibid.* **84**, 1848 (2000).
84. J. B. Mehl and W. Zimmerman, *Phys. Rev. Lett.* **14**, 815 (1965).
85. H. Kojima *et al.*, *Phys. Rev. Lett.* **27**, 714 (1971).
86. P. Sindzingre, D. M. Ceperley, and M. L. Klein, *Phys. Rev. Lett.* **67**, 1871 (1991).
87. Y. Kwon and K. B. Whaley, (2002), in progress.

CHAPTER 4

STRUCTURE AND DYNAMICS OF THE BULK LIQUID AND BULK MIXTURES

M. Saarela
Dept. of Physical Sciences, P.O. Box 3000, FIN-90014 University of Oulu, Finland
E-mail: Mikko.Saarela@oulu.fi

V. Apaja
Institute for Theoretical Physics, Johannes Kepler University
A-4040 Linz, Austria
E-mail: Vesa.Apaja@jku.at

J. Halinen
Dept. of Physical Sciences, P.O. Box 3000, FIN-90014 University of Oulu, Finland
E-mail: Jani.Halinen@oulu.fi

1. Introduction

The first theoretical model of dilute quantum fluid mixtures was proposed by Landau and Pomeranchuck [1] over fifty years ago. The model, which went relatively unnoticed at the time, was based on a quasi–particle picture of the ^{3}He impurity atom moving in superfluid ^{4}He. However, the discovery of the finite solubility of ^{3}He in ^{4}He near absolute zero made in 1965 launched intensive experimental and theoretical efforts to better understand the thermodynamic properties of this quantum fluid mixture. Excellent review articles on these developments, containing discussion of quantities such as the equation of state, heat of mixing, osmotic pressure, both zero– and second–sound velocities, and so on, as a function of the ^{3}He concentration and pressure, have been published by Edwards and collaborators, [2] by Baym and Pethick, [3] and by Ouboter and Yang [4]. Bardeen, Baym, and Pines [5] (BBP) formulated a theory of effective interaction between the quasi–particles to explain transport properties. The BBP model has since then been further refined by including corrections due to the momentum dependence of the quasi–particle interaction, [2,5] as well as improvements beyond the Hartree–Fock approximation. [6] The most complete treatment of these effects is given in the pseudopotential model [7–9] by Pines *et al.* Their effective interaction between the helium isotopes combines aspects like the short–range repulsion, Pauli principle corrections, and exact sum rules with experimental information such as the density and concentration dependence of the speed of zero sound in the mixture.

A truly microscopic theory starts from a Hamiltonian containing the two–body interaction between pairs of particles. In calculating the energetics, pressure, chemical potentials, *etc.*, and structural properties like distribution and structure functions, the many–body effects are incorporated through the use of cluster expansions and summations. The convergence rate of such methods depends, of course, on the physical quantities one chooses to calculate; when making predictions on physical observables one must be aware that some predictions of the theory will be better than others. In the microscopic treatment of the ^{3}He–^{4}He mixture one faces the problem of two different length scales: the short–ranged interactions between the individual helium atoms, and a weak effective interaction between ^{3}He quasi–particles giving rise in three dimensions to pairing [5,10] and in two dimensions to weakly bound dimers. [11,12] The energy differences between the mixed and separated phases are minute: accurate results are therefore needed for both the pure fluid and the mixture.

Our review concentrates on the Jastrow–Feenberg variational approach which is today the method of choice for microscopic studies of strongly correlated quantum fluids. One reason for this is that the theory describes both short– and long–ranged correlations correctly, and even the simplest version of the optimized (Fermi)–hypernetted–chain ((F)HNC/EL) approximation used to sum up the many–body diagrams is, for many purposes, sufficient to obtain a reasonable qualitative or, in dilute systems, even a quantitative understanding of the essential physics. Woo, Massey, and Tan [13] were the first to develop a microscopic theory of binary quantum mixtures based on the Jastrow–Feenberg wave function. They ignored, however, the fermion character of the ^{3}He component which is crucial for the stability of the mixture. These calculations were later improved through the use of HNC summations, [14] generalized to include the fermionic nature of the mixed particles. [15–18] For strongly correlated systems like the helium liquids, it is essential to include also the three–body correlations into the trial wave function if quantitatively correct results are to be obtained, as well as to add the elementary diagrams into the HNC summation. [19–24] Again, this is even more true in the case of a dilute mixture where small energy differences are important. Energy derivatives, such as the chemical potentials, having a central role in the theory also gain important corrections from these terms. [25–28]

We start this review by formulating the theory of quantum fluid mixtures in terms of the full set of FHNC/EL equations, and indicate places where improvement like triplet–correlation functions and elementary diagrams must be added into the description. Special attention is paid to the stability of the mixture by studying the behavior of collective excitations. Going beyond the variational theory, we will also briefly review the calculation of ring diagrams within the theory of correlated basis functions (CBF). This extension of the theory becomes necessary because, as we will show, the FHNC approximation leads to an incorrect prediction of the critical concentration at which the mixture becomes locally unstable against phase separation. In fact, we must even go beyond this calculation and include CBF corrections to all orders. This will be done by adopting ideas from the parquet–diagram theory and generalizing them for mixtures.

The second part of this review discusses the dynamics of bulk liquids. The preliminary work on the collective excitations of ^{4}He was done by Landau and Bijl in the 1940s. [29,30] Landau proposed that there are two separate collective excitation modes in liquid ^{4}He: phonons, thought of as collective density (sound) modes having linear dispersion, and rotons, assumed to be a collective rotation of the fluid having a separate dispersion curve. Later on, he joined these excitations into a single collective mode dispersion curve continuous in the wave vector k; phonons and rotons were then interpreted as the low– and high–k regions of the same collective excitation. Between them we have the so–called maxon region. This was consistent with the dispersion curve for excitations in a dilute Bose gas derived by Bogoliubov in his seminal paper. [31] A vast number of experiments has been performed since then: for a review, we refer to the book by Glyde. [32]

The microscopic variational description of the dynamics of strongly correlated fluids was initiated by Feynman who suggested a trial excited–state wave function. [33] Specifically, he wrote the wave function of an excited state $\Psi_{\mathbf{k}}$ of momentum $\hbar\mathbf{k}$ as a product $\Psi_{\mathbf{k}} = \rho_{\mathbf{k}}\Psi_0$ of the ground–state wave function Ψ_0 and of a density–fluctuation (or phonon–creation) operator $\rho_{\mathbf{k}} = \Sigma_j \exp(i\mathbf{k}\cdot\mathbf{r}_j)$, offering thus a microscopic explanation for phonons and rotons as collective density excitations. The proposed wave function leads to the dispersion relation

$$\varepsilon_F(k) = \frac{\hbar^2 k^2}{2mS(k)}\,, \tag{1.1}$$

where m is the particle mass and $S(k)$ the ground–state static structure function. The dispersion relation of Eq. (1.1) provides an upper bound for the lowest–lying excitation and is exact in the long–wavelength limit, but at shorter wavelengths it grossly overestimates the experimental spectrum: for example, in the roton region the computed energy is twice as large as the experimentally observed value. Owing to this discrepancy, the theory was subsequently supplemented by Feynman and Cohen [34] to include so–called backflow corrections which increased the flexibility of the wave function and, thus, lowered the roton energy significantly towards measured values. The term backflow is used here to describe the correlated motion of neighboring particles around a given reference atom. Feynman and Cohen assumed that the particles move in a dipolar flow field, behaving in a sense like a smoke ring.

Following Feynman's original arguments, the method of correlated basis functions was developed by Feenberg and his collaborators. [19,35–38] In the CBF approach the excited–state wave function $\Psi_{\mathbf{k}}$ is written as $\Psi_{\mathbf{k}} = F_{\mathbf{k}}\Psi_0$, and the excitation operator $F_{\mathbf{k}}$ is further expressed as a polynomial in the density–fluctuation operators $\{\rho_{\mathbf{k}}\}$. Thus, in the lowest order we have the usual Feynman form for the excited states, and terms beyond the linear one introduce the backflow effects.

Attempts to calculate the dynamic structure function were also made. [39]

The latter part of this work reviews further developments in the microscopic description of the dynamic structure, based on the equations–of–motion method. [40–43] In brief, the method commences from a Hamiltonian which contains an infinitesimal external interaction driving excitations into the system. The least–action principle can then be utilized to derive equations of motion for the fluctuations in the one– and two–body densities. The time dependence of the external interaction also creates currents in the system, solved, together with the density fluctuations, from the equations of motion. We also demonstrate how the linear response function and, hence, the dynamic structure function can be obtained from the equations.

2. Variational theory of quantum fluid mixtures

In this section, we review the microscopic variational theory of a mixture of two quantum fluids. Besides the dilute mixture of ^{3}He and ^{4}He atoms, we also have in mind the mixture of electrons and holes in semiconductors.

The theory starts from an empirical, non-relativistic Hamiltonian

$$H = -\sum_{\alpha}\sum_{i=1}^{N_\alpha} \frac{\hbar^2}{2m_\alpha}\nabla_i^2 + \frac{1}{2}\sum_{\alpha,\beta}\sum_{i,j}^{N_\alpha,N_\beta}{}' V^{\alpha\beta}(|\mathbf{r}_i - \mathbf{r}_j|)\,, \tag{2.1}$$

where the two–body interaction $V^{\alpha\beta}(r)$ is either known like the Coulomb interaction between charged particles, or experimentally determined as the case is for the interaction [44] between helium atoms. In our notation the Greek indices $\alpha, \beta, \ldots \in \{1,2\}$ refer to the type of a particle (^{3}He or ^{4}He atom), and Latin subscripts $i, j, \ldots$ refer to the individual particles. The number of particles of each species is N_α, and $N = N_1 + N_2$ is the total number of particles in the system. In terms of the concentration x of particle type 1, we have

$$N_1 = xN, \qquad N_2 = (1-x)N\,. \tag{2.2}$$

The prime on the summation in Eq. (2.1) indicates that no two pairs (i,α), (j,β) can be the same.

Along with the Hamiltonian, a central ingredient in the Jastrow–Feenberg variational method [19,35,45] is an ansatz for the ground–state wave function,

$$\begin{aligned}\Psi_0(\{\mathbf{r}_i^{(\alpha)}\}) &= e^{\frac{1}{2}U(\{\mathbf{r}_i^{(\alpha)}\})}\Phi(\{\mathbf{r}_i^{(\alpha)}\}) \\ U(\{\mathbf{r}_i^{(\alpha)}\}) &= \frac{1}{2!}\sum_{\alpha\beta}\sum_{i,j}^{N_\alpha,N_\beta}{}' u^{\alpha,\beta}(\mathbf{r}_i,\mathbf{r}_j) + \frac{1}{3!}\sum_{\alpha,\beta,\gamma}\sum_{i,j,k}^{N_\alpha,N_\beta,N_\gamma}{}' u^{\alpha\beta\gamma}(\mathbf{r}_i,\mathbf{r}_j,\mathbf{r}_k)\,.\end{aligned} \tag{2.3}$$

Here the shorthand notation $(\{\mathbf{r}_i^{(\alpha)}\})$ in the list of arguments refers to the full list of coordinates $(\mathbf{r}_1^{(1)}, ..., \mathbf{r}_{N_1}^{(1)}, \mathbf{r}_1^{(2)}, ..., \mathbf{r}_{N_2}^{(2)})$, and $\Phi_0(\{\mathbf{r}_i^{(\alpha)}\})$ is a product of Slater determinants of plane waves ensuring the required antisymmetry of the fermion part of the wave function. The functions $u^{\alpha\beta}(\mathbf{r}_i, \mathbf{r}_j)$ and $u^{\alpha\beta\gamma}(\mathbf{r}_i, \mathbf{r}_j, \mathbf{r}_k)$ represent pair and triplet correlations; the superscripts determine the type of correlation. Both the pair and the triplet correlation functions are determined from the variational principles [20,21,23] (Euler or Euler–Lagrange equations)

$$\frac{\delta E}{\delta u^{\alpha\beta}(\mathbf{r}_i, \mathbf{r}_j)} = 0 \tag{2.4}$$

and

$$\frac{\delta E}{\delta u^{\alpha\beta\gamma}(\mathbf{r}_i, \mathbf{r}_j, \mathbf{r}_k)} = 0\,, \tag{2.5}$$

where

$$E = \frac{\langle \Psi_0 | H | \Psi_0 \rangle}{\langle \Psi_0 | \Psi_0 \rangle} \tag{2.6}$$

is the variational energy–expectation value.

Key quantities in the theory are, besides the correlation functions, the partial densities

$$\rho_\alpha = N_\alpha / \Omega \tag{2.7}$$

with Ω the total volume of the system, the two– and three–body densities

$$\rho^{\alpha\beta}(\mathbf{r}, \mathbf{r}') = \sum_{i,j}^{N_\alpha, N_\beta}{}' \frac{\langle \Psi_0 | \delta(\mathbf{r}_i^{(\alpha)} - \mathbf{r}) \delta(\mathbf{r}_j^{(\beta)} - \mathbf{r}') | \Psi_0 \rangle}{\langle \Psi_0 | \Psi_0 \rangle} \tag{2.8}$$

$$\rho^{\alpha\beta\gamma}(\mathbf{r}, \mathbf{r}', \mathbf{r}'') = \sum_{i,j,k}^{N_\alpha, N_\beta, N_\gamma}{}' \frac{\langle \Psi_0 | \delta(\mathbf{r}_i^{(\alpha)} - \mathbf{r}) \delta(\mathbf{r}_j^{(\beta)} - \mathbf{r}') \delta(\mathbf{r}_k^{(\gamma)} - \mathbf{r}'') | \Psi_0 \rangle}{\langle \Psi_0 | \Psi_0 \rangle}\,, \tag{2.9}$$

and the pair– and triplet–distribution functions

$$g^{\alpha\beta}(\mathbf{r}, \mathbf{r}') = \frac{\rho^{\alpha\beta}(\mathbf{r}, \mathbf{r}')}{\rho_\alpha \rho_\beta} \tag{2.10}$$

$$g^{\alpha\beta\gamma}(\mathbf{r}, \mathbf{r}', \mathbf{r}'') = \frac{\rho^{\alpha\beta\gamma}(\mathbf{r}, \mathbf{r}', \mathbf{r}'')}{\rho_\alpha \rho_\beta \rho_\gamma}\,. \tag{2.11}$$

Again, the number of coordinates and/or species labels tells the type of a density or distribution function. In the translationally invariant system under consideration here, all two–point functions (such as the pair–distribution function) depend only on the distance between the two particles,

$$g^{\alpha\beta}(\mathbf{r}_i, \mathbf{r}_j) \equiv g^{\alpha\beta}(|\mathbf{r}_i - \mathbf{r}_j|) \equiv g^{\alpha\beta}(r_{ij})\,. \tag{2.12}$$

The static structure functions $S^{\alpha\beta}(k)$ are related to the pair–distribution functions through the Fourier transform

$$S^{\alpha\beta}(k) = \delta_{\alpha\beta} + \sqrt{\rho_\alpha \rho_\beta} \int d^3r \left[g^{\alpha\beta}(r) - 1 \right] e^{i\mathbf{k}\cdot\mathbf{r}}\,. \tag{2.13}$$

2.1. *Exact Euler equation for the pair–distribution function*

In any approximate theory, utmost care must be exercised to guarantee that approximations for the variational problems (2.4) and (2.5) have physically meaningful solutions. The main structural properties of strongly correlated quantum fluids are determined by two–body correlations, so we concentrate here on them. For detailed discussion on the triplet–correlation functions, we point the reader to Refs. 24 and 46.

Based on this, the safest way to proceed is to derive the two–body Euler equation from the exact energy–expectation value. Firstly, we make use of the Jackson–Feenberg identity [36]

$$e^{\frac{1}{2}U}\nabla^2 e^{\frac{1}{2}U} = \frac{1}{2}(\nabla^2 e^U + e^U\nabla^2) + \frac{1}{4}e^U\left[\nabla,[\nabla,U]\right] - \frac{1}{4}\left[\nabla,[\nabla,e^U]\right], \tag{2.14}$$

and write the total energy in the form

$$\frac{E}{N} = \frac{T_F}{N} + \sum_{\alpha\beta}\frac{\rho_\alpha\rho_\beta}{2\rho}\int d^3r g^{\alpha\beta}(r)V_{JF}^{\alpha\beta}(r) + \frac{T_{JF}}{N} + \frac{T_{(3)}}{N}. \tag{2.15}$$

The first term gives the sum of the kinetic energies of the free fermion components in the mixture

$$\frac{T_F}{N} = \frac{3}{10}\hbar^2\sum_\alpha\frac{k_{F\alpha}^2}{m_\alpha}\frac{N_\alpha}{N}, \tag{2.16}$$

with the Fermi wave number $k_{F\alpha}$. In the electron–hole mixture both constituents are fermions, in helium mixtures only the ^{3}He component. Here $V_{JF}^{\alpha\beta}(r)$ is what is known as the Jackson–Feenberg effective interaction,

$$V_{JF}^{\alpha\beta}(r) = V^{\alpha\beta}(r) - \left[\frac{\hbar^2}{8m_\alpha} + \frac{\hbar^2}{8m_\beta}\right]\nabla^2 u^{\alpha\beta}(r), \tag{2.17}$$

and T_{JF} in Eq. (2.15) is a fermionic contribution originating from the last term in the Jackson–Feenberg identity (2.14),

$$T_{JF} = \sum_{\alpha=1}^{2}\frac{\hbar^2}{8m_\alpha}\sum_{i=1}^{N_\alpha}\frac{\langle\Phi_0|(\nabla_i^2 e^U)|\Phi_0\rangle}{\langle\Phi_0|e^U|\Phi_0\rangle}. \tag{2.18}$$

The genuine two–body term consists of closed exchange loops and can be written as

$$\frac{T_{JF}}{N} = \sum_\alpha\frac{\hbar^2\rho_\alpha^2}{8m_\alpha\rho}\int d^3r\,\Gamma_{dd}^{\alpha\alpha}(r)\nabla^2 g_F^{\alpha\alpha}(r), \tag{2.19}$$

where $g_F^{\alpha\alpha}(r)$ is the pair–distribution function of the non–interacting system,

$$\begin{aligned} g_F^{\alpha\alpha}(r) &= 1 - \frac{1}{2}\ell^2(rk_{F\alpha}) \\ \ell(x) &\equiv \frac{3}{x}j_1(x), \end{aligned} \tag{2.20}$$

with the spherical Bessel function $j_1(x)$, and $\Gamma_{dd}^{\alpha\alpha}(r)$ is the direct–direct correlation function; its explicit form will be derived in the next section. Finally, $T_{(3)}/N$ includes all the terms containing triplet–correlation functions and other higher–order terms such as three–point diagrams with exchange lines. [24,46]

Starting from the expression (2.15) for the energy, we see that the Euler equation for the two–body correlation function consist of two parts, one originating from the explicit appearance of $u^{\alpha\beta}(r)$ in the kinetic–energy terms, and the other from the implicit dependence of the two–body functions on the correlation functions. The Euler equation then takes the general form

$$\left[\frac{\hbar^2}{8m_\alpha}+\frac{\hbar^2}{8m_\beta}\right]\nabla^2 g^{\alpha\beta}(r)=g'^{\alpha\beta}(r)\,, \tag{2.21}$$

where

$$\begin{aligned} g'^{\alpha\beta}(r) = &\frac{1}{\rho_\alpha\rho_\beta}\sum_{\gamma,\delta}\rho_\gamma\rho_\delta\int d^3r'\frac{\delta g^{\gamma\delta}(r')}{\delta u^{\alpha\beta}(r)}V_{JF}^{\gamma\delta}(r') \\ &+\frac{2}{\rho_\alpha\rho_\beta}\frac{\delta}{\delta u^{\alpha\beta}(r)}\left[\frac{T_{JF}}{N}+\frac{T_{(3)}}{N}\right]. \end{aligned} \tag{2.22}$$

The function $g'^{\alpha\beta}(r)$, which is essentially the solution of a linearized set of HNC equations, is most conveniently constructed by introducing the prime–derivative technique of Campbell and Feenberg. [19,20] The pair–correlation function is generalized into a function which depends on a parameter λ,

$$u^{\alpha\beta}(r;\lambda)\equiv u^{\alpha\beta}(r)+\lambda\left[V^{\alpha\beta}(r)-\left(\frac{\hbar^2}{8m_\alpha}+\frac{\hbar^2}{8m_\beta}\right)\nabla^2 u^{\alpha\beta}(r)\right]. \tag{2.23}$$

The generalized pair–distribution function $g^{\alpha\beta}(r;\lambda)$ is defined in exactly the same way as the pair–distribution function in Eq. (2.8). It is then easy to verify that

$$g'^{\alpha\beta}(r)=\left.\frac{\partial}{\partial\lambda}g^{\alpha\beta}(r;\lambda)\right|_{\lambda=0}+\frac{2}{\rho_\alpha\rho_\beta}\frac{\delta}{\delta u^{\alpha\beta}(r)}\left[\frac{T_{JF}}{N}\right]. \tag{2.24}$$

This demonstrates how $g'^{\alpha\beta}(r)$ can be constructed from $g^{\alpha\beta}(r)$ without the need for a new derivation of the variational derivatives appearing in the Euler equation. Finally, the variational derivative in the last term is obtained in a diagrammatic expansion of the pair–distribution function $g^{\alpha\beta}(r)$, by replacing, in turn, every connected pair of exchange lines $\ell(r_{ij}k_{F\alpha})\ell((r_{ij}k_{F\alpha})$ by $(\hbar^2/4m_\alpha)\nabla_i^2[\ell(r_{ij}k_{F\alpha})\ell((r_{ij}k_{F\alpha})]$.

2.1.1. *Fermi–hypernetted–chain equations*

It has been known for quite some time [19,21] that the HNC hierarchy of approximations preserves the properties of the exact Euler equation at every level of its implementation. One desirable property is that the Euler equations (2.4) and (2.5) cease to yield solutions describing a uniform system if the system is physically unstable. [47,48] The (F)HNC equations describe the self–consistent summation of two types of diagrams: chain connections and parallel connections. For both pure systems and mixtures, these equations have been derived and discussed in the literature; [14,16,18,24,25,49–51] we therefore review here only the essential steps of the derivation.

In the FHNC method the pair–distribution function is decomposed as

$$g^{\alpha\beta}(r) = 1 + \Gamma_{dd}^{\alpha\beta}(r) + \Gamma_{de}^{\alpha\beta}(r) + \Gamma_{ed}^{\alpha\beta}(r) + \Gamma_{ee}^{\alpha\beta}(r)\,. \tag{2.25}$$

Each $\Gamma_{\{ii\}}^{\alpha\beta}(r)$ represents certain subclass of diagrams, labeled by the subscript $ii = \{dd, de, ed, ee\}$. Each of these functions is further decomposed into a sum of nodal $N_{\{ii\}}^{\alpha\beta}$ and non-nodal or direct $X_{\{ii\}}^{\alpha\beta}$ diagrams,

$$\Gamma_{\{ii\}}^{\alpha\beta} = N_{\{ii\}}^{\alpha\beta} + X_{\{ii\}}^{\alpha\beta}\,. \tag{2.26}$$

A nodal diagram can be cut into two disconnected pieces by removing one internal point (node), hence the diagrams can be summed up by solving convolution integral equations. As a result, they can be expressed in momentum space as products of the Fourier transforms

$$\tilde{f}^{\alpha\beta}(\mathbf{k}) = \sqrt{\rho_\alpha\rho_\beta}\int d^3r\; e^{i\mathbf{k}\cdot\mathbf{r}}\; f^{\alpha\beta}(\mathbf{r})\,. \tag{2.27}$$

In constructing the products, all possible products of $\tilde{\Gamma}_{\{ii\}}^{\alpha\beta}$ and $\tilde{X}_{\{ii\}}^{\alpha\beta}$ are summed together, recalling also that exchange loops cannot connect different types of particles, and that exchange loops cannot overlap. [18] Using the 2×2 matrix notation

$$\tilde{\mathbf{f}} \equiv \begin{pmatrix} \tilde{f}^{11}(\mathbf{k}) & \tilde{f}^{12}(\mathbf{k}) \\ \tilde{f}^{21}(\mathbf{k}) & \tilde{f}^{22}(\mathbf{k}) \end{pmatrix}, \tag{2.28}$$

we get a set of matrix equations

$$\begin{aligned}
\tilde{\mathbf{N}}_{dd} &= \tilde{\mathbf{\Gamma}}_{dd}\tilde{\mathbf{X}}_{dd} + \tilde{\mathbf{\Gamma}}_{dd}\tilde{\mathbf{X}}_{ed} + \tilde{\mathbf{\Gamma}}_{de}\tilde{\mathbf{X}}_{dd} \\
\tilde{\mathbf{N}}_{de} &= \tilde{\mathbf{\Gamma}}_{dd}\tilde{\mathbf{X}}_{de} + \tilde{\mathbf{\Gamma}}_{dd}\tilde{\mathbf{X}}_{ee} + \tilde{\mathbf{\Gamma}}_{de}\tilde{\mathbf{X}}_{de} \\
\tilde{\mathbf{N}}_{ed} &= \tilde{\mathbf{\Gamma}}_{ed}\tilde{\mathbf{X}}_{dd} + \tilde{\mathbf{\Gamma}}_{ed}\tilde{\mathbf{X}}_{ed} + \tilde{\mathbf{\Gamma}}_{ee}\tilde{\mathbf{X}}_{dd} \\
\tilde{\mathbf{N}}_{ee} &= \tilde{\mathbf{\Gamma}}_{ed}\tilde{\mathbf{X}}_{de} + \tilde{\mathbf{\Gamma}}_{ed}\tilde{\mathbf{X}}_{ee} + \tilde{\mathbf{\Gamma}}_{ee}\tilde{\mathbf{X}}_{de} \\
\tilde{\mathbf{N}}_{cc} &= (\tilde{\mathbf{\Gamma}}_{cc} - \mathbf{L})\tilde{\mathbf{X}}_{cc}\,.
\end{aligned} \tag{2.29}$$

Here $\mathbf{L}$ is a diagonal matrix with elements giving the occupation probabilities of single–particle states in an ideal Fermi system, *i.e.* $L^{\alpha\beta} = \delta_{\alpha\beta}$ in case $k < k_{F_\alpha}$ and zero otherwise. Matrices $\tilde{\mathbf{N}}_{dd}$ and $\tilde{\mathbf{N}}_{ee}$ are symmetric, $\tilde{\mathbf{N}}_{ed}$ is the transpose of $\tilde{\mathbf{N}}_{de}$, and $\tilde{\mathbf{N}}_{cc}$ is diagonal; the same holds for $\tilde{\mathbf{\Gamma}}_{\{ii\}}$ and $\tilde{\mathbf{X}}_{\{ii\}}$.

Eqs. (2.29) can be solved for the matrices $\tilde{\mathbf{\Gamma}}_{\{ii\}}$,

$$\tilde{\mathbf{\Gamma}}_{dd}^{-1} = -\tilde{\mathbf{A}} + \tilde{\mathbf{B}}^T\tilde{\mathbf{X}}_{dd}^{-1}\tilde{\mathbf{B}} \tag{2.30}$$

$$\tilde{\Gamma}_{de} = -\mathbf{I} - \tilde{\Gamma}_{dd} + \tilde{\Gamma}_{dd}\tilde{\mathbf{B}}^T\tilde{\mathbf{X}}_{dd}^{-1} \tag{2.31}$$

$$\tilde{\Gamma}_{ed} = -\mathbf{I} - \tilde{\Gamma}_{dd} + \tilde{\mathbf{X}}_{dd}^{-1}\tilde{\mathbf{B}}\tilde{\Gamma}_{dd} \tag{2.32}$$

$$\tilde{\Gamma}_{ee} = \mathbf{I} + \tilde{\Gamma}_{dd} - \tilde{\Gamma}_{dd}\tilde{\mathbf{B}}^T\tilde{\mathbf{X}}_{dd}^{-1} - \tilde{\mathbf{X}}_{dd}^{-1}\tilde{\mathbf{B}}\tilde{\Gamma}_{dd} + (\tilde{\mathbf{B}}^T)^{-1}\tilde{\mathbf{A}}\tilde{\Gamma}_{dd}\tilde{\mathbf{B}}^T\tilde{\mathbf{X}}_{dd}^{-1}, \tag{2.33}$$

where we have used the notation

$$\begin{aligned}\tilde{\mathbf{A}} &= \mathbf{I} + \tilde{\mathbf{X}}_{ee}\\ \tilde{\mathbf{B}} &= \mathbf{I} - \tilde{\mathbf{X}}_{de}\end{aligned} \tag{2.34}$$

with the unit matrix **I**. Inserting the sum

$$\tilde{\Gamma}_{dd} + \tilde{\Gamma}_{de} + \tilde{\Gamma}_{ed} + \tilde{\Gamma}_{ee} = -\mathbf{I} + (\tilde{\mathbf{B}}^T)^{-1}\tilde{\mathbf{A}}\tilde{\Gamma}_{dd}\tilde{\mathbf{B}}^T\tilde{\mathbf{X}}_{dd}^{-1} \tag{2.35}$$

into Eqs. (2.25) and (2.13), we get the static structure function matrix

$$\mathbf{S} = (\tilde{\mathbf{B}}^T)^{-1}\tilde{\mathbf{A}}\tilde{\Gamma}_{dd}\tilde{\mathbf{B}}^T\tilde{\mathbf{X}}_{dd}^{-1}. \tag{2.36}$$

The inverse $\tilde{\mathbf{X}}_{dd}^{-1}$ can be solved from Eq. (2.30) in terms of $\tilde{\Gamma}_{dd}^{-1}$, and we get

$$\mathbf{S} = (\tilde{\mathbf{B}}^T)^{-1}\tilde{\mathbf{A}}\left(\mathbf{I} + \tilde{\Gamma}_{dd}\tilde{\mathbf{A}}\right)\tilde{\mathbf{B}}^{-1}. \tag{2.37}$$

Finally, we solve this equation for $\tilde{\Gamma}_{dd}$,

$$\tilde{\Gamma}_{dd} = \tilde{\mathbf{A}}^{-1}\tilde{\mathbf{B}}^T\mathbf{S}\tilde{\mathbf{B}}\tilde{\mathbf{A}}^{-1} - \tilde{\mathbf{A}}^{-1}, \tag{2.38}$$

and insert the result into Eq. (2.36) to obtain an expression for $\tilde{\mathbf{X}}_{dd}$ in terms of the structure function,

$$\tilde{\mathbf{X}}_{dd} = \tilde{\mathbf{B}}\tilde{\mathbf{A}}^{-1}\tilde{\mathbf{B}}^T - \mathbf{S}^{-1}. \tag{2.39}$$

The coordinate–space equations take care of the correct short–range behavior of the correlation functions. The HNC summation of the nodal diagrams $N_{dd}^{\alpha\beta}(r)$ and elementary diagrams $E_{dd}^{\alpha\beta}(r)$ defines the components of the direct–direct correlation matrix $\Gamma_{dd}^{\alpha\beta}(r)$,

$$\Gamma_{dd}^{\alpha\beta}(r) = e^{\left[u^{\alpha\beta}(r) + N_{dd}^{\alpha\beta}(r) + E_{dd}^{\alpha\beta}(r)\right]} - 1. \tag{2.40}$$

This function determines the short–range behavior of the other $\Gamma_{\{ii\}}^{\alpha\beta}(r)$–functions through the hypernetting equations,

$$\begin{aligned}\Gamma_{de}^{\alpha\beta}(r) &= \left[1 + \Gamma_{dd}^{\alpha\beta}(r)\right]\left[N_{de}^{\alpha\beta}(r) + E_{de}^{\alpha\beta}(r)\right]\\ \Gamma_{ee}^{\alpha\beta}(r) &= \left[1 + \Gamma_{dd}^{\alpha\beta}(r)\right]\left\{N_{ee}^{\alpha\beta}(r) + E_{ee}^{\alpha\beta}(r) - 2\left[\mathcal{L}^{\alpha\beta}(r)\right]^2\right.\\ &\quad\left. + \left[N_{ed}^{\alpha\beta}(r) + E_{ed}^{\alpha\beta}(r)\right]\left[N_{de}^{\alpha\beta}(r) + E_{de}^{\alpha\beta}(r)\right]\right\}\\ \Gamma_{cc}^{\alpha\beta}(r) &= \left[1 + \Gamma_{dd}^{\alpha\beta}(r)\right]N_{cc}^{\alpha\beta}(r) - \frac{1}{2}\ell(k_{F\alpha}r)\delta_{\alpha\beta}\Gamma_{dd}^{\alpha\beta}(r),\end{aligned} \tag{2.41}$$

with

$$\mathcal{L}^{\alpha\beta}(r) = \frac{1}{2}\ell(k_{F\alpha}r)\delta_{\alpha\beta} - N_{cc}^{\alpha\beta}(r) - E_{cc}^{\alpha\beta}(r). \tag{2.42}$$

Our aim was to introduce the relationships between the pair–correlation functions and the pair–distribution functions, and so, inserting the above results into Eq. (2.25), we finally get the pair–distribution functions in the respective channels,

$$g^{\alpha\beta}(r) = \left[1 + \Gamma^{\alpha\beta}(r)\right] \left\{-2\left[\mathcal{L}^{\alpha\beta}(r)\right]^2 + C^{\alpha\beta}(r)\right\} \tag{2.43}$$

with

$$\begin{aligned} C^{\alpha\beta}(r) = &\left[1 + N_{de}^{\alpha\beta}(r) + E_{de}^{\alpha\beta}(r)\right]\left[1 + N_{ed}^{\alpha\beta}(r) + E_{ed}^{\alpha\beta}(r)\right] \\ &+ N_{ee}^{\alpha\beta}(r) + E_{ee}^{\alpha\beta}(r) . \end{aligned} \tag{2.44}$$

2.1.2. *Single–loop approximation*

In what comes to ^{3}He–^{4}He mixtures, we are mostly interested in the dilute limit where the mixture is stable against phase separation. This focuses our calculations to concentrations of no more than ten percent. For such small concentrations, the simplest version of the FHNC equations, [49,52] called the single–loop approximation and denoted FHNC/0, is quite adequate in describing correlations between the ^{3}He particles. It is also accurate for weakly correlated systems like the electron–hole mixture in the high–density limit. The single–loop exchange approximation amounts in

$$\tilde{\mathbf{X}}_{ee} = \mathbf{S_F} - \mathbf{I}, \quad \tilde{\mathbf{X}}_{de} = 0 \tag{2.45}$$

which immediately implies that

$$\tilde{\mathbf{A}} = \mathbf{S_F}; \quad \tilde{\mathbf{B}} = \mathbf{I} . \tag{2.46}$$

The static structure function of the noninteracting mixture is given by the diagonal matrix

$$\mathbf{S_F} \equiv \begin{pmatrix} S_F^{11}(k) & 0 \\ 0 & S_F^{22}(k) \end{pmatrix} , \tag{2.47}$$

where each $S_F^{\alpha\alpha}(k)$ is the static structure function of the corresponding noninteracting Fermi system,

$$S_F^{\alpha\alpha}(k) = \begin{cases} \frac{3k}{4k_F^\alpha} - \frac{k^3}{16(k_F^\alpha)^3}, & \text{if } k < 2k_F^\alpha \\ 1, & \text{otherwise.} \end{cases} \tag{2.48}$$

For bosons (like ^{4}He) $S_F^{\alpha\alpha}(k) \equiv 1$. Using these definitions, we can write the matrices $\mathbf{S}$ in Eq. (2.37) and $\tilde{\mathbf{X}}_{dd}$ in Eq. (2.38) in the form

$$\mathbf{S} = \mathbf{S_F} + \mathbf{S_F}\tilde{\mathbf{\Gamma}}_{dd}\mathbf{S_F} \tag{2.49}$$

$$\tilde{\mathbf{X}}_{dd} = \mathbf{S_F}^{-1} - \mathbf{S}^{-1} \tag{2.50}$$

and calculate the sums of nodal diagrams

$$\tilde{\mathbf{N}}_{de} = \tilde{\mathbf{\Gamma}}_{de} = \tilde{\mathbf{\Gamma}}_{dd}(\mathbf{S_F} - \mathbf{I})$$

$$\tilde{\mathbf{N}}_{ed} = \tilde{\mathbf{\Gamma}}_{ed} = (\mathbf{S_F} - \mathbf{I})\tilde{\mathbf{\Gamma}}_{dd}$$

$$\tilde{\mathbf{N}}_{ee} = \tilde{\mathbf{\Gamma}}_{ee} - \mathbf{S_F} + \mathbf{I} = (\mathbf{S_F} - \mathbf{I})\tilde{\mathbf{\Gamma}}_{dd}(\mathbf{S_F} - \mathbf{I})\,. \tag{2.51}$$

Furthermore, if we set $\tilde{\mathbf{N}}_{cc} = 0$ and ignore the contribution of elementary diagrams, we get a first approximation for the function $C^{\alpha\beta}(r)$ in Eq. (2.44) in momentum space

$$\tilde{C}^{\alpha\beta}(k) = S_F^{\alpha\alpha}\tilde{\Gamma}_{dd}^{\alpha\beta}(k)S_F^{\beta\beta} - \tilde{\Gamma}_{dd}^{\alpha\beta}(k) + (S_F^{\alpha\alpha} - 1)\left[\tilde{\Gamma}_{dd}^{\alpha\beta}(k)\right]^2\left(S_F^{\beta\beta} - 1\right) + 1\,. \tag{2.52}$$

2.1.3. *Euler equations in the single–loop approximation*

The stage is now set for the Euler equations (2.21) and (2.22) to be brought into numerically solvable form. In the following, the method is reviewed in the single–loop approximation starting from Eq. (2.50), but the extension to the complete set of FHNC equations of Eq. (2.39) is straightforward and leads to exactly the same structure of the Euler equations. [46]

Again, it is more convenient to work in momentum space and use the matrix representation we introduced. We define the free particle kinetic–energy matrix

$$\mathbf{H_1} \equiv \begin{pmatrix} t_\alpha & 0 \\ 0 & t_\beta \end{pmatrix}, \tag{2.53}$$

with $t_\alpha = \hbar^2 k^2/(2m_\alpha)$, and write the Euler equation (2.21) in matrix form,

$$\mathbf{S}' = -\frac{1}{4}\left[\mathbf{H_1}(\mathbf{S} - \mathbf{I}) + (\mathbf{S} - \mathbf{I})\mathbf{H_1}\right]\,. \tag{2.54}$$

Here $\mathbf{S}'$ is the matrix obtained by Fourier transforming $g'^{\alpha\beta}$, *i.e.*

$$S'^{\alpha\beta}(k) = \sqrt{\rho_\alpha\rho_\beta}\int d^3r e^{i\mathbf{k}\cdot\mathbf{r}} g'^{\alpha\beta}(r)\,. \tag{2.55}$$

To take the contributions from T_{JF} into account, we have to resort to a diagrammatic expansion. These contributions are obtained by replacing, in turn, in the diagrammatic expansion of the pair–distribution functions $g^{\alpha\beta}(r)$ every connected exchange loop $\ell^2(|\mathbf{r}_i - \mathbf{r}_j|\ k_F^\alpha)$ with a differentiated exchange loop $\frac{\hbar^2}{8m_\alpha}\nabla^2\ell^2(|\mathbf{r}_i - \mathbf{r}_j|\ k_F^\alpha)$. In momentum space, the equivalent statement is that we must replace, in turn, every occurrence of $S_F^{\alpha\alpha}(k)$ with $-\frac{\hbar^2}{4m_\alpha}k^2[S_F^{\alpha\alpha}(k) - 1]$, or in matrix representation

$$\mathbf{S_F}' = -\frac{1}{4}\left[\mathbf{H_1}(\mathbf{S_F} - \mathbf{I}) + (\mathbf{S_F} - \mathbf{I})\mathbf{H_1}\right]\,. \tag{2.56}$$

Again, we use this result, together with Eq. (2.54), to do the "priming" of the direct correlation function $\tilde{\mathbf{X}}_{dd}$ in Eq. (2.50),

$$\tilde{\mathbf{X}}'_{dd} = \mathbf{S}^{-1}\mathbf{S}'\mathbf{S}^{-1} - \mathbf{S_F}^{-1}\mathbf{S_F}'\mathbf{S_F}^{-1}\,. \tag{2.57}$$

We can write this also in the form

$$\tilde{\mathbf{X}}'_{dd} = \frac{1}{4}\left(\tilde{\mathbf{X}}_{dd}\mathbf{H_1} + \mathbf{H_1}\tilde{\mathbf{X}}_{dd}\right) + \frac{1}{2}\left(\mathbf{S}^{-1}\mathbf{H_1}\mathbf{S}^{-1} - \mathbf{S_F}^{-1}\mathbf{H_1}\mathbf{S_F}^{-1}\right). \quad (2.58)$$

The next task is to carry out the "priming operation" on the HNC equation, Eq. (2.40). We obtain

$$\Gamma'^{\alpha\beta}_{dd}(r) = \left[1 + \Gamma^{\alpha\beta}_{dd}(r)\right]\left\{V^{\alpha\beta}(r) + N'^{\alpha\beta}_{dd}(r) + E'^{\alpha\beta}_{dd}(r) - \left[\frac{\hbar^2}{8m_\alpha} + \frac{\hbar^2}{8m_\beta}\right]\nabla^2 u^{\alpha\beta}(r)\right\}. \quad (2.59)$$

This immediately gives

$$\begin{aligned} X'^{\alpha\beta}_{dd}(r) &= \Gamma'^{\alpha\beta}_{dd}(r) - N'^{\alpha\beta}_{dd}(r) \\ &= \left[1 + \Gamma^{\alpha\beta}_{dd}(r)\right]\left\{V^{\alpha\beta}(r) + E'^{\alpha\beta}_{dd}(r) - \left[\frac{\hbar^2}{8m_\alpha} + \frac{\hbar^2}{8m_\beta}\right]\nabla^2 u^{\alpha\beta}(r)\right\} \\ &\quad + \Gamma^{\alpha\beta}_{dd}(r) N'^{\alpha\beta}_{dd}(r). \end{aligned} \quad (2.60)$$

If we now use the HNC equation (2.40) to eliminate the pair correlations $u^{\alpha\beta}(r)$ from the equation,

$$u^{\alpha\beta}(r) = \ln\left[1 + \Gamma^{\alpha\beta}_{dd}(r)\right] - E^{\alpha\beta}_{dd}(r) - N^{\alpha\beta}_{dd}(r), \quad (2.61)$$

we find, after a few manipulations,

$$\begin{aligned} X'^{\alpha\beta}_{dd}(r) &= \left[1 + \Gamma^{\alpha\beta}_{dd}(r)\right]\left[V^{\alpha\beta}(r) + \Delta V^{\alpha\beta}(r)\right] + \left[\frac{\hbar^2}{2m_\alpha} + \frac{\hbar^2}{2m_\beta}\right]\left|\nabla\sqrt{1 + \Gamma^{\alpha\beta}_{dd}(r)}\right|^2 \\ &\quad + \Gamma^{\alpha\beta}_{dd}(r) w^{\alpha\beta}_I(r) - \left[\frac{\hbar^2}{8m_\alpha} + \frac{\hbar^2}{8m_\beta}\right]\nabla^2 X^{\alpha\beta}_{dd}(r). \end{aligned} \quad (2.62)$$

Here we have used the abbreviation

$$\Delta V^{\alpha\beta}(r) = E'^{\alpha\beta}_{dd}(r) + \left[\frac{\hbar^2}{8m_\alpha} + \frac{\hbar^2}{8m_\beta}\right]\nabla^2 E^{\alpha\beta}_{dd}(r) \quad (2.63)$$

for the contribution from triplet correlations and elementary diagrams, and introduced the so-called "induced potential"

$$w^{\alpha\beta}_I(r) = N'^{\alpha\beta}_{dd}(r) + \left[\frac{\hbar^2}{8m_\alpha} + \frac{\hbar^2}{8m_\beta}\right]\nabla^2 N^{\alpha\beta}_{dd}(r). \quad (2.64)$$

Finally, we define a static, effective particle–hole interaction $V^{\alpha\beta}_{p-h}(r)$ as

$$V^{\alpha\beta}_{p-h}(r) = X'^{\alpha\beta}_{dd}(r) + \left[\frac{\hbar^2}{8m_\alpha} + \frac{\hbar^2}{8m_\beta}\right]\nabla^2 X^{\alpha\beta}_{dd}(r). \quad (2.65)$$

This can be solved from Eq. (2.62), and the result is

$$\begin{aligned} V^{\alpha\beta}_{p-h}(r) &= \left[1 + \Gamma^{\alpha\beta}_{dd}(r)\right]\left[V^{\alpha\beta}(r) + \Delta V^{\alpha\beta}(r)\right] + \left[\frac{\hbar^2}{2m_\alpha} + \frac{\hbar^2}{2m_\beta}\right]\left|\nabla\sqrt{1 + \Gamma^{\alpha\beta}_{dd}(r)}\right|^2 \\ &\quad + \Gamma^{\alpha\beta}_{dd}(r) w^{\alpha\beta}_I(r). \end{aligned} \quad (2.66)$$

At this point, $V^{\alpha\beta}_{p-h}(r)$ serves only as a convenient quantity with which to formulate the Euler equations and to design a numerical algorithm to solve them, but later on we will draw a connection to the linear response function.

Defining the momentum–space matrix $\tilde{\mathbf{V}}_{p-h} \equiv \left(\tilde{V}^{\alpha\beta}_{p-h}(k)\right)$ we can combine Eq. (2.65) with Eq. (2.58), and get the particle-hole potentials in terms of the structure functions,

$$\mathbf{S}^{-1}\mathbf{H_1}\mathbf{S}^{-1} - \mathbf{S_F}^{-1}\mathbf{H_1}\mathbf{S_F}^{-1} = 2\tilde{\mathbf{V}}_{p-h}\,. \tag{2.67}$$

This is the final form of the Euler equation in momentum space, a generalization of the Euler equation derived in the "paired–phonon analysis" (PPA). [20–22] A structurally identical Euler equation has also been derived in applying the variational approach to nonuniform systems. [43,53] The equation makes no assumption on the inclusion of multi–body correlations and/or elementary diagrams. The only approximation that has been made is the single–loop approximation for the exchange diagrams. One can make more sophisticated approximations for these exchange contributions, only to arrive at an essentially identical form, the only difference being that the interaction matrix $\tilde{\mathbf{V}}_{p-h}$ will be defined differently. [46]

We still need to calculate the induced interaction. This derivation is done in momentum space. Once again, we use the matrix notation for $w_I^{\alpha\beta}(r)$,

$$\tilde{\mathbf{w}}_\mathbf{I} = -\frac{1}{4}\left[\mathbf{H_1}\tilde{\mathbf{N}}_{dd} + \tilde{\mathbf{N}}_{dd}\mathbf{H_1}\right] + \tilde{\mathbf{N}}'_{dd}\,. \tag{2.68}$$

The right–hand side can be divided into two parts. The first part contains the direct correlation function $\tilde{\mathbf{X}}$ and is equal to the particle–hole potential $\tilde{\mathbf{V}}_{p-h}$, the second part depends on the distribution functions $\tilde{\mathbf{\Gamma}}$,

$$\tilde{\mathbf{w}}_\mathbf{I} = -\tilde{\mathbf{V}}_{p-h} - \frac{1}{4}\left[\mathbf{H_1}\tilde{\mathbf{\Gamma}}_{dd} + \tilde{\mathbf{\Gamma}}_{dd}\mathbf{H_1}\right] + \tilde{\mathbf{\Gamma}}'_{dd}\,. \tag{2.69}$$

Using then Eq. (2.49), together with the Euler equations (2.54) and Eq. (2.56), we can calculate $\tilde{\mathbf{\Gamma}}'$ and insert it back into Eq. (2.69). The result can be written in the form

$$\tilde{\mathbf{w}}_\mathbf{I} = -\tilde{\mathbf{V}}_{p-h} - \frac{1}{2}\left[\mathbf{S_F}^{-1}\mathbf{H_1}\tilde{\mathbf{\Gamma}}_{dd} + \tilde{\mathbf{\Gamma}}_{dd}\mathbf{H_1}\mathbf{S_F}^{-1}\right]\,. \tag{2.70}$$

For a given initial guess of $\tilde{\mathbf{\Gamma}}_{dd}$ and a given choice of the elementary diagrams and three–body correlation functions, Eqs. (2.63), (2.66), (2.67), and (2.70) form a closed set of equations that can be solved iteratively until convergence is reached.

The iteration method suggested by the PPA is the most effective procedure known to solve the Euler equations numerically. Physically, the momentum–space equations are closely related to the excitation spectrum of collective modes and, therefore, of independent interest. In particular, instabilities of the variational problem are reflected in the "softening" of the predicted collective modes, which will eventually preclude solving the Euler equation in cases where the system is locally unstable. Nevertheless, it is of interest to study the Euler equation also in coordinate space, because one should expect that the pair–distribution function is determined, for small distances, by a two–body Schrödinger equation. Related to this, problems are to be anticipated: it has been pointed out repeatedly in the literature (Refs. 49, 54, 55) that there is no way to formulate an Euler equation for a Fermi system that is consistent both in coordinate and momentum space without including infinite sets of "elementary" exchange diagrams.

In the case of a very dilute fermion component, for example one or two ^{3}He impurities in ^{4}He, we can set $\mathbf{S_F} \equiv \mathbf{I}$, and Eq. (2.70) can be rewritten in coordinate space as

$$\left(\frac{\hbar^2}{2m_\alpha}+\frac{\hbar^2}{2m_\beta}\right)\nabla^2\sqrt{1+\Gamma^{\alpha\beta}(r)} = \left[V^{\alpha\beta}(r)+\Delta V^{\alpha\beta}(r)+w_I^{\alpha\beta}(r)\right]\sqrt{1+\Gamma^{\alpha\beta}(r)}\,. \tag{2.71}$$

This has the desired form of a two–body Schrödinger equation. The induced potential that takes into account the many–body effects in this "two–body" problem can also be derived without any references to variational wave functions as we will show later on.

2.2. *Variational energy in the HNC approximation*

Let us go back to Eq. (2.15) and evaluate the ground–state energy using our optimized correlation functions. We restrict the evaluation to two–body terms and to those elementary diagrams that can be constructed from two–body correlations alone. Starting from Eq. (2.15), we can write the correlation energy per particle as

$$\begin{aligned}\frac{E_{(2)}}{N} &= \sum_{\alpha\beta}\frac{\rho_\alpha\rho_\beta}{2\rho}\int d^3r g^{\alpha\beta}(r)\left\{V^{\alpha\beta}(r)-\left[\frac{\hbar^2}{8m_\alpha}+\frac{\hbar^2}{8m_\beta}\right]\nabla^2u^{\alpha\beta}(r)\right\}\\ &+\sum_\alpha\frac{\hbar^2\rho_\alpha^2}{8m_\alpha\rho}\int d^3r\Gamma_{dd}^{\alpha\alpha}(r)\nabla^2 g_F^{\alpha\alpha}(r)\,.\end{aligned} \tag{2.72}$$

The pair–correlation functions are eliminated through the use of Eq. (2.61), and writing the full pair–distribution functions in the form of Eq. (2.43). Then

$$\begin{aligned}\frac{E_{(2)}}{N} &= \sum_{\alpha\beta}\frac{\rho_\alpha\rho_\beta}{2\rho}\int d^3r\Bigg\{g^{\alpha\beta}(r)\left[V^{\alpha\beta}(r)+\left[\frac{\hbar^2}{8m_\alpha}+\frac{\hbar^2}{8m_\beta}\right]\nabla^2E_{dd}^{\alpha\beta}(r)\right]\\ &+\left[g_F^{\alpha\beta}(r)+C^{\alpha\beta}(r)\right]\left[\frac{\hbar^2}{2m_\alpha}+\frac{\hbar^2}{2m_\beta}\right]\left|\nabla\sqrt{1+\Gamma_{dd}^{\alpha\beta}(r)}\right|^2\Bigg\}\\ &+\sum_{\alpha\beta}\frac{\rho_\alpha\rho_\beta}{2\rho}\left[\frac{\hbar^2}{8m_\alpha}+\frac{\hbar^2}{8m_\beta}\right]\int d^3r\left\{g^{\alpha\beta}(r)\nabla^2N_{dd}^{\alpha\beta}(r)-C^{\alpha\beta}(r)\nabla^2\Gamma_{dd}^{\alpha\beta}(r)\right\}.\end{aligned} \tag{2.73}$$

It is possible, although complicated, to derive the Euler equation directly from this approximate energy expression. The FHNC equations could be treated as constraints included in the optimization procedure through Lagrange multipliers. However, the derivation of the Euler equation from the *exact* energy functional allows us to verify that the approximations one might subsequently choose to use do not change the character of the equation nor its solution. To be specific, the only point where approximations enter in our derivations is the correction term $\Delta V^{\alpha\beta}(r)$, c.f. Eq. (2.66), to the particle–hole interaction. This term is short–ranged. Hence, the inclusion of elementary diagrams and triplet correlations has only the effect of introducing an additive correction to the bare interaction. It may cause quantitative changes in the results, but including them will not change the analytic structure of the Euler equation (2.67).

2.3. *Collective excitations and stability*

In studying the stability of quantum fluid mixtures, we distinguish between *global* and *local* instabilities. A global instability means that there exists a phase of lower energy; the system can reach it through a *macroscopic* perturbation of its configuration. When the chemical potential of a particle in the mixture becomes higher than in the pure phase, the mixture will phase separate, but it takes a *finite* external perturbation and latent heat to initiate the transition. Theoretically, one can detect such an instability only by comparing the ground–state energies of the two different phases.

A local instability occurs when the system becomes unstable against *infinitesimal* fluctuations about its equilibrium configuration. Such an instability is indicated by the softening of a collective excitation and should also be reflected in a divergence of the theory. Experimentally, it is very difficult to get close to this type of an instability because it involves generating an over–saturated mixture. Theoretically, this instability is interesting because it provides a consistency test for the theoretical description.

The necessary condition for the existence of a solution to the Euler equation (2.67) is obtained by diagonalizing the Feynman–like matrix

$$\text{diag}\left(\mathbf{H_1 S^{-1}}\right) = \mathbf{D}^{1/2}\,, \tag{2.74}$$

and requiring that the eigenvalues are real. This is equivalent to require that the eigenvalues of the matrix

$$\mathbf{H_1}\left[\,2\tilde{\mathbf{V}}_{p-h} + \mathbf{S_F}^{-1}\mathbf{H_1}\mathbf{S_F}^{-1}\right] \tag{2.75}$$

are positive. The diagonal elements of $\mathbf{D}^{1/2}$ are the Feynman approximations of the two collective–excitation modes in the mixture; the zeroth and second sound in ^{3}He–^{4}He mixtures or the optical and acoustic modes in the electron–hole liquid. As discussed in the introduction, in the one–component Bose system the Feynman approximation is exact in the long–wavelength limit, provided that the structure function $S(k)$ is known exactly, and the stability criterion is satisfied in the density regime where the fluid is stable against density fluctuations. The second eigenvalue appearing in a dilute mixture is related to concentration fluctuations. For the stability of the mixed phase it is sufficient to require the positivity of the determinant

$$\det\left[\tilde{\mathbf{V}}_{p-h} + \frac{1}{2}\mathbf{S_F}^{-1}\mathbf{H_1}\mathbf{S_F}^{-1}\right] > 0\,. \tag{2.76}$$

In the ^{3}He–^{4}He mixture, this reduces in the long–wavelength limit, $k \to 0+$, to

$$\det\left[\tilde{\mathbf{V}}_{p-h}(0+)\right] + \frac{4}{9}\frac{\hbar^2 k_{F3}^2}{m_3}\,\tilde{V}^{44}_{\rm p-h}(0+) > 0 \tag{2.77}$$

where the indices 3 and 4 refer to ^{3}He and ^{4}He, respectively.

From the hydrodynamic point of view, a similar stability condition can be derived from the density derivatives. The mixture is stable against *infinitesimal* changes in density and concentration if the second–derivative matrix of the energy with respect to concentration and particle density is positive definite. We write the total energy as a function of the ^{4}He and ^{3}He densities,

$$E = T_F + E_c[\rho_3, \rho_4]\,, \tag{2.78}$$

where E_c is the correlation energy and T_F the kinetic energy of the noninteracting fermion component. We then calculate the incompressibility matrix

$$\left(\frac{\sqrt{\rho_\alpha\rho_\beta}}{\Omega}\frac{\partial^2 E}{\partial\rho_\alpha\partial\rho_\beta}\right)_{\alpha\beta} = \begin{pmatrix} \hat{V}^{33}(0+) + \frac{1}{3}\frac{\hbar^2 k_{F3}^2}{m_3} & \hat{V}^{34}(0+) \\ \hat{V}^{34}(0+) & \hat{V}^{44}(0+) \end{pmatrix}, \tag{2.79}$$

with

$$\hat{V}^{\alpha\beta}(0+) \equiv \frac{\sqrt{\rho_\alpha\rho_\beta}}{\Omega}\frac{\partial^2 E_c}{\partial\rho_\alpha\partial\rho_\beta}\,. \tag{2.80}$$

Hydrodynamic stability requires then that

$$\det\left[\hat{\mathbf{V}}(\mathbf{0}+)\right] + \frac{1}{3}m_3 c_2^2\,\hat{V}^{44}(0+) > 0\,. \tag{2.81}$$

A comparison of Eqs. (2.79) and (2.77) suggests the identification $\tilde{V}^{\alpha\beta}_{\rm p-h}(0+) = \hat{V}^{\alpha\beta}(0+)$. In fact, the $\hat{V}^{\alpha\beta}(0+)$ are the same as the long–wavelength limits of the particle–hole interactions $\tilde{V}^{\alpha\beta}_{\rm p-h}(0+)$ in case the Jastrow–Feenberg wave function is optimized for all n–body correlations, and if all elementary diagrams are included: in other words, in an exact theory. The positivity of the matrix (2.79) is clearly the correct stability condition; the discrepancy in the free Fermi gas term is due to the so–called "mean–spherical approximation" leading to the term $\frac{1}{2}\mathbf{S_F}^{-1}\mathbf{H_1}\mathbf{S_F}^{-1}$. This discrepancy can be cured by including correlated basis functions, as discussed in the next section.

3. Correlated basis functions

The Jastrow–Feenberg variational theory for bosons is a systematic method which leads, in principle, to the exact wave function. This is not the case for Fermi systems as we saw by comparing the two ways of analyzing the stability of a mixture — the positivity of the second derivative of the energy, Eq. (2.81), and the existence of a solution to the HNC/FHNC Euler equations, Eq. (2.77). The Jastrow–Feenberg wave function (2.3) replaces the particle–hole propagator with a "collective" or "mean–spherical" approximation (MSA). [56,57] This is already seen in the weakly–interacting limit and has, among others, the consequence that the wave function (2.3) does not reproduce the correct high–energy limit of the correlation energy of the electron gas.

The most complete analysis between variational and perturbative many–body theories has been worked out, in the case of a one–component Bose system, in the parquet–diagram theory. [58–61] It has been shown that the HNC/EL theory is equivalent to a self–consistent, approximate summation of ring and ladder diagrams. Similar analysis holds also for mixtures and Fermi systems.

The formally systematic way to go beyond the Jastrow–Feenberg theory is provided by the CBF theory. [19,35] The theory extends the use of the correlation operator $\exp\left[\frac{1}{2}U(\{\mathbf{r}_i^{(\alpha)}\})\right]$ to generate a nonorthogonal basis

$$|\Psi_{\mathbf{m}}\rangle = \mathrm{I}_{\mathbf{mm}}^{-1/2} \exp\left[\frac{1}{2}U(\{\mathbf{r}_i^{(\alpha)}\})\right] \Big|\Phi_{\mathbf{m}}\Big\rangle \tag{3.1}$$

$$\mathrm{I}_{\mathbf{mm}} = \Big\langle\Phi_{\mathbf{m}}\Big| \exp\left[U(\{\mathbf{r}_i^{(\alpha)}\})\right] \Big|\Phi_{\mathbf{m}}\Big\rangle \tag{3.2}$$

of the Hilbert space, where $\{|\Phi_{\mathbf{m}}\rangle\}$ is a complete set of Slater determinants. The relation of the variational theory to the CBF theory is not immediately obvious, and the actual calculations are quite tedious as they require both an (F)HNC analysis [62] of the effective interactions defined by the CBF theory *and* an analysis of ring diagrams in correlated basis functions [63] to all orders. Nevertheless, the result is quite plausible: the effect of summing all ring diagrams in the CBF theory is simply to remove the collective approximation.

The random–phase approximation (RPA) sums the ring diagrams for the density–density response function which, for a mixture, is given by a 2×2 matrix,

$$\boldsymbol{\chi}(k,\omega) = \boldsymbol{\chi}_0(k,\omega) + \boldsymbol{\chi}_0(k,\omega)\tilde{\mathbf{V}}_{\mathrm{p-h}}(k)\boldsymbol{\chi}(k,\omega)\,. \tag{3.3}$$

Here $\boldsymbol{\chi}_0(k,\omega)$ is the density–density response function of the noninteracting system and $\tilde{\mathbf{V}}_{p-h}(k)$ serves as an energy–independent, local approximation for the particle–hole interaction. Considering the ^{3}He–^{4}He mixture, we have

$$\chi_0(k,\omega) = \begin{pmatrix} \chi_0^{33}(k,\omega) & 0 \\ 0 & \chi_0^{44}(k,\omega) \end{pmatrix}. \tag{3.4}$$

For noninteracting bosons, $\chi_0(k,\omega)$ is given by

$$\chi_0^{44}(k,\omega) = \frac{2t_4(k)}{(\hbar\omega + i\eta)^2 - t_4^2(k)} \tag{3.5}$$

with the free particle kinetic energy t_4, see Eq. (2.53). For fermions, $\chi_0(k,\omega)$ is the Lindhard function. The "mean–spherical approximation" is to replace it with

$$\chi_{0,\mathrm{MSA}}^{33}(k,\omega) = \frac{2t_3(k)}{(\hbar\omega + i\eta)^2 - \hbar^2\omega_3^2(k)}\,. \tag{3.6}$$

Here

$$\hbar\omega_3(k) \equiv t_3(k)/S_F(k) \tag{3.7}$$

is an effective collective energy, determined in such a way that $\chi^{33}_{0,\mathrm{MSA}}$ satisfies the first two energy–weighted sum rules.

Once the response function is known, the static structure function can be calculated from the dynamic structure function

$$\mathbf{S}(k,\omega) = -\frac{1}{\pi}\Im m\,\chi(k,\omega) \tag{3.8}$$

through the fluctuation–dissipation theorem

$$\mathbf{S}(k) = \int_0^\infty d(\hbar\omega)\,\mathbf{S}(k,\omega)\,. \tag{3.9}$$

The use of $\chi^{33}_{0,\mathrm{MSA}}(k,\omega)$ instead of the Lindhard function in Eq. (3.3) leads immediately to the Euler equation (2.67) of the (F)HNC/EL theory. [24] Thus, we have located the problem with the stability criterion of the fermionic part into the mean–spherical approximation used in the FHNC/EL theory. This can be cured through the use of the ordinary RPA equations of motion with the Lindhard function. Doing this, however, adds frequency as a new variable into the Euler equation. The frequency integral (3.9) must be calculated in every iteration cycle in solving the Euler equation (2.67).

At the same level of approximations, one can calculate the correction to the ground–state energy due to the Lindhard function. This is done by adding to the ground–state energy a correction due to all ring diagrams and then subtracting the same set of diagrams, but evaluated with the collective approximation (3.6) replacing the Lindhard function. The coupling–constant integration gives then the results [24]

$$\Delta E_{\mathrm{RPA}} = E_{\mathrm{RPA}} - E^{\mathrm{MSA}}_{\mathrm{RPA}} = \frac{1}{2\rho}\Im m \int \frac{d^3k\,d(\hbar\omega)}{(2\pi)^4} \ln\left[D(k,\omega)/D_{\mathrm{MSA}}(k,\omega)\right] \tag{3.10}$$

with the notation

$$\begin{aligned} D(k,\omega) = &\left[1-\chi^{33}_0(k,\omega)\tilde V^{33}_{\mathrm{p-h}}(k)\right]\left[1-\chi^{44}_0(k,\omega)\tilde V^{44}_{\mathrm{p-h}}(k)\right] \\ &-\chi^{33}_0(k,\omega)\chi^{44}_0(k,\omega)[\tilde V^{34}_{\mathrm{p-h}}(k)]^2 \end{aligned} \tag{3.11}$$

and the same for $D_{\mathrm{MSA}}(k,\omega)$ with $\chi^{33}_{0,\mathrm{MSA}}(k,\omega)$].

4. Results for dilute ^{3}He–^{4}He mixtures in 2D and 3D

4.1. *Pure liquid ^{4}He: a performance test*

The variational theory has been used extensively to study one–component Bose fluids, [20,22,23,67–69] which form a limiting case for the formalism we presented. In

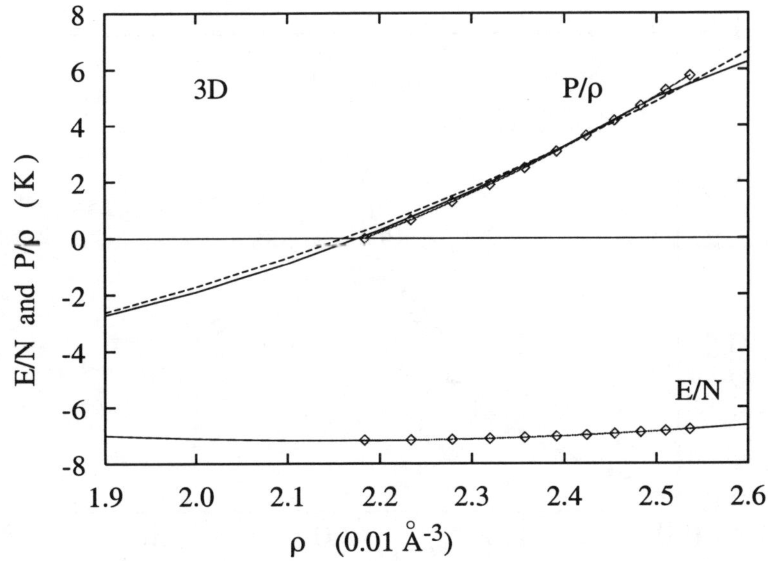

Fig. 1. Energy/particle and pressure/density as a function of density for liquid ^{4}He in three dimensions as obtained from the current theory (solid lines), compared with the experiments of Ref. 2 (lines with diamonds). The dashed lines for the pressure are taken from a cubic polynomial fit to the calculated energy per particle.

such a system, the Euler equation (2.67) can be formulated in coordinate space as a Schrödinger–like equation

$$\left[-\frac{\hbar^2}{m}\nabla^2 + V(r) + \Delta V(r) + w_{\rm I}(r)\right]\sqrt{g(r)} = 0\,, \tag{4.1}$$

and the induced potential (2.70) takes the form

$$\begin{aligned} w_{\rm I}(k) &= -V_{\rm p-h}(k) - \frac{\hbar^2 k^2}{2m}\left[S(k)-1\right] \\ &= -\frac{\hbar^2 k^2}{2m}\left[S(k)-1\right] - \frac{\hbar^2 k^2}{4m}\left[\frac{1}{S^2(k)}-1\right]. \end{aligned} \tag{4.2}$$

Again, $\Delta V(r)$ is the additive correction term arising from triplet correlations and elementary diagrams. To return to the points we made earlier, this coordinate–space formulation of the Euler equation can readily be identified with the boson Bethe–Goldstone equation which sums the dominant diagrams in the strong–coupling limit. On the other hand, the Euler equation (2.67) can also be written in momentum space in terms of the static structure function $S(k)$,

$$S(k) = \frac{k}{\sqrt{k^2 + \frac{4m}{\hbar^2}V_{\rm p-h}(k)}}\,. \tag{4.3}$$

This equation is formally identical to the boson–RPA expression for $S(k)$. However, and as we have seen, the current theory goes well beyond that simple description

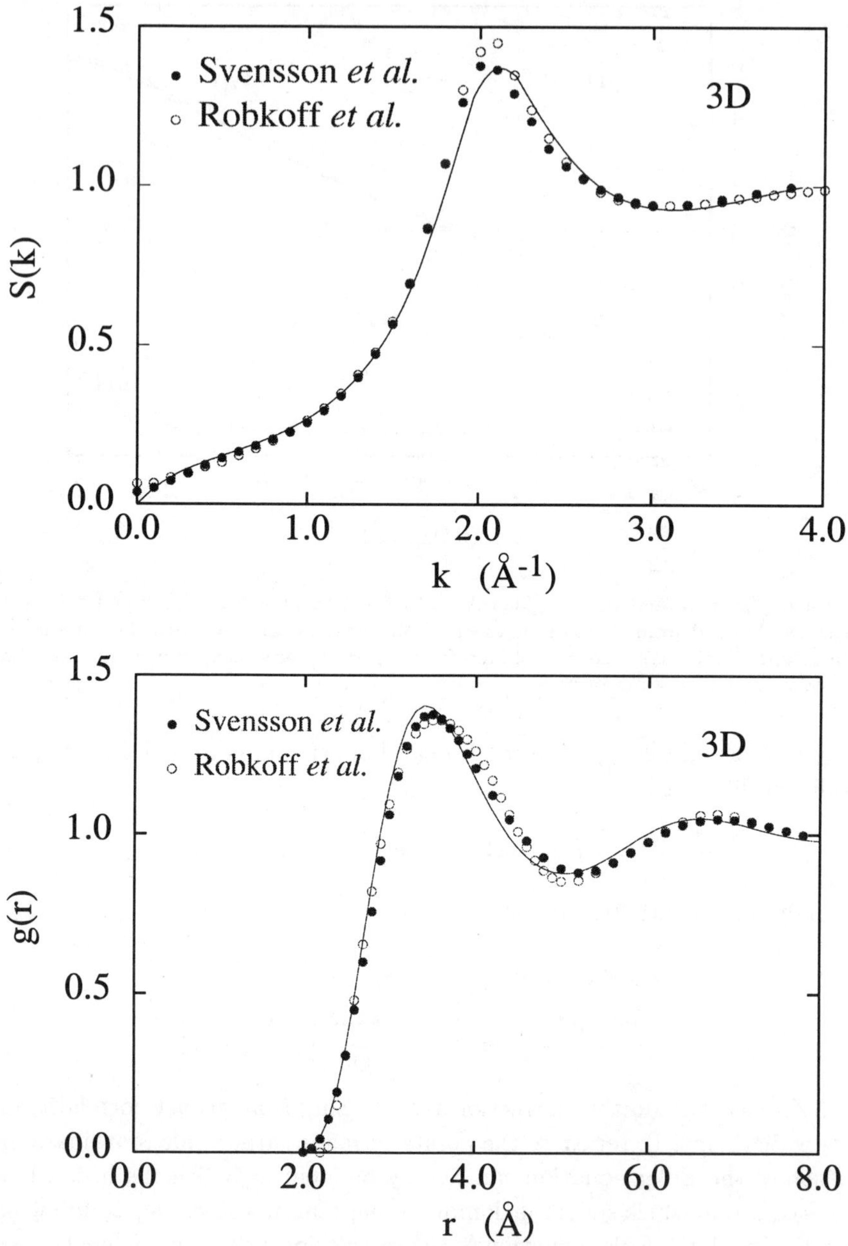

Fig. 2. Calculated static structure and pair–distribution functions in 3D. Our results (solid lines) are given at the density $\rho = 0.022\text{Å}^{-3}$ (near equilibrium), and compared with the experimental data by Svensson *et al.* [64] (dots) and Robkoff *et al.* [65] (circles).

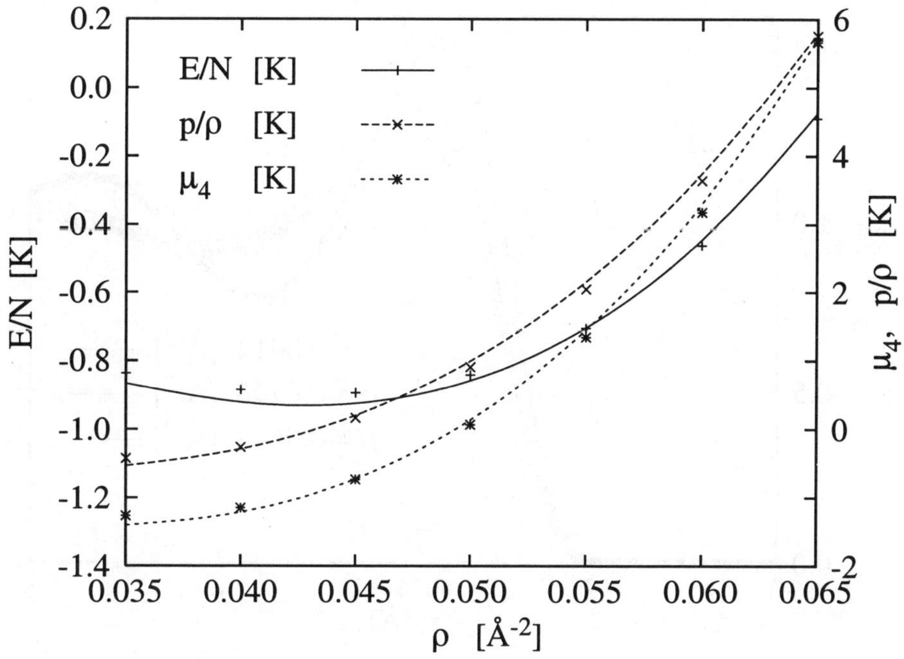

Fig. 3. HNC/EL results for the energy per particle E/N, (solid line, left scale), pressure per density P/ρ (long–dashed line, right scale), and chemical potential μ_4 (short–dashed line, right scale) in 2D. Also shown are the DMC results of Ref. 66 for the energy (+–symbols), pressure per density (crosses), and chemical potential (stars). All energies are given in Kelvins.

by supplementing the RPA with a microscopic theory of the effective (or "particle–hole") interaction,

$$V_{\text{p-h}}(r) = g(r)\left[V(r) + \Delta V(r)\right] + \frac{\hbar^2}{m}\left|\nabla\sqrt{g(r)}\right|^2 + \left[g(r) - 1\right] w_{\text{I}}(r)\,. \tag{4.4}$$

Consequently, we conclude that the theory sums *both* important sets of diagrams self–consistently.

We already mentioned that a rewarding and physically important feature of the HNC/EL equations is that there are no solutions to the equations when the system under consideration is unstable; for example, there are no uniform solutions to be found when some other state becomes energetically more favorable than the uniform ground state we started from. Having the low–density limit of liquid ^{4}He in mind, this is the case at the so–called spinodal instability where the speed of sound becomes zero (compressibility diverges), or using other terms, the second derivative of the energy with respect to density becomes zero. Physically speaking, this means that lowering the density further would cause the liquid to break into droplets. Having identified

$$\lim_{k\to 0} V_{\text{p-h}}(k) = mc^2\,, \tag{4.5}$$

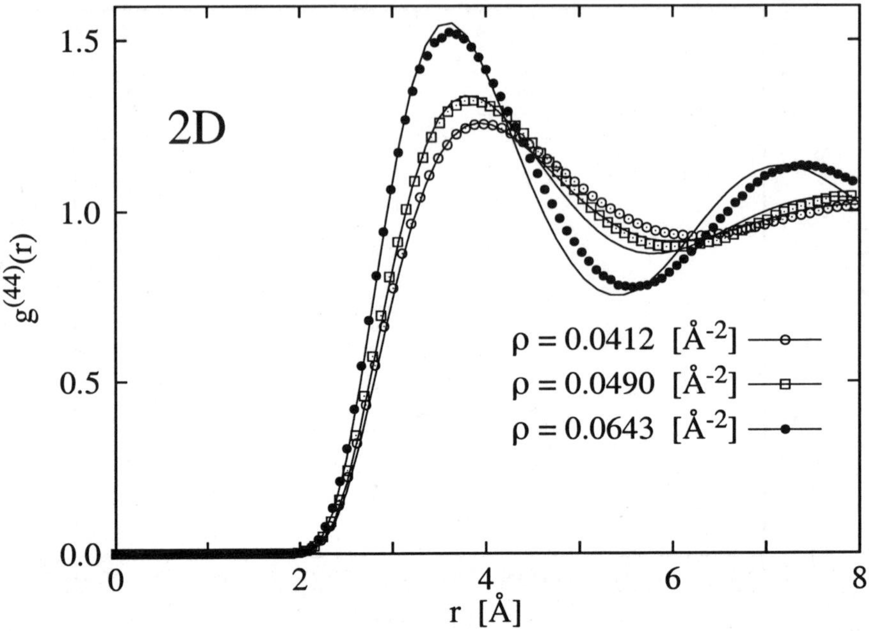

Fig. 4. HNC/EL results for the pair–distribution function $g^{44}(r)$ (solid lines), compared with DMC data [66] (markers) at the densities $\rho = 0.0412$ Å^{-2}, $\rho = 0.0490$ Å^{-2}, and $\rho = 0.0643$ Å^{-2}.

we see directly from Eq. (4.3) that the existence of a real solution for $S(k)$ requires that $c > 0$.

Once the ground–state structure (and thus also the equation of state) is known, we can proceed to calculate other thermodynamic quantities of interest. For example, the pressure of the system is obtained from the derivative of the energy per particle with respect to density,

$$\frac{P}{\rho} = \rho \left[\frac{d(E/N)}{d\rho} \right]_N , \tag{4.6}$$

hence also the pressure can be expressed entirely in terms of ground–state structure functions. The chemical potential of the system is defined as $\mu = E/N + P/\rho$.

Figs. 1 and 2 show some of the results obtained using the HNC/EL formalism for three–dimensional liquid ^{4}He, compared with experimental energies and pressures. The agreement is excellent, but here we must point out that the calculated four- and five-body elementary diagrams are scaled with a single free parameter in such a way that the energy at the saturation density agrees with experiments. [24] (Otherwise the qualitative agreement would remain, but the quantitative agreement would not be quite this good. The validity of this scaling has been discussed in length in Ref. 23). In Figs. 3 and 4, we show the corresponding results for two–dimensional liquid ^{4}He, and compare them with the Monte Carlo results by Giorgini, Boronat, and Casulleras. [66] Again, the agreement is rather rewarding, and here there are no free parameters left, the scaling of the elementary diagrams being fixed by the three–dimensional results.

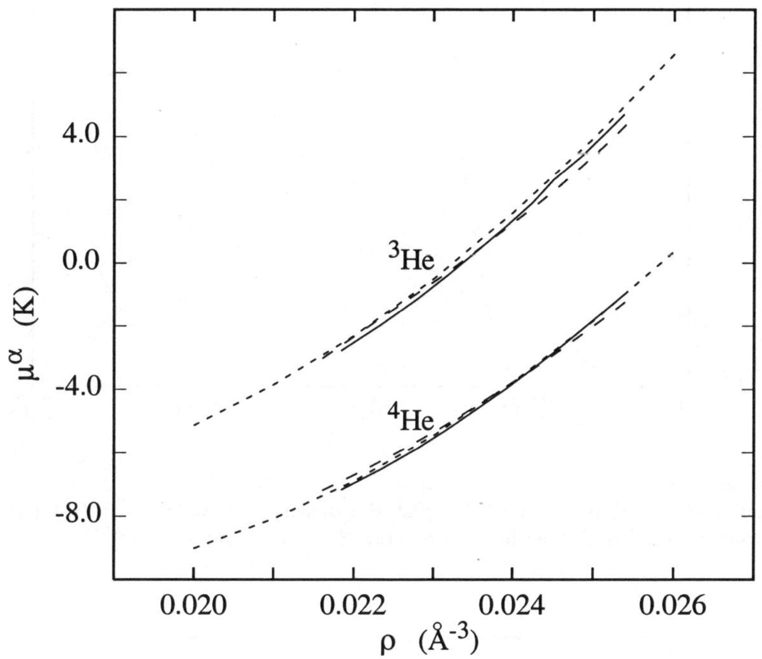

Fig. 5. Chemical potentials of ^{4}He and a ^{3}He impurity in ^{4}He as a function of density. The short–dashed lines correspond to results obtained from a calculation of the analytic derivatives of the energy/particle with respect to density. The long–dashed lines give the small–concentration limit of the mixture calculation, and the solid lines correspond to the experimental results by Ebner and Edwards. [2]

4.2. *Single–impurity limit*

Having a more general analysis of ^{3}He–^{4}He mixtures in mind, it is nevertheless instructive to study also the low–concentration limit, even so that one examines the properties of only a single impurity, or calculates the effective interactions between isolated pairs of impurities. In the limit of a single ^{3}He atom in liquid ^{4}He, the trial wave function is obtained from the mixture wave function by omitting all correlation functions containing two or more ^{3}He indices, and by setting the Slater determinant equal to one. As a convention, we denote the impurity coordinate with $\mathbf{r}_0$, in distinction from the ^{4}He coordinates $\mathbf{r}_i$ with $1 \le i \le N_4 \equiv N$. The wave function describing the system consisting of N ^{4}He particles (the background) and one ^{3}He impurity is then

$$\Psi^I_{N+1}(\mathbf{r}_0, \mathbf{r}_1, ..., \mathbf{r}_N) = \exp \frac{1}{2} \Big[\sum_{j=1}^{N} u^{34}(\mathbf{r}_0, \mathbf{r}_j) + \frac{1}{2!} \sum_{i,j=1}^{N}{}' u^{44}(\mathbf{r}_i, \mathbf{r}_j)$$

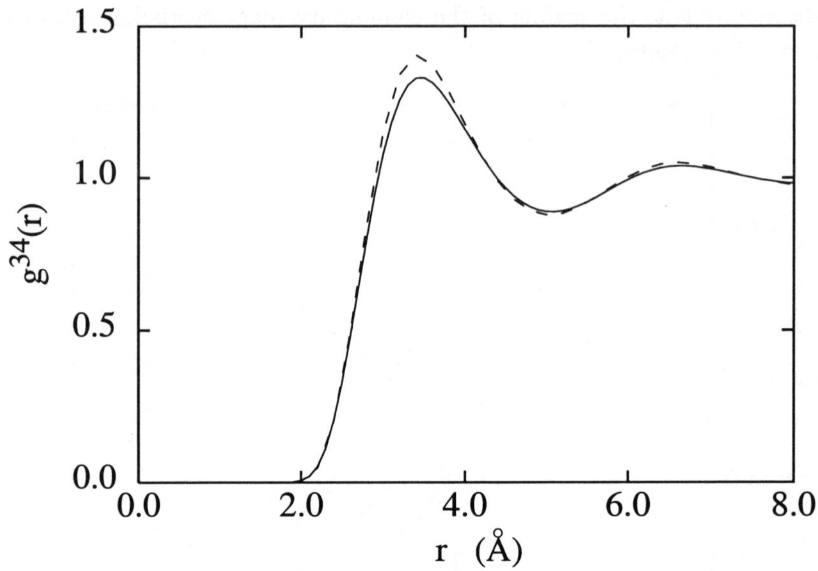

Fig. 6. Radial distribution function $g^{34}(r)$ at the density $\rho = 0.022$ Å^{-3} (solid line) for 0.1 percent concentration. The dashed line represents $g^{44}(r)$ at the same density.

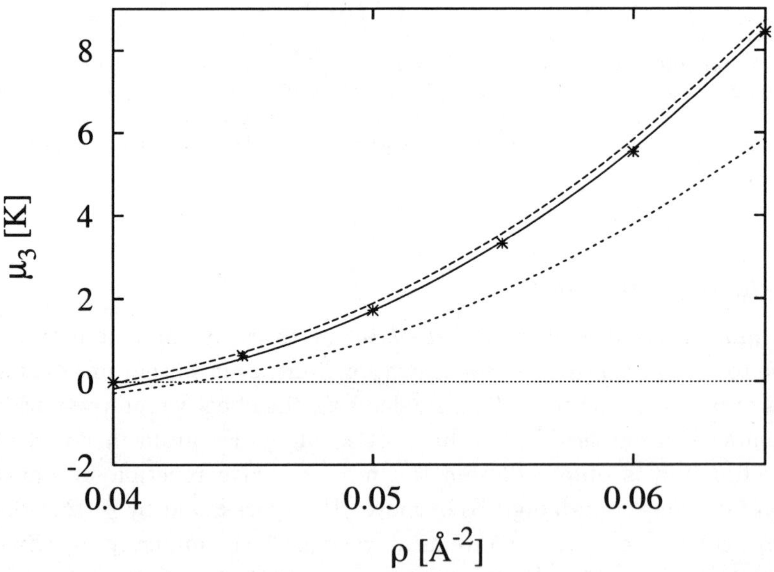

Fig. 7. Chemical potentials of the ^{3}He impurity as a function of density. The result of the full calculation (solid line) is compared with the average correlation approximation (long–dashed line). The stars give the average correlation approximation obtained from Monte Carlo simulations. [66] Also plotted in the figure is the pressure/density (short–dashed line).

$$+\frac{1}{2!}\sum_{i,j=1}^{N}{}' u^{344}(\mathbf{r}_0,\mathbf{r}_j,\mathbf{r}_k)+\frac{1}{3!}\sum_{i,j,k=1}^{N}{}' u^{444}(\mathbf{r}_i,\mathbf{r}_j,\mathbf{r}_k)\Big]. \tag{4.7}$$

Indices 3 and 4 refer to ^{3}He and ^{4}He, respectively and the prime in the summation means that no summation indices can be the same. The chemical potential of the impurity is given by

$$\mu_3^I = \frac{\langle\Psi_{N+1}^I|H_{N+1}^I|\Psi_{N+1}^I\rangle}{\langle\Psi_{N+1}^I|\Psi_{N+1}^I\rangle} - \frac{\langle\Psi_N|H_N|\Psi_N\rangle}{\langle\Psi_N|\Psi_N\rangle} \equiv E_{N+1}^I - E_N\,. \tag{4.8}$$

Here, H_N corresponds to the Hamiltonian of the background liquid with N particles and H_{N+1}^I the Hamiltonian of the whole $(N+1)$–body system.

The correlations between the impurity and background atoms are again determined variationally. Because the background energy does not depend on the presence of the impurity, solving the Euler equation for the impurity is equivalent to minimizing the chemical potential

$$\frac{\delta\mu_3^I}{\delta u^{34}(\mathbf{r}_0,\mathbf{r}_1)} = 0\,, \qquad \frac{\delta\mu_3^I}{\delta u^{344}(\mathbf{r}_0,\mathbf{r}_1,\mathbf{r}_2)} = 0\,. \tag{4.9}$$

The structure function can again be calculated from the pair–distribution function, but now the Fourier transform must be taken a bit differently,

$$S^{34}(k) = \rho_4 \int d^3r e^{i\mathbf{k}\cdot\mathbf{r}}\left[g^{34}(r) - g^{34}(\infty)\right]\,. \tag{4.10}$$

Note that the normalization factor is ρ_4 and not $\sqrt{\rho_3\rho_4}$ as for the mixture. The value of the impurity structure function at the origin gives the so–called volume excess factor, [5]

$$S^{34}(0+) = -\beta\,. \tag{4.11}$$

The two–body Euler equation for the impurity can be obtained directly from the mixture Euler equation (2.67)by setting $S_F(k) = S^{33}(k) = g^{33}(r) = 1$,

$$\Gamma^{\alpha\beta}(r) = g^{\alpha\beta}(r) - 1\,, \tag{4.12}$$

and recalling the above change in normalization. For future discussion, we prefer, however, to be slightly more flexible and allow for a dilute gas of *weakly correlated* fermions. The low–concentration limit of Eqs. (2.67) and (2.70) is then taken in such a way that all quantities contain no more than one (or two for the two-impurity case) *dynamical* impurity correlations, but they may still be correlated *statistically* to arbitrary orders. This implies, among others, that $S^{34}(k) = S_F(k)\tilde{\Gamma}^{34}(k)$. The rigorous single–impurity limit is then obtained by setting $S_F(k) = 1$.

To see how this works, we start from the mixture Euler equation (2.67). The single–impurity limit implies that there cannot be two dynamically correlated ^{3}He atoms present, hence we can set $\Gamma^{33}(k) = 0$ and $S^{33}(k) = S_F(k)$. The (34)–channel equation then reads explicitly

$$\begin{aligned}\tilde{\Gamma}^{34}(k) &= -2\frac{\tilde{V}^{34}_{\rm p-h}(k)S^{44}(k)}{\hbar\omega_3(k)+\epsilon_4(k)}\left[1-\frac{S_F(k)(\tilde{\Gamma}^{34}(k))^2}{S^{44}(k)}\right]^2\\ &\approx -2\frac{\tilde{V}^{34}_{\rm p-h}(k)S^{44}(k)}{\hbar\omega_3(k)+\epsilon_4(k)}\,,\end{aligned} \tag{4.13}$$

where $\epsilon_4(k) = \hbar^2k^2/\left[2m_4S^{44}(k)\right]$ is the Feynman excitation energy, and the last step follows from the dictum to omit all contributions that contain more than one correlated impurity atom. Alternatively, one may start from Eq. (3.3) which reads, in the single–impurity limit,

$$\chi^{34}(k,\omega) = \chi_0^{33}(k,\omega)\tilde{V}^{34}_{\rm p-h}(k)\chi^{44}(k,\omega)\,, \tag{4.14}$$

use the "mean–spherical approximation" (3.6) for the Lindhard function, and calculate $S^{34}(k)$ through frequency integration, Eq. (3.9). This leads to the same result (4.13). In the strictly low–concentration limit $S_F(k) = 1$, $S^{34}(k) = \tilde{\Gamma}^{34}(k)$, we recover the familiar single–impurity Euler equation [28]

$$S^{34}(k) = -2\frac{\tilde{V}^{34}_{\rm p-h}(k)S^{44}(k)}{t_3(k)+\epsilon_4(k)}\,, \tag{4.15}$$

with the particle–hole potential (2.66) and the induced potential,

$$\tilde{w}_{\rm I}^{34}(k) = -\frac{1}{2}\frac{S^{34}(k)(S^{44}(k)-1)}{S^{44}(k)}\left[t_3(k)+t_4(k)+\epsilon_4(k)\right]. \tag{4.16}$$

Again, Eqs. (4.15), (4.16), and (2.66) form a closed system of equations which can be solved by iteration, given a practical choice for the contributions of the elementary diagrams and triplet correlations to the particle–hole potential, $\Delta V^{34}(\mathbf{r}_0,\mathbf{r}_1)$.

Using the PPA equation (4.3) for the background liquid

$$\tilde{V}^{44}_{\rm p-h}(k) = \frac{\hbar^2k^2}{4m_4}\left\{\left[\frac{1}{S^{44}(k)}\right]^2-1\right\} \tag{4.17}$$

we find, in the limit $k\to 0+$,

$$S^{34}(0+) = -\frac{\tilde{V}^{34}_{\rm p-h}(0+)}{\tilde{V}^{44}_{\rm p-h}(0+)} = -\beta\,. \tag{4.18}$$

The relation to the volume excess factor in the last equality comes from the identification of the interaction $\tilde{V}^{34}_{\rm p-h}(0+)$ as the effective ^{3}He–^{4}He interaction used by Bardeen, Baym, and Pines. [5]

Our results for the three–dimensional case are shown in Figs. 5 and 6. The chemical potentials agree nicely with the experimental results of Ref. 2, even though here the triplet correlations and elementary diagrams are very important for quantitative results. [28]

In two dimensions, there are no experimental or simulation results available to be used to make a comparison. That is why we make use of the "average correlation approximation" (ACA) to calculate a rigorous upper bound for the impurity chemical potential as follows: let $\Psi_{N+1}(\mathbf{r}_0,\ldots,\mathbf{r}_N)$ be the ground–state wave function of $N+1$ ^{4}He atoms, and $\Psi^I_{N+1}(\mathbf{r}_0,\ldots,\mathbf{r}_N)$ the ground state of N ^{4}He atoms and one ^{3}He impurity, approximated, for example, by the variational function (4.7). Then, the impurity chemical potential μ_3^I can be estimated as [70,71]

$$\begin{aligned}\mu_3^I &\equiv \langle\Psi^I_{N+1}|\,H^I_{N+1}\,|\Psi^I_{N+1}\rangle - \langle\Psi_N|\,H_N\,|\Psi_N\rangle\\ &\le \langle\Psi_{N+1}|\,H^I_{N+1}\,|\Psi_{N+1}\rangle - \langle\Psi_N|\,H_N\,|\Psi_N\rangle\\ &= \left(\frac{m_4}{m_3}-1\right)\langle T_4\rangle + \mu_4 \equiv \mu_3^{ACA}\,,\end{aligned} \tag{4.19}$$

where $\langle T_4\rangle$ is the kinetic energy per particle of the ^{4}He component, and μ_4 is the ^{4}He chemical potential. Both quantities can be obtained from simulations [66] without recourse to semi–analytic theories, but naturally also within our theoretical framework. That being said, we can assure ourselves of the accuracy of our calculations in two ways: firstly, by asserting that our ACA estimate for the ^{3}He chemical potential agrees sufficiently well with what one would obtain from Ref. 66. Secondly, the relevant quantity is only the *increase* of the ^{3}He binding due to relaxing the ACA, and the energy differences gained by relaxing the ACA should be more accurate than the absolute energies.

In Fig. 7, we compare the result obtained from full calculation of the ^{3}He chemical potential with the ACA result. The gain in energy due to relaxing the ACA is about 0.2 – 0.3 K in the whole density range. As expected, Monte Carlo results for the ACA are slightly below our approximate result, and we can thus conclude that our full calculation gives reliable results also in two dimensions.

4.3. *Two–impurity limit*

Let us then turn to the effective interaction between pairs of impurities. Again, one can start from a variational wave function that contains two impurities, and then calculate the correlations by minimizing the second–order energy difference

$$\Delta^{(2)}E \equiv E^{II}_{N+2} - 2E^I_{N+1} + E_N\,. \tag{4.20}$$

However, this energy difference is only of secondary interest; the physically more relevant quantity is the effective interaction between impurities which can be obtained, in the same manner as outlined above, from the induced interaction (2.70). We proceed again by allowing first for no more than two *dynamically* correlated impurities. Because we are interested in the two–impurity limit of the coordinate–space Schrödinger–like equation (2.71) which contains already a common factor $\Gamma^{33}(r)$, we must not allow for dynamically correlated pairs of impurities in the induced potential. For calculating the induced potential, we must interpret the particle–hole interaction matrix appearing in Eq. (2.70) as being expressed by the PPA equation

(2.67). Expanding the (33)–component of $\tilde{w}_{\rm I}^{33}(k)$ then leads to

$$\tilde{w}_{\rm I}^{33}(k) = -\frac{[\tilde{\Gamma}^{34}(k)]^2}{2S^{44}(k)}\,[2\hbar\omega_3(k) + \epsilon_4(k)]\,. \tag{4.21}$$

As a further approximation, one may also ignore the *statistical* correlations by setting $S_F(k) = 1$. By doing so, we arrive at

$$\tilde{w}_{\rm I}^{33}(k) = -\frac{[S^{34}(k)]^2}{2S^{44}(k)}\,[2t_3(k) + \epsilon_4(k)]\,, \tag{4.22}$$

which is the induced interaction derived, among others, by Owen. [72]

An alternative derivation, providing an interpretation of the Jastrow–Feenberg theory in terms of Green's functions, is offered by the localizing approximations of parquet–diagram theory. [58-60] The energy–dependent effective interaction between two impurities that is mediated by the exchange of phonons is generally

$$\begin{aligned}\tilde{V}_{\rm eff}(k,\omega) &= \tilde{V}^{33}_{\rm p-h}(k) + \tilde{V}^{34}_{\rm p-h}(k)\,\chi^{44}(k,\omega)\,\tilde{V}^{34}_{\rm p-h}(k)\\ &= \tilde{V}^{33}_{\rm p-h}(k) + \tilde{V}^{34}_{\rm p-h}(k)\,\frac{2t_4(k)}{(\hbar\omega)^2-(\epsilon_4(k))^2}\,\tilde{V}^{34}_{\rm p-h}(k)\,.\end{aligned} \tag{4.23}$$

The prescription from parquet–theory to make this energy–dependent interaction local is as follows: first, construct the RPA static structure function

$$S^{33}_{\rm RPA}(k) = -\int_0^\infty \frac{d(\hbar\omega)}{\pi}\,\Im m\left[\chi_0^{33}(k,\omega) + \chi_0^{33}(k,\omega)\tilde{V}_{\rm eff}(k,\omega)\chi_0^{33}(k,\omega)\right]. \tag{4.24}$$

Then, construct the *ladder approximation* for the same quantity in terms of a different and yet unspecified local effective interaction, say $\tilde{V}_{\rm L}(k)$

$$S^{33}_{\rm ladder}(k) = -\int_0^\infty \frac{d(\hbar\omega)}{\pi}\,\Im m\left[\chi_0^{33}(k,\omega) + \chi_0^{33}(k,\omega)\,\tilde{V}_{\rm L}(k)\chi_0^{33}(k,\omega)\right]. \tag{4.25}$$

Now choose an average frequency $\bar{\omega}(k)$ such that these two forms of the static structure function are *identical* for

$$\tilde{V}_{\rm L}(k) = \tilde{V}_{\rm eff}(k,\bar{\omega}(k))\,. \tag{4.26}$$

The calculation can be carried out in closed form in the collective approximation, leading to

$$V_{\rm L}(k) = V^{33}_{\rm p-h}(k) - 2\left[\tilde{V}^{34}_{\rm p-h}(k)\right]^2 S^{44}(k)\frac{2\hbar\omega_3(k)+\epsilon_4(k)}{(\hbar\omega_3(k)+\epsilon_4(k))^2} \tag{4.27}$$

from which we conclude that

$$\tilde{w}_{\rm I}^{33}(k) = V_{\rm L}(k) - V^{33}_{\rm p-h}(k)\,. \tag{4.28}$$

This is seen to be identical with the one given in Eq. (4.21).

4.3.1. *Bound states in two-dimensional mixtures*

Bashkin [11] has pointed out that ^{3}He atoms form weakly bound dimers in a dilute ^{3}He–^{4}He mixture in case their motion is restricted to (quasi–)two dimensions. In particular, this dimerization should be observable in a low–density 2D mixture. The effective ^{3}He–^{3}He interaction is a plausible candidate for producing such bound states. To justify the existence of these states, we must show that the sum of the bare interaction, the induced interaction, and the correction originating from elementary diagrams and triplets appearing in Eq. (2.71) is sufficiently attractive. [12,73]

It is relatively easy to see that the elementary diagram contribution falls of at least as steeply as r^{-6} at large separations, in other words, in the way of the bare interaction. A similar proof is possible, although quite tedious, for the correction from triplet correlations. To calculate the large–distance behavior of the induced–interaction term, we study the long–wavelength limit. The line of thought is the following: firstly, we see from Eqs. (2.66) that the particle–hole interactions $V^{44}_{\rm p-h}(r)$ and $V^{34}_{\rm p-h}(r)$ are short–ranged functions that fall off at least as r^{-6} as $r \to \infty$, hence

$$\tilde{V}^{\alpha\beta}_{\rm p-h}(k) = a_{\alpha\beta} + b_{\alpha\beta}k^2 + O(k^4)\,, \tag{4.29}$$

where we can identify $a_{44} = m_4c^2$ and $a_{34} = \beta m_4 c^2$, *cf.* Eq. (4.18), with c the speed of sound. Inserting these expansions into Eq. (4.22) yields a small–k expansion

$$\tilde{w}^{33}_{\rm I}(k) = -\beta^2 m_4 c^2 + \text{const.} \times k^2 - \frac{\beta^2\hbar^3 m_4}{4m_3^3 c}k^3 + \mathcal{O}(k^4)\,, \tag{4.30}$$

from which we conclude that

$$\tilde{w}^{33}_{\rm I}(r) \sim -\frac{9\beta^2\hbar^3 m_4}{8\pi\rho m_3^3 c}r^{-5} + \mathcal{O}(r^{-6}) \quad \text{as r} \to \infty\,. \tag{4.31}$$

In other words, the *phonon–induced* interaction dominates the effective interaction at large separations, being also attractive. In Fig. 8, we show the binding energy of the dimer as a function of density in 2D. The density $\rho = 0.043\text{Å}^{-2}$ is the saturation density of 2D liquid ^{4}He.

A similar phenomenon also takes place between charged impurities in the electron gas. The induced potential is responsible of the perfect screening at large distances. At intermediate distances it over-screens the Coulomb repulsion of like charges and at low enough densities the over–screening becomes so attractive that impurities form a bound state.

Things look different when the effect of Fermi statistics of the dilute component is included. Firstly, the "Schrödinger equation" (2.71) for the coordinate–space correlations is no longer rigorously valid as pointed out above. Even ignoring this, it is no longer possible to prove that $w^{33}_{\rm I}(r)$ is attractive and dominates at large distances. Using Eqs. (4.21) and (4.13) instead of (4.22) and (4.15), one finds that the small–k expansion of $\tilde{w}^{33}_{\rm I}(k)$ contains only even powers up to k^4. To elaborate, the phonon–induced interaction is effectively screened by Fermi statistics and when the concentration increases then the bound state disappears and a homogeneous atomic mixture becomes stable.

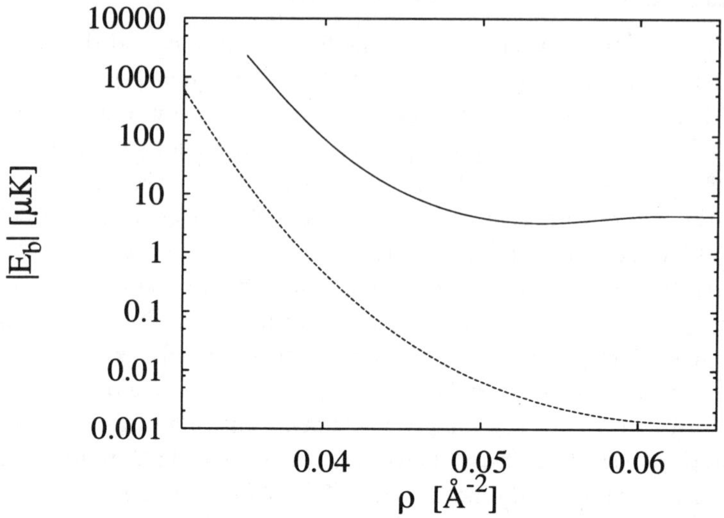

Fig. 8. Binding energy of a dimer of ^{3}He atoms within two–dimensional ^{4}He on a logarithmic scale as a function of density. The solid line gives the result of the full calculation, the dashed line shows the result from the HNC approximation, omitting the corrections $\Delta V^{33}(r)$ from the elementary diagrams and triplet correlations.

4.4. *Finite–concentration mixtures in 2D*

The examination of the single– and two–impurity limits and the comparison of the energetics and effective interactions has lead us to two conclusions: firstly, at slightly positive pressures, the chemical potential of a single ^{3}He impurity in the mixture is less than that in two–dimensional liquid ^{3}He at the same pressure as shown in Fig. 9. This means that the low–concentration mixture will not phase separate into disjoint regions of ^{4}He and ^{3}He. Secondly, the effective interaction between pairs of ^{3}He impurities is sufficiently attractive for the impurities to form dimers as shown in Fig. 8. Both of these conclusions were independent of the theoretical tools and based on either simulation data or rigorous upper bounds. Among others, our analysis demonstrates that an atomic mixture of ^{3}He and ^{4}He is, in two dimensions and at sufficiently low concentrations, unstable against dimerization, but stable against phase separation. [73]

As the concentration is increased, one expects that the dimers dissolve and an atomic mixture is formed. A fully microscopic theory is needed to extract the properties of such a mixture. The accuracy of these calculations for ^{4}He has been demonstrated above; we therefore only need to estimate the concentrations for which we expect the calculations for the ^{3}He component to be reliable. We have used the version of the (F)HNC/EL theory explained in detail in Ref. 24. The method implies, for the fermion channels, a simplified version of the FHNC/EL equation which is not sufficiently accurate for a quantitative description of pure ^{3}He in three dimensions. We can indirectly assess the expected accuracy of our calculations as follows: in

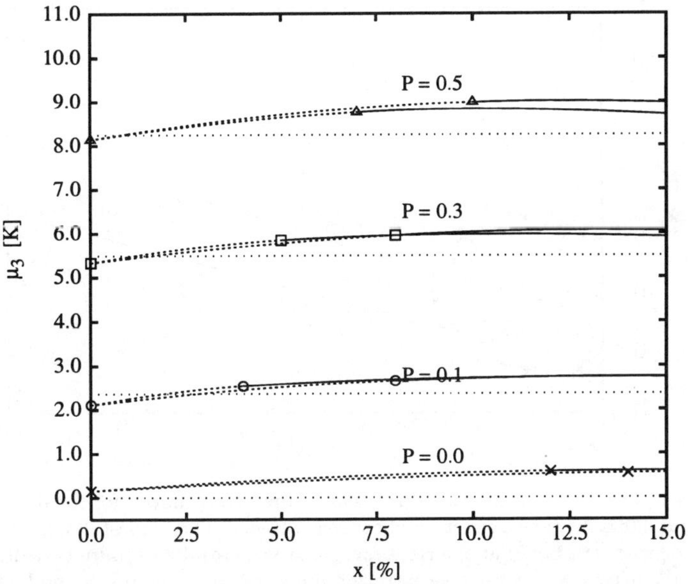

Fig. 9. Chemical potential of the ^{3}He component in two–dimensional mixture is shown, as a function of concentration x, for the pressures 0.0, 0.1, 0.3, and 0.5 dyn/cm as obtained from the (F)HNC/EL calculation and from the CBF–corrected calculation. The results have been smoothly interpolated through the unstable region to match the $x = 0$ value. The interpolated region is drawn as dashed lines and ended by markers. The CBF–corrected chemical potentials lie slightly higher. The four dotted horizontal lines show the chemical potentials of the pure ^{3}He gas at the indicated pressures as given in Ref. 75.

the three–dimensional mixture, we have demonstrated the accuracy of our results for the energy up to 20 Atm pressure, corresponding to a density of approximately 0.026 Å^{-3}, and up to 10 percent concentration. The average distance between ^{3}He atoms is, at this density, about 4.5 Å. This translates to a ^{3}He density of 0.016 Å^{-2} in two dimensions, which is about 25 percent of the solidification density of 0.065 Å^{-2}. A conservative estimate would then lead us to expect that our calculations are as accurate as those of Ref. 24 up to about 25 percent concentration.

Results for the chemical potential of the ^{3}He component $\mu_3(\rho, x)$, equation of state $E(\rho, x)/N$, and the speed of second sound mc_2^2 are shown in Figs. 9, 10, and 11. The most interesting result is the concentration dependence of the speed of second sound. As pointed out above, it is unavoidable that the system exhibits a *local* instability at low concentrations (coming from above). Fig. 11 shows, for three different densities, the long–wavelength limit of mc_2^2 as a function of concentration as obtained both from the (F)HNC/EL calculation and the CBF–corrected calculations. Our estimates for the critical concentration, as a function of density, are shown in Fig. 12. With all the caution that should be exercised when relying on extrapolations, we assert that the FHNC/EL approximation predicts a critical concentration of about one to two percent below which the atomic mixture is un-

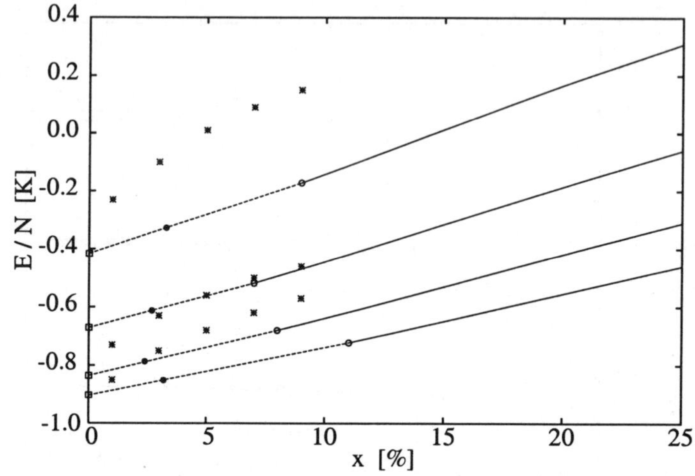

Fig. 10. Energy per particle of the two–dimensional mixture is shown, for the densities $\rho = 0.045$, 0.050, 0.055, and 0.060 Å^{-2}, as a function of concentration x. The highest–lying curve corresponds to the highest density, the boxes at the right margin are the results for pure two–dimensional ^{4}He. The unstable area between the zero–concentration limit and the estimated spinodal point for phase separation is dashed, the estimated spinodal concentrations are indicated by heavy dots, whereas the endpoint of our numerical calculations is indicated by an open circle. Also shown are the (F)HNC results of Ref. 74 for $\rho = 0.045$, 0.50, and 0.60 Å^{-2} (stars).

stable. The better CBF calculation predicts a somewhat steeper functional form of $mc_2^2(x)$ and, hence, a higher critical concentration. Because of this steeper behavior, it is numerically even harder to get close to the spinodal point. The CBF results have been derived from an extrapolation reaching much farther and have, therefore, larger uncertainties. This is also clearly visible in Fig. 12 where the results are more scattered.

5. Dynamic structure of quantum fluids

In examining the microscopic properties of many–body systems, functions describing the interdependence of observables take the central stage. We have seen that in the case of ground–state and static properties such functions are the distribution functions, in particular the pair–distribution function and its Fourier transform, the static structure function. In what comes to the excited states and dynamics, this role is played by the density–density response function and especially its imaginary part, the dynamic structure function. In the previous section, we introduced these quantities in connection with the collective excitations and the random–phase approximation of quantum fluid mixtures. In this section, we push the description beyond the collective excitations in a systematic way using variational techniques. For simplicity, we limit the discussion to one–component fluids.

Let us assume that a quantum fluid is driven out of the ground state by an infinitesimal external perturbation $\tilde{U}_{\rm ext}(k,\omega)$, with a given frequency ω and wave

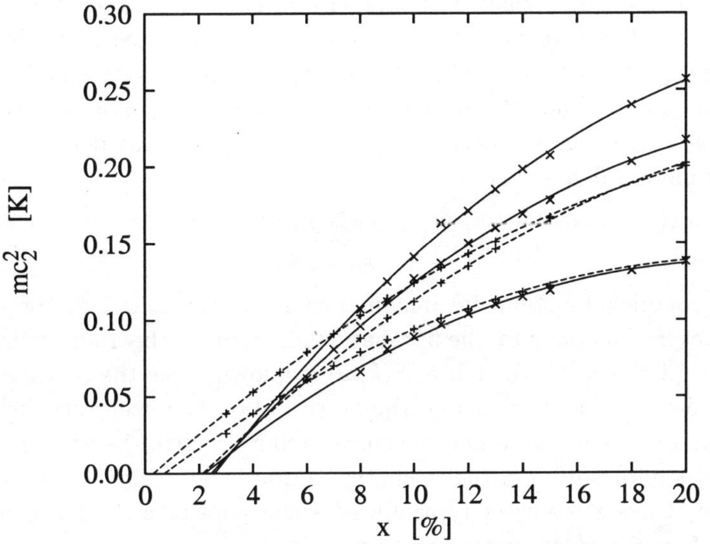

Fig. 11. The speed of second sound in the two–dimensional mixture is shown, for the densities $\rho = 0.045$, 0.055, and 0.065 Å^{-2}, as a function of concentration x as obtained from the (F)HNC/EL calculation (+–symbols) and the CBF–corrected calculation (crosses). Also shown are the fits to these data, determining the estimates for the lowest concentration that is stable against concentration fluctuations (dashed and solid lines, respectively).

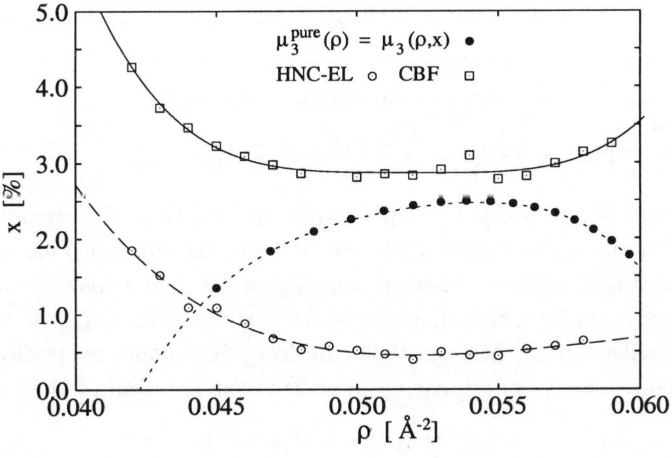

Fig. 12. Extrapolated critical concentration at which the atomic mixture becomes unstable against *infinitesimal* concentration fluctuations. The squares and circles show the (actually extrapolated) values obtained from the CBF and FHNC/EL calculations, respectively; the long–dashed and the solid line give a smooth interpolation of these results. The short–dashed line marked with filled dots gives the upper bound for the concentration below which the mixture is globally stable.

number k. It induces a change in the density, $\delta\tilde{\rho}_1(k,\omega)$, of an originally homogeneous system. The response of the system can be assumed to be linear, because macroscopic perturbations used in the measurements are very small on the scale of microscopic forces. The information of the dynamic properties of the system, like the possible excitation and decay modes, is contained in the density–density linear response function defined as

$$\chi(k,\omega) = \frac{\delta\tilde{\rho}_1(k,\omega)}{\rho_0 \tilde{U}_{\rm ext}(k,\omega)} . \tag{5.1}$$

This is a complex function; its imaginary part is connected to the dynamic and static structure functions by the fluctuation–dissipation theorem as already pointed out in Eqs. (3.8) and (3.9). Hence, $S(k,\omega)$ is known, once the relation between the one–body density fluctuations and the perturbation has been established.

The dynamic structure function is measured in (neutron) scattering experiments and provides information of the strength, lifetime, and dispersion of excitations. At low temperatures it consists typically of a sharp peak and of a broad contribution. [76,32] It is therefore customary to write

$$S(k,\omega) = Z(k)\delta(\hbar\omega - \hbar\omega_0(k)) + S_{\rm mp}(k,\omega) . \tag{5.2}$$

This immediately suggests that the density–density response function can be written in the form familiar from the RPA, Eq. (3.3),

$$\chi(k,\omega) = S(k)\left[\frac{1}{\hbar\omega - \varepsilon(k) - \Sigma(k,\omega)} - \frac{1}{\hbar\omega + \varepsilon(k) + \Sigma^*(k,-\omega)}\right] . \tag{5.3}$$

Here $\varepsilon(k)$ is the energy of a single collective mode and $\Sigma(k,\omega)$ is the complex self–energy. Recalling Plemelj's formula for singular integrands

$$\lim_{\eta\to 0}\frac{1}{\omega - \omega' + i\eta} = \mathcal{P}\frac{1}{\omega - \omega'} - i\pi\delta(\omega - \omega') , \tag{5.4}$$

where $\mathcal{P}$ denotes the Cauchy principal value, we see that the dynamic structure function, *i.e.* the imaginary part of the linear response function, has contributions from the poles of the response function, which give the sharp peak of the elementary excitation modes and from the imaginary part of the self–energy, which corresponds to excitations with a finite lifetime that can decay into other excitations.

The real pole $\omega = \omega_0(k)$ appearing in (5.2) is the solution of the equation

$$\hbar\omega_0(k) = \varepsilon(k) + \Sigma(k,\omega_0(k)) \tag{5.5}$$

where $\Sigma(k,\omega_0(k))$ is real, and the strength of the pole $Z(k)$ can be evaluated from the derivative of the self–energy,

$$Z(k) = S(k)\left[1 - \left.\frac{d\Sigma(k,\omega)}{d(\hbar\omega)}\right|_{\omega=\omega_0}\right]^{-1} . \tag{5.6}$$

The second term in (5.2), $S_{\rm mp}(k,\omega)$, is the multi–phonon background, *i.e.* the contribution in which a neutron probing the system exchanges energy with two or more excitations. In addition, the relative weight, $Z(k)/S(k)$, gives the efficiency of single–collective–excitation scattering processes, as seen from the (zeroth–moment) sum rule

$$\int_0^\infty S(k,\omega)d(\hbar\omega) = S(k)\,. \tag{5.7}$$

In other words, it gives the fraction of available scattering processes going through a single collective mode at a given wave vector. If the only excitation in the system were the collective mode $\varepsilon(k)$, like in the Feynman approximation, then the ratio $Z(k)/S(k)$ would be one.

5.1. *Equations–of–motion method*

5.1.1. *Least–action principle*

After these general considerations, let us implement these concepts within the framework of the variational theory. In the presence of a weak, time–dependent perturbation the ground–state wave function, $\Psi_0(\mathbf{r}_1,\ldots\mathbf{r}_N)$, must be allowed more flexibility. A plausible way to do this is to allow the correlation functions in Eq. (2.3) to pick up time dependence,

$$\Psi(\mathbf{r}_1,\ldots\mathbf{r}_N;t) = e^{-iEt/\hbar}\Phi(\mathbf{r}_1,\ldots\mathbf{r}_N;t) \tag{5.8}$$

with

$$\Phi(\mathbf{r}_1,\ldots\mathbf{r}_N;t) = \frac{1}{\sqrt{\mathcal{N}(t)}}\, e^{\frac{1}{2}\delta U(\mathbf{r}_1,\ldots\mathbf{r}_N;t)}\Psi_0(\mathbf{r}_1,\ldots\mathbf{r}_N)\,. \tag{5.9}$$

The phase factor with the ground–state energy E removes the ground–state contributions, and the normalization factor contains the ratio of the ground–state and perturbed–state normalizations,

$$\mathcal{N}(t) = \frac{1}{\langle\Psi_0|\Psi_0\rangle}\int d\mathbf{r}_1\ldots d\mathbf{r}_N|\Psi_0(\mathbf{r}_1,\ldots\mathbf{r}_N)|^2 e^{\Re e[\delta U(\mathbf{r}_1,\ldots\mathbf{r}_N;t)]}\,. \tag{5.10}$$

The complex–valued excitation operator

$$\delta U(\mathbf{r}_1,\ldots\mathbf{r}_N;t) = \sum_i \delta u_1(\mathbf{r}_i;t) + \sum_{i<j}\delta u_2(\mathbf{r}_i,\mathbf{r}_j;t) + \cdots \tag{5.11}$$

represents fluctuations in the correlation functions due to the perturbation. Here, we truncate the expansion to the two–body level. Contrary to the ground–state calculation, the one–body function $\delta u_1(\mathbf{r}_i;t)$ must now be included into the description as the dynamics will normally break the translational invariance. On the other hand, restricting the time dependence to the one–body component only would lead directly to the Feynman theory of excitations. The time–dependent two–body component is significant in situations where the external field excites fluctuations of wavelengths comparable to the interparticle distance, as has been explicitly demonstrated for liquid ^{4}He [38,40–42,77] and for charged bosonic systems. [69] Finally, we point out that the wave function (5.9) represents a ground state perturbed by an external time–dependent potential, and not a true excited state.

The time evolution of the correlation functions is governed by the least–action principle [78,79]

$$\delta \mathcal{S} = \delta \int_{t_0}^{t} dt \mathcal{L}(t) \equiv \delta \int_{t_0}^{t} dt \left\langle \Psi(t) \left| H(t) - i\hbar \frac{\partial}{\partial t} \right| \Psi(t) \right\rangle = 0\,, \tag{5.12}$$

where the Hamiltonian

$$H(t) = H_0 + \hat{U}_{\rm ext}(t) \tag{5.13}$$

now contains, besides the ground–state Hamiltonian H_0, also the time–dependent operator $\hat{U}_{\rm ext}(t) = \sum_{i=1}^{N} U_{\rm ext}(\mathbf{r}_i; t)$, which introduces the external, infinitesimal disturbance of the system. The least–action priciple provides us the equations connecting the one–body density fluctuations to the perturbation, needed to calculate the dynamic response.

Recalling the ground–state Schrödinger equation, we can reformulate the integrand as

$$\begin{aligned}
\mathcal{L}(t) &= \left\langle \Phi(t) \left| H - E - \frac{\hbar}{2}\left(i\frac{\partial}{\partial t} + h.c. \right) \right| \Phi(t) \right\rangle \\
&= \frac{1}{\mathcal{N}(t)} \left\langle \Psi_0 \left| e^{\frac{1}{2}\delta U^*(t)} [H_0, e^{\frac{1}{2}\delta U(t)}] \right| \Psi_0 \right\rangle \\
&\quad + \left\langle \Phi(t) \left| -\frac{\hbar}{2}\left(i\frac{\partial}{\partial t} + h.c. \right) + \hat{U}_{\rm ext}(t) \right| \Phi(t) \right\rangle .
\end{aligned} \tag{5.14}$$

The potential–energy operator commutes with $\delta U(t)$, so only the kinetic–energy operator contributes to the commutator. Elementary manipulations, partial integration, and evaluation of the time derivative result finally in

$$\mathcal{L}(t) = \left\langle \Phi(t) \left| \frac{\hbar^2}{8m} \sum_{j=1}^{N} |\nabla_j \; \delta U(t)|^2 + \frac{1}{2}\hbar \, \Im m[\delta \dot{U}(t)] + \hat{U}_{\rm ext}(t) \right| \Phi(t) \right\rangle . \tag{5.15}$$

The point in the least–action principle is to search for the correlation functions $\delta u_1(\mathbf{r}_1; t)$, $\delta u_2(\mathbf{r}_1, \mathbf{r}_2; t)$, *etc.*, which minimize the action integral (5.12). By performing the variation with respect to a general correlation function $\delta u_n(\mathbf{r}_1, \ldots, \mathbf{r}_n; t)$, which depends on n coordinates and time, and using the form we derived above for the integrand, we can write the least–action principle in the form

$$\begin{aligned}
\int dr_{n+1} \ldots dr_N \Bigg\{ &\frac{\hbar^2}{4m} \sum_{j=1}^{n} \nabla_j \cdot [|\Psi|^2 \nabla_j \; \delta U(t)] + |\Psi|^2 \left[-\frac{i\hbar}{2} \delta \dot{U}(t) + \hat{U}_{\rm ext}(t) \right] \\
&- |\Psi|^2 \left\langle \Psi \left| -\frac{i\hbar}{2} \delta \dot{U}(t) + \hat{U}_{\rm ext}(t) \right| \Psi \right\rangle \Bigg\} = 0\,.
\end{aligned} \tag{5.16}$$

The last term originates from the variation of the normalization integral (5.10).

5.2. *Continuity equations*

The previous discussion moved on a formally transparent but, unfortunately, still non–practicable level. In the following, we discuss the implications of the least–action principle: how to calculate $S(k,\omega)$ by reducing the problem to one of solving two coupled and linear equations of motion, taking the form of continuity equations, for the correlation functions.

In order to simplify Eq. (5.16), we make use of the familiar definition

$$\rho_n(\mathbf{r}_1,\ldots\mathbf{r}_n;t) = \frac{N!}{(N-n)!}\int d^3r_{n+1}\ldots d^3r_N|\Psi(\mathbf{r}_1,\ldots\mathbf{r}_N;t)|^2 \tag{5.17}$$

for the n–particle density. As we are working within the linear response theory, we can separate the time–dependent and time–independent parts in the density, and write

$$\bar{\rho}_n(\mathbf{r}_1,\ldots\mathbf{r}_n;t) = \rho_n(\mathbf{r}_1,\ldots\mathbf{r}_n) + \delta\rho_n(\mathbf{r}_1,\ldots\mathbf{r}_n;t)\,. \tag{5.18}$$

Expanding to the first order in $\delta U(t)$, we get

$$\delta\rho_n(\mathbf{r}_1,\ldots\mathbf{r}_n;t) = \frac{N!}{(N-n)!}\int dr_{n+1}\ldots dr_N|\Psi_0|^2\Big[\delta U(t) - \langle\Psi_0\,|\delta U(t)|\,\Psi_0\rangle\Big]\,. \tag{5.19}$$

In doing this, we have generalized the definition for complex density fluctuations; the physical density is given by the real part. Similarly, we define the complex n–particle current linear in $\delta U(t)$

$$\mathbf{j}_n(\mathbf{r}_1,\ldots\mathbf{r}_n;t) = $$
$$= \frac{\hbar}{2mi}\frac{N!}{(N-n)!}\int d^3r_{n+1}\ldots d^3r_N|\Psi_0(\mathbf{r}_1,\ldots\mathbf{r}_N)|^2\sum_{j=1}^{n}\nabla_j\delta U(\mathbf{r}_1,\ldots\mathbf{r}_N;t)\,.$$

Our starting point was a homogeneous system in its ground state with constant one–body density, $\rho_1(\mathbf{r}) = \rho_0$. It is then convenient to introduce the n–particle distribution function

$$g_n(\mathbf{r}_1,\ldots,\mathbf{r}_n) = \frac{1}{(\rho_0)^n}\rho_n(\mathbf{r}_1,\ldots\mathbf{r}_n)\,, \tag{5.20}$$

which gives the probability to find n particles at positions $\mathbf{r}_1,\ldots,\mathbf{r}_n$.

Using these definitions, we can write the general form of the least–action principle (5.16) in terms of the one– and two–particle continuity equations [77,41,42]

$$\nabla_1\cdot\mathbf{j}_1(\mathbf{r}_1;t) + \delta\dot{\rho}_1(\mathbf{r}_1;t) = D_1(\mathbf{r}_1;t) \tag{5.21}$$
$$\nabla_1\cdot\mathbf{j}_2(\mathbf{r}_1,\mathbf{r}_2;t) + \text{same for }\; (1\leftrightarrow 2) + \delta\dot{\rho}_2(\mathbf{r}_1,\mathbf{r}_2;t) = D_2(\mathbf{r}_1,\mathbf{r}_2;t)\,. \tag{5.22}$$

These are the coupled, linear equations of motion that we aimed to derive.

To proceed into a more detailed analysis of the ingredients that we have here, let us first examine the term with the time derivative. Inserting the definition (5.11) of the excitation operator (truncated after the two–body term) into the definition (5.19), we get for the one–body density

$$\begin{aligned}\delta\rho_1(\mathbf{r}_1;t) = \rho_0\ \delta u_1(\mathbf{r}_1;t) + \rho_0^2 \int d^3r_2[g_2(\mathbf{r}_1,\mathbf{r}_2)-1]\delta u_1(\mathbf{r}_2;t) \\ +\rho_0^2 \int d^3r_2\Big\{g_2(\mathbf{r}_1,\mathbf{r}_2)\delta u_2(\mathbf{r}_1,\mathbf{r}_2;t) \\ +\frac{\rho_0}{2}\int d^3r_3\,[g_3(\mathbf{r}_1,\mathbf{r}_2,\mathbf{r}_3)-g_2(\mathbf{r}_2,\mathbf{r}_3)]\,\delta u_2(\mathbf{r}_2,\mathbf{r}_3;t)\Big\}.\end{aligned} \tag{5.23}$$

Additionally, definition (5.19) tells us directly that the particle number is conserved in the fluctuations,

$$\int d^3r\delta\rho_1(\mathbf{r}) = 0\,, \tag{5.24}$$

and that the fluctuations satisfy the sequential relation

$$\int d^3r_2\delta\rho_2(\mathbf{r}_1,\mathbf{r}_2;t) = (N-1)\delta\rho_1(\mathbf{r}_1;t)\,. \tag{5.25}$$

The one– and two–body currents can be extracted from Eq. (5.20),

$$\mathbf{j}_1(\mathbf{r}_1;t) = \frac{\hbar\rho_0}{2mi}\Big\{\nabla_1\delta u_1(\mathbf{r}_1;t) + \rho_0\int d^3r_2 g_2(\mathbf{r}_1,\mathbf{r}_2)\nabla_1\delta u_2(\mathbf{r}_1,\mathbf{r}_2;t)\Big\} \tag{5.26}$$

$$\begin{aligned}\mathbf{j}_2(\mathbf{r}_1,\mathbf{r}_2;t) = \frac{\hbar\rho_0^2}{2mi}\Big\{g_2(\mathbf{r}_1,\mathbf{r}_2)[\nabla_1\delta u_1(\mathbf{r}_1;t)+\nabla_1\delta u_2(\mathbf{r}_1,\mathbf{r}_2;t)] \\ +\rho_0\int d^3r_3 g_3(\mathbf{r}_1,\mathbf{r}_2,\mathbf{r}_3)\nabla_1\delta u_2(\mathbf{r}_1,\mathbf{r}_3;t)\Big\}.\end{aligned} \tag{5.27}$$

Also the currents satisfy the sequential condition

$$\int d^3r_2\ \mathbf{j}_2(\mathbf{r}_1,\mathbf{r}_2;t) = (N-1)\ \mathbf{j}_1(\mathbf{r}_1;t)\,. \tag{5.28}$$

This confirms that the two continuity equations are indeed not independent.

Finally, the external potential appears only in the functions

$$D_1(\mathbf{r}_1;t) = \frac{2\rho_0}{i\hbar}\Big\{U_{\rm ext}(\mathbf{r}_1;t) + \rho_0\int d^3r_2\Big[g_2(\mathbf{r}_1,\mathbf{r}_2)-1\Big]U_{\rm ext}(\mathbf{r}_2;t)\Big\} \tag{5.29}$$

$$\begin{aligned}D_2(\mathbf{r}_1,\mathbf{r}_2;t) = \frac{2\rho_0^2}{i\hbar}\Big\{g_2(\mathbf{r}_1,\mathbf{r}_2)\,[U_{\rm ext}(\mathbf{r}_1;t)+U_{\rm ext}(\mathbf{r}_2;t)] \\ +\rho_0\int d^3r_3\Big[g_3(\mathbf{r}_1,\mathbf{r}_2,\mathbf{r}_3)-g_2(\mathbf{r}_1,\mathbf{r}_2)\Big]U_{\rm ext}(\mathbf{r}_3;t)\Big\}.\end{aligned} \tag{5.30}$$

Up to now, we have formulated the problem of finding out the dynamics in terms of a Hamiltonian, a trial wave function, and the least–action principle, and transformed the requirement of least action into two coupled continuity equations. What we still need to do is to find a way to actually solve the equations. Assuming that all ground–state quantities are known, the continuity equations still contain four unknown functions, namely $\delta u_1(\mathbf{r};t)$ and $\delta u_2(\mathbf{r}_1,\mathbf{r}_2;t)$, and the time derivatives of $\delta\rho_1(\mathbf{r};t)$ and $\delta\rho_2(\mathbf{r}_1,\mathbf{r}_2;t)$. Clearly these are not all independent, but connected by the definition (5.19).

Before introducing various approximation schemes developed to work out the solutions, let us define the Fourier transforms. In a homogeneous system the fluctuations are small, and it is possible and more convenient to work in the momentum and frequency space. We define the one–body Fourier transform and its inverse as

$$\begin{aligned}\mathcal{F}[f(\mathbf{r};t)] &= \rho_0 \int d^3r\, dt\; e^{-i(\mathbf{k}\cdot\mathbf{r}-\omega t)} f(\mathbf{r};t) = \tilde{f}(\mathbf{k};\omega)\\ \mathcal{F}^{-1}[\tilde{f}(\mathbf{k};\omega)] &= \int \frac{d^3k\, d\omega}{(2\pi)^4\rho_0} e^{i(\mathbf{k}\cdot\mathbf{r}-\omega t)} \tilde{f}(\mathbf{k};\omega) = f(\mathbf{r};t)\,,\end{aligned} \tag{5.31}$$

and the two–body Fourier transforms in the form

$$\begin{aligned}\mathcal{F}[f(\mathbf{r}_1,\mathbf{r}_2;t)] &= \rho_0^2 \int d^3r_1 d^3r_2\; dt\; e^{-i(\mathbf{k}\cdot\mathbf{R}+\mathbf{p}\cdot\mathbf{r}-\omega t)} f(\mathbf{r}_1,\mathbf{r}_2;t)\\ \mathcal{F}^{-1}[\tilde{f}(\mathbf{k},\mathbf{p};t)] &= \int \frac{d^3k d^3p\; d\omega}{(2\pi)^7\rho_0^2} e^{i(\mathbf{k}\cdot\mathbf{R}+\mathbf{p}\cdot\mathbf{r}-\omega t)} \tilde{f}(\mathbf{k},\mathbf{p};\omega)\,.\end{aligned} \tag{5.32}$$

Here $\mathbf{R} = (\mathbf{r}_1 + \mathbf{r}_2)/2$ is the center–of–mass vector and $\mathbf{r} = \mathbf{r}_1 - \mathbf{r}_2$ the relative position vector; $\mathbf{k}$ and $\mathbf{p}$ are the center–of–mass and relative momenta, respectively.

5.3. *Feynman approximation*

The simplest physically consistent approximation is the Feynman approximation where we limit the time dependence to the one–body correlation function by setting $\delta u_2(\mathbf{r}_1,\mathbf{r}_2;t) = 0$. In this case, we need to solve only the first continuity equation (5.21). The one–body current reduces to

$$\mathbf{j}_1(\mathbf{r}_1;t) = \frac{\hbar\rho_0}{2mi}\nabla_1\delta u_1(\mathbf{r}_1;t)\,, \tag{5.33}$$

and the time–dependent part of the density is simply

$$\delta\rho_1(\mathbf{r}_1;t) = \rho_0\delta u_1(\mathbf{r}_1;t) + \rho_0^2\int d^3r_2\Big[g_2(\mathbf{r}_1,\mathbf{r}_2) - 1\Big]\delta u_1(\mathbf{r}_2;t)\,. \tag{5.34}$$

The Fourier transforms can readily be calculated, resulting in

$$\delta\tilde{\rho}_1(\mathbf{k};\omega) = S(k)\rho_0\delta\tilde{u}_1(\mathbf{k};\omega)\,. \tag{5.35}$$

Inserting these results into the continuity equation (5.21), along with the Fourier transform of D_1, we end up with

$$-\frac{\hbar k^2}{2mS(k)}\delta\tilde{\rho}_1(\mathbf{k};\omega) + \omega\delta\tilde{\rho}_1(\mathbf{k};\omega) = \frac{2\rho_0}{\hbar}S(k)\tilde{U}_{\text{ext}}(\mathbf{k},\omega)\,. \tag{5.36}$$

The external potential is included in the Hamiltonian, which must be Hermitian. Thus the potential is real function. We assume the perturbation to be harmonic,

$$U_{\text{ext}}(\mathbf{r}_i;t) = U_e(\mathbf{r}_i)\cos(\omega' t)e^{\eta t}, \qquad \eta \to 0+\,, \tag{5.37}$$

and switch on the time dependence adiabatically using a small, positive parameter η, which can be set equal to zero at the end of the calculations. The Fourier transform of the perturbation is

$$\tilde{U}_{\text{ext}}(\mathbf{k},\omega) = \frac{1}{2}\tilde{U}_e(\mathbf{k})\left[\delta(\omega-\omega'-i\eta)+\delta(\omega+\omega'-i\eta)\right] . \tag{5.38}$$

Inserting this into Eq. (5.36), we can solve for the density fluctuations,

$$\delta\tilde{\rho}(\mathbf{k},\omega) = \rho_0 S(k)\tilde{U}_e(\mathbf{k})\left[\frac{\delta(\omega-\omega'-i\eta)}{\hbar\omega'-\varepsilon_F(k)+i\eta}+\frac{\delta(\omega+\omega'-i\eta)}{-\hbar\omega'-\varepsilon_F(k)+i\eta}\right] \tag{5.39}$$

where we have again used the notation

$$\varepsilon_F(k) = \frac{\hbar^2k^2}{2mS(k)} . \tag{5.40}$$

Our aim is to calculate the linear response function (5.1) and the dynamic structure function (3.8). For that we need the transformation back to the time–coordinate

$$\delta\rho(\mathbf{k};t) = \rho_0 S(k)\tilde{U}_e(\mathbf{k})e^{\eta t}\left[\frac{e^{-i\omega' t}}{\hbar\omega'-\varepsilon_F(k)+i\eta}+\frac{e^{i\omega' t}}{-\hbar\omega'-\varepsilon_F(k)+i\eta}\right] , \tag{5.41}$$

because the physical density needed for the linear response must be a real function

$$\begin{aligned}\Re e[\delta\rho(\mathbf{k};t)] = \rho_0 S(k)\tilde{U}_e(\mathbf{k})\ \frac{1}{2}e^{\eta t}\Big\{ & e^{-i\omega' t}\left[\frac{1}{\hbar\omega'-\varepsilon_F(k)+i\eta}-\frac{1}{\hbar\omega'+\varepsilon_F(k)+i\eta}\right] \\ & + e^{i\omega' t}\left[\frac{1}{\hbar\omega'-\varepsilon_F(k)-i\eta}-\frac{1}{\hbar\omega'+\varepsilon_F(k)-i\eta}\right]\Big\} .\end{aligned} \tag{5.42}$$

This is symmetric in ω' and can be written as

$$\Re e[\delta\rho(\mathbf{k};t)] = \frac{1}{2}e^{\eta t}\left[\delta\tilde{\rho}_1(k,\omega')e^{-i\omega' t}+\delta\tilde{\rho}_1^*(k,\omega')e^{i\omega' t}\right] . \tag{5.43}$$

Similarly, we can write the external potential (5.37) in the form

$$\tilde{U}_{\text{ext}}(k;t) = \frac{1}{2}e^{\eta t}\tilde{U}_e(k)\left[e^{-i\omega' t}+e^{i\omega' t}\right] . \tag{5.44}$$

We are studying the time evolution of the ground state, so the system can be driven only with positive frequencies, $\omega' > 0$. Using this fact, we can write the linear response function as

$$\chi(k,\omega') = \frac{\delta\tilde{\rho}_1(k,\omega')}{\rho_0\tilde{U}_e(k)} = S(k)\left[\frac{1}{\hbar\omega'-\varepsilon_F(k)+i\eta}-\frac{1}{\hbar\omega'+\varepsilon_F(k)+i\eta}\right] . \tag{5.45}$$

The pole defines the collective excitation mode of the system, known as the Feynman approximation

$$\hbar\omega' = \varepsilon_F(k) = \frac{\hbar^2k^2}{2mS(k)} . \tag{5.46}$$

In the limit $\omega' = 0$ we obtain the static response function

$$\chi(k,0) = -\frac{4mS^2(k)}{\hbar^2k^2} . \tag{5.47}$$

In liquid ^{4}He (and, in fact, also in systems where the particles interact through the screened Coulomb, or Yukawa, interaction), the excitation mode is linear in the long–wavelength limit and proportional to the speed of sound c,

$$\varepsilon_F \to \hbar kc \ , \ \text{as} \ \ k \to 0 \,. \tag{5.48}$$

The structure function is also linear at low momenta,

$$S(k) \to \frac{\hbar k}{2mc} \ , \ \text{as} \ \ k \to 0 \,, \tag{5.49}$$

and the inverse of the static response function determines the incompressibility

$$-\chi^{-1}(k,0) \to mc^2 \,. \tag{5.50}$$

Due to the long range of the Coulomb interaction, the long wavelength limits in charged systems depend strongly on the dimensionality of the system. For example, there is a gap in the three–dimensional plasmon spectrum contrary to the two–dimensional case. That is to say, in a three–dimensional system the energy of the collective excitation (plasmon) does not go to zero at low momenta but to a constant known as the plasma frequency.

Everything put together, we have now solved our original problem: the imaginary part of the density–density response function determines the dynamic structure function (3.8)

$$S(k,\omega) = S(k)\delta(\hbar\omega - \varepsilon_F(k)) \,. \tag{5.51}$$

Analyzing this in a bit more detail, we see that the Feynman approximation is a single–pole approximation, and that the strength of the pole is given by $Z(k) = S(k)$. The fact that $S(k,\omega)$ consists of a single non–decaying excitation branch is also where the Feynman approximation misses much of the physics. For example, compared to the experimental phonon–roton (or "single–mode") spectrum, the Feynman approximation gives results that seem to be qualitatively correct — and even in quantitative agreement at very low momentum transfers — but in the roton region the excitation energy is too large by a factor of two. Additionally, the single–mode spectrum does not have an upper limit with increasing wave number, contrary to what is observed. Again, we return to these points when we discuss our results for the dynamic structure.

The elementary excitation modes of the system can also be obtained directly by setting $U_{\text{ext}} = 0$ in the continuity equations. Working still within the Feynman approximation and using the results (5.33) and (5.34), we get the differential equation

$$\begin{aligned} \frac{\hbar\rho_0}{2mi}\nabla_1^2\delta u_1(\mathbf{r}_1;\omega) - i\omega\Big[\rho_0\delta u_1(\mathbf{r}_1;\omega) \\ + \rho_0^2\int d^3r_2\,[g(\mathbf{r}_1,\mathbf{r}_2)-1]\,\delta u_1(\mathbf{r}_2;\omega)\Big] = 0\,. \end{aligned} \tag{5.52}$$

This has the solution (5.46) with

$$\delta u_1(\mathbf{r}_1;\omega) = e^{i\mathbf{k}\cdot\mathbf{r}_1}\,. \tag{5.53}$$

The excitation operator now takes the Feynman form,

$$\delta U = \sum_j \delta u_1(\mathbf{r}_j;\omega) = \sum_j e^{i\mathbf{k}\cdot\mathbf{r}_j}\,. \tag{5.54}$$

The operator in Eq. (5.11) can then be viewed as a generalization of this phonon–creation operator.

5.4. *CBF approximation*

Let us now look for ways of improving upon the Feynman approximation by allowing the two–body correlation function to vary in time and including the second continuity equation, Eq. (5.22), into our analysis. The approximations we make here result in the correlated basis functions form for the dynamic response function and the self–energy. [38,77,80] In discussing the Feynman approximation, we saw how fluctuations in the one–body correlation function describe single–excitation states. Fluctuations in the two–body correlation function then represent a correlated pair of excitations, coupled to single–excitation states via the equations of motion. Hence, excitations can decay into two excitations of lower energy, provided that energy and momentum are conserved or vice versa, *i.e.* excitations can have a finite lifetime. This is something that is required from a meaningful description of the dynamic structure, and the physics we would like the approximations to retain.

5.4.1. *Convolution approximation*

In the evaluation of the two–body current (5.27), we must first approximate the three–body distribution function. This subject has been extensively discussed in the literature. [19] The approach we take is basically what is known as the convolution approximation, but we also include a special set of diagrams containing the triplet–correlation function $u_3(\mathbf{r}_1,\mathbf{r}_2,\mathbf{r}_3)$. The diagrams are shown in Fig. 13. [22–24] In brief, this amounts in writing the triplet distribution function as

$$\begin{aligned} g_3(\mathbf{r}_1,\mathbf{r}_2,\mathbf{r}_3) &= 1 + h(\mathbf{r}_1,\mathbf{r}_2) + h(\mathbf{r}_1,\mathbf{r}_3) + h(\mathbf{r}_2,\mathbf{r}_3) \\ &\quad + h(\mathbf{r}_1,\mathbf{r}_2)h(\mathbf{r}_1,\mathbf{r}_3) + h(\mathbf{r}_1,\mathbf{r}_2)h(\mathbf{r}_2,\mathbf{r}_3) \\ &\quad + h(\mathbf{r}_1,\mathbf{r}_3)h(\mathbf{r}_2,\mathbf{r}_3) + \int d^3r_4 h(\mathbf{r}_1,\mathbf{r}_4)h(\mathbf{r}_2,\mathbf{r}_4)h(\mathbf{r}_3,\mathbf{r}_4) \\ &\quad + \text{terms with triplet correlation functions.} \end{aligned} \tag{5.55}$$

Here we have also introduced the short–hand notation $h(\mathbf{r}_1,\mathbf{r}_2) = g_2(\mathbf{r}_1,\mathbf{r}_2) - 1$. In the momentum space, this simplifies to

$$\begin{aligned} \mathcal{F}[g_3(\mathbf{r}_1,\mathbf{r}_2,\mathbf{r}_3) - 1] &= S(k_1)S(k_2)S(k_3) \\ &\quad \times [1 + u_3(\mathbf{k}_1,\mathbf{k}_2,\mathbf{k}_3)] - 1\,. \end{aligned} \tag{5.56}$$

We now ignore the triplet correlations for a moment, but return to them later on.

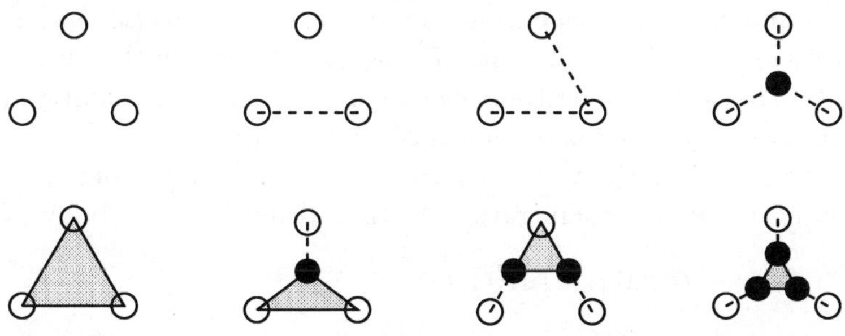

Fig. 13. Convolution approximation for $g_3(\mathbf{r}_1, \mathbf{r}_2, \mathbf{r}_3)$. The dots represent particle coordinates; the filled dots are coordinates that are integrated over. Dashed lines correspond to the functions $h(\mathbf{r}_1, \mathbf{r}_2)$, and triangles to the triplet–correlation functions $u_3(\mathbf{r}_1, \mathbf{r}_2, \mathbf{r}_3)$. There are three topologically identical combinations for the diagrams of the second, third, fifth, and sixth kind, just with different particle coordinates.

5.4.2. *Approximating the two–body continuity equation*

In the following, we evaluate the two–body terms in the two–body continuity equation. The simplest term to approximate is the driving term $D_2(\mathbf{r}_1, \mathbf{r}_2; t)$, given in Eq. (5.30). Introducing the convolution approximation (as already stated, without the triplet–correlation functions), we can write D_2 in the form

$$\begin{aligned} D_2(\mathbf{r}_1, \mathbf{r}_2; t) = \rho_0 \Big\{ & g_2(\mathbf{r}_1, \mathbf{r}_2)\left[D_1(\mathbf{r}_1; t) + D_1(\mathbf{r}_2; t)\right] \\ & + \frac{2\rho_0^2}{i\hbar} \int d^3r_3 U_{\rm ext}(\mathbf{r}_3; t) \\ & \times \left[h(\mathbf{r}_1, \mathbf{r}_3)h(\mathbf{r}_2, \mathbf{r}_3) + \int d^3r_4 h(\mathbf{r}_1, \mathbf{r}_4)h(\mathbf{r}_2, \mathbf{r}_4)h(\mathbf{r}_3, \mathbf{r}_4) \right] \Big\} . \end{aligned} \tag{5.57}$$

Recalling the expression (5.29) for the one–body driving term D_1, we immediately see that the last two lines can be combined, and as a result, we obtain the two–body driving term entirely in terms of $D_1(\mathbf{r}; t)$,

$$\begin{aligned} D_2(\mathbf{r}_1, \mathbf{r}_2; t) = \rho_0 \Big\{ & g_2(\mathbf{r}_1, \mathbf{r}_2)\left[D_1(\mathbf{r}_1; t) + D_1(\mathbf{r}_2; t)\right] \\ & + \rho_0 \int d^3r_3 h(\mathbf{r}_1, \mathbf{r}_3)h(\mathbf{r}_2, \mathbf{r}_3) D_1(\mathbf{r}_3; t) \Big\} . \end{aligned} \tag{5.58}$$

The next term to be approximated is the time derivative of the two–body density fluctuation. Using Eqs. (5.19) and (5.23), and making the convolution approximation results in

$$\begin{aligned} \delta\rho_2(\mathbf{r}_1, \mathbf{r}_2; t) = \rho_0 \Big\{ & g_2(\mathbf{r}_1, \mathbf{r}_2)\left[\delta\rho_1(\mathbf{r}_1; t) + \delta\rho_1(\mathbf{r}_2; t)\right] \\ & + \rho_0 \int d^3r_3 h(\mathbf{r}_1, \mathbf{r}_3)h(\mathbf{r}_2, \mathbf{r}_3) \delta\rho_1(\mathbf{r}_3; t) \Big\} \\ & + \rho_0^2 g_2(\mathbf{r}_1, \mathbf{r}_2)\delta u_2(\mathbf{r}_1, \mathbf{r}_2; t) + \mathcal{F}[\delta u_2] . \end{aligned} \tag{5.59}$$

Here we have eliminated the dependence on the correlation function $\delta u_1(\mathbf{r};t)$ in favor of the density $\delta\rho_1(\mathbf{r};t)$. The functional $\mathcal{F}[\delta u_2]$ includes rest of the terms containing $\delta u_2(\mathbf{r}_1,\mathbf{r}_2;t)$. They could be written down explicitly using the definition (5.19), but for discussing the CBF approximation this is immaterial.

Let us then consider the two–body current. It contains not only the one–body current, but also structure coming from the time–dependent two–body correlations,

$$\mathbf{j}_2(\mathbf{r}_1,\mathbf{r}_2;t) = \rho_0 g_2(\mathbf{r}_1,\mathbf{r}_2)\,\mathbf{j}_1(\mathbf{r}_1;t)$$

$$+\frac{\hbar\rho_0^2}{2mi}\Big\{g_2(\mathbf{r}_1,\mathbf{r}_2)\nabla_1\delta u_2(\mathbf{r}_1,\mathbf{r}_2;t)+\rho_0\int d^3r_3 \tag{5.60}$$

$$\times\big[g_3(\mathbf{r}_1,\mathbf{r}_2,\mathbf{r}_3)-g_2(\mathbf{r}_1,\mathbf{r}_2)g_2(\mathbf{r}_1,\mathbf{r}_3)\big]\nabla_1\delta u_2(\mathbf{r}_1,\mathbf{r}_3;t)\Big\}. \tag{5.61}$$

This expression is identical to Eq. (5.27), we have merely rewritten it differently.

Up to this stage, we have just rearranged the terms appearing in the two–body continuity equation within the convolution approximation. The central step of our derivation is to bring the two–body equation in a numerically tractable form. We adopt the general strategy of the uniform–limit approximation. [19] The approximation has turned out to be quite successful in the calculation of optimal static three–body correlations. [21–23] Its essence is to consider all products of two or more two–body functions small in coordinate space, but not necessarily in momentum space. This allows for long–range effects like phonons.

In the specific case we are dealing with here, the approximation amounts in taking $g_2(\mathbf{r}_1,\mathbf{r}_2)\delta u_2(\mathbf{r}_1,\mathbf{r}_2;t)\approx\delta u_2(\mathbf{r}_1,\mathbf{r}_2;t)$, and similarly for $\nabla_1\delta u_2(\mathbf{r}_1,\mathbf{r}_2;t)$. While it places more emphasis on the structure of $\delta u_2(\mathbf{r}_1,\mathbf{r}_2;t)$, it is physically appealing simply because it removes the *redundant* relevant short–range structure shared by both $g_2(\mathbf{r}_1,\mathbf{r}_2)$ and $\delta u_2(\mathbf{r}_1,\mathbf{r}_2;t)$. Invoking the equivalent uniform limit for the three–body distribution function, the terms in Eq. (5.60) depending on $\delta u_2(\mathbf{r}_1,\mathbf{r}_2;t)$ read

$$\frac{\hbar\rho_0^2}{2mi}\Big\{g_2(\mathbf{r}_1,\mathbf{r}_2)\nabla_1\delta u_2(\mathbf{r}_1,\mathbf{r}_2;t)+\rho_0\int d^3r_3$$

$$\times[g_3(\mathbf{r}_1,\mathbf{r}_2,\mathbf{r}_3)-g_2(\mathbf{r}_1,\mathbf{r}_3)g_2(\mathbf{r}_1,\mathbf{r}_2)]\nabla_1\delta u_2(\mathbf{r}_1,\mathbf{r}_3;t)\Big\}$$

$$\approx\frac{\hbar\rho_0^3}{2mi}\int d^3r_3\,[\delta(\mathbf{r}_3-\mathbf{r}_2)+h(\mathbf{r}_3,\mathbf{r}_2)]\,\nabla_1\delta u_2(\mathbf{r}_1,\mathbf{r}_3;t)\,.$$

We now have the necessary elements to construct the approximate two–body continuity equation. Collecting the above terms together, we have

$$\nabla_1\cdot\Big\{g_2(\mathbf{r}_1,\mathbf{r}_2)\,\mathbf{j}_1(\mathbf{r}_1;t)+\frac{\hbar\rho_0^2}{2mi}\int d^3r_3\,[\delta(\mathbf{r}_3-\mathbf{r}_2)+h(\mathbf{r}_3,\mathbf{r}_2)]\,\nabla_1\delta u_2(\mathbf{r}_1,\mathbf{r}_3;t)\Big\}$$

$$+\text{same with }\ (1\leftrightarrow 2)$$

$$=g_2(\mathbf{r}_1,\mathbf{r}_2)\Big[D_1(\mathbf{r}_1;t)-\delta\dot\rho_1(\mathbf{r}_1;t)+D_1(\mathbf{r}_2;t)-\delta\dot\rho_1(\mathbf{r}_2;t)\Big]$$

$$+\rho_0 \int d^3r_3 h(\mathbf{r}_1,\mathbf{r}_3)h(\mathbf{r}_2,\mathbf{r}_3)\left[D_1(\mathbf{r}_3;t) - \delta\dot{\rho}_1(\mathbf{r}_3;t)\right]$$
$$+\rho_0 \delta\dot{u}_2(\mathbf{r}_1,\mathbf{r}_2;t)\,. \tag{5.62}$$

As already mentioned, from the terms containing the time derivative $\delta\dot{u}_2(\mathbf{r}_1,\mathbf{r}_2;t)$ we have retained the leading term in accordance with the uniform–limit approximation, and ignored those collected in $\mathcal{F}[\delta\dot{u}_2]$. Again, we can rewrite this using the one–body continuity equation to replace the one–body quantities with one–body currents to arrive at our final approximate form,

$$\begin{aligned}
&\left\{\frac{\hbar\rho_0^2}{2mi}\int d^3r_3\left[\delta(\mathbf{r}_3-\mathbf{r}_2)+h(\mathbf{r}_3,\mathbf{r}_2)\right]\nabla_1^2\delta u_2(\mathbf{r}_1,\mathbf{r}_3;t)\right.\\
&\qquad\left.+\text{same with }\ (1\leftrightarrow 2)\right\} - \rho_0\delta\dot{u}_2(\mathbf{r}_1,\mathbf{r}_2;t) \qquad (5.63)\\
&= \mathbf{j}_1(\mathbf{r}_1;t)\cdot\nabla_1 g_2(\mathbf{r}_1,\mathbf{r}_2) + \mathbf{j}_1(\mathbf{r}_2;t)\cdot\nabla_2 g_2(\mathbf{r}_1,\mathbf{r}_2)\\
&\qquad + \rho_0\int d^3r_3 h(\mathbf{r}_1,\mathbf{r}_3)h(\mathbf{r}_2,\mathbf{r}_3)\nabla_3\cdot\mathbf{j}_1(\mathbf{r}_3;t)\,.
\end{aligned}$$

The last step is to decouple the equations of motion. We can do this by approximating the one–body current, given in Eq. (5.26), by the Feynman current,

$$\begin{aligned}
\mathbf{j}(\mathbf{r};t) &= \frac{\hbar\rho_0}{2mi}\nabla_1\delta u_1(\mathbf{r};\mathbf{t}) \qquad (5.64)\\
&= \frac{\hbar\rho_0}{2mi}\nabla_1\left[\delta\rho_1(\mathbf{r}_1;t) - \rho_0\int d^3r_2 X(\mathbf{r}_1,\mathbf{r}_2)\delta\rho_1(\mathbf{r}_2;t)\right],
\end{aligned}$$

in other words, by dropping the terms with the fluctuating two–body correlation function from the equation. Here $X(\mathbf{r}_1,\mathbf{r}_2)$ is the direct correlation function, familiar from the ground–state calculations.

We have now expressed the fluctuating two–body function in closed form as a functional of one–body quantities alone. To better see this, let us put everything together in momentum space. The second continuity equation reads

$$\left[\hbar\omega - \varepsilon_{\mathrm{F}}\left(\left|\tfrac{\mathbf{k}}{2}+\mathbf{p}\right|\right) - \varepsilon_{\mathrm{F}}\left(\left|\tfrac{\mathbf{k}}{2}-\mathbf{p}\right|\right)\right] S\left(\left|\tfrac{\mathbf{k}}{2}+\mathbf{p}\right|\right) S\left(\left|\tfrac{\mathbf{k}}{2}-\mathbf{p}\right|\right)\delta u_2(\mathbf{k},\mathbf{p};\omega)$$
$$+\varepsilon_{\mathrm{F}}(k)\sigma_{\mathbf{k}}(\mathbf{p})\delta\tilde{\rho}_1(\mathbf{k};\omega) = 0\,. \tag{5.65}$$

Here $\sigma_{\mathbf{k}}(\mathbf{p})$ is just a short–hand notation collecting together terms containing the ground–state structure,

$$\begin{aligned}
\sigma_{\mathbf{k}}(\mathbf{p}) &= -\frac{1}{k^2}\left[\mathbf{k}\cdot\left(\tfrac{\mathbf{k}}{2}+\mathbf{p}\right) S\left(\left|\tfrac{\mathbf{k}}{2}-\mathbf{p}\right|\right) + (\mathbf{p}\leftrightarrow -\mathbf{p})\right]\\
&\quad + S\left(\left|\tfrac{\mathbf{k}}{2}+\mathbf{p}\right|\right) S\left(\left|\tfrac{\mathbf{k}}{2}-\mathbf{p}\right|\right)\left[1+\tilde{u}_3\left(\tfrac{\mathbf{k}}{2}+\mathbf{p},\tfrac{\mathbf{k}}{2}-\mathbf{p},-\mathbf{k}\right)\right]. \qquad (5.66)
\end{aligned}$$

We have also written down the triplet correlation function $\tilde{u}_3$ which we did not carry along in the derivations. The reader can easily verify that adding them brings only length into the equations, and that everything finally results in the form given above.

It is now evident that we can solve directly the fluctuating two–body correlation function in terms of the one–body density fluctuation,

$$\frac{\delta u_2(\mathbf{k},\mathbf{p};\omega)}{\delta\tilde{\rho}_1(\mathbf{k};\omega)} = -\frac{\varepsilon_{\rm F}(k)\sigma_{\mathbf{k}}(\mathbf{p})\left[S\left(\left|\frac{\mathbf{k}}{2}+\mathbf{p}\right|\right)S\left(\left|\frac{\mathbf{k}}{2}-\mathbf{p}\right|\right)\right]^{-1}}{\hbar\omega-\varepsilon_{\rm F}\left(\left|\frac{\mathbf{k}}{2}+\mathbf{p}\right|\right)-\varepsilon_{\rm F}\left(\left|\frac{\mathbf{k}}{2}-\mathbf{p}\right|\right)}\,. \tag{5.67}$$

We will need this expression to construct the self–energy when we solve the one–body equation to get our hands finally on the dynamic response function.

5.4.3. *Solving the one–body equation: dynamic response*

Having worked out the two–body part, let us now return to the first continuity equation (5.21) with the one–body current (5.26) and one–body density (5.23), and see how the terms are spelled out within the convolution approximation. Again, we want to change variables from $\delta u_1(\mathbf{r}_1;t)$ to $\delta\rho_1(\mathbf{r}_1;t)$. By using the "renormalized" correlation function

$$\begin{aligned}\delta\bar{u}_1(\mathbf{r}_1;t) &= \delta u_1(\mathbf{r}_1;t)+\rho_0\int d^3r_2 g_2(\mathbf{r}_1,\mathbf{r}_2)\delta u_2(\mathbf{r}_1,\mathbf{r}_2;t)\\ &+\frac{1}{2}\rho_0^2\int d^3r_2d^3r_3h(\mathbf{r}_2,\mathbf{r}_1)h(\mathbf{r}_3,\mathbf{r}_1)\delta u_2(\mathbf{r}_2,\mathbf{r}_3;t)\,,\end{aligned} \tag{5.68}$$

we can write the one–body density fluctuation (5.23) as

$$\delta\rho_1(\mathbf{r}_1;t) = \rho_0\;\delta\bar{u}_1(\mathbf{r}_1;t)+\rho_0^2\int d^3r_2h(\mathbf{r}_1,\mathbf{r}_2)\delta\bar{u}_1(\mathbf{r}_2;t)\,. \tag{5.69}$$

We can easily solve this for $\delta\bar{u}_1(\mathbf{r}_1;t)$. The result is

$$\rho_0\delta\bar{u}_1(\mathbf{r}_1;t) = \delta\rho_1(\mathbf{r}_1;t)-\rho_0\int d^3r_2X(\mathbf{r}_1,\mathbf{r}_2)\delta\rho_1(\mathbf{r}_2;t)\,. \tag{5.70}$$

The one–body correlation function is then simply

$$\begin{aligned}\rho_0\delta u_1(\mathbf{r}_1;t) &= \delta\rho_1(\mathbf{r}_1;t)-\rho_0\int d^3r_2X(\mathbf{r}_1,\mathbf{r}_2)\delta\rho_1(\mathbf{r}_2;t)\\ &-\rho_0^2\int d^3r_2g_2(\mathbf{r}_1,\mathbf{r}_2)\delta u_2(\mathbf{r}_1,\mathbf{r}_2;t)\\ &-\frac{1}{2}\rho_0^3\int d^3r_2d^3r_3h(\mathbf{r}_2,\mathbf{r}_1)h(\mathbf{r}_3,\mathbf{r}_1)\delta u_2(\mathbf{r}_2,\mathbf{r}_3;t)\,.\end{aligned} \tag{5.71}$$

Taking the gradient and inserting the result into equation (5.26), we get the one–body current

$$\begin{aligned}\mathbf{j}_1(\mathbf{r}_1;t) = \frac{\hbar}{2mi}\Bigg\{&\nabla_1\left[\delta\rho_1(\mathbf{r}_1;t)-\rho_0\int d^3r_2X(\mathbf{r}_1,\mathbf{r}_2)\delta\rho_1(\mathbf{r}_2;t)\right]\\ &-\rho_0^2\int d^3r_2\delta u_2(\mathbf{r}_1,\mathbf{r}_2;t)\nabla_1g_2(\mathbf{r}_1,\mathbf{r}_2)\\ &-\frac{1}{2}\rho_0^3\int d^3r_2d^3r_3\nabla_1\left[h(\mathbf{r}_2,\mathbf{r}_1)h(\mathbf{r}_3,\mathbf{r}_1)\right]\delta u_2(\mathbf{r}_2,\mathbf{r}_3;t)\Bigg\}\,.\end{aligned} \tag{5.72}$$

This is all the adjusting we need to do for the one–body continuity equation.

To connect with the derivations we made in the previous section, we write the equation (5.21) in momentum space,

$$\begin{aligned}&[\hbar\omega - \varepsilon_{\rm F}(k)]\delta\tilde{\rho}_1(\mathbf{k};\omega)\\&\quad + \frac{\hbar^2 k^2}{4m}\int \frac{d^3p}{(2\pi)^3\rho_0}\sigma_{\mathbf{k}}(\mathbf{p})\delta u_2(\mathbf{k},\mathbf{p};\omega) = 2\rho_0 S(k)\tilde{U}_{ext}(k,\omega)\,,\end{aligned} \tag{5.73}$$

where $\sigma_{\mathbf{k}}(\mathbf{p})$ is exactly the same short–hand notation as in the two–body equation, and given in Eq. (5.66). From this, we can solve the density fluctuations

$$\delta\tilde{\rho}_1(\mathbf{k};\omega) = \frac{2\rho_0 S(k)\tilde{U}_{ext}(k,\omega)}{\hbar\omega - \varepsilon_{\rm F}(k) - \Sigma(\mathbf{k},\omega)} \tag{5.74}$$

with the self–energy defined as

$$\Sigma(\mathbf{k},\omega) = -\frac{\hbar^2 k^2}{4m}\int \frac{d^3p}{(2\pi)^3\rho_0}\sigma_{\mathbf{k}}(\mathbf{p})\frac{\delta u_2(\mathbf{k},\mathbf{p};\omega)}{\delta\tilde{\rho}_1(\mathbf{k};\omega)}\,. \tag{5.75}$$

The ratio $\delta u_2(\mathbf{k},\mathbf{p};\omega)/\delta\tilde{\rho}_1(\mathbf{k};\omega)$ is exactly what we got out of the two–body equation (5.67) in closed form.

Using the expression given in Eq. (5.37) for the external potential and following the derivations of the Feynman approximation, we can write the linear response function in the form of Eq. (5.3),

$$\chi(k,\omega) = S(k)\left[\frac{1}{\hbar\omega - \varepsilon(k) - \Sigma(k,\omega)} - \frac{1}{\hbar\omega + \varepsilon(k) + \Sigma^*(k,-\omega)}\right]\,. \tag{5.76}$$

This completes the solution. All that is left to do are the numerical integrations. Notice that we can rearrange Eq. (5.75) a bit. Inserting our solution for $\delta u_2/\delta\tilde{\rho}_1$, changing the variables $\frac{\mathbf{k}}{2}+\mathbf{p} \to -\mathbf{p}$ and $\frac{\mathbf{k}}{2}-\mathbf{p} \to -\mathbf{q}$, and introducing the Dirac delta function to ensure momentum conservation $\mathbf{k} = -\mathbf{p} - \mathbf{q}$, the self–energy correction takes the form incountered in the literature, [38]

$$\Sigma^{\rm CBF}(k,\omega) = \frac{1}{2}\int \frac{d^3p\, d^3q}{(2\pi)^3\rho_0}\delta(\mathbf{k}+\mathbf{p}+\mathbf{q})\frac{|V_3(\mathbf{k};\mathbf{p},\mathbf{q})|^2}{\hbar\omega - \varepsilon_{\rm F}(p) - \varepsilon_{\rm F}(q)}\,. \tag{5.77}$$

The three–excitation ("three–phonon") coupling matrix element,

$$\begin{aligned}V_3(\mathbf{k};\mathbf{p},\mathbf{q}) &= -\frac{\hbar^2}{2m}\frac{1}{\sqrt{S(p)S(q)S(k)}}\\&\quad\times\left[\mathbf{k}\cdot\mathbf{p}\tilde{S}(q) + \mathbf{k}\cdot\mathbf{q}\tilde{S}(p) + k^2 S(p)S(q)\left(1+\tilde{u}_3(\mathbf{k},\mathbf{p},\mathbf{q})\right)\right]\\&= \frac{\hbar^2}{2m}\sqrt{\frac{S(p)S(q)}{S(k)}}[\mathbf{k}\cdot\mathbf{p}\tilde{X}(p) + \mathbf{k}\cdot\mathbf{q}\tilde{X}(q) - k^2\tilde{u}_3(\mathbf{k},\mathbf{p},\mathbf{q})]\end{aligned} \tag{5.78}$$

is given in terms of the static structure function $S(k)$, the direct correlation function $\tilde{X}(k) = 1 - S(k)^{-1}$, and the three–body correlation function $\tilde{u}_3$. These are all quantities that we know from the ground–state calculations.

Referring to Eq. (5.4), the imaginary part of the self–energy is determined by the poles at

$$\hbar\omega - \varepsilon_{\rm F}(p) - \varepsilon_{\rm F}(q) = 0\,. \tag{5.79}$$

The critical value is

$$\hbar\omega_{\rm crit}(k) \equiv \min_{\mathbf{q}} \left[\varepsilon(q) + \varepsilon(|\mathbf{k}+\mathbf{q}|)\right]\,. \tag{5.80}$$

Above that energy the self–energy is complex. Moreover, it follows from Eq. (5.77) that

$$\frac{d\Sigma(k,\omega)}{d\omega} < 0 \qquad \text{for} \qquad \hbar\omega < \hbar\omega_{\rm crit}(k)\,. \tag{5.81}$$

Thus the self–energy contribution lowers the excitation energy of the collective mode below the reference single–excitation mode. Above the critical energy $\hbar\omega > \hbar\omega_{\rm crit}(k)$, excitations can decay into two Feynman excitations. Of course, in actual physical systems the decay thresholds are determined by the collective modes themselves. This deficiency reflects, for example, in the fact that the calculated two–roton limit is not twice the roton energy: instead, it is given by twice the energy of the Feynman roton.

To summarize this part, we went beyond the Feynman approximation by allowing for fluctuating two–body correlations. From the physical point of view, this means adding three–phonon scattering processes into the description: an excitation can decay into two, provided that energy and momentum are conserved. Thus, we can study, for example, the anomalous dispersion of phonons in liquid ^{4}He. This added flexibility also lowers the excitation energies considerably towards measured values. We introduced first the convolution approximation to get rid of the three–body distribution function, then the uniform–limit approximation to be able to solve the two–body equation. This gave us the fluctuating two–body correlation function in terms of the fluctuating one–body density. We were then able to solve the first continuity equation and extract the dynamic response function.

In the spirit of the ground–state calculation, we might next look for ways to change variables from δu_2 to the fluctuating pair–distribution function δg_2. Namely, simple analysis shows that had we chosen to optimize both the correlation functions and not replaced δu_1 with the one–body density fluctuation, we would have ended up with the free–particle spectrum as our reference spectrum. This suggests that we can do quantitatively better and bring the results into an even closer agreement with the experimental data by applying the least–action principle to observable quantities.

5.5. *Beyond the CBF approximation ("full optimization")*

5.5.1. *Continuity equations revisited*

Recalling Eq. (5.19) for the n–particle density, we can calculate general expressions for the gradients of the one– and two–body densities appearing in the continuity equations,

$$\begin{aligned}\nabla_1 \delta\rho_1(\mathbf{r}_1;t) &= \rho_0 \nabla_1 \delta u_1(\mathbf{r}_1;t) \\ &+\rho_0^2 \int d^3r_2 g_2(\mathbf{r}_1,\mathbf{r}_2)\nabla_1 \delta u_2(\mathbf{r}_1,\mathbf{r}_2;t) \\ &+ \int d^3r_2 \delta\rho_2(\mathbf{r}_1,\mathbf{r}_2;t)\nabla_1 u_2(\mathbf{r}_1,\mathbf{r}_2)\end{aligned} \tag{5.82}$$

and

$$\begin{aligned}\nabla_1 \delta\rho_2(\mathbf{r}_1,\mathbf{r}_2;t) &= \delta\rho_2(\mathbf{r}_1,\mathbf{r}_2;t)\nabla_1 u_2(\mathbf{r}_1,\mathbf{r}_2) \\ &+\rho_0 g_2(\mathbf{r}_1,\mathbf{r}_2)\nabla_1 \left[\delta u_1(\mathbf{r}_1;t) + \delta u_2(\mathbf{r}_1,\mathbf{r}_2;t)\right] \\ &+\rho_0^3 \int d^3r_3 g_3(\mathbf{r}_1,\mathbf{r}_2,\mathbf{r}_3)\nabla_1 \delta u_2(\mathbf{r}_1,\mathbf{r}_3;t) \\ &+ \int d^3r_3 \delta\rho_3(\mathbf{r}_1,\mathbf{r}_2,\mathbf{r}_3;t)\nabla_1 u_2(\mathbf{r}_1,\mathbf{r}_3)\,.\end{aligned} \tag{5.83}$$

These are exact within the chosen wave function. For the above equations, we have assumed that the ground state is constructed only of one– and two–body correlations. Again, including the triplet correlations only adds some length to the already lengthy expressions.

The terms depending on $\delta u_1(\mathbf{r};t)$ and $\delta u_2(\mathbf{r}_1,\mathbf{r}_2;t)$ in Eqs. (5.82) and (5.83) appear also in the expressions for the one– and two–body currents in Eqs. (5.26) and (5.27). This similarity stems from the parallel ways they are derived from the modified Jastrow–Feenberg wave function. We can thus eliminate them and write the currents solely in terms of fluctuating densities. This gives

$$\mathbf{j}_1(\mathbf{r}_1;t) = \frac{\hbar}{2mi}[\nabla_1 \delta\rho_1(\mathbf{r}_1;t) - \int d^3r_2 \delta\rho_2(\mathbf{r}_1,\mathbf{r}_2;t)\nabla u_2(\mathbf{r}_1,\mathbf{r}_2)] \tag{5.84}$$

$$\begin{aligned}\mathbf{j}_2(\mathbf{r}_1,\mathbf{r}_2;t) &= \frac{\hbar}{2mi}[\nabla_1 \delta\rho_2(\mathbf{r}_1,\mathbf{r}_2;t) - \delta\rho_2(\mathbf{r}_1,\mathbf{r}_2;t)\nabla_1 u_2(\mathbf{r}_1,\mathbf{r}_2) \\ &- \int d^3r_3 \delta\rho_3(\mathbf{r}_1,\mathbf{r}_2,\mathbf{r}_3;t)\nabla_1 u_2(\mathbf{r}_1,\mathbf{r}_3)]\,.\end{aligned} \tag{5.85}$$

Here Eq. (5.85) contains the three–body density variation, $\delta\rho_3(\mathbf{r}_1,\mathbf{r}_2,\mathbf{r}_3;t)$. Following the same reasoning as in deriving the CBF approximation, we must formulate it in terms of one– and two–body density fluctuations.

From the definition of the n–particle density (5.17), we can derive the so–called Born–Green–Yvon (BGY) equations simply by applying the gradient operator,

$$\begin{aligned}\nabla_1 g_2(\mathbf{r}_1,\mathbf{r}_2) &= g_2(\mathbf{r}_1,\mathbf{r}_2)\nabla_1 u_2(\mathbf{r}_1,\mathbf{r}_2) \\ &+\rho_0 \int d^3r_3 g_3(\mathbf{r}_1,\mathbf{r}_2,\mathbf{r}_3)\nabla_1 u_2(\mathbf{r}_1,\mathbf{r}_3)\,.\end{aligned} \tag{5.86}$$

We need this relation to construct the continuity equations: this is exactly the equation that gives us the tool to change variables from the fluctuating two–body correlation function to the fluctuating pair–distribution function δg_2. Additionally, because we started from a homogeneous ground state, the gradient with respect to one–body density vanishes provided that

$$\rho_0 \int d^3r_2 g_2(\mathbf{r}_1,\mathbf{r}_2)\nabla_1 u_2(\mathbf{r}_1,\mathbf{r}_2) = 0\,. \tag{5.87}$$

Making these substitutions, and writing

$$\delta g_2(\mathbf{r}_1,\mathbf{r}_2) = g_2(\mathbf{r}_1,\mathbf{r}_2)\delta(\mathbf{r}_1,\mathbf{r}_2;t) \tag{5.88}$$

(similarly for δg_3), we get

$$\mathbf{j}_1(\mathbf{r}_1;t) = \frac{\hbar}{2mi}\Big\{\nabla_1\delta\rho_1(\mathbf{r}_1;t) \tag{5.89}$$

$$-\rho_0\int d^3r_2 g_2(\mathbf{r}_1,\mathbf{r}_2)\left[\delta\rho_1(\mathbf{r}_2;t)+\rho_0\delta(\mathbf{r}_1,\mathbf{r}_2;t)\right]\nabla_1 u_2(\mathbf{r}_1,\mathbf{r}_2)\Big\}$$

$$\mathbf{j}_2(\mathbf{r}_1,\mathbf{r}_2;t) = \frac{\hbar\rho_0}{2mi}\Big\{g_2(\mathbf{r}_1,\mathbf{r}_2)\nabla_1\left[\delta\rho_1(\mathbf{r}_1;t)+\rho_0\delta(\mathbf{r}_1,\mathbf{r}_2;t)\right]$$

$$-\rho_0\int d^3r_3\; g_3(\mathbf{r}_1,\mathbf{r}_2,\mathbf{r}_3)\Big[\delta\rho_1(\mathbf{r}_3;t)+\rho_0\delta(\mathbf{r}_1,\mathbf{r}_3;t)$$

$$+\rho_0\delta(\mathbf{r}_2,\mathbf{r}_3;t)+\rho_0\delta(\mathbf{r}_1,\mathbf{r}_2,\mathbf{r}_3;t)\Big]\nabla_1 u_2(\mathbf{r}_1,\mathbf{r}_3)\Big\}\,. \tag{5.90}$$

The problem we are facing now is identical to the one we dealt with when deriving the CBF approximation: the currents contain three unknown quantities, including the fluctuating three–body distribution function. A natural truncation would be to set $\delta(\mathbf{r}_1,\mathbf{r}_2,\mathbf{r}_3;t) = 0$, because we have so far ignored also the fluctuating three–body correlation functions. This approximation is consistent with making the so–called Kirkwood superposition approximation for the triplet distribution function, in which we set

$$g_3(\mathbf{r}_1,\mathbf{r}_2,\mathbf{r}_3) \approx g_2(\mathbf{r}_1,\mathbf{r}_2)g_2(\mathbf{r}_1,\mathbf{r}_3)g_2(\mathbf{r}_2,\mathbf{r}_3)\,. \tag{5.91}$$

Compared to the convolution approximation we made in constructing the CBF equations, this is a real–space factorization in terms of three pair–distribution functions, whereas the convolution approximation was a momentum–space factorization using three static structure functions. The convolution approximation works better at low momenta, the superposition approximation at high momenta. It is also worthwhile to point out that in this approximation we can take the ground–state pair–distribution function as a common factor in the two–body current, hence in liquid ^{4}He the current won't exhibit any spurious flow inside the hard–core radius. The approximation has also the benefit that after linearization the three–body density fluctuation can be expressed simply as a sum of one– and two–body density fluctuations.

Within these approximations and notations, we can write the one–body continuity equation as

$$\frac{\hbar}{2mi}\nabla_1\cdot\Big\{\nabla_1\delta\rho_1(\mathbf{r}_1;t)$$

$$-\rho_0 \int d^3r_2 g_2(\mathbf{r}_1,\mathbf{r}_2)\nabla_1 u_2(\mathbf{r}_1,\mathbf{r}_2)\big[\delta\rho_1(\mathbf{r}_2;t)+\rho_0\delta(\mathbf{r}_1,\mathbf{r}_2;t)\big]\Bigg\} \tag{5.92}$$
$$+\delta\dot{\rho}_1(\mathbf{r}_1;t) = D_1(\mathbf{r}_1;t)$$

and the two–body equation takes the somewhat lengthy form

$$\begin{aligned}
&\frac{\hbar\rho_0}{2mi}\nabla_1\cdot\Bigg\{h(\mathbf{r}_1,\mathbf{r}_2)\nabla_1\Big[\delta\rho_1(\mathbf{r}_1;t)+\rho_0\delta(\mathbf{r}_1,\mathbf{r}_2;t)\Big]+\rho_0\nabla_1\delta(\mathbf{r}_1,\mathbf{r}_2;t)\\
&\quad -\rho_0\int d^3r_3\big[g_2(\mathbf{r}_1,\mathbf{r}_2)g_2(\mathbf{r}_2,\mathbf{r}_3)-1\big]\\
&\quad \times\big[\delta\rho_1(\mathbf{r}_3;t)+\rho_0\delta(\mathbf{r}_1,\mathbf{r}_3;t)+\rho_0\delta(\mathbf{r}_2,\mathbf{r}_3;t)\big]g_2(\mathbf{r}_1,\mathbf{r}_3)\nabla_1 u_2(\mathbf{r}_1,\mathbf{r}_3)\\
&\quad -\rho_0^2\int d^3r_3\delta(\mathbf{r}_2,\mathbf{r}_3;t)g_2(\mathbf{r}_1,\mathbf{r}_3)\nabla_1 u_2(\mathbf{r}_1,\mathbf{r}_3)\Bigg\} + \text{ same with } (1\leftrightarrow 2)\\
&\quad +h(\mathbf{r}_1,\mathbf{r}_2)\big[\delta\dot{\rho}_1(\mathbf{r}_1;t)+\delta\dot{\rho}_1(\mathbf{r}_2;t)+\rho_0\dot{\delta}(\mathbf{r}_1,\mathbf{r}_2;t)\big]\\
&= \left[D_2(\mathbf{r}_1,\mathbf{r}_2;t)-D_1(\mathbf{r}_1;t)-D_1(\mathbf{r}_2;t)\right].
\end{aligned} \tag{5.93}$$

5.5.2. *Solving the continuity equations in momentum space*

Above, we reformulated the continuity equations by taking $\delta\rho_1$ and δg_2 as our functions to be optimized through the use of the BGY equations. We also applied the superposition approximation to express the three–body density fluctuation in terms of one– and two–body quantities. To solve these two coupled equations, we again transform them into momentum space. The one–body continuity equation (5.92) takes the familiar form

$$\hbar\omega-\varepsilon(k)-\Sigma(k,\omega)=\frac{2\rho_0 S(k)\tilde{U}_{ext}(k,\omega)}{\delta\tilde{\rho}_1(\mathbf{k};\omega)}, \tag{5.94}$$

and the linear response function can be calculated from Eq. (5.3), except that now the single–mode spectrum $\varepsilon(k)$ is not the bare Feynman mode, but contains a correction term which becomes significant at higher momenta.

We can see this by picking out the terms from the one–body continuity equation that transform into $\varepsilon(k)$, and calculating the Fourier transform

$$k^2Q(k)=\mathcal{F}\left[\nabla_1^2\delta\rho_1(\mathbf{r}_1;t)-\rho_0\int d^3r_2\nabla_1\cdot g_2(\mathbf{r}_1,\mathbf{r}_2)\nabla_1 u_2(\mathbf{r}_1,\mathbf{r}_2)\delta\rho_1(\mathbf{r}_2;t)\right]. \tag{5.95}$$

We can evaluate this using the HNC equation

$$g_2(\mathbf{r}_1,\mathbf{r}_2)=\exp\left[u_2(\mathbf{r}_1,\mathbf{r}_2)+N(\mathbf{r}_1,\mathbf{r}_2)+E(\mathbf{r}_1,\mathbf{r}_2)\right], \tag{5.96}$$

which gives

$$\begin{aligned}
g_2(\mathbf{r}_1,\mathbf{r}_2)\nabla_1 u_2(\mathbf{r}_1,\mathbf{r}_2) &= \nabla_1 g_2(\mathbf{r}_1,\mathbf{r}_2)+g_2(\mathbf{r}_1,\mathbf{r}_2)\big[\nabla_1 N(\mathbf{r}_1,\mathbf{r}_2)+\nabla_1 E(\mathbf{r}_1,\mathbf{r}_2)\big]\\
&= \nabla_1 X(\mathbf{r}_1,\mathbf{r}_2)-\nabla_1 E(\mathbf{r}_1,\mathbf{r}_2)\\
&\quad -h(\mathbf{r}_1,\mathbf{r}_2)\big[\nabla_1 N(\mathbf{r}_1,\mathbf{r}_2)+\nabla_1 E(\mathbf{r}_1,\mathbf{r}_2)\big].
\end{aligned} \tag{5.97}$$

Here $N(\mathbf{r}_1, \mathbf{r}_2)$ is the sum of nodal diagrams and $E(\mathbf{r}_1, \mathbf{r}_2)$ the sum of elementary diagrams known from the ground–state calculation.

In the long–wavelength limit, the leading term is the Fourier transform of the direct correlation function, given by

$$\tilde{X}(k) = 1 - \frac{1}{S(k)} \,. \tag{5.98}$$

We can then write

$$Q(k) = 1 - \frac{1}{S(k)} + I(k) \,, \tag{5.99}$$

where $I(k)$ contains rest of the terms and stands for the integral

$$I(k) = -E(k) - \int \frac{d^3q}{(2\pi)^3 \rho k^2} \mathbf{k}\cdot\mathbf{q}\big[S(|\mathbf{k}-\mathbf{q}|) - 1\big]\big[N(q) + E(q)\big] \,. \tag{5.100}$$

This leads to the expression

$$\varepsilon(k) = \frac{\hbar^2 k^2}{2mS(k)}[1 - S(k)I(k))] \tag{5.101}$$

for the reference spectrum. We see that the correction term, $S(k)I(k)$, to the Feynman spectrum is positive; therefore the correction lowers the reference energy. Note also that had we included the three–body correlations, we should add the term

$$\frac{\rho}{2}\int d^3r g_3(\mathbf{r}_1, \mathbf{r}_2, \mathbf{r}_3)\nabla_1 u_3(\mathbf{r}_1, \mathbf{r}_2, \mathbf{r}_3) \tag{5.102}$$

into the elementary diagrams to retain consistency.

In this improved theory the self–energy $\Sigma(k, \omega)$ is given by the integral

$$\Sigma(k,\omega) = -\frac{\hbar^2}{2m}\int \frac{d^3p}{(2\pi)^3\rho}\, \mathbf{k}\cdot\left(\tfrac{\mathbf{k}}{2} + \mathbf{p}\right) Q\left(\left|\tfrac{\mathbf{k}}{2} + \mathbf{p}\right|\right) \rho_0 \frac{\delta(\mathbf{k},\mathbf{p};\omega)}{\delta\tilde{\rho}_1(\mathbf{k};\omega)} \,. \tag{5.103}$$

In the CBF approximation we were able to solve the second continuity equation for $\delta u_2/\delta\tilde{\rho}_1$ explicitly, and to insert that into the self–energy integral. Now the second continuity equation is an integral equation for $\delta(\mathbf{k},\mathbf{p};\omega)/\delta\tilde{\rho}_1(\mathbf{k};\omega)$ and it cannot be solved in closed form.

The singularity structure of both the self–energy and the second continuity equation can be made more transparent by introducing the notation

$$\beta_{\mathbf{k},\omega}(\mathbf{p}) \equiv \left[\hbar\omega - \varepsilon\left(\left|\tfrac{\mathbf{k}}{2} + \mathbf{p}\right|\right) - \varepsilon\left(\left|\tfrac{\mathbf{k}}{2} - \mathbf{p}\right|\right)\right] \rho_0 \frac{\delta(\mathbf{k},\mathbf{p};\omega)}{\delta\rho_1(\mathbf{k};\omega)} \,.$$

The self–energy now reads

$$\Sigma(k,\omega) = -\frac{\hbar^2}{2m}\int \frac{d\mathbf{p}}{(2\pi)^3\rho_0} \frac{\mathbf{k}\cdot\left(\frac{\mathbf{k}}{2} + \mathbf{p}\right) Q\left(\left|\frac{\mathbf{k}}{2} + \mathbf{p}\right|\right)\beta_{\mathbf{k},\omega}(\mathbf{p})}{\hbar\omega - \varepsilon\left(\left|\frac{\mathbf{k}}{2} + \mathbf{p}\right|\right) - \varepsilon\left(\left|\frac{\mathbf{k}}{2} - \mathbf{p}\right|\right)} \,. \tag{5.104}$$

The function $\beta_{\mathbf{k},\omega}(\mathbf{p})$ is to be solved self–consistently from the second continuity equation, which is now an integral equation for $\beta_{\mathbf{k},\omega}(\mathbf{p})$,

$$\beta_{\mathbf{k},\omega}(\mathbf{p}) = [\hbar\omega - \varepsilon(k)]\, M_{\mathbf{k}}(\mathbf{p}) + N_{\mathbf{k}}(\mathbf{p}) \tag{5.105}$$

$$+ \int \frac{d^3q}{(2\pi)^3\rho} \frac{\beta_{\mathbf{k},\omega}(\mathbf{q})\,[K_{\mathbf{k}}(\mathbf{p},\mathbf{q}) - \hbar\omega s(|\mathbf{p}-\mathbf{q}|)]}{\hbar\omega - \varepsilon\left(|\frac{\mathbf{k}}{2}+\mathbf{q}|\right) - \varepsilon\left(|\frac{\mathbf{k}}{2}-\mathbf{q}|\right)}\,.$$

With the aid of the notations

$$s(\mathbf{k}) = S(\mathbf{k}) - 1 \tag{5.106}$$

$$s^{\mathrm{D}}(\mathbf{k}) = \delta(\mathbf{k}) + s(\mathbf{k})\,, \tag{5.107}$$

we can write

$$M_{\mathbf{k}}(\mathbf{p}) = \frac{1}{S(k)} \int \frac{d\mathbf{q}}{(2\pi)^d\rho} s(|\tfrac{\mathbf{k}}{2}+\mathbf{q}|)s(|\tfrac{\mathbf{k}}{2}-\mathbf{q}|)s^{\mathrm{D}}\left(|\mathbf{p}-\mathbf{q}|\right) \tag{5.108}$$

$$N_{\mathbf{k}}(\mathbf{p}) = -\frac{\varepsilon(k)}{k^2}\mathbf{k}\cdot\left(\tfrac{\mathbf{k}}{2}+\mathbf{p}\right) s\left(|\tfrac{\mathbf{k}}{2}+\mathbf{p}|\right) \tag{5.109}$$

$$-\frac{\hbar^2}{2m}\int \frac{d\mathbf{q}}{(2\pi)^d\rho}\left(\tfrac{\mathbf{k}}{2}+\mathbf{p}\right)\cdot\left(\tfrac{\mathbf{k}}{2}+\mathbf{q}\right) Q(|\tfrac{\mathbf{k}}{2}+\mathbf{q}|)s(|\tfrac{\mathbf{k}}{2}-\mathbf{q}|)s^{D}(|\mathbf{p}-\mathbf{q}|)$$

$$+(\mathbf{p}\leftrightarrow -\mathbf{p})$$

$$K_{\mathbf{k}}(\mathbf{p},\mathbf{q}) = \frac{\hbar^2}{2m}\Bigg[\left(\tfrac{\mathbf{k}}{2}+\mathbf{p}\right)\cdot\left(\tfrac{\mathbf{k}}{2}+\mathbf{q}\right)\left[1 - Q\left(|\tfrac{\mathbf{k}}{2}+\mathbf{q}|\right)\right] s(|\mathbf{p}-\mathbf{q}|) \tag{5.110}$$

$$-\mathbf{k}\cdot\left(\tfrac{\mathbf{k}}{2}+\mathbf{q}\right) Q\left(|\tfrac{\mathbf{k}}{2}+\mathbf{q}|\right) N_{\mathbf{k}}(\mathbf{p})$$

$$-\int \frac{d\mathbf{q}'}{(2\pi)^d\rho}\Bigg[\left(\tfrac{\mathbf{k}}{2}+\mathbf{p}\right)\cdot(\mathbf{q}-\mathbf{q}')Q(|\mathbf{q}-\mathbf{q}'|)s\left(|\tfrac{\mathbf{k}}{2}-\mathbf{q}'|\right)$$

$$-\left(\tfrac{\mathbf{k}}{2}+\mathbf{p}\right)\cdot\left(\tfrac{\mathbf{k}}{2}+\mathbf{q}'\right) Q\left(|\tfrac{\mathbf{k}}{2}+\mathbf{q}'|\right) s(|\mathbf{q}-\mathbf{q}'|)\Bigg]\Bigg] + (\mathbf{p}\leftrightarrow -\mathbf{p})\,.$$

To repeat the point we made above, the single–mode reference spectrum entering the energy denominator in Eq. (5.105) is not the Feynman spectrum (1.1) but the spectrum of Eq. (5.101), which lies closer to the experimental result. Therefore, this approach accounts better for the energetics of the excitations, especially at high momenta. The fact that we call this the "fully optimized" solution comes from the aspect that we have now replaced both the fluctuating abstract correlation functions with experimentally observable quantities.

5.6. *Results: dynamic structure and related applications*

5.6.1. *Phonon–roton spectrum in liquid ^{4}He*

Having formulated the problem of solving the dynamics of bosonic quantum fluids in a number of ways, let us next see how the different approximations do in the very basic task of accounting for the phonon–roton spectrum in liquid ^{4}He. This is shown in Figs. 14 and 15. We can also study the density dependence of the spectrum; this is done in Fig. 16.

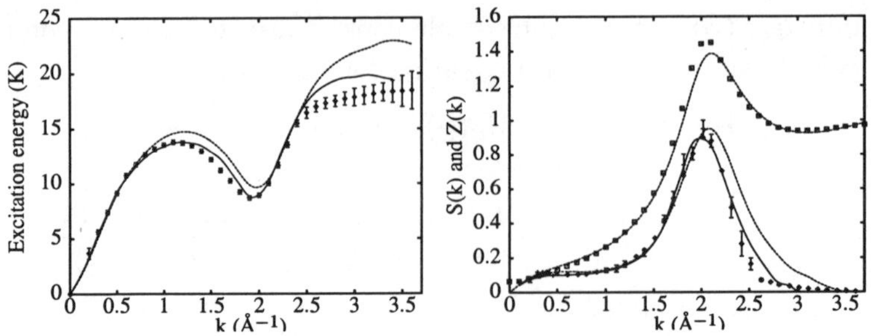

Fig. 14. Left figure compares the phonon–roton spectrum obtained from the fully optimized theory calculated using the experimental (solid line) and the variational (dashed line) $S(k)$ with measurements [76] at the 3D saturation density. The experimental [65] (squares) and variational (dashed line) structure functions are shown on the right, together with the pole strength $Z(k)$ as obtained from the experimental (solid line) and variational (dashed line) $S(k)$, and in the measurements of Ref.~76

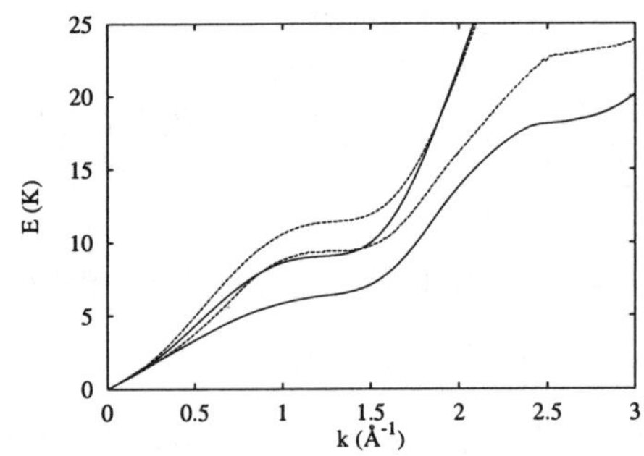

Fig. 15. Phonon–roton spectrum at the 2D equilibrium density 0.0421 $Å^{-2}$ in different approximations. The dashed curves give the Feynman spectrum (upper curve) and the CBF result (lower curve). The solid lines represent the reference spectrum (upper curve) and the result obtained from the fully optimized theory (lower curve).

Fig. 15 displays the Feynman spectrum, the CBF result, and the result obtained from the fully optimized theory for two–dimensional ^{4}He at the saturation density 0.0421 $Å^{-2}$. As asserted above, the Feynman approximation works reasonably well at low momenta (becoming exact in the limit $k \to 0$), but overshoots already the roton minimum by a factor of two. Adding three–phonon processes in the CBF approximation improves the situation considerably, although the result still lies rather far from the fully optimized solution [83] that is in a reasonable agreement with the shadow wave function calculations of Grisenti and Reatto [84] near the roton minimum.

It is worth noting that the phonon–roton spectrum terminates (its strength goes to zero) at high momenta when the mode merges into the two–roton continuum. We return to this in more detail when we discuss the full dynamic structure function. Additionally, the phonon dispersion is anomalous at momenta around $k \sim 0.5$ $Å^{-1}$: a phonon can decay into two of lower energy (the dispersion curve "bends upwards",

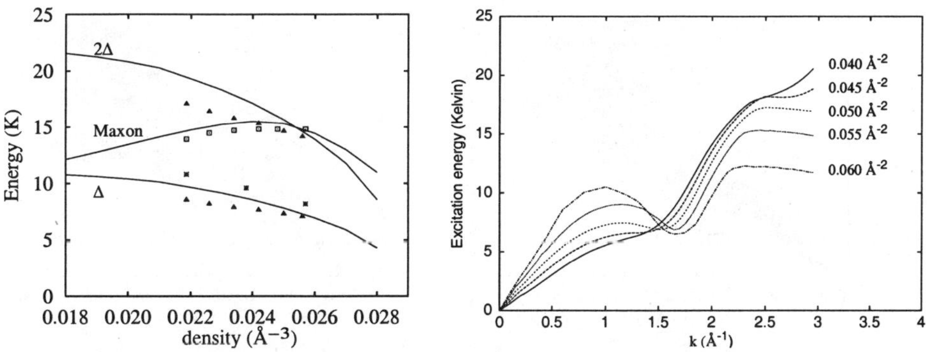

Fig. 16. On the left: Density dependence of the roton and maxon excitation energies in 3D. The squares represent the measured maxon energies of Ref. 81 and the triangles give the one– and two–roton energies of Ref. 82. The stars show the Brillouin–Wigner perturbation theory (CBF–BW) results of Chang and Campbell. [38] The solid lines correspond to the results obtained from the fully optimized theory. The theoretical maxon energy crosses the two–roton energy between 0.024 and 0.025 $Å^{-3}$, and at higher densities the theoretical maxon curve drops down along with the two–roton curve. On the right: Phonon–roton mode in 2D as obtained from the fully optimized theory at densities indicated in the figure.

i.e. is convex). The existence of a real solution to the CBF equation in such a situation has been discussed extensively in Ref. 77 for ^{4}He and in Ref. 69 for the charged Bose gas. In the figures, we have indicated the lowest–lying solution to the CBF or fully optimized equations, but it should be noted that this solution may actually have negligible strength — the main strength may be concentrated around a complex pole (decaying mode). This is the case for anomalous phonon dispersion; measurements of the pole strength $Z(k)$ give under such circumstances the integrated strength of the Landau damped phonon. The strength also vanishes when the phonon–roton mode crosses the two–roton threshold. This can also be seen as a "kink" in Fig. 15 at $k \sim 2.5$ $Å^{-1}$.

To summarize all these figures, we can say with confidence that the microscopic variational method gives at the fully optimized level results that agree remarkably well with the available experimental data.

5.6.2. *Dynamic structure function*

The results for the dynamic structure function we show here have been calculated in the CBF approximation. Such calculations have also been reported in the literature. [77,80] Although the theory leads to excitations energies somewhat too high as compared with the experimental data, the overall qualitative features of the spectrum are well reproduced. The most apparent shortcoming is the one we already pointed out, namely that excitations decay into two Feynman excitations, not into two excitations with self–energy–corrected energies. As a result, the two–roton limit appears at an energy higher than the limit taken from the phonon–roton spectrum given by the theory itself. Yet, the results satisfy the lowest energy weighted sum

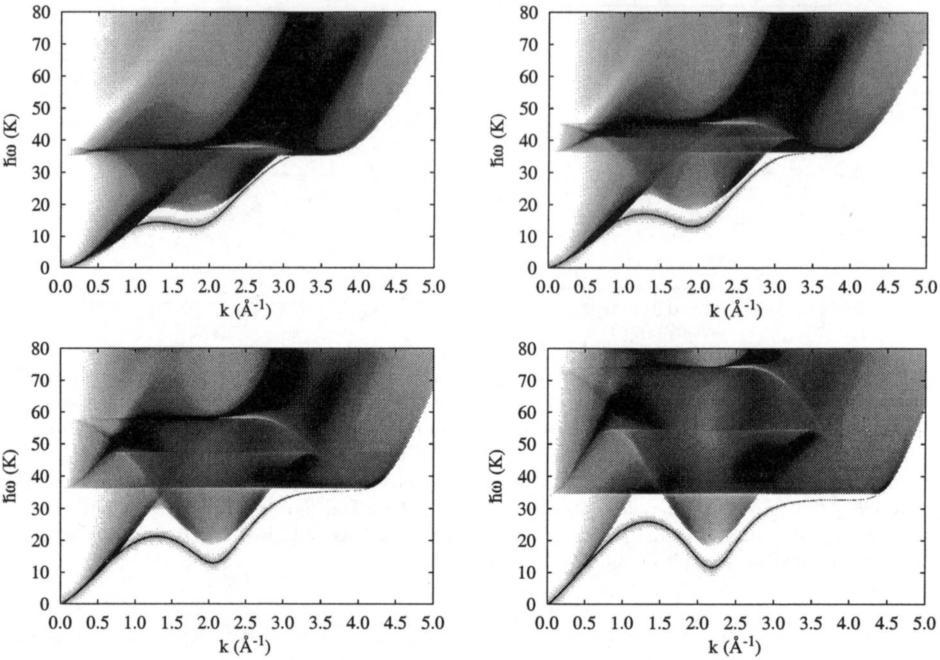

Fig. 17. $S(k,\omega)$ in 3D at 0.017 Å^{-3} near the low–density spinodal point, at 0.0200 Å^{-3}, at 0.0240 Å^{-3}, and near the freezing density at 0.0280 Å^{-3}.

rules

$$S(k) = \int_{-\infty}^{\infty} d(\hbar\omega) S(k,\omega) \tag{5.111}$$

$$\frac{\hbar^2 k^2}{2m} = \int_{-\infty}^{\infty} d(\hbar\omega)\ \hbar\omega\ S(k,\omega) \tag{5.112}$$

$$\frac{1}{2mc^2} = \lim_{k\to 0} \int_{-\infty}^{\infty} d(\hbar\omega) \frac{S(k,\omega)}{\hbar\omega} \tag{5.113}$$

exactly. Also the numerical implementation preserves this property to better than 1%. In fact, this feature of the CBF theory was first proven by Jackson. [85]

The phonon spectrum becomes soft (the speed of sound drops to zero) at the so–called spinodal point at low densities. Lowering the density below that would lead to the liquid breaking into droplets. Near the spinodal point the dispersion has a strong anomaly (the phonon mode can decay into phonons of lower energy) reaching down to very low momenta, shown in Fig. 17 at the density of 0.0170 Å^{-3}, as the ground state is on the verge of becoming unstable due to diverging compressibility. At high momenta the collective mode merges into the continuum at the two–roton threshold. At lower densities this happens at lower momenta than at the equilibrium density 0.0218 Å^{-3}, shown Fig. 18, where the single–excitation pole loses rapidly strength but stays below the two–roton continuum in consonance

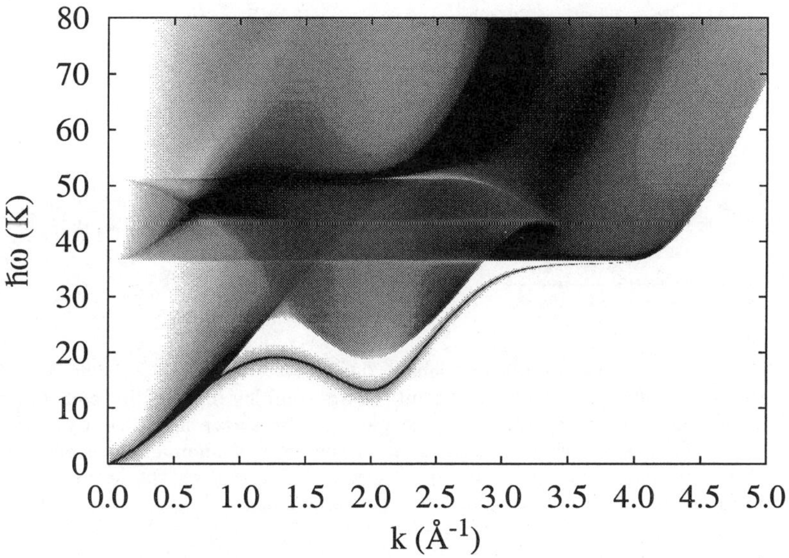

Fig. 18. $S(k, \omega)$ in 3D at the equilibrium density 0.0218 Å^{-3}. For convenience, we have convolved the spectra with a Lorentzian. This causes the weak spurious weight seen below the phonon–roton curve. Additionally, the strength scale in the figures has been chosen to be $S(k, \omega)^{1/4}$ to enhance structures of low intensity.

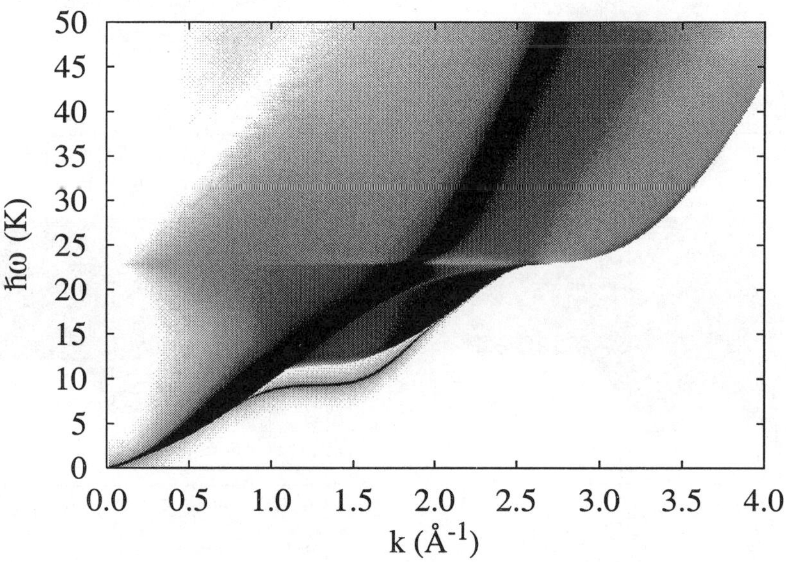

Fig. 19. $S(k, \omega)$ in 2D at the equilibrium density 0.0421 Å^{-3}.

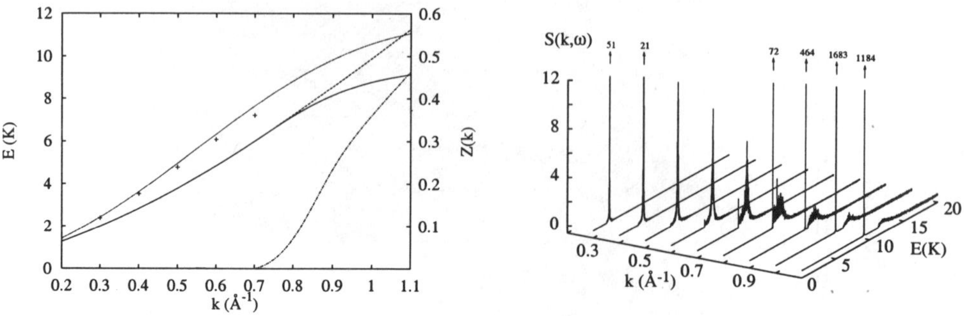

Fig. 20. $S(k,\omega)$ in 2D at the equilibrium density 0.0421 Å^{-3}. Left figure shows the CBF result (solid line), the critical limit above which all solutions are complex (dashed line coinciding with the CBF result), and the Feynman spectrum (upper curve). Also shown are a few complex solutions around which the main strength is concentrated in the case of anomalous dispersion (crosses, compare with the peaks in the right figure). The lowest curve gives the strength $Z(k)$ of the phonon–roton mode. On the right, we plot $S(k,\omega)$ for momenta 0.3, 0.4, ..., 1.1 Å^{-1}.

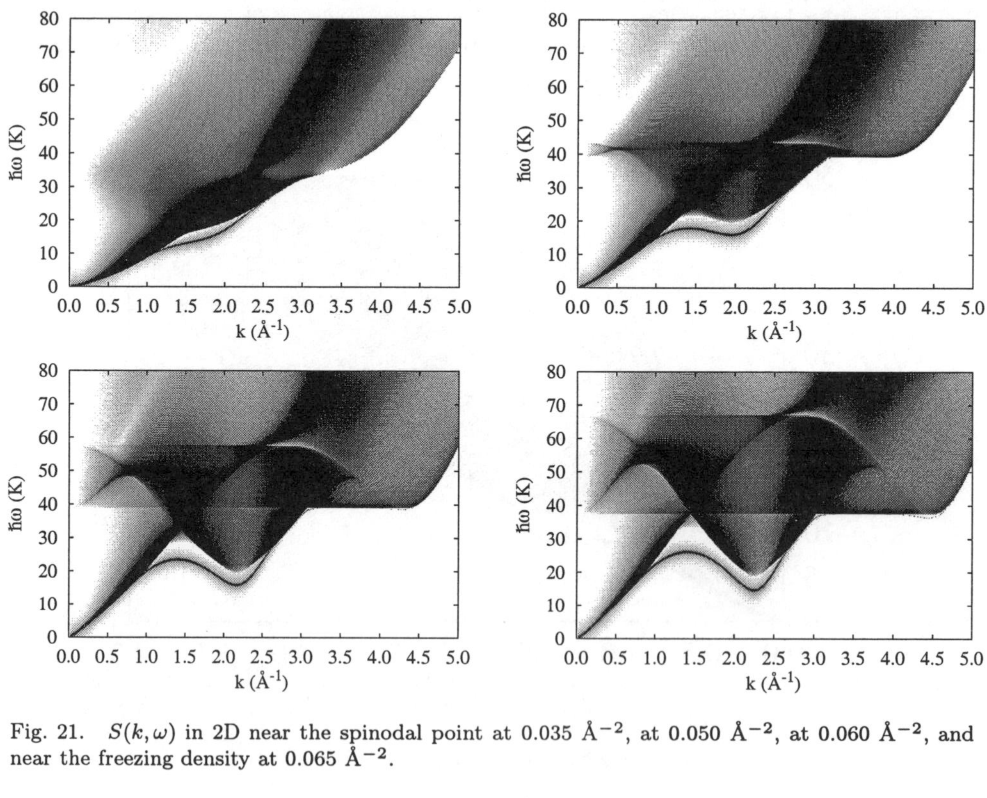

Fig. 21. $S(k,\omega)$ in 2D near the spinodal point at 0.035 Å^{-2}, at 0.050 Å^{-2}, at 0.060 Å^{-2}, and near the freezing density at 0.065 Å^{-2}.

with the results of Ref. 86. The sharp features in the continuum are roton–roton, roton–maxon and maxon–maxon resonances.

The anomalous dispersion is even more pronounced in 2D, shown in Figs. 19 and 20. Especially, Fig. 20 demonstrates how a second peak (corresponding to the phonon–roton mode) develops next to the peak (the decaying phonon mode) containing the main strength in that region. The phonon–roton branch emerges from the continuum at $k \sim 0.8$ Å^{-1}, terminating again just below $k \sim 2.0$ Å^{-1} (Fig. 19). Contrary to the 3D case, no high frequency scattering is present in the results at the equilibrium density. This can be understood from the absence of van Hove singularities in the two–body density of states at this density. This situation changes as we move towards higher densities.

5.6.3. *Transition currents*

The continuity equations define the one– and two–body currents related to the fluctuations in the one– and two–body densities. These currents represent particle motion in the perturbed ground state, being thence of first order in the density fluctuations. In contrast to these "transition currents", the currents in an excited state would be of second order because the excited state itself is of first order. In the following, we study the currents at wave numbers and frequencies along the phonon–roton curve in liquid ^{4}He to shed some light on the microscopic processes involved. We are especially interested in the roton excitation.

The two–body current, given in Eq. (5.90), can be regrouped as

$$\mathbf{j}_2(\mathbf{r}_1, \mathbf{r}_2; t) = \rho_0 g_2(|\mathbf{r}_1 - \mathbf{r}_2|) \Big[\mathbf{j}_1(\mathbf{r}_1; t) + \frac{\hbar\rho_0}{2mi} \mathbf{T}(\mathbf{r}_1, \mathbf{r}_2; t)\Big] , \tag{5.114}$$

where we have just picked out the ground–state pair–distribution function as a common factor, and separated the one–body current from the rest of the terms. Here $\mathbf{T}(\mathbf{r}_1, \mathbf{r}_2; t)$ depends on the fluctuations $\delta(\mathbf{r}_2, \mathbf{r}_3; t)$ in the pair–distribution function,

$$\begin{aligned}\mathbf{T}(\mathbf{r}_1, \mathbf{r}_2; t) = {} & \nabla_1 \delta(\mathbf{r}_1, \mathbf{r}_2; t) - \rho_0 \int d\mathbf{r}_3 \delta(\mathbf{r}_2, \mathbf{r}_3; t) g_2(\mathbf{r}_1, \mathbf{r}_3) \nabla_1 u_2(\mathbf{r}_1, \mathbf{r}_3) \\ & -\rho_0 \int d\mathbf{r}_3 h(\mathbf{r}_2, \mathbf{r}_3 g_2(\mathbf{r}_1, \mathbf{r}_3) \nabla_1 u_2(\mathbf{r}_1, \mathbf{r}_3) \\ & \times \left[\frac{1}{\rho_0} \delta\rho_1(\mathbf{r}_3; t) + \delta(\mathbf{r}_1, \mathbf{r}_3; t) + \delta(\mathbf{r}_2, \mathbf{r}_3; t)\right] .\end{aligned} \tag{5.115}$$

We present our results for the real part of the two–body current of Eq. (5.114) in a mixed representation as a function of the center–of–mass momentum $\mathbf{k}$ and the relative coordinate $\mathbf{r}$, that is,

$$\begin{aligned}\mathbf{j}_2(\mathbf{k}, \mathbf{r}, \omega) = {} & \rho_0 g_2(r) \Big[\mathbf{j}_1(\mathbf{k}, \omega) \cos\left(\tfrac{1}{2}\mathbf{k} \cdot \mathbf{r}\right) \\ & + \frac{\hbar\rho_0}{2mi} \tilde{\mathbf{T}}(\mathbf{k}, \mathbf{r}, \omega)\Big] .\end{aligned} \tag{5.116}$$

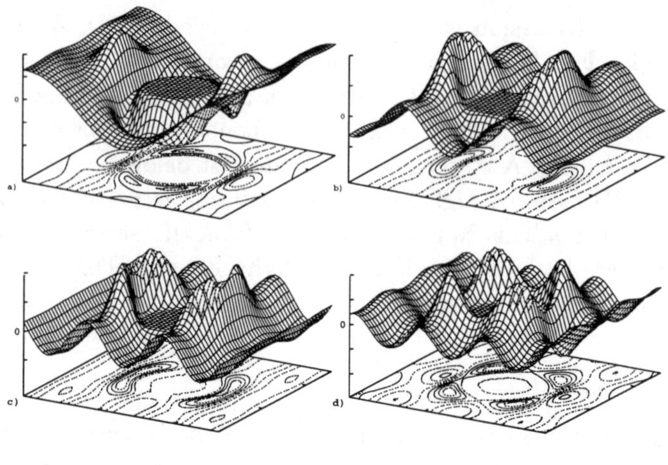

Fig. 22. z component of the two–body current (a) at the maxon region $k = 1.0\ ^{-1}$, (b) near the roton minimum $k = 2.0\ ^{-1}$, (c) at $k = 2.5\ ^{-1}$, and (d) in the asymptotic region $k = 3.0\ ^{-1}$. The direction of $\mathbf{k}$ is along the x axis. (Originally from Ref. 87.)

The radial distribution function of the ground state $g_2(r)$ is proportional to the probability of finding a particle at the distance r from a given reference particle. Let this reference particle be located at the origin. Because of the strongly repulsive core of the ^{4}He–^{4}He interaction, other particles are repelled outside the radius of about 2 Å . This "correlation hole" can be clearly seen in Figs. 22 and 23. We also separate the oscillating sound–wave–like behavior in which the particles move towards and away from each other with wavelength determined by the center–of–mass motion in the cosine term from the more complicated flow patterns collected into $\tilde{\mathbf{T}}(\mathbf{k}, \mathbf{r}, \omega)$.

In Fig. 22, we have plotted the z component of the two–body current for four typical cases, corresponding to center–of–mass motion with wave numbers k=1.0 Å^{-1} (maxon), 2.0 Å^{-1} (roton), 2.5Å^{-1} and 3.0 Å^{-1} (asymptotic region). Besides the center–of–mass oscillations, there are additional oscillations due to the interparticle correlations; the most pronounced feature in Fig. 22 is the nearest–neighbor peak. From Fig. 22.a it is evident that in the maxon region these oscillations are out of phase, whereas in the roton region (Fig. 22.b) they are in phase. This explains why the roton region is energetically favorable and why the minimum corresponds to the wave number of the peak of the static structure function. The latter aspect is, of course, already well–known from the Feynman description and the dispersion relation $\hbar^2 k^2/[2mS(k)]$. At wavelengths shorter than the size of the correlation hole the simple wave pattern breaks down (see Figs. 22.c and 22.d).

A more detailed view into the structure of the current is obtained from Fig. 23, where we have subtracted the center–of–mass oscillations. The arrows show the

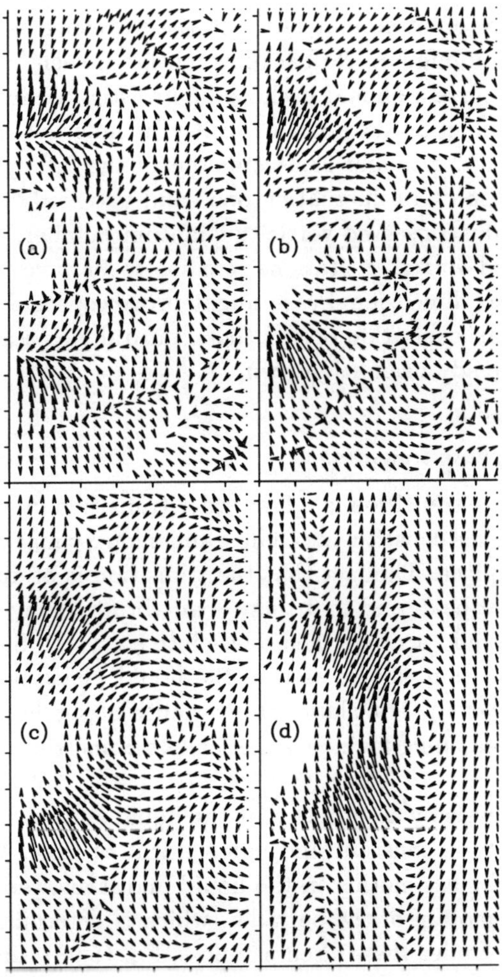

Fig. 23. Short–range part of the two–body current (a) at the maxon region $k = 1.0\ ^{-1}$, (b) near the roton minimum $k = 2.0\ ^{-1}$, (c) at $k = 2.5\ ^{-1}$, and (d) in the asymptotic region $k = 3.0\ ^{-1}$. The direction of **k** is upwards and the tick-mark spacing is 1.0 . (Originally from Ref. 87.)

direction of the current. Due to the cylindrical symmetry, only the xz–plane with $x \geq 0$ is displayed. In the maxon and roton regions (Figs. 23.a and 23.b), oscillations of the radial distribution function dominate — although some interesting topological structures could also be identified. The pattern, however, changes completely at 2.5 Å^{-1}: a clear "backflow loop" forms around each atom (Fig. 23.c). The radius of the circulation seen in the figure is of the order of atomic radius. (It can be compared with the white area in the figures.) At k >2.5 Å^{-1}, the loop gets elongated with increasing wave number and forms a tube–like structure with a diameter of atomic size. A typical case at 3.0 Å^{-1} is shown in Fig. 23.d.

To summarize, we have seen that at the roton minimum the size of the correlation hole created by an atom matches with the wavelength of the center–of–mass motion — hence the minimum. No backflow motion is seen, although the topological structure of the current should perhaps be investigated further. The perturbed ground–state wave function does not explicitly include quantized vortices: [88] the structures seen in Figs. 23.c and 23.d come out of the full optimization of the action integral with respect to fluctuating one– and two–body correlation functions. They do not carry any conserved vorticity quantum number. The relation between these excitations and vortex excitations remains to be investigated in the future. In this light, the two–dimensional case might prove especially useful, because the vortex–antivortex pair excitation occurs there naturally as a low–lying mode. Such calculations have already been attempted. [83]

5.6.4. *Liquid–solid phase transition*

The liquid–solid phase transition is a first order transition in which the close–range correlations between the constituents change and a long–range order penetrating the whole system takes place. The fact that the transition is of first order also implies the existence of latent heat, and that there is an abrupt change in the density when the particles in the liquid arrange themselves into a crystal. Some signatures of the emerging phase transition can be seen in the liquid phase by studying two–body structures which break the translational invariance of the liquid phase. [42,83,89–92] We can do this by searching for non–spherically symmetric solutions to our continuity equations.

We start our analysis by assuming that the liquid state is homogeneous and the single–particle density is constant and independent of time. We also assume that there is no center–of–mass motion in the liquid and, thus, the fluctuations depend only on the relative motion of the particles, which is determined by the two–body continuity equation. Under these circumstances, Eq. (5.93) can be written as

$$\begin{aligned}
&\frac{\hbar^2}{2m}\nabla_1\cdot\Big(g_2(\mathbf{r}_1,\mathbf{r}_2)\Big\{\nabla_1\delta(\mathbf{r}_1,\mathbf{r}_2)\\
&\qquad -\rho_0\int d^3r_3 g_2(\mathbf{r}_2,\mathbf{r}_3)g_2(\mathbf{r}_1,\mathbf{r}_3)\nabla_1 u_2(\mathbf{r}_1,\mathbf{r}_3)\Big[\delta(\mathbf{r}_1,\mathbf{r}_3)+\delta(\mathbf{r}_2,\mathbf{r}_3)\Big]\Big\}\Big)\\
&\qquad +(1\leftrightarrow 2)\\
&= \hbar\omega\; g_2(\mathbf{r}_1,\mathbf{r}_2)\delta(\mathbf{r}_1,\mathbf{r}_2)\,. \qquad (5.117)
\end{aligned}$$

A convenient way to solve Eq. (5.117) is to Fourier transform it into momentum space, with $F\{\delta(\mathbf{r}_1,\mathbf{r}_2)\}=\delta(\mathbf{p})$, and separate the angle and momentum dependencies in the fluctuation by expanding $\delta(\mathbf{p})$ in terms of either Legendre polynomials (3D) or in terms of exponentials (2D). It turns out that the different symmetry components separate, and we end up with a set of decoupled equations, one for each symmetry component.

To demonstrate the approach, we present the result obtained by solving the equation for two–dimensional liquid ^{4}He, as this case is somewhat easier to visu-

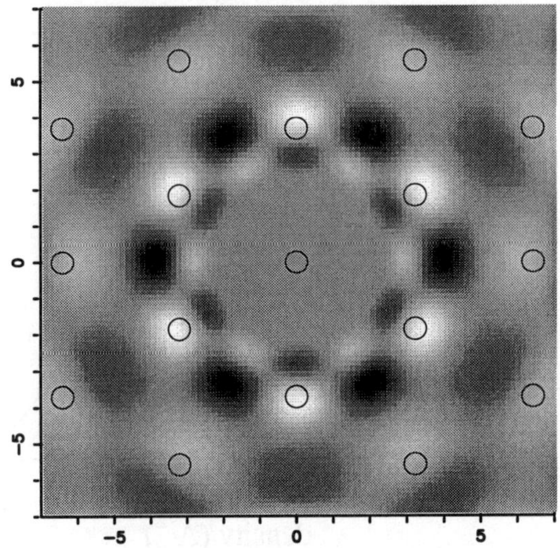

Fig. 24. Change in the relative two–body structure corresponding to the first mode (with hexagonal symmetry) to become soft at ~0.069 Å^{-2}. Lighter shade indicates increased probability to find a particle at given coordinates. The gray area in the middle is the correlation hole. Superimposed (rings) are the lattice sites in the hexagonal crystal at the same density.

alize than the three–dimensional one. The first mode to become soft in the long–wavelength limit has hexagonal symmetry. The critical density where the energy of this mode becomes zero is around 0.069 Å^{-2}, in fair agreement with Monte Carlo results. [66,93,94] An analysis of the ground–state optimization process has brought us to the conclusion that this instability should be identified with freezing. [92] At some higher density also the spherically symmetric mode becomes unstable, and no liquid solution can be found after that. We identify this density as the melting density of the crystal.

This example demonstrates how the method can be extended further to obtain detailed information of the liquid–solid phase transition in strongly–correlated quantum fluids by studying the stability of the liquid ground state against small changes in two–body structure, or more precisely, against non–spherically symmetric fluctuations in the interparticle correlations. This fact, together with the observation that the equations optimizing the ground state have solutions only for locally stable geometries, bring us to the conclusion that the relevant structural phase transitions are indeed embedded into the equations in a natural way, without external constraints.

6. Summary

The Jastrow–Feenberg variational theory and its extension to correlated basic functions provides a versatile tool in obtaining microscopic understanding of the many–body structure and processes present in strongly–correlated quantum fluids and

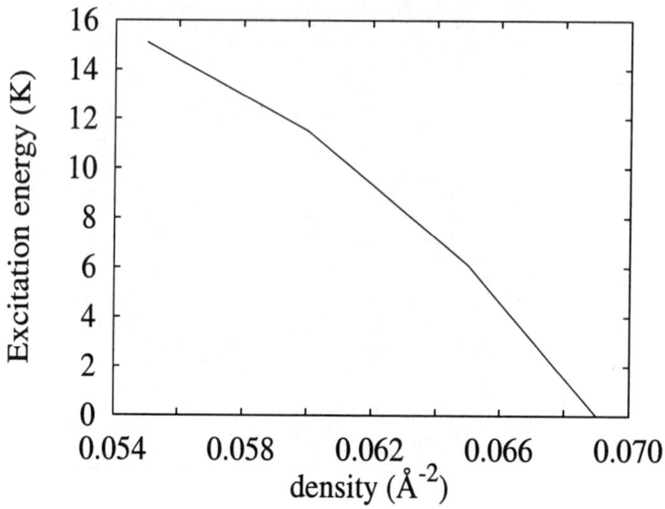

Fig. 25. Energy of the mode with hexagonal symmetry as a function of density.

fluid mixtures, competing in accuracy with the best available Monte Carlo calculations and being in keeping with the available experimental data, as we have shown for bulk liquid ^{4}He and bulk ^{3}He–^{4}He mixtures. Moreover, the theory has a built–in sensitivity to the local stability of the system through collective excitations. For example, at low densities the liquid is unstable against softening of the sound mode. We also saw how the ground–state tools can be used to predict new phenomena and explain experiments, not just to reproduce the already existing data. In the two–dimensional mixture a close analysis of the basic physics and model–independent simulation data hinted towards the existence of a very interesting phase diagram: at low ^{3}He concentrations, typically below two percent, the mixture is energetically preferable over a phase–separated system, and the mixture does not exist in an atomic form, but rather in the form of ^{3}He dimers within the ^{4}He background. The existence of these states follows from the long–wavelength properties of the phonon–mediated interaction. At high densities, a precursor of the solidification is clearly seen in the two–body structure. These observations show that the relevant structural phase transitions are indeed naturally embedded into the equations optimizing the system.

We also introduced a way to deal with excitations within the variational approach by allowing the correlations to fluctuate both in space and time. By extending the time dependence to two–body correlations and, hence, bringing the three–phonon processes into the description, we went beyond the conventional Feynman theory of excitations, first in the CBF approximation which already gave reasonable results for the dynamic structure, and then in the fully optimized approach which brought the results for the phonon–roton spectrum in liquid ^{4}He into quantitative agreement with experimental data. We presented an analysis of the dynamic structure function and made a clear distinction between the real elementary-excitation mode and the broader complex mode, which can decay into elementary excitations. Liquid ^{4}He displays anomalous dispersion at low momenta where the complex mode dominates; this is why phonons below 8 Kelvins decay rapidly into low-energy phonons. We also showed how the formalism allows us to obtain detailed information of the particle currents. In this way, we were able to shed light on the microscopic structure of the roton excitation, interpreting it as a favourable matching of the center-of-mass oscillations and the oscillations following from the interparticle correlations.

Acknowledgments

We thank E. Krotscheck, K. Schörhuber for many dicussion in preparing this material. The work was supported, in part, by the Academy of Finland under project 163358 (M.S. and J.H.), and by the Austrian Science fund under grants No. P12832-TPH and P11098-PHY (V.A.).

References

1. L. D. Landau and I. Pomeranchuk, in *Collected Papers of L. D. Landau*, edited by D. T. Haar (Gordon and Breach, New York, NY, 1967), pp. 469–470, [*Sov. Phys. Dokl.* **59**, 669 (1948)].
2. C. Ebner and D. O. Edwards, *Physics Reports* **2**, 77 (1971).
3. G. Baym and C. Pethick, in *The Physics of Liquid and Solid Helium, Part II*, edited by K. H. Bennemann and J. B. Ketterson (Wiley, New York, 1978), Vol. XXIX, pp. 123–175.
4. R. de Bruyn Ouboter and C. N. Yang, *Physica* **144B**, 127 (1986).
5. J. Bardeen, G. Baym, and D. Pines, *Phys. Rev.* **156**, 207 (1967).
6. H. H. Fu and C. J. Pethick, *Phys. Rev. B* **14**, 3837 (1976).
7. C. H. Aldrich and D. Pines, *J. Low Temp. Phys.* **25**, 677 (1976).
8. W. Hsu and D. Pines, *J. Stat. Phys.* **38**, 273 (1985).
9. W. Hsu, D. Pines, and C. H. Aldrich, *Phys. Rev. B* **32**, 7179 (1985).
10. E. Krotscheck *et al.*, *Phys. Rev. B* **58**, 12282 (1998).
11. E. P. Bashkin, *Sov. Phys. JETP* **51**, 181 (1980).
12. E. Krotscheck, J. Paaso, M. Saarela, and K. Schörkhuber, *Phys. Rev. Lett.* **85**, 2344 (2000).
13. W. E. Massey, C. W. Woo, and H.-T. Tan, *Phys. Rev. A* **1**, 519 (1970).
14. K. E. Kürten and C. E. Campbell, *Phys. Rev. B* **26**, 124 (1982).
15. M. D. Miller, *Phys. Rev. B* **17**, 1139 (1978).
16. A. Fabrocini and A. Polls, *Phys. Rev. B* **25**, 4533 (1982).
17. T. Chakraborty, A. Kallio, L. J. Lantto, and P. Pietiläinen, *Phys. Rev. B* **27**, 3061 (1983).
18. L. J. Lantto, *Phys. Rev. B* **36**, 5160 (1987).
19. E. Feenberg, *Theory of Quantum Fluids* (Academic, New York, 1969).
20. C. E. Campbell and E. Feenberg, *Phys. Rev.* **188**, 396 (1969).
21. C. E. Campbell, *Phys. Lett. A* **44**, 471 (1973).
22. C. C. Chang and C. E. Campbell, *Phys. Rev. B* **15**, 4238 (1977).

23. E. Krotscheck, *Phys. Rev. B* **33**, 3158 (1986).
24. E. Krotscheck and M. Saarela, *Physics Reports* **232**, 1 (1993).
25. A. Fabrocini and A. Polls, *Phys. Rev. B* **30**, 1200 (1984).
26. J. Boronat, A. Fabrocini, and A. Polls, *J. Low Temp. Phys.* **74**, 347 (1989).
27. M. Saarela, in *Recent Progress in Many Body Theories*, edited by Y. Avishai (Plenum, New York, 1990), Vol. 2, pp. 337–346.
28. M. Saarela and E. Krotscheck, *J. Low Temp. Phys.* **90**, 415 (1993).
29. L. Landau, *J. Phys. U.S.S.R.* **5**, 71 (1941).
30. A. Bijl, *Physica* **7**, 869 (1940).
31. N. N. Bogoliubov, *J. Phys. U.S.S.R* **9**, 23 (1947).
32. H. Glyde, *Excitations in liquid and solid helium* (Oxford University Press, Oxford, 1994).
33. R. P. Feynman, *Phys. Rev.* **94**, 262 (1954).
34. R. P. Feynman and M. Cohen, *Phys. Rev.* **102**, 1189 (1956).
35. J. W. Clark, in *Progress in Particle and Nuclear Physics*, edited by D. H. Wilkinson (Pergamon Press Ltd., Oxford, 1979), Vol. 2, pp. 89–199.
36. H. W. Jackson and E. Feenberg, *Ann. Phys. (NY)* **15**, 266 (1961).
37. H. W. Jackson and E. Feenberg, *Rev. Mod. Phys.* **34**, 686 (1962).
38. C. C. Chang and C. E. Campbell, *Phys. Rev. B* **13**, 3779 (1976).
39. S. Manousakis and V. R. Pandharipande, *Phys. Rev. B* **30**, 5062 (1984).
40. M. Saarela, *Phys. Rev. B* **33**, 4596 (1986).
41. M. Saarela and J. Suominen, in *Condensed Matter Theories*, edited by J. S. Arponen, R. F. Bishop, and M. Manninen (Plenum, New York, 1988), Vol. 3, pp. 157–165.
42. J. Suominen and M. Saarela, in *Condensed Matter Theories*, edited by J. Keller (Plenum, New York, 1989), Vol. 4, p. 377.
43. E. Krotscheck, *Phys. Rev. B* **31**, 4258 (1985).
44. R. A. Aziz *et al.*, *J. Chem. Phys.* **70**, 4330 (1979).
45. C. E. Campbell, in *Progress in Liquid Physics*, edited by C. A. Croxton (Wiley, London, 1977), Chap. 6, pp. 213–308.
46. E. Krotscheck, *J. Low Temp. Phys.* **119**, 103 (2000).
47. A. D. Jackson, A. Lande, and L. J. Lantto, *Nucl. Phys. A* **317**, 70 (1979).
48. A. D. Jackson, *Ann. Review Nucl. Part. Sci* **33**, 105 (1983).
49. E. Krotscheck, *J. Low Temp. Phys.* **27**, 199 (1977).
50. E. Krotscheck, R. A. Smith, J. W. Clark, and R. M. Panoff, *Phys. Rev. B* **24**, 6383 (1981).
51. E. Krotscheck, *Ann. Phys. (NY)* **155**, 1 (1984).
52. E. Krotscheck and M. L. Ristig, *Phys. Lett. A* **48**, 17 (1974).
53. E. Krotscheck, Q.-X. Qian, and W. Kohn, *Phys. Rev. B* **31**, 4245 (1985).
54. E. Krotscheck, *Phys. Rev. A* **15**, 397 (1977).
55. J. C. Owen, *Phys. Rev. B* **23**, 2169 (1981).
56. G. Ripka, *Physics Reports* **56**, 1 (1979).
57. R. F. Bishop and K. H. Lührmann, *Phys. Rev. B* **26**, 5523 (1982).
58. A. D. Jackson, A. Lande, and R. A. Smith, *Physics Reports* **86**, 55 (1982).
59. A. D. Jackson, A. Lande, and R. A. Smith, *Phys. Rev. Lett.* **54**, 1469 (1985).
60. E. Krotscheck, A. D. Jackson, and R. A. Smith, *Phys. Rev. A* **33**, 3535 (1986).
61. A. Lande and R. A. Smith, *Phys. Rev. A* **45**, 913 (1992).
62. E. Krotscheck and J. W. Clark, *Nucl. Phys. A* **328**, 73 (1979).
63. E. Krotscheck, *Phys. Rev. A* **26**, 3536 (1982).
64. E. C. Svensson, V. F. Sears, A. D. B. Woods, and P. Martel, *Phys. Rev. B* **21**, 3638 (1980).

65. H. N. Robkoff and R. B. Hallock, *Phys. Rev. B* **24**, 159 (1981).
66. S. Giorgini, J. Boronat, and J. Casulleras, *Phys. Rev. B* **54**, 6099 (1996).
67. L. J. Lantto and P. J. Siemens, *Phys. Lett. B* **68**, 308 (1977).
68. A. Kallio and R. A. Smith, *Phys. Lett. B* **68**, 315 (1977).
69. V. Apaja *et al.*, *Phys. Rev. B* **55**, 12925 (1997).
70. G. Baym, *Phys. Rev. Lett.* **17**, 952 (1966).
71. J. Boronat, A. Fabrocini, and A. Polls, *Phys. Rev. B* **39**, 2700 (1989).
72. J. C. Owen, *Phys. Rev. Lett.* **47**, 586 (1981).
73. E. Krotscheck, J. Paaso, M. Saarela, and K. Schörkhuber, *Phys. Rev. B* **64**, 054504 (2001).
74. C. Um, J. Kahng, and T. F. George, *J. Low Temp. Phys.* **112**, 399 (1998).
75. J. Boronat, 1999, private communication.
76. R. A. Cowley and A. D. B. Woods, *Can. J. Phys.* **49**, 177 (1971).
77. B. E. Clements *et al.*, *Phys. Rev. B* **50**, 6958 (1994).
78. A. K. Kerman and S. E. Koonin, *Ann. Phys. (NY)* **100**, 332 (1976).
79. P. Kramer and M. Saraceno, *Geometry of the time-dependent variational principle in quantum mechanics*, Vol. 140 of *Lecture Notes in Physics* (Springer, Berlin, Heidelberg, and New York, 1981).
80. B. E. Clements, E. Krotscheck, and C. J. Tymczak, *Phys. Rev. B* **53**, 12253 (1996).
81. E. H. Graf, V. J. Minkiewicz, H. B. Möller, and L. Passell, *Phys. Rev. A* **10**, 1748 (1974).
82. O. W. Dietrich, E. H. Graf, C. H. Huang, and L. Passell, *Phys. Rev. A* **5**, 1377 (1972).
83. J. Halinen, V. Apaja, K. A. Gernoth, and M. Saarela, *J. Low Temp. Phys.* **121**, 531 (2000).
84. R. E. Grisenti and L. Reatto, *J. Low Temp. Phys.* **109**, 477 (1997).
85. H. W. Jackson, *Phys. Rev. A* **9**, 964 (1974).
86. B. Fåk and J. Bossy, *J. Low Temp. Phys.* **112**, 1 (1998).
87. V. Apaja and M. Saarela, *Phys. Rev. B* **57**, 5358 (1998).
88. Saarela and F. V. Kusmartsev, *Phys. Lett. A* **202**, 317 (1995).
89. A. D. Jackson, B. K. Jennings, A. Lande, and R. A. Smith, *Phys. Rev. B* **24**, 105 (1981).
90. L. Castillejo, A. D. Jackson, B. K. Jennings, and R. A. Smith, *Phys. Rev. B* **20**, 3631 (1979).
91. V. Apaja, J. Halinen, and M. Saarela, *J. Low Temp. Phys.* **113**, 909 (1998).
92. V. Apaja, J. Halinen, and M. Saarela, *Physica B* **284-288**, 29 (2000).
93. P. A. Whitlock, G. V. Chester, and M. H. Kalos, *Phys. Rev. B* **38**, 2418 (1988).
94. M. C. Gordillo and D. M. Ceperley, *Phys. Rev. B* **58**, 6447 (1998).

CHAPTER 5

A MICROSCOPIC VIEW OF CONFINED QUANTUM LIQUIDS

V. Apaja and E. Krotscheck
Institute for Theoretical Physics, Johannes Kepler University
A-4040 Linz, Austria
E-mail: Eckhard.Krotscheck@jk.uni-linz.ac.at

We describe a quantitative microscopic theory of non-uniform quantum liquids to explore the structure and the dynamics of inhomogeneous quantum liquids. Such systems are manifested, most prominently, in thin films of ^{4}He adsorbed to a substrate or in helium clusters; most recent developments also look at quasi-one-dimensional systems like nano-tubes, or quasi-two-dimensional systems between hectorite sheets.
The *physical problem* at hand is particularly well suited for highlighting the interconnections between *decisive experiments* and *properly executed theory.* We base our theoretical methods on the time-honored Jastrow-Feenberg variational theory, but we shall try to take, whenever appropriate, the "view from the top" and try to highlight the physical content behind the sometimes lengthy equations.

1. Introduction

This Chapter discusses the physics of inhomogeneous quantum liquids, specifically ^{4}He, looking at these systems from the manifestly microscopic point of view. The ability to deal with non-uniform quantum many-particle systems in a systematic and quantitative manner, and in particular without recourse to uncontrolled phenomenological input that can be biased by the theorist's perceptions, opens a wide range of possibilities:

First, inhomogeneous quantum liquids display vastly more complex behavior than homogeneous ones. Confined systems have therefore enjoyed much experimental attention over the past two decades. Quantum liquid surfaces and clusters, recently also quasi-one- and quasi-two-dimensional quantum liquids, may nowadays be generated in nano-tubes or between hectorite sheets. Quantum liquid films and mixture films let us study the issues of *dimensionality,* specifically the transition from a two– to three–dimensional geometry. Experiments have revealed many phenomena, such as layer modes, "dispersionless" excitations, and transient superfluidity. The skill to produce small droplets of ^{4}He brought under examination, for the first time, small systems that resemble in many aspects finite nuclei. But unlike in the nuclear many–body problem, the Hamiltonian is not only known, but also sufficiently simple such that one can examine almost the complete dynamics in a both experimentally and theoretically well–controlled situation.

Second, the theory of inhomogeneous quantum liquids is a technically difficult many–body problem. Inhomogeneity and confinement literally adds a new dimension to the problem and puts more demands on the internal consistency and calculational efficiency of a theory. In fact, a significant part of theoretical physics is indeed about developing approximation methods, as well as examining whether they are — at least qualitatively — able to reproduce what an "exact" theory (meaning, in our case, a hypothetical exact solution of the many–body Schrödinger equation) would predict. Attacking the technical problems and examining the validity of one's approaches is what comprises *many-body theory.* The application of these theories is what one does in *many-body physics.* The study of a physically rich system, like inhomogeneous quantum liquid, will always tell us about the nature of physical phenomena within our theoretical methods; it can highlight the strengths just as much as it can expose its weaknesses.

2. HNC–EL Theory for Inhomogeneous Bose Systems

The Hypernetted Chain Euler-Lagrange (HNC-EL) theory for the homogeneous liquid has been discussed in Chapter 4 of this volume, [1] hence we only need to look at the generalization of the theory to an inhomogeneous geometry. The Feenberg trial wave function

$$\Psi_N(\mathbf{r}_1,\ldots,\mathbf{r}_N) = \exp\frac{1}{2}\left[\sum_i u_1(\mathbf{r}_i) + \sum_{i<j} u_2(\mathbf{r}_i,\mathbf{r}_j) + \sum_{i<j<k} u_3(\mathbf{r}_i,\mathbf{r}_j,\mathbf{r}_k)\right] . \quad (2.1)$$

already allows for the symmetry breaking which we now explore. As in the bulk liquid, we seek to functionally minimize the expectation value of the N-body Hamiltonian

$$H_N = \sum_{i=1}^{N}\left[-\frac{\hbar^2}{2m}\nabla_i^2 + U_{\rm ext}(\mathbf{r}_i)\right] + \sum_{i<j} V(\mid \mathbf{r}_i - \mathbf{r}_j \mid) \quad (2.2)$$

in the space of all trial functions permitted in the *ansatz* (2.1). The essential new property that must be dealt with is the breaking of translational invariance and isotropy of the system. Such symmetry breakings may be induced by an external "substrate" potential $U_{\rm ext}(\mathbf{r}_i)$, but it may also occur spontaneously like in the formation of quantum liquid droplets to be discussed later in this Chapter.

The first task is to manipulate the expectation value of the Hamiltonian in a way that makes it accessible to numerical treatment. Clearly one now has a non–trivial one–body correlation function $u_1(\mathbf{r}_1)$. Furthermore, symmetry breaking affects *all* correlation functions, *i.e.*, the pair correlation function u_2 is not just a function of the distance between two particles, but also depends on the direction and on where in the system the particles are. The strategy is to replace the correlation functions

u_n by the physically observable n-body densities or distribution functions. The n–body density is defined as

$$\rho_n(\mathbf{r}_1,\ldots,\mathbf{r}_n) = \frac{N!}{(N-n)!}\frac{\int d^3r_{n+1}\ldots d^3r_N\Psi_0^2(\mathbf{r}_1,\ldots,\mathbf{r}_N)}{\int d^3r_1\ldots d^3r_N\Psi_0^2(\mathbf{r}_1,\ldots,\mathbf{r}_N)}\,, \tag{2.3}$$

and by calculating $\nabla_1\rho_1(\mathbf{r}_1)$ using the Feenberg wave function in Eq. (2.3) one obtains the first equation in the Born–Green–Yvon (BGY) hierarchy,

$$\boldsymbol{\nabla}_1\rho_1(\mathbf{r}_1) = \rho_1(\mathbf{r}_1)\boldsymbol{\nabla}_1 u_1(\mathbf{r}_1) + \rho_1(\mathbf{r}_1)\int d^3r_2\rho_1(\mathbf{r}_2)g(\mathbf{r}_1,\mathbf{r}_2)\boldsymbol{\nabla}_1 u_2(\mathbf{r}_1,\mathbf{r}_2)\,. \tag{2.4}$$

Eq. (2.4) is exact for a wave function containing pair correlations only; a third term is added to the right hand side if triplet correlations u_3 are included.

Eq. (2.4) can be used to eliminate $u_1(\mathbf{r})$. For the pair correlation function $u_2(\mathbf{r}_1,\mathbf{r}_2)$, we take the same route that has worked well in the homogeneous case and use the HNC hierarchy of integral equations for inhomogeneous systems to derive a relationship between $u_2(\mathbf{r}_1,\mathbf{r}_2)$ and the pair distribution function $g(\mathbf{r}_1,\mathbf{r}_2)$. The cleanest, but somewhat lengthy diagrammatic derivation of the HNC equations for a non-uniform system may be found in Ref. 2; a somewhat heuristic derivation, tuned to the application to quantum liquids, was given in Ref. 3. We shall rely heavily on the latter article for the basics of our theory; the reader is advised to consult this article in particular for a derivation of the relevant equations.

Square roots of the density appear frequently as factors, we therefore introduce the "tilde notation",

$$\tilde{C}(\mathbf{r}_1) \equiv \sqrt{\rho_1(\mathbf{r}_1)}C(\mathbf{r}_1) \tag{2.5}$$

$$\tilde{A}(\mathbf{r}_1,\mathbf{r}_2) \equiv \sqrt{\rho_1(\mathbf{r}_1)}A(\mathbf{r}_1,\mathbf{r}_2)\sqrt{\rho_1(\mathbf{r}_2)}\,. \tag{2.6}$$

We also mark convolution products with an asterisk. The unit operator and the inverse will generally be interpreted in the sense of a convolution product. Using this notation the HNC equations assume the form

$$g(\mathbf{r}_1,\mathbf{r}_2) = \exp\left[u_2(\mathbf{r}_1,\mathbf{r}_2) + N(\mathbf{r}_1,\mathbf{r}_2) + E(\mathbf{r}_1,\mathbf{r}_2)\right] \tag{2.7}$$

$$\tilde{N}(\mathbf{r}_1,\mathbf{r}_2) = \left[\tilde{h} * \tilde{X}\right](\mathbf{r}_1,\mathbf{r}_2) \tag{2.8}$$

$$h(\mathbf{r}_1,\mathbf{r}_2) = g(\mathbf{r}_1,\mathbf{r}_2) - 1 \tag{2.9}$$

$$X(\mathbf{r}_1,\mathbf{r}_2) = h(\mathbf{r}_1,\mathbf{r}_2) - N(\mathbf{r}_1,\mathbf{r}_2)\,, \tag{2.10}$$

where $N(\mathbf{r}_1,\mathbf{r}_2)$ are the chain (nodal) diagrams and $X(\mathbf{r}_1,\mathbf{r}_2)$ are the non-nodal diagrams. Here $E(\mathbf{r}_1,\mathbf{r}_2)$ is the infinite series of "elementary" diagrams which can be expressed as multi-dimensional integrals involving $\rho_1(\mathbf{r})$ and $g(\mathbf{r}_1,\mathbf{r}_2)$. It may also contain triplet correlations. We shall also use the coordinate–space static structure function

$$S(\mathbf{r},\mathbf{r}') = \delta(\mathbf{r},\mathbf{r}') + \tilde{h}(\mathbf{r},\mathbf{r}')\,, \tag{2.11}$$

is related to the direct correlation function $X(\mathbf{r}_1,\mathbf{r}_2)$ through

$$S(\mathbf{r}_1,\mathbf{r}_2) = \left[\mathbf{1} - \tilde{X}\right]^{-1}(\mathbf{r}_1,\mathbf{r}_2)\,. \tag{2.12}$$

2.1. *Variational energy expectation value and Euler equations*

Neglecting triplet correlations for the time being, one can express the variational energy expectation value

$$E_{\rm var}[u_n] = \frac{1}{I}\int d^3r_1 \ldots d^3r_N \Psi_0(\mathbf{r}_1,\ldots,\mathbf{r}_N) H \Psi_0(\mathbf{r}_1,\ldots,\mathbf{r}_N)\ . \tag{2.13}$$

entirely in terms of the one body functions $u_1(\mathbf{r}_1)$ and $\rho_1(\mathbf{r}1)$ and the two–body functions $u_2(\mathbf{r}_1,\mathbf{r}_2)$ and $g(\mathbf{r}_1,\mathbf{r}_2)$. As usual, the expectation value of the kinetic energy operator is transformed using the Jackson–Feenberg identity and expressed in terms of the one– and the two–body densities:

$$\begin{aligned} T[u_n] = &-\int d^3r_1 \rho_1(\mathbf{r}_1)\frac{\hbar^2}{8m}\nabla_1^2 u_1(\mathbf{r}_1) \\ &-\frac{1}{2}\int d^3r_1 d^3r_2 \rho_2(\mathbf{r}_1,\mathbf{r}_2)\frac{\hbar^2}{8m}[\nabla_1^2+\nabla_2^2]u_2(\mathbf{r}_1,\mathbf{r}_2)\ . \end{aligned} \tag{2.14}$$

The energy expectation value is still a functional of four unknown functions, $u_1(\mathbf{r}_1)$ and $u_2(\mathbf{r}_1,\mathbf{r}_2)$, as well as $\rho_1(\mathbf{r}_1)$ and $\rho_2(\mathbf{r}_1,\mathbf{r}_2)$. We can use the exact equation (2.4) to eliminate the one-body factor $u_1(\mathbf{r}_1)$ in the energy expectation value and consider the physically observable $\rho_1(\mathbf{r})$ as the independent one-body function. This elimination of a quantity that we do not need to know, namely $u_1(\mathbf{r}_1)$, in favour of a quantity that we would like to know, namely $\rho_1(\mathbf{r})$, is not only convenient but also essential for a number of reasons. It is *convenient* that we do not need to calculate $\rho_1(\mathbf{r}_1)$ from $u_1(\mathbf{r}_1)$. It is also *convenient* that the only remaining approximate relationship is the choice of the elementary diagrams $E(\mathbf{r}_1,\mathbf{r}_2)$. But the elimination is *essential* to guarantee that the solution of the two–body Euler-Lagrange equation in an inhomogeneous geometry satisfies the "cluster property" $u_2(\mathbf{r}_1,\mathbf{r}_2) \to 0$ for $|\mathbf{r}_1-\mathbf{r}_2| \to \infty$ for *any* given $u_1(\mathbf{r}_1)$. Specifying $u_1(\mathbf{r}_1)$ beforehand, one might have chosen a non-optimal one-body function, and the two–body function would have to compensate for that by violating the "cluster property". This would lead to a breakdown of the HNC resummation procedure. Thus, the technical *importance* of reformulating the variational problem in terms of the physics observables $\rho_1(\mathbf{r})$ and $g(\mathbf{r}_1,\mathbf{r}_2)$ is that it removes this difficulty.

After u_1 has been eliminated using (2.4), the energy expectation value can be written in a form suitable for further manipulations:

$$\begin{aligned} E_{\rm var}[\rho_1,g,u_2] &= \frac{\hbar^2}{2m}\int d^3r_1 \left|\boldsymbol{\nabla}_1\sqrt{\rho_1(\mathbf{r}_1)}\right|^2 + \int d^3r_1\rho_1(\mathbf{r}_1)U_{\rm ext}(\mathbf{r}_1) \\ &+\frac{1}{2}\int d^3r_1 d^3r_2 \rho_1(\mathbf{r}_1)\rho_1(\mathbf{r}_2)g(\mathbf{r}_1,\mathbf{r}_2)v_{JF}(\mathbf{r}_1,\mathbf{r}_2) \\ &= T_{\rm var} + E_{\rm ext} + E_{\rm c}\ . \end{aligned} \tag{2.15}$$

Here $v_{\rm JF}(\mathbf{r}_1,\mathbf{r}_2)$ is the generalized Jackson–Feenberg interaction introduced in Ref. 4 to an inhomogeneous geometry,

$$v_{\rm JF}(\mathbf{r}_1,\mathbf{r}_2) \equiv V(|\mathbf{r}_1-\mathbf{r}_2|)$$

$$-\frac{\hbar^2}{8m}\left[\frac{1}{\rho_1(\mathbf{r}_1)}\nabla_1\rho_1(\mathbf{r}_1)\cdot\nabla_1+\frac{1}{\rho_1(\mathbf{r}_2)}\nabla_2\rho_1(\mathbf{r}_2)\cdot\nabla_2\right]u_2(\mathbf{r}_1,\mathbf{r}_2)\ . \tag{2.16}$$

Two new quantities appear in (2.15). The first one,

$$T_{\text{var}}=\frac{\hbar^2}{2m}\int d^3r_1\left|\nabla_1\sqrt{\rho_1(\mathbf{r}_1)}\right|^2 \tag{2.17}$$

is the kinetic energy of a non-interacting, inhomogeneous Bose system whose single-particle wave function is $\sqrt{\rho_1(\mathbf{r}_1)}$. The second term,

$$E_{\text{ext}}=\int d^3r_1\rho_1(\mathbf{r}_1)U_{\text{ext}}(\mathbf{r}_1) \tag{2.18}$$

is the energy of the system in the external field $U_{\text{ext}}(\mathbf{r}_1)$. This physics of the interactions is contained in the third term, E_c, which we shall call the "correlation energy".

The most straightforward (although not the most general) way of deriving an Euler equation is to use the HNC relationship (2.7) to eliminate $u_2(\mathbf{r}_1,\mathbf{r}_2)$ in favor of $g(\mathbf{r}_1,\mathbf{r}_2)$ from the variational energy functional. In analogy to the bulk system [1] one finds

$$E_c = K+V+E_Q \tag{2.19}$$

$$K = \frac{\hbar^2}{2m}\int d^3r_1d^3r_2\rho_1(\mathbf{r}_1)\rho_1(\mathbf{r}_2)\left[\left|\nabla_1\sqrt{g(\mathbf{r}_1,\mathbf{r}_2)}\right|^2+\left|\nabla_2\sqrt{g(\mathbf{r}_1,\mathbf{r}_2)}\right|^2\right] \tag{2.20}$$

$$V = \frac{1}{2}\int d^3r_1d^3r_2\rho_1(\mathbf{r}_1)\rho_1(\mathbf{r}_2)g(\mathbf{r}_1,\mathbf{r}_2)V(|\mathbf{r}_1-\mathbf{r}_2|) \tag{2.21}$$

$$E_Q = -\frac{1}{4}\int d^3r_1d^3r_2\left[\tilde{N}(\mathbf{r}_1,\mathbf{r}_2)+\tilde{E}(\mathbf{r}_1,\mathbf{r}_2)\right]H_1(\mathbf{r}_1)\tilde{h}(\mathbf{r}_1,\mathbf{r}_2)\ , \tag{2.22}$$

where

$$H_1(\mathbf{r}_1)=-\frac{\hbar^2}{2m}\frac{1}{\sqrt{\rho(\mathbf{r}_1)}}\nabla_1\cdot\rho_1(\mathbf{r}_1)\nabla_1\frac{1}{\sqrt{\rho(\mathbf{r}_1)}}\ . \tag{2.23}$$

The operator $H_1(\mathbf{r}_1)$ can be thought of as the kinetic energy operator generalized to inhomogeneous systems. The functions $\rho(\mathbf{r}_1)$ and $g(\mathbf{r}_1,\mathbf{r}_2)$ are determined by minimization of the total energy. Let us first look at the optimization of pair correlations. Ignoring, for the time being, the elementary contribution $\tilde{E}(\mathbf{r}_1,\mathbf{r}_2)$, one arrives at the Euler-Lagrange equation [3]

$$\left[\tilde{X}*H_1*\tilde{X}\right](\mathbf{r}_1,\mathbf{r}_2)-\left[H_1(\mathbf{r}_1)+H_1(\mathbf{r}_2)\right]\tilde{X}(\mathbf{r}_1,\mathbf{r}_2)=2\tilde{V}_{\text{p-h}}(\mathbf{r}_1,\mathbf{r}_2) \tag{2.24}$$

or, using the relationship (2.12)

$$\left[S^{-1}*H_1*S^{-1}-H_1\right](\mathbf{r}_1,\mathbf{r}_2)=2\tilde{V}_{\text{p-h}}(\mathbf{r}_1,\mathbf{r}_2)\ . \tag{2.25}$$

The "particle-hole interaction" $V_{\text{p-h}}(\mathbf{r}_1,\mathbf{r}_2)$ appearing in Eq. (2.24) is

$$\begin{aligned}V_{\text{p-h}}(\mathbf{r}_1,\mathbf{r}_2) &= g(\mathbf{r}_1,\mathbf{r}_2)\left[V(|\mathbf{r}_1-\mathbf{r}_2|)+\Delta V_e(\mathbf{r}_1,\mathbf{r}_2)\right] \\ &+\frac{\hbar^2}{2m}\left[\left|\nabla_1\sqrt{g(\mathbf{r}_1,\mathbf{r}_2)}\right|^2+\left|\nabla_2\sqrt{g(\mathbf{r}_1,\mathbf{r}_2)}\right|^2\right]+h(\mathbf{r}_1,\mathbf{r}_2)w_I(\mathbf{r}_1,\mathbf{r}_2)\end{aligned} \tag{2.26}$$

with the induced interaction

$$\tilde{w}_I(\mathbf{r}_1,\mathbf{r}_2) = -\frac{1}{2}\left[\left[H_1(\mathbf{r}_1)+H_1(\mathbf{r}_2)\right]\tilde{N}(\mathbf{r}_1,\mathbf{r}_2) + \left[\tilde{X} * H_1 * \tilde{X}\right](\mathbf{r}_1,\mathbf{r}_2)\right] \ . \qquad (2.27)$$

As in the homogeneous case, $\Delta V_e(\mathbf{r}_1,\mathbf{r}_2)$ denotes the correction from elementary diagrams and triplet correlations.

Finally, we must also derive an Euler equation for the one–body density. The energy is minimized subject to the constraint of fixed particle number

$$N = \int d^3r_1 \rho_1(\mathbf{r}_1) \ , \qquad (2.28)$$

i.e., we carry out the variation

$$\frac{1}{2}\frac{\delta(E_{\rm var}-\mu N)}{\delta\sqrt{\rho_1(\mathbf{r}_1)}} = \left[-\frac{\hbar^2}{2m}\nabla_1^2 + U_{\rm ext}(\mathbf{r}_1) + V_{\rm H}(\mathbf{r}_1) - \mu\right]\sqrt{\rho_1(\mathbf{r}_1)} = 0 \ , \qquad (2.29)$$

where

$$V_{\rm H}(\mathbf{r}_1) = \frac{\delta E_c}{\delta\rho_1(\mathbf{r}_1)} \qquad (2.30)$$

is a generalized Hartree–potential, which depends implicitly on the density and the pair distribution function. The chemical potential μ enters the theory as a Lagrange parameter to enforce the constraint (2.28). Explicit forms of the effective one-body potential $V_H(\mathbf{r})$ have been derived in Ref. 5; the form presented here is best suited for an iterative solution with excellent convergence properties in numerical applications:

$$\begin{aligned}
V_{\rm H}(\mathbf{r}_1) &= V_{\rm H}^{(1)}(\mathbf{r}_1) + V_{\rm H}^{(2)}(\mathbf{r}_1) \\
V_{\rm H}^{(1)}(\mathbf{r}_1) &= \int d^3r_2\rho_1(\mathbf{r}_2)\left[V_{\rm p-h}(\mathbf{r}_1,\mathbf{r}_2) - \frac{1}{2}h(\mathbf{r}_1,\mathbf{r}_2)w_{\rm I}(\mathbf{r}_1,\mathbf{r}_2)\right] \qquad (2.31)\\
V_{\rm H}^{(2)}(\mathbf{r}_1) &= -\frac{\hbar^2}{16m\rho_1(\mathbf{r}_1)}\nabla_1\cdot\left[\rho_1(\mathbf{r}_1)\nabla_1\int d^3r_2\rho_1(\mathbf{r}_2)h(\mathbf{r}_1,\mathbf{r}_2)N(\mathbf{r}_1,\mathbf{r}_2)\right] \ .
\end{aligned}$$

One might interpret $V_{\rm H}^{(1)}(\mathbf{r}_1)$ as a "volume term" and $V_{\rm H}^{(2)}(\mathbf{r}_1)$ as a "curvature term", because the former survives in the bulk liquid whereas the latter vanishes. There are also corrections from elementary diagrams; these terms are not spelled out here for brevity.

2.2. *Normal–Mode Analysis*

The key equation of our theory is the Euler equation (2.24) for the inhomogeneous system. In the bulk case, the equation can be trivially solved in momentum space, but in the inhomogeneous case Eq. (2.24) is a second-order, non-linear partial integro-differential equation in a large number of variables. We review the most efficient algorithm for solving this equation numerically. This algorithm is closely related to the linear–response theory of the non-uniform system; the quantities introduced for the numerical solution of the optimization problem are therefore of independent physical interest.

Consider the eigenvalue problem

$$\int d^3r_2 H_1(\mathbf{r}_1)\left[H_1(\mathbf{r}_1)\delta(\mathbf{r}_1-\mathbf{r}_2)+2\tilde{V}_{\rm p-h}(\mathbf{r}_1,\mathbf{r}_2)\right]\phi^{(n)}(\mathbf{r}_2)=\epsilon_n^2\phi^{(n)}(\mathbf{r}_1) \qquad (2.32)$$

and the adjoint equation

$$\int d^3r_2 \left[H_1(\mathbf{r}_1)\delta(\mathbf{r}_1-\mathbf{r}_2)+2\tilde{V}_{\rm p-h}(\mathbf{r}_1,\mathbf{r}_2)\right]H_1(\mathbf{r}_2)\psi^{(n)}(\mathbf{r}_2)=\epsilon_n^2\psi^{(n)}(\mathbf{r}_1)\ . \qquad (2.33)$$

The eigenvectors of the two equations are related by

$$\phi^{(n)}(\mathbf{r}_1)=\frac{1}{\epsilon_n}H_1(\mathbf{r}_1)\psi^{(n)}(\mathbf{r}_1)\ , \qquad (2.34)$$

where the normalization factor is a matter of convenience. The states $\psi^{(n)}$ and $\phi^{(n)}$ then satisfy the orthogonality and closure relations

$$\langle\psi^{(n)}\,|H_1|\,\psi^{(m)}\rangle=\epsilon_n\delta_{m,n}. \qquad (2.35)$$

$$\sum_n \psi^{(n)}(\mathbf{r}_1)\phi^{(n)}(\mathbf{r}_2)=\delta(\mathbf{r}_1-\mathbf{r}_2) \qquad (2.36)$$

It is then straightforward to verify that the solutions $\tilde{X}(\mathbf{r}_1,\mathbf{r}_2)$ of (2.24) and $S(\mathbf{r}_1,\mathbf{r}_2)$ of (2.25) can be constructed by the normal-mode decomposition

$$\tilde{X}(\mathbf{r}_1,\mathbf{r}_2)=\delta(\mathbf{r}_1-\mathbf{r}_2)-\sum_n \psi^{(n)}(\mathbf{r}_1)\psi^{(n)}(\mathbf{r}_2) \qquad (2.37)$$

$$S(\mathbf{r}_1,\mathbf{r}_2)=\sum_n \phi^{(n)}(\mathbf{r}_1)\phi^{(n)}(\mathbf{r}_2)\ . \qquad (2.38)$$

Given $S(\mathbf{r}_1,\mathbf{r}_2)$, one can proceed through an iteration path that is identical to the one for the bulk liquid. The necessary operations are just matrix products and do not require any specific numerical techniques. The eigenvalues ϵ_n and the eigenfunctions $\psi^{(n)}(\mathbf{r}_1)$ and $\phi^{(n)}(\mathbf{r}_1)$ will play central roles in many further applications of the theory.

2.3. *Atomic Impurities*

Next we formulate our theory for isolated atomic impurities in a non–uniform bosonic background. Since impurity particles interact with each other by exchanging elementary excitations of the host system, the study of impurity properties also provides useful tests of our understanding of these excitations.

The Hamiltonian of the $N+1$ particle system consisting of N ^{4}He atoms and one impurity is

$$H^I_{N+1} = -\frac{\hbar^2}{2m_I}\nabla_0^2+U^I_{\rm ext}(\mathbf{r}_0)+\sum_{i=1}^N V^I(|\mathbf{r}_0-\mathbf{r}_i|)+H_N\,. \qquad (2.39)$$

We adopt the convention that the coordinate $\mathbf{r}_0$ refers to the impurity particle and coordinates $\mathbf{r}_i$, with $i=1\ldots N$, to the background particles. Note that the substrate potentials $U_{\rm sub}(\mathbf{r}_i)$ appearing in the background Hamiltonian H_N, and $U^I_{\rm ext}(\mathbf{r}_0)$, as well as the interactions $V^I(|\mathbf{r}_0 - \mathbf{r}_j|)$ and $V(|\mathbf{r}_i - \mathbf{r}_j|)$, can be different functions for different particle species.

The generalization of the wave function (2.1) for an inhomogeneous N-particle Bose system with a single impurity atom is

$$\Psi^I_{N+1}(\mathbf{r}_0, \mathbf{r}_1, \ldots, \mathbf{r}_N) \tag{2.40}$$
$$= \exp \frac{1}{2}\left[u^I_1(\mathbf{r}_0) + \sum_{1\le i\le N} u^I_2(\mathbf{r}_0, \mathbf{r}_i) + \sum_{1\le i<j\le N} u^I_3(\mathbf{r}_0, \mathbf{r}_i, \mathbf{r}_j)\right] \Psi_N(\mathbf{r}_1, \ldots, \mathbf{r}_N)\,.$$

The energy necessary for (or gained by) adding one impurity atom into the system is the impurity chemical potential $\mu^I \equiv E^I_{N+1} - E_N$, where E^I_{N+1} is the energy of the system containing one impurity and N background atoms, and E_N is the energy of the unperturbed background system. Here E^I_{N+1} is to be understood as the energy expectation value of the Hamiltonian (2.39) with respect to the wave function (2.40).

It is plausible to write the impurity chemical potential μ^I in the form

$$\mu^I = T^I + V^I_{\rm ext} + E^I_c\,. \tag{2.41}$$

Here the first term is the kinetic energy of a single non-interacting impurity with the ground-state wave function $\sqrt{\rho^I_1(\mathbf{r}_0)}$,

$$T^I = \frac{\hbar^2}{2m_I}\int d^3r_0 \left|\nabla\sqrt{\rho^I_1(\mathbf{r}_0)}\right|^2\,, \tag{2.42}$$

and the second one is the energy of the impurity due to the external potential,

$$V^I_{\rm ext} = \int d^3r_0 U^I_{\rm ext}(\mathbf{r}_0)\rho^I_1(\mathbf{r}_0)\,. \tag{2.43}$$

The correlation energy E^I_c contains the many–body effects, which can be written as a functional of four quantities: the impurity density $\rho^I_1(\mathbf{r}_0)$, the background density $\rho_1(\mathbf{r}_1)$, the impurity-background pair-distribution function $g^I(\mathbf{r}_0, \mathbf{r}_1)$ and the background pair-distribution function $g(\mathbf{r}_1, \mathbf{r}_2)$,

When minimizing the impurity energy, one must keep in mind that all background quantities are changed by the presence of the impurity by terms of the order of $1/N$; these changes give rise to quantitatively important rearrangement effects. The details of the derivation are given in Ref. 6, the final results are similar to those of the ground–state theory of the background liquid.

The impurity density is calculated by minimizing the chemical potential (2.41) with respect to $\sqrt{\rho^I_1(\mathbf{r}_0)}$. This leads to the usual Hartree equation

$$-\frac{\hbar^2}{2m_I}\nabla^2_0\sqrt{\rho^I_1(\mathbf{r}_0)} + \left[U^I_{\rm ext}(\mathbf{r}_0) + V_H(\mathbf{r}_0)\right]\sqrt{\rho^I_1(\mathbf{r}_0)} = \mu^I\sqrt{\rho^I_1(\mathbf{r}_0)}\,, \tag{2.44}$$

with the self-consistent one-body Hartree potential for the single impurity,

$$V_H(\mathbf{r}_0) = \frac{\delta E_c}{\delta\rho^I_1(\mathbf{r}_0)}\,. \tag{2.45}$$

The impurity chemical potential μ^I appears as the Lagrange multiplier to ensure the normalization $\int d^3r_0 \rho_1^I(\mathbf{r}_0) = 1$.

The two-body Euler equation is derived by variation of the impurity chemical potential with respect to $\sqrt{g^I(\mathbf{r}_0, \mathbf{r}_1)}$. After a number of algebraic manipulations that are virtually identical to those used in the derivation of the Euler equations for the background, [5,7] one arrives at (*cf.* Eq. (2.24))

$$[\tilde{X}^I * H_1 * \tilde{X}](\mathbf{r}_0, \mathbf{r}_1) - [H_1^I(\mathbf{r}_0) + H_1(\mathbf{r}_1)]\tilde{X}^I(\mathbf{r}_0, \mathbf{r}_1) = 2\tilde{V}^I_{\mathrm{p-h}}(\mathbf{r}_0, \mathbf{r}_1) , \qquad (2.46)$$

where the one-impurity Hamiltonian is (*cf.* Eq. (2.23))

$$H_1^I(\mathbf{r}_0) = -\frac{\hbar^2}{2m_I} \frac{1}{\sqrt{\rho_1^I(\mathbf{r}_0)}} \nabla_0 \cdot \rho_1^I(\mathbf{r}_0) \nabla_0 \frac{1}{\sqrt{\rho_1^I(\mathbf{r}_0)}} . \qquad (2.47)$$

The direct correlation function for the impurity $\tilde{X}^I(\mathbf{r}_0, \mathbf{r}_1)$ is related to the impurity pair-distribution function through the Ornstein-Zernike relation,

$$\sqrt{\rho_1^I(\mathbf{r}_0)\rho_1(\mathbf{r}_1)}[g^I(\mathbf{r}_0, \mathbf{r}_1) - 1] \equiv S^I(\mathbf{r}_0, \mathbf{r}_1) = \int d^3r_2 \tilde{X}^I(\mathbf{r}_0, \mathbf{r}_2) S(\mathbf{r}_2, \mathbf{r}_1) . \qquad (2.48)$$

The expressions for the *impurity particle-hole interaction* $\tilde{V}^I_{\mathrm{p-h}}(\mathbf{r}_0, \mathbf{r}_1)$ and the Hartree potential $V_H(\mathbf{r}_0)$ are given in Ref. 6.

A convenient way to solve the two-body equation (2.46) is again by normal–mode decomposition. We first solve the eigenvalue problem

$$H_1^I(\mathbf{r}_0)\eta^{(o)}(\mathbf{r}_0) = t_o \eta^{(o)}(\mathbf{r}_0) , \qquad (2.49)$$

which is equivalent with the calculation of the eigenvalues and eigenfunctions of the Hartree equation (2.44) with the spectrum shifted by the amount μ^I.

The matrix representation of the impurity correlation functions can be obtained now in the basis of the *impurity states* $\eta^{(o)}$ and the Feynman phonon states $\phi^{(m)}$,

$$\tilde{X}^I(\mathbf{r}_0, \mathbf{r}_1) = -2 \sum_{om} \eta^{(o)}(\mathbf{r}_0) \frac{\langle \eta^{(o)} | \tilde{V}^I_{\mathrm{p-h}} | \phi^{(m)} \rangle}{t_o + \epsilon_m} \psi^{*(m)}(\mathbf{r}_1) , \qquad (2.50)$$

and

$$S^I(\mathbf{r}_0, \mathbf{r}_1) = -2 \sum_{om} \eta^{(o)}(\mathbf{r}_0) \frac{\langle \eta^{(o)} | \tilde{V}^I_{\mathrm{p-h}} | \phi^{(m)} \rangle}{t_o + \epsilon_m} \phi^{*(m)}(\mathbf{r}_1) . \qquad (2.51)$$

3. Theory of Excitations

The features of many–body systems that are primarily accessible to experiments are excitations: They determine the outcome of thermodynamic, sound, and scattering experiments. The ground state and the excited states are determined by the same Hamiltonian; a look at the excitations will therefore also aid in an assessment of the physical content and interpretation of the ground–state theory. For example, the static structure function $S(k)$ is *a priori* a ground–state property, but it can also be obtained, through the fluctuation–dissipation theorem, [8] from the dynamic structure function $S(k,\omega)$ by frequency integration. Self–consistency requires that these structure functions are identical. We will also see how the eigenvalues ϵ_n of Eq. (2.33) can be related to the excitation energies of the system.

The basic principles of the theory of excited states of inhomogeneous liquids parallel those of homogeneous ones (see Chapter 4 in this book). There are now two static potentials in the Hamiltonian: The pair potential between particles, and the external potential that makes the liquid inhomogeneous in the first place. In addition, we introduce a small, time–dependent perturbation that momentarily drives the quantum liquid out of its ground state and causes the n-body correlations in the wave function presented in Eq. (2.1) to acquire time dependence. The excited state is

$$|\Psi(t)\rangle = \frac{e^{-iE_0 t/\hbar} e^{(1/2)\delta U(t)} |\Psi_0\rangle}{\left[\langle\Psi_0|e^{\delta U(t)}|\Psi_0\rangle\right]^{1/2}} , \tag{3.1}$$

where E_0 is the ground–state energy, $|\Psi_0\rangle$ is the ground state, and

$$\delta U(t) = \sum_i \delta u_1(\mathbf{r}_i, t) + \sum_{i<j} \delta u_2(\mathbf{r}_i, \mathbf{r}_j, t) + \ldots \tag{3.2}$$

is the complex *excitation operator*. The components of the time–dependent correlation operator $\delta U(t)$ are determined by the action principle:

$$\delta S = \delta \int_{t_0}^{t} dt \mathcal{L}(t) = 0 , \tag{3.3}$$

where the Lagrangian is given by [9]

$$\begin{aligned}
\mathcal{L}(t) &= \left\langle \Psi(t) \middle| H(t) - E_{\rm var} - i\hbar\frac{\partial}{\partial t} \middle| \Psi(t) \right\rangle \\
&= \frac{1}{8}\left\langle \Psi_0 \middle| [\delta U^*, [T, \delta U]] \middle| \Psi_0 \right\rangle + \left\langle \Psi(t) \middle| \sum_i U_{ext}(\mathbf{r}_i; t) \middle| \Psi(t) \right\rangle \\
&- \frac{i\hbar}{8}\left[\left\langle \Psi_0 \middle| \delta\dot{U}\left[\delta U^* - \langle\Psi_0|\delta U^*|\Psi_0\rangle\right] \middle| \Psi_0 \right\rangle - c.c. \right] .
\end{aligned} \tag{3.4}$$

If the state has no time dependence, Eq. (3.3) reduces to the familiar Rayleigh-Ritz variational principle. One normally assumes that $|\Psi_0\rangle$ is the exact ground state, but if the excitation operator δU is truncated at δu_2 it is sufficient to assume that the correlations up to and including u_4 have been optimized. By keeping terms to leading (second) order in the dynamical correlations, the resulting Euler equations can be cast in the form of coupled *equations of motion* (EOM). The conjugate variable to the time is the excitation energy $\hbar\omega$. In general it is complex; the real part is the excitation's dispersion and the imaginary part its inverse lifetime.

The truncation of the sequence of time–dependent correlation functions defines the level of approximation in which we treat the excitations. We start with the simplest case of just time–dependent one–body correlations referred to, hereafter, as the *Feynman approximation.* [10]

3.1. *Feynman Theory of Excitations and the Static Structure Function*

The original Feynman theory of collective excitations [10] is obtained by restricting the excitation operator (3.2) to the one–body component $\delta u_1(\mathbf{r}_i;t)$. In this case the Lagrangian can be expressed in terms of one– and two–body densities,

$$\mathcal{L}(t) = \frac{\hbar^2}{8m}\int d^3r_1\rho_1(\mathbf{r}_1)\left|\nabla\delta u_1(\mathbf{r}_1;t)\right|^2 - \frac{i\hbar}{8}\int d^3r_1\dot\rho_1(\mathbf{r}_1;t)\delta u_1^*(\mathbf{r}_1;t)$$
$$+\int d^3r_1U_{\rm ext}(\mathbf{r}_1;t)\Re e\rho_1(\mathbf{r}_1;t) \qquad (3.5)$$

where, to first order in the time–dependent functions,

$$\rho_1(\mathbf{r}_1;t) = \rho_1(\mathbf{r}_1)+\delta\rho_1(\mathbf{r}_1;t)\ ,$$
$$\delta\rho_1(\mathbf{r}_1;t) = \rho_1(\mathbf{r}_1)\delta u_1(\mathbf{r}_1;t)+\int d^3r_2\left[\rho_2(\mathbf{r}_1,\mathbf{r}_2)-\rho_1(\mathbf{r}_1)\rho_1(\mathbf{r}_2)\right]\delta u_1(\mathbf{r}_2;t). \qquad (3.6)$$

The *complex* function $\delta\rho_1(\mathbf{r}_1;t)$ should be regarded as an abbreviation of the functional expansion (3.6), while the physical density fluctuation is its real part.

Assuming harmonic time dependence, $\delta u_1(\mathbf{r};t) = \delta u_1(\mathbf{r}_1)e^{i\omega t}$, and defining

$$\psi(\mathbf{r}_1) = \sqrt{\rho_1(\mathbf{r}_1)}\delta u_1(\mathbf{r}_1), \qquad (3.7)$$

the action principle (3.3) can be cast into the generalized eigenvalue problem

$$H_1\psi^{(n)}(\mathbf{r}_1) \equiv -\frac{\hbar^2}{2m}\frac{1}{\sqrt{\rho_1(\mathbf{r})}}\nabla\cdot\left(\rho_1(\mathbf{r})\nabla\frac{\psi^{(n)}(\mathbf{r}_1)}{\sqrt{\rho_1(\mathbf{r})}}\right)$$
$$= \hbar\omega_n\int d^3r'S(\mathbf{r}_1,\mathbf{r}_2)\psi^{(n)}(\mathbf{r}_2) \qquad (3.8)$$

with the coordinate–space representations of the kinetic energy operator H_1 (2.23) and the static structure function $S(\mathbf{r}_1,\mathbf{r}_2)$ (2.11). The bulk limit of the eigenvalue problem (3.8) is the Feynman dispersion relation $\hbar\omega(k) = \hbar^2k^2/[2mS(k)]$. The adjoint states

$$\phi^{(m)}(\mathbf{r}_1) \equiv \frac{1}{\hbar\omega_n}H_1\psi^{(n)}(\mathbf{r}_1) \qquad (3.9)$$

are related to the physical density fluctuations in state n (*cf.* Refs. 11, 12) via

$$\delta\rho_1^{(n)}(\mathbf{r}_1) = \sqrt{\rho_1(\mathbf{r}_1)}\phi^{(n)}(\mathbf{r}_1). \qquad (3.10)$$

The eigenstates $\psi^{(n)}$ and their adjoint states $\phi^{(n)}$ are related to the Feynman excitation functions $\delta u_1(\mathbf{r}_1)$ through Eq. (3.7). These *Feynman states* appeared also in the ground–state theory as solutions of the eigenvalue equations (2.32) and (2.33).

Within the Feynman approximation, which corresponds here to the random phase approximation (RPA), we construct from these states the density-density response function:

$$\chi^{RPA}(\mathbf{r}_1,\mathbf{r}_2,\omega) = \sum_{mn}\delta\rho_1^{(m)}(\mathbf{r}_1)\left[G_{mn}^{RPA}(\omega)+G_{mn}^{RPA}(-\omega)\right]\delta\rho_1^{(n)}(\mathbf{r}_2)\ , \qquad (3.11)$$

where

$$G_{mn}^{RPA}(\omega) = \frac{\delta_{mn}}{\hbar\left[\omega - \omega_n + i\epsilon\right]} \tag{3.12}$$

is the Green's function for a free Feynman phonon. Eq. (3.11) describes the propagation of density fluctuations from point $\mathbf{r}_2$ to $\mathbf{r}_1$.

We can now calculate the static structure function from the response function (3.11) by frequency integration:

$$\begin{aligned} S(\mathbf{r}_1,\mathbf{r}_2) &= -\frac{1}{\sqrt{\rho_1(\mathbf{r}_1)\rho_1(\mathbf{r}_2)}}\int_0^\infty \frac{d(\hbar\omega)}{\pi}\Im m\chi^{RPA}(\mathbf{r}_1,\mathbf{r}_2;\omega) \\ &= \sum_n \phi^{(n)}(\mathbf{r}_1)\phi^{(n)}(\mathbf{r}_2)\,. \end{aligned} \tag{3.13}$$

This is exactly the same expression which we obtained from the optimization of the ground–state correlations, Eq. (2.38). In the latter calculation excitations are not involved in any explicit way. Thus, we have demonstrated the aforementioned internal consistency of the theory.

3.2. *Multiphonon Excitations*

The Feynman theory can be improved by letting both u_1 and u_2 depend on time. We start again from the Lagrangian (3.4). Inserting the explicit form of our time–dependent correlations allows us to rewrite the double-commutator term in (3.4) as

$$\begin{aligned} \frac{1}{8}\left\langle \Psi_0\left|[\delta U^*,[T,\delta U]]\right|\Psi_0\right\rangle = \frac{\hbar^2}{8m}\Bigg\{ &\int d^3r\rho_1(\mathbf{r}_1)\left|\nabla\delta u_1(\mathbf{r};t)\right|^2 \\ + &\int d^3r_1d^3r_2\rho_2(\mathbf{r}_1,\mathbf{r}_2)\left[\nabla_1\delta u_1(\mathbf{r}_1;t)\cdot\nabla_1\delta u_2^*(\mathbf{r}_1,\mathbf{r}_2;t) + c.c. + \left|\nabla_1\delta u_2(\mathbf{r}_1,\mathbf{r}_2;t)\right|^2\right] \\ + &\int d^3r_1d^3r_2d^3r_3\rho_3(\mathbf{r}_1,\mathbf{r}_2,\mathbf{r}_3)\nabla_1\delta u_2(\mathbf{r}_1,\mathbf{r}_2)\cdot\nabla_1\delta u_2^*(\mathbf{r}_1,\mathbf{r}_3)\Bigg\}\,. \end{aligned} \tag{3.14}$$

The time–derivative term in the Lagrangian reads

$$\begin{aligned} &\left\langle\Psi_0\left|\delta\dot U\left[\delta U^* - \left\langle\Psi_0\left|\delta U^*\right|\Psi_0\right\rangle\right]\right|\Psi_0\right\rangle \\ &= \int d^3r\dot\rho_1(\mathbf{r};t)\delta u_1^*(\mathbf{r};t) + \frac{1}{2}\int d^3r_1d^3r_2\dot\rho_2(\mathbf{r}_1,\mathbf{r}_2;t)\delta u_2^*(\mathbf{r}_1,\mathbf{r}_2;t)\,. \end{aligned} \tag{3.15}$$

The action principle (3.3) leads to two equations of motion, which have the form of continuity equations,

$$\nabla_1\cdot\mathbf{j}_1(\mathbf{r}_1;t) + i\dot\rho_1(\mathbf{r}_1;t) = D_1(\mathbf{r}_1;t) \tag{3.16}$$

$$(\nabla_1+\nabla_2)\cdot\mathbf{j}_2(\mathbf{r}_1,\mathbf{r}_2,;t) + i\dot\rho_2(\mathbf{r}_1,\mathbf{r}_2;t) = D_2(\mathbf{r}_1,\mathbf{r}_2;t)\,, \tag{3.17}$$

where we have defined the one– and two–body currents

$$\mathbf{j}_1(\mathbf{r}_1;t) = \frac{\hbar}{2mi}\left\{\rho_1(\mathbf{r}_1)\boldsymbol{\nabla}_1\delta u_1(\mathbf{r}_1;t) + \int d^3r_2\rho_2(\mathbf{r}_1,\mathbf{r}_2)\boldsymbol{\nabla}_1\delta u_2(\mathbf{r}_1,\mathbf{r}_2;t)\right\} \quad (3.18)$$

$$\mathbf{j}_2(\mathbf{r}_1,\mathbf{r}_2;t) = \frac{\hbar}{2mi}\left\{\rho_2(\mathbf{r}_1,\mathbf{r}_2)\boldsymbol{\nabla}_1\left[\delta u_1(\mathbf{r}_1;t) + \delta u_2(\mathbf{r}_2,\mathbf{r}_2;t)\right]\right.$$
$$\left. + \int d^3r_3\rho_3(\mathbf{r}_1,\mathbf{r}_2,\mathbf{r}_3)\boldsymbol{\nabla}_1\delta u_2(\mathbf{r}_1,\mathbf{r}_3;t)\right\} + \text{ same for } (1 \leftrightarrow 2) \quad (3.19)$$

and the driving terms

$$D_1(\mathbf{r}_1;t) = \frac{2}{\hbar}\left[\rho_1(\mathbf{r}_1)U_{ext}(\mathbf{r}_1;t)\right. \quad (3.20)$$
$$\left. + \int d^3r_2[\rho_2(\mathbf{r}_1,\mathbf{r}_2) - \rho_1(\mathbf{r}_1)\rho_1(\mathbf{r}_2)]U_{ext}(\mathbf{r}_2;t)\right]$$

$$D_2(\mathbf{r}_1,\mathbf{r}_2;t) = \frac{2}{\hbar}\rho_2(\mathbf{r}_1,\mathbf{r}_2)\left[U_{ext}(\mathbf{r}_1;t) + U_{ext}(\mathbf{r}_2;t)\right] \quad (3.21)$$
$$+\frac{2}{\hbar}\int d^3r_3\left[\rho_3(\mathbf{r}_1,\mathbf{r}_2,\mathbf{r}_3) - \rho_2(\mathbf{r}_1,\mathbf{r}_2)\rho_1(\mathbf{r}_3)\right]U_{ext}(\mathbf{r}_3;t)\ .$$

Eqs. (3.16) and (3.17) are the starting point for the equations–of–motion method for the calculation of collective excitations in quantum liquids. Different implementations [13–16] differ by the approximations used for the three– and four–body densities.

An important point is that the one– and two–body equations are *not* independent: The one-body equation (3.16) results from the two-body equation in the asymptotic limit $|\mathbf{r}_1 - \mathbf{r}_2| \to \infty$. Moreover, using the sequential relations

$$\int d^3r_n\rho_n(\mathbf{r}_1,\ldots,\mathbf{r}_n) = (N-n)\rho_{n-1}(\mathbf{r}_1,\ldots,\mathbf{r}_{n-1}) \quad (3.22)$$

it is easily seen that the one-body equation also results when Eq. (3.17) is integrated over one coordinate, say, $\mathbf{r}_2$. This sequential property is preserved if one carries out a cumulant analysis for the three-body distribution function. [17] It turns out, that the external potential can be eliminated from the second continuity equation, showing that the time-dependence of the *pair correlations* is driven by the one-body current alone. [17] Thus it is appropriate to consider the equations–of–motion method as a *systematic* approach to introduce current–current coupling effects into the theory of excitations. Note that at this point we haven't made any approximations apart from assuming that the time-dependence of the wave function is described appropriately by an one-body and a two-body component and that it is legitimate to linearize the exponential excitation operator.

3.3. *Convolution Approximation*

We now need to make a specific approximation to all the three–body distribution functions and densities, as well as for the connection between the fluctuating pair–correlation function and the time–dependent part of the two–body distribution functions. We will refer to the approximation scheme we have chosen as the *convolution approximation* since it is diagrammatically equivalent to the approximation used by Chang and Campbell [13] for the bulk system.

The convolution approximation is equivalent to the *uniform limit approximation* [4] which assumes that the two-body quantities under consideration are small in coordinate space, but not necessarily in momentum space. This allows for long–range effects like phonons. Accordingly, two simplifications are made: We approximate

$$g(\mathbf{r}_1,\mathbf{r}_2)\nabla_1\delta u_2(\mathbf{r}_1,\mathbf{r}_2;t) \approx \nabla_1\delta u_2(\mathbf{r}_1,\mathbf{r}_2;t) \approx \nabla_1\delta X(\mathbf{r}_1,\mathbf{r}_2;t)\ , \tag{3.23}$$

and

$$\int d^3r_3\rho_1(\mathbf{r}_3)\left[g_3(\mathbf{r}_1,\mathbf{r}_2,\mathbf{r}_3) - g(\mathbf{r}_1,\mathbf{r}_3)g(\mathbf{r}_1,\mathbf{r}_2)\right]\nabla_1\delta u_2(\mathbf{r}_1,\mathbf{r}_3;t) \tag{3.24}$$

$$\approx \int d^3r_3\rho_1(\mathbf{r}_3)\nabla_1\delta u_2(\mathbf{r}_1,\mathbf{r}_3;t)h(\mathbf{r}_3,\mathbf{r}_2) \approx \int d^3r_3\rho_1(\mathbf{r}_3)\nabla_1\delta X(\mathbf{r}_1,\mathbf{r}_3;t)h(\mathbf{r}_3,\mathbf{r}_2)\ .$$

Here $\delta X(\mathbf{r}_1,\mathbf{r}_3;t)$ describes fluctuations of the direct correlation function defined in Eq. (2.12). This approximation is actually less dramatic as it may seem at first glance; a careful diagrammatic analysis of the three-body distribution function shows that, by expressing $\delta u_2(\mathbf{r}_1,\mathbf{r}_3;t)$ in terms of $\delta X(\mathbf{r}_1,\mathbf{r}_3;t)$, a large number of diagrams are eliminated that would contribute if one worked in terms of $\delta u_2(\mathbf{r}_1,\mathbf{r}_3;t)$. Unfortunately, a complete elimination of $\delta u_2(\mathbf{r}_1,\mathbf{r}_3;t)$ in favor of either $\delta X(\mathbf{r}_1,\mathbf{r}_3;t)$ or $\delta g(\mathbf{r}_1,\mathbf{r}_3;t)$ does not lead to any simplifications and appears to be impractical in the non-uniform geometry.

In order to avoid explicit density factors, it is again useful to introduce the tilde notation, Eq. (2.6), and mark convolution products with an asterisk. Besides the Feynman states $\psi^{(n)}(\mathbf{r})$ and their adjoints $\phi^{(n)}(\mathbf{r}_1)$, we will also need the abbreviations

$$\zeta^{(n)}(\mathbf{r}_1) = \frac{\phi^{(n)}(\mathbf{r}_1) - \psi^{(n)}(\mathbf{r}_1)}{\sqrt{\rho_1(\mathbf{r}_1)}} \tag{3.25}$$

and

$$\xi^{(n)}(\mathbf{r}_1) = \sqrt{\rho_1(\mathbf{r}_1)}\phi^{(n)}(\mathbf{r}_1). \tag{3.26}$$

From here on, it is advantageous to work entirely in the space defined by the Feynman wave functions and to express the one- and two-body fluctuations as

$$\begin{aligned}\frac{\delta\rho_1(\mathbf{r};t)}{\sqrt{\rho_1(\mathbf{r}_1)}} &= \sum_m \varphi_m(t)\phi^{(m)}(\mathbf{r}_1)\\ \delta\tilde{X}(\mathbf{r}_1,\mathbf{r}_2;t) &= \sum_{mn}\delta X_{mn}(t)\psi^{(m)}(\mathbf{r}_1)\psi^{(n)}(\mathbf{r}_2)\end{aligned} \tag{3.27}$$

$$\tilde{U}_{\rm ext}(\mathbf{r};t) = \sum_m U_{\rm ext}^{(m)}(t)\psi^{(m)}(\mathbf{r}_1).$$

In this basis, the static structure function has the form (2.38), and the direct correlation function is

$$\tilde{X}(\mathbf{r}_1,\mathbf{r}_2) = \delta(\mathbf{r}_1-\mathbf{r}_2) - \sum_m \psi^{(m)}(\mathbf{r}_1)\psi^{(m)}(\mathbf{r}_2). \tag{3.28}$$

The EOM (3.16) and (3.17) projected on the Feynman states $\psi^{(m)}$ turn into the one-body equation

$$\left[i\hbar\frac{\partial}{\partial t} - \hbar\omega_m\right]\varphi_m = \frac{1}{2}\sum_{st} V_{st}^{(m)}\delta X_{st} + 2U_{\rm ext}^{(m)}(t)\ , \tag{3.29}$$

and to the two-body equation

$$\left[i\hbar\frac{\partial}{\partial t} - \hbar\omega_m - \hbar\omega_n\right]\delta X_{nm} = \sum_s V_{mn}^{(s)}\varphi_s\ . \tag{3.30}$$

Here

$$V_{mn}^{(s)} = \frac{\hbar^2}{2m}\int d^3r \frac{\psi^{(s)}(\mathbf{r}_1)}{\sqrt{\rho_1(\mathbf{r}_1)}}\boldsymbol{\nabla}\cdot\left[\sqrt{\rho_1(\mathbf{r}_1)}\tilde{\mathbf{W}}_{mn}(\mathbf{r}_1)\right]\ , \tag{3.31}$$

with

$$\tilde{\mathbf{W}}_{mn}(\mathbf{r}_1) = \phi^{(m)}(\mathbf{r}_1)\boldsymbol{\nabla}\zeta^{(n)}(\mathbf{r}_1) + \phi^{(n)}(\mathbf{r}_1)\boldsymbol{\nabla}\zeta^{(m)}(\mathbf{r}_1) + \sqrt{\rho_1(\mathbf{r}_1)}\boldsymbol{\nabla}X_{mn}(\mathbf{r}_1) \tag{3.32}$$

which contains the triplet term

$$X_{mn}(\mathbf{r}_1) = \frac{1}{\sqrt{\rho_1(\mathbf{r}_1)}}\int d^3r_1 d^3r_2 \phi^{(m)}(\mathbf{r}_1)\phi^{(n)}(\mathbf{r}_2)\tilde{X}_3(\mathbf{r}_1,\mathbf{r}_2,\mathbf{r}_1)\ . \tag{3.33}$$

We use for the triplet function $X_3(\mathbf{r}_1,\mathbf{r}_2,\mathbf{r}_3)$ the optimized form obtained in Ref. 7.

To get the expression for the response function, we now make a harmonic expansion of the external field and the fluctuations

$$\begin{aligned} U_{\rm ext}^{(m)}(t) &= U_{\rm ext}^{(m)}\left[e^{-i\omega t} + e^{i\omega t}\right], \\ \varphi_m(t) &= x_m e^{-i\omega t} + y_m e^{i\omega t}, \\ \delta X_{mn}(t) &= x_{mn}e^{-i\omega t} + y_{mn}e^{i\omega t}\ , \end{aligned} \tag{3.34}$$

where we can assume that x_m, y_m and x_{mn}, y_{mn} are real. Defining

$$T_{mn,st}(\omega) = \hbar\left[\omega_n + \omega_m - \omega\right]\delta_{ms}\delta_{nt} \tag{3.35}$$

and separating the portions with positive and negative frequency allows us to formally solve for the one- and two-body equations for the x_m, y_m and the x_{mn} and y_{mn}

$$\begin{aligned} x_{mn} &= -\sum_{r,pq}\left[T^{-1}(\omega)\right]_{mn,pq}\tilde{V}_{pq}^{(r)}x_r\ , \\ y_{mn} &= -\sum_{r,pq}\left[T^{-1}(-\omega)\right]_{mn,pq}\tilde{V}_{pq}^{(r)}y_r\ , \end{aligned}$$

$$\begin{aligned}\hbar(\omega-\omega_s)x_s &= \frac{1}{2}\sum_{mn}\tilde{V}^{(s)}_{mn}x_{mn} + 2U^{(s)}_{\rm ext}\,, \\ \hbar(-\omega-\omega_s)y_s &= \frac{1}{2}\sum_{mn}\tilde{V}^{(s)}_{mn}y_{mn} + 2U^{(s)}_{\rm ext}\,,\end{aligned} \tag{3.36}$$

and, finally, to obtain the full solutions of the problem:

$$\sum_t E_{st}(\omega)x_t = 2U^{(s)}_{ext}\,, \qquad \sum_t E_{st}(-\omega)y_t = 2U^{(s)}_{ext}\,,$$

with

$$E_{st}(\omega) = \hbar(\omega-\omega_s)\delta_{st} + \frac{1}{2}\sum_{mn,pq}\tilde{V}^{(s)}_{mn}\left[T^{-1}(\omega)\right]_{mn,pq}\tilde{V}^{(t)}_{pq}. \tag{3.37}$$

Evidently, we can identify $E_{st}(\omega)$ with the inverse of the one-phonon Greens' function

$$G_{st}(\omega) = \left[\hbar(\omega-\omega_s+i\epsilon)\delta_{st} + \Sigma_{st}(\omega)\right]^{-1} \tag{3.38}$$

with the self energy

$$\Sigma_{st}(\omega) = \frac{1}{2}\sum_{mn,pq}\tilde{V}^{(s)}_{mn}\left[T^{-1}(\omega)\right]_{mn,pq}\tilde{V}^{(t)}_{pq}. \tag{3.39}$$

We are now ready to calculate the real part of the density fluctuations.

$$\begin{aligned}\Re e\ r_s(t) &= \frac{1}{2}\left[x_s+y_s\right]\left[e^{-i\omega t}+e^{i\omega t}\right] \\ &= \sum_t\left[G_{st}(\omega)+G_{st}(-\omega)\right]\tilde{U}^{(t)}_{\rm ext}\left[e^{-i\omega t}+e^{i\omega t}\right].\end{aligned} \tag{3.40}$$

The term in the square bracket can be identified with the density-density response function $\chi_{st}(\omega)$ in Feynman space. We shall call this the CBF response function, after the first derivation [18] of this form within the theory of correlated basis functions (CBF). It has the same form as the RPA response function (3.11), only the Greens' function G^{RPA} is replaced with

$$G^{CBF}_{mn}(\omega) = \left[\hbar[\omega-\omega_m+i\epsilon]\delta_{mn} + \Sigma_{mn}(\omega)\right]^{-1}\ . \tag{3.41}$$

The CBF density–density then response reads

$$\begin{aligned}&\chi^{\rm CBF}(\mathbf{r}_1,\mathbf{r}_2;\omega) \\ &\quad= \sqrt{\rho_1(\mathbf{r}_1)}\sum_{m,n}\left[\phi^*_m(\mathbf{r}_1)\left[G^{\rm CBF}_{mn}(\omega)+G^{\rm CBG}_{mn}(-\omega)\right]\phi_n(\mathbf{r}_2)\right]\sqrt{\rho_1(\mathbf{r}_2)}\,,\end{aligned} \tag{3.42}$$

with

$$\Sigma^{\rm CBF}_{st}(\omega) = \frac{1}{2}\sum_{mn}\frac{\tilde{V}^{(s)}_{mn}\tilde{V}^{(t)}_{mn}}{\hbar(\omega_m+\omega_n-\omega)}, \tag{3.43}$$

cf. Eqs. (3.35,3.39). The expression for the "three-phonon" vertex function $V^{(t)}_{mn}$ can be found in Ref. 17. It contains the three-body vertex of the ground state theory, which is necessary for the correct density dependence of the roton energy in the bulk liquid. [13] The *normal modes* of the system are given, as usual, by the singularities of the response function, in other words by the solutions of the generalized eigenvalue problem

$$\hbar\omega_n\varphi_n - \frac{1}{2}\sum_{stm}\frac{V_{st}^{(m)}V_{st}^{(n)}}{\hbar(\omega_s+\omega_s-\omega)}\varphi_m = \hbar\omega\varphi_n \,. \tag{3.44}$$

Eq. (3.44) has evidently the structure of a Brillouin-Wigner perturbation formula. We will use the CBF-BW approximation for the numerical parts of this paper.

3.4. *Impurity Dynamics*

It is tempting to identify the higher–lying eigenstates of the "Hartree-equation" (2.44) with the excited states of the impurity. This is indeed legitimate to the extent that one is permitted to make a *static approximation* for the impurity features, and we shall see below in more detail the approximations this entails. However, such a simplification misses two important features:

- If the momentum is a good quantum number, low–lying excited states are properly discussed in terms of an *effective mass.* In the present case, an effective mass can be associated with the motion of an impurity particle *parallel* to the surface. More specifically, this is the "hydrodynamic effective mass" which is caused by the coupling of the impurity motion to the excitations of the background liquid. If, however, the one–body operator determining the impurity states is independent of the momentum of the particle parallel to the surface, the effective mass is necessarily equal to the bare mass. This is the case in the simple, stationary impurity theory, where one has

$$t_o = \epsilon_o + \frac{\hbar^2 q_\parallel^2}{2m_I}\,, \tag{3.45}$$

  where ϵ_o are the eigenvalues of Eq. (2.49) for $q_\parallel = 0$.
- The effective Hartree potential $V_H(z)$ is real, *i.e.* all "excitations" defined by the local equation (2.44) have an infinite lifetime. This is also a consequence of the static approximation. A more refined theory should, more appropriately, describe resonances and allow for their decay by the coupling to the low–lying background excitations of the host film.

Hence, to describe the full physics of impurity motion, one must supplement the ground–state theory by a dynamic theory that takes these effects into account.

The natural generalization of the variational approach to excited states is to allow for a time dependence of the correlation functions $u_n(\mathbf{r}_0,\ldots,\mathbf{r}_n;t)$. We begin again by separating the *kinematic* from the *dynamic* correlations and write the time dependent variational wave function in the form

$$\phi(t) = \frac{1}{\sqrt{\langle\psi^I \mid \psi^I\rangle}} e^{-iE_{N+1}^I t/\hbar}\psi^I(\mathbf{r}_0,\mathbf{r}_1,\ldots\mathbf{r}_N;t)\,, \tag{3.46}$$

where $\psi^I(\mathbf{r}_0, \mathbf{r}_1, ...\mathbf{r}_N; t)$ contains the dynamic, time–dependent correlations between the impurity particle and the background. Consistent with the general strategy of variational methods, we include the time dependence in the one–particle *and* two–particle impurity–background correlations, *i.e.* we write

$$\psi^I(\mathbf{r}_0, \mathbf{r}_1, ...\mathbf{r}_N; t) = \exp\frac{1}{2}\Big[\delta u_1(\mathbf{r}_0; t) + \sum_{1\le i\le N} \delta u_2(\mathbf{r}_0, \mathbf{r}_i; t)\Big]\Psi^I_{N+1}(\mathbf{r}_0, \mathbf{r}_1, ..., \mathbf{r}_N)\ . \tag{3.47}$$

The time–independent part remains the same as defined in Eq. (2.40). The time–dependent correlations are determined by searching for a stationary state of the action integral

$$\begin{aligned} \mathcal{S} &= \int_{t_0}^{t} \mathcal{L}(t)dt \\ \mathcal{L}(t) &= \langle\phi(t)|H^I_{N+1} - i\hbar\frac{\partial}{\partial t}|\phi(t)\rangle\ , \end{aligned} \tag{3.48}$$

where H^I_{N+1} is the Hamiltonian (2.39) of the impurity-background system. Details of the derivation of a set of useful equations of motion for the impurity are given in the Appendix of Ref. 6. As one would expect, the final result is readily identified with a Greens' function expression, where the three–body vertex function describes an impurity atom scattering off a phonon, and is given in terms of ground–state quantities.

It is again convenient to work in the basis defined by the phonon states $\phi^{(m)}(\mathbf{r}_1)$ and the *impurity states* $\eta^{(o)}(\mathbf{r}_0)$ defined through the eigenvalue problem (2.49). We expand the time dependent part of the one–body density as

$$\frac{\delta\rho_1(\mathbf{r}_0; t)}{\sqrt{\rho^I_1(\mathbf{r}_0)}} = e^{-i\omega t}\sum_p r_p \eta^{(p)}(\mathbf{r}_0)\ , \tag{3.49}$$

and write the first equation of motion (3.16) in the following form:

$$\hbar\omega r_o = \sum_p \left[\delta_{op} t_o + \Sigma_{op}(\omega)\right] r_p\ , \tag{3.50}$$

where

$$\Sigma_{op}(\omega) = -\sum_{qm} \frac{W^{*(o)}_{mq} W^{(p)}_{mq}}{\hbar\omega_m + t_q - \hbar\omega} \tag{3.51}$$

is interpreted as the impurity's self-energy. The impurity-impurity–phonon vertex function occurring in the self-energy (3.51),

$$W^{(p)}_{mo} = \int d^3r_0 \frac{\eta^{*(o)}(\mathbf{r}_0)\eta^{(p)}(\mathbf{r}_0)}{\sqrt{\rho^I_1(\mathbf{r}_0)}}\left[\frac{1}{2}(t_o - t_p + \hbar\omega_m)\tilde{X}_m(\mathbf{r}_0) + \tilde{V}_m(\mathbf{r}_0)\right]\ , \tag{3.52}$$

has the form of an effective interaction within the theory of correlated basis functions [4,19,20] (CBF). The quantities $\tilde{X}_m(\mathbf{r}_0)$ and $\tilde{V}_m(\mathbf{r}_0)$ are defined as

$$\tilde{X}_m(\mathbf{r}_0) = \sum_p X_{pm}\eta^{(p)}(\mathbf{r}_0) = \int d^3r_1 \tilde{X}^I(\mathbf{r}_0, \mathbf{r}_1)\phi^{(m)}(\mathbf{r}_1)$$

$$= \int d^3r_1 \tilde{S}^I(\mathbf{r}_0, \mathbf{r}_1)\psi^{(m)}(\mathbf{r}_1) \; ,$$

$$\tilde{V}_m(\mathbf{r}_0) = \sum_p V_{pm}\eta^{(p)}(\mathbf{r}_0) = \int d^3r_1 \tilde{V}^I_{\mathrm{p-h}}(\mathbf{r}_0, \mathbf{r}_1)\phi^{(m)}(\mathbf{r}_1) \; . \tag{3.53}$$

The structure of Eqs. (3.50) and (3.51) is of the expected form of an energy–dependent Hartree–equation with a self-energy correction involving the energy loss or gain of the impurity particle by coupling to the excitations of the background system. More general forms can be derived in a similar manner, for example, within our theory by eliminating some of the approximations made, or in Greens' functions based theories by directly summing more Feynman diagrams. The most prominent approximation evident from Eq. (3.51) is the occurrence of bare single-particle energies and the Feynman excitation spectrum in the energy denominator. In more advanced theories these should be replaced with the proper self-energy of the impurity particle and the proper phonon propagator. These corrections can be important, [17,21] but we shall refrain here from such improvements (which are difficult to include fully self-consistently) to keep our formulation of the theory as transparent as possible.

Before we turn to the applications of our theory, we would like to stress that *dynamic correlations are properties of the excited state, and not of the ground state.* This requirement is satisfied by the present theory. To see this, inserting

$$\eta^{(0)}(z_0) = \sqrt{\rho^I_1(z_0)} \; . \tag{3.54}$$

into the matrix element (3.52) leads to

$$W^{(0)}_{mq} = -\frac{1}{2}\int d^3r_0 \eta^{*(q)}(\mathbf{r}_0)\left[t_q - H^I_1(\mathbf{r}_0)\right]\tilde{X}_m(\mathbf{r}_0) = 0 \tag{3.55}$$

and, hence, $\Sigma_{o,p=0}(\omega) = \Sigma_{o=0,p}(\omega) = 0$. Note that $o = 0$ and $p = 0$ here means that we take both the lowest state of Eq. (3.50) *and* the parallel momentum equal to zero. Hence the lowest–lying solution of the *dynamic* theory is identical to the ground–state solution of the static theory. This shows that the ground state is not renormalized by dynamic correlations.

Next, we *expand* the self-energy for small energies and momenta. If we are interested in the effect on the ground state only, it is sufficient to consider $\Sigma_{(0,\mathbf{q}_\parallel),(0,\mathbf{q}_\parallel)}(\omega = 0)$ for small momenta $\mathbf{q}_\parallel$ parallel to the surface. The ground–state solution of the impurity state reads

$$\eta^{(0,q\|)}(\mathbf{r}_0) = \sqrt{\rho^I_1(z_0)}e^{-iq_\parallel \cdot r_\parallel} \tag{3.56}$$

Using this form we calculate the matrix elements in the self-energy to leading order in $\mathbf{q}_\parallel$,

$$W^{(0,\mathbf{q}_\parallel)}_{mp} = -\frac{\hbar^2}{2m_I}\mathbf{q}_o \cdot \mathbf{q}_\parallel \delta(\mathbf{q}_o - \mathbf{q}_p + \mathbf{q}_\parallel) X_{om} \; , \tag{3.57}$$

where $X_{om} = \left(\eta^{(o)} \mid \tilde{X}^I(\mathbf{r}_0,\mathbf{r}_1) \mid \phi^{(m)}\right)$. Consequently, the self-energy for small momenta and energies is

$$\begin{aligned}\Sigma_{(0,\mathbf{q}_\parallel),(0,\mathbf{q}_\parallel)}(\omega=0) &= -\left(\frac{\hbar^2}{2m_I}\right)^2 \sum_m \int \frac{d^2q_o}{(2\pi)^2} \left(\mathbf{q}_o \cdot \mathbf{q}_\parallel\right)^2 \frac{\left|X_{(0,q_o)m}\right|^2}{\hbar\omega_m + t_o} + O(q_\parallel^4)\\ &= -\frac{q_\parallel^2}{2}\left(\frac{\hbar^2}{2m_I}\right)^2 \sum_m \int \frac{d^2q_o}{(2\pi)^2} q_o^2 \frac{\left|X_{(0,q_o)m}\right|^2}{\hbar\omega_m + t_o} + O(q_\parallel^4)\ . \quad (3.58)\end{aligned}$$

This defines an expansion for the lowest excitation energy of the impurity as a function of momentum,

$$\hbar\omega(q_\parallel) = \frac{\hbar^2 q_\parallel^2}{2m_I} + \Sigma_{(0,\mathbf{q}_\parallel),(0,\mathbf{q}_\parallel)}(\omega=0) = \frac{\hbar^2 q_\parallel^2}{2m_I^*} + O(q_\parallel^4)\ , \tag{3.59}$$

and the effective mass,

$$\frac{m_I}{m_I^*} = 1 - \frac{\hbar^2}{4m_I} \sum_m \int \frac{d^2q_o}{(2\pi)^2} q_o^2 \frac{\left|X_{(0,q_o)m}\right|^2}{\hbar\omega_m + t_o}\ . \tag{3.60}$$

Inserting the matrix elements of the PPA equation (2.50) into Eq. (3.58) shows that our result is indeed identical to the effective mass calculated in Ref. 22, however, the present theory is more general and is now also applicable to higher–lying excited states.

For higher lying excitations, the phase–space integrals of Eq. (3.51) have a pole and must be interpreted as principal value integrals, *i.e.* they are complex. Therefore, all excited states acquire a finite lifetime, as they should, by being able to decay into the ground state through the emission of ripplons or phonons. Hence, *dynamic correlations define both the effective mass and, via the same mechanism, the lifetime of the impurity states.*

3.5. *Thermodynamics*

A number of interesting issues arise at finite temperatures. A quantitatively successful microscopic theory for bulk liquid ^{4}He at finite, but low temperatures was formulated in Refs. 23-25. Their approach was to calculate the trial Helmholtz free energy corresponding to a proposed ansatz for the statistical density matrix ρ_t. The relevant minimum principle for the trial density matrix in this formulation is the Gibbs-Delbrück-Moliére principle for the Helmholtz free energy F, *i.e.*,

$$F \le F_t \equiv \mathcal{TV}(H\rho_t) + k_B T\ \mathcal{TV}(\rho_t \ln \rho_t)\ , \tag{3.61}$$

where the subscript t signifies *trial*, and the two terms are the trial internal energy and entropy term ($-TS = -k_B T\ \mathcal{TV}(\rho_t \ln \rho_t)$), respectively. Here T is the temperature and H is the Hamiltonian (2.2). Similar to the ground–state theory, the density matrix is written in a variational form:

$$\rho_t(\mathbf{r}_1,\ldots,\mathbf{r}_N;\mathbf{r}'_1,\ldots,\mathbf{r}'_N)$$
$$= \Psi_t(\mathbf{r}_1,\ldots,\mathbf{r}_N)Q_t(\mathbf{r}_1,\ldots,\mathbf{r}_N;\mathbf{r}'_1,\ldots,\mathbf{r}'_N)\Psi^*{}_t(\mathbf{r}'_1,\ldots,\mathbf{r}'_N)\,. \tag{3.62}$$

The *incoherence factor* can now be written as a Feenberg function, provided that the $w_{\alpha\beta}$ vanish in the limit that $|\mathbf{r}_i - \mathbf{r}_j| \to \infty$

$$Q_t(\mathbf{r}_1,\ldots,\mathbf{r}_N;\mathbf{r}'_1,\ldots,\mathbf{r}'_N) \tag{3.63}$$
$$= \exp\left\{\sum_{i,j} w_{1,1}(\mathbf{r}_i,\mathbf{r}'_j) + \sum_{i,j<k}\left[w_{1,2}(\mathbf{r}_i;\mathbf{r}'_j,\mathbf{r}'_k) + w_{2,1}(\mathbf{r}_j,\mathbf{r}_k;\mathbf{r}'_i)\right] + \ldots\right\}.$$

The result is much more plausible than the individual steps leading to it and we just state that the final entropy is

$$S_{\mathrm{var}} = \sum_m \left[(1+n_m)\ln(1+n_m) - n_m \ln n_m\right]\ , \tag{3.64}$$

where

$$n_m = \frac{1}{\exp(\beta\epsilon_m) - 1} \tag{3.65}$$

may be identified with the "quasiparticle occupation number" of a state characterized by the quantum number(s) m. This identification of n_m is at this point a matter of observation and formal analogy, but not of physical insight. We should also point out that the form (3.64) of the entropy is derived in the "uniform limit approximation", which basically keeps only excitations that can be characterized by a single momentum quantum number, and ignores any coupling of these excitations. In that sense, the approximations leading to Eqs. (3.64) and (3.65) stop short of where the power of the HNC–EL theory begins, namely to be able to do *more* than that.

The variational free energy functional is then

$$F_{\mathrm{var}} = T_{\mathrm{var}} + E_{\mathrm{ext}} + E_{\mathrm{c}} + \sum_m \epsilon^*_m n_m\left[n_m+1\right] - TS_{\mathrm{var}}$$
$$\equiv U_{\mathrm{var}} - TS_{\mathrm{var}}\ , \tag{3.66}$$
$$\epsilon^*_m = \epsilon_m \tanh\left[\frac{\beta}{2}\epsilon_m\right]\,. \tag{3.67}$$

The first three terms in F_{var} are the same as in the zero-temperature theory, they have been defined in Eqs. (2.17), (2.18) and (2.19). The fourth term in F_{var} describes the coupling between the internal energy and the quasiparticle excitations.

The remaining manipulation are straightforward: Convenient independent functions are the pair distribution function $g(\mathbf{r}_1,\mathbf{r}_2)$ and the occupation number n_m. Corrections to the two Euler-Lagrange equations arise from the new term in the free energy upon varying with respect to $\sqrt{\rho_1(\mathbf{r}_1)}$ and $g(\mathbf{r}_1,\mathbf{r}_2)$. The one–body equation acquires a term additive to the Hartree-potential,

$$\left[-\frac{\hbar^2}{2m}\nabla^2 + U_{\mathrm{ext}}(\mathbf{r}_1) + V_H(\mathbf{r}_1) + \delta V_H^{(T)}\right]\sqrt{\rho_1(\mathbf{r}_1)} = \mu\sqrt{\rho_1(\mathbf{r}_1)} \tag{3.68}$$

where $V_H(\mathbf{r}_1)$ is the effective potential defined in (2.31), and the temperature–induced correction is.

$$\delta V_H^{(T)}(\mathbf{r}_1) = \sum_m n_m \left[n_m + 1\right] \frac{\delta \epsilon_m^*}{\delta \rho_1(\mathbf{r}_1)} \quad . \tag{3.69}$$

In addition to this explicit temperature term, the two-body equation (2.25) contains an implicit temperature dependence through $\rho_1(\mathbf{r}_1)$ and $g(\mathbf{r}_1, \mathbf{r}_2)$. One has

$$\left[S^{-1} * H_1 * S^{-1} - H_1\right](\mathbf{r}_1, \mathbf{r}_2) = 2\left[\tilde{V}_{p-h}(\mathbf{r}_1, \mathbf{r}_2) + \tilde{v}^*(\mathbf{r}_1, \mathbf{r}_2)\right] \tag{3.70}$$

with

$$\begin{aligned} \tilde{v}^*(\mathbf{r}_1, \mathbf{r}_2) &= \frac{2}{\sqrt{\rho_1(\mathbf{r}_1)\rho_1(\mathbf{r}_2)}} \sum_m \frac{\delta \epsilon_m^*}{\delta g(\mathbf{r}_1, \mathbf{r}_2)} n_m \left[n_m + 1\right] \\ &= -2 \sum_m n_m \left[n_m + 1\right] \epsilon_m^* \psi^{(m)}(\mathbf{r}_1) \psi^{(m)}(\mathbf{r}_2) \ . \end{aligned} \tag{3.71}$$

Here $V_{\mathrm{p-h}}(\mathbf{r}_1, \mathbf{r}_2)$ is the particle-hole interaction (2.26) which depends explicitly on the temperature only through the correction term $v^*(\mathbf{r}_1, \mathbf{r}_2)$ to the induced potential $w_{\mathrm{I}}(\mathbf{r}_1, \mathbf{r}_2)$. The Euler-Lagrange equation (3.70) for the static structure function $S(\mathbf{r}_1, \mathbf{r}_2)$ is again conveniently solved by the normal mode method (2.33). For the eigenstates $\psi^{(m)}(\mathbf{r}_1)$ of (2.33) one adopts the normalization

$$\langle \psi^{(m)} \mid H_1 \mid \psi^{(n)} \rangle = \epsilon_n^* \delta_{n,m} \ , \tag{3.72}$$

and defines the adjoint states as

$$\phi^{(n)}(\mathbf{r}_1) = \frac{1}{\epsilon_n^*} H_1(\mathbf{r}_1) \psi^{(n)}(\mathbf{r}_1) \, . \tag{3.73}$$

Then, the temperature–dependent static structure function is again given by (2.38).

A fully self-consistent implementation of the finite temperature theory is needed close to a phase transition, because the temperature dependence of the excitation spectrum becomes important. The most obvious example for such a problem is the calculation of the critical point in bulk ^{4}He; we will encounter a related problem in the film geometry in connection with the temperature–dependence of "layering transitions". Computations are significantly simpler in the low–temperature expansion, which assumes that the excitation spectrum is temperature independent. Comparison between the different implementations will then tell us whether the low–temperature expansions were valid or not.

4. Structure of Inhomogeneous Quantum Liquids

The systems chosen here as examples of inhomogeneous liquids are families of quantum liquid films, and spherically symmetric clusters of ^{4}He atoms.

Quantum films are ^{4}He liquids adsorbed to more–or–less attractive plane substrates. A variation of these systems are helium liquids between two "walls" as they are, for example, in hectorite. [26,27,28] These systems provide an opportunity for studying the structure of "nearly" two–dimensional quantum liquids and, as the thickness of such an adsorbed film increases, the transition from an "essentially two–dimensional" to an "essentially three–dimensional" geometry. Structure, excitations, and growth of liquid ^{4}He films adsorbed to weakly attractive substrates has therefore been a matter of considerable experimental and theoretical research. It is fascinating to uncover the underlying mechanisms that drive an outwardly *simple-appearing* system to display a complicated growth scenario, [29,30] highly-complex neutron scattering spectra, [31,32] a multitude of thermodynamic phases, [33] transient superfluidity, [34] and many more.

For a many–body theorist quantum films are ideal, because each of the above characteristics is a direct consequence of many–body correlations inherent to a hard-core boson liquid in a confined geometry. For our present purposes, quantum liquid films provide an experimentally relevant possibility for applying microscopic many–body theory, for seeing how the theory works, and for demonstrating how the system under consideration "chooses its own shape".

4.1. *General Properties of Quantum Films*

The interaction between the substrate and the ^{4}He liquid is normally described by a substrate potential, whose shape is obtained by averaging Lennard–Jones interactions between the helium atoms and the atoms of the substrate over a half space. [35] These potentials are $3-9$ potentials characterized by their *range* C_3 and their *well depth* D, the values of C_3 and D are chosen to reproduce the experimental binding energies of single ^{4}He atoms to the substrate.

$$U_{\rm ext}(z) = \left[\frac{4C_3^3}{27D^2}\right]\frac{1}{z^9} - \frac{C_3}{z^3}\ . \tag{4.1}$$

Some typical substrate potentials for alkali metals are shown in Fig. 1. A slightly more complicated potential, which will be considered repeatedly throughout this article, describes a graphite substrate with two solid layers of ^{4}He on top; this potential is the steepest. The substrate potential consists of three terms,

$$U_{\rm ext}(z) = U_0(z+z_0) + \sum_{i=1}^{2} U_i(z+z_i) \tag{4.2}$$

with

$$U_0(z) = e\left[\frac{1}{15}\left(\frac{s}{z}\right)^9 - \frac{1}{2}\left(\frac{s}{z}\right)^3\right] \tag{4.3}$$

where the strength e has been chosen to match the asymptotic strength of the graphite-helium interaction, [36] and $e\,s^3/2 = 186$ meV. The two layers of solid helium are modeled by averaging Lennard-Jones potentials over a plane,

$$U_i(z) = 4\pi\epsilon n_i\sigma^2\left[\frac{1}{5}\left(\frac{\sigma}{z}\right)^{10} - \frac{1}{2}\left(\frac{\sigma}{z}\right)^4\right] \tag{4.4}$$

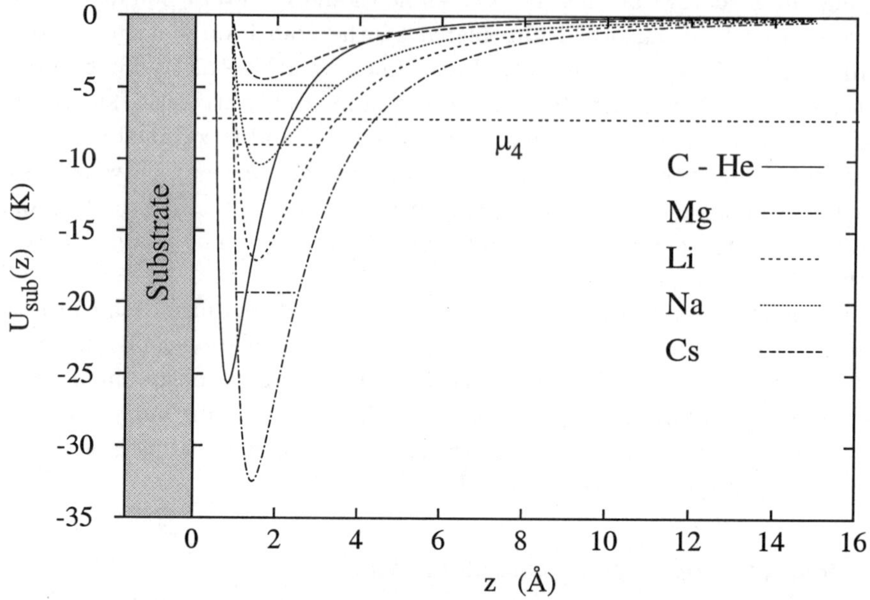

Fig. 1. A comparison of the substrate potentials defining the graphite/solid helium model (solid line), and the Mg (long/short dashed line), Li (short–dashed line), Na (dotted line), and Cs (long-dashed line) substrates. The horizontal lines of the same linetype as the potential give the binding energy of a single impurity atom in that field. The parametrization of the alkali metal potentials is taken from Ref. 37. Also shown is the chemical potential of bulk ^{4}He.

with $\epsilon = 10.22$ K and $\sigma = 2.556$ Å. The surface densities n_i $(i = 1, 2)$ were taken to be the experimental values [31] $n_1 = 0.115$ Å^{-2} and $n_2 = 0.093$ Å^{-2}. The offsets z_i are such that the distance between individual solid layers and between the first solid layer and the substrate is about 3.3 Å.

A recurrent issue in the following discussions will be the *stability* of a given configuration. For that purpose, we have to mention some results that will be derived later in Section 5.1, since stability is closely connected to the energetics of long–wavelength excitations.

Thermodynamic stability requires that the chemical potential $\mu(n)$ should be a *monotonically increasing* function of the surface coverage n,

$$n = \int_0^\infty dz \rho_1(z) \,, \tag{4.5}$$

where $\rho_1(\mathbf{r}) = \rho_1(z)$ is the density profile. It is illustrative to look at the *low-coverage* and the *high–coverage* limit of the chemical potential. A strongly–bound monolayer film can be reasonably well approximated by a two–dimensional model. In that case, the chemical potential of a ^{4}He atom is

$$\mu(n) \approx e_0 + \mu_{2D}(n) \,, \tag{4.6}$$

where e_0 is the binding energy of a single ^{4}He atom to the substrate, and $\mu_{2D}(n)$ the chemical potential of the two–dimensional liquid. As the film grows, $\mu(n)$ will tend towards its bulk value as [38]

$$\mu(n) = \mu_\infty - \frac{C_3}{d^3} = \mu_\infty - \frac{C_3 \rho_\infty^3}{n^3},$$

where $\mu_\infty \approx -7.15\,\mathrm{K}$ is the chemical potential of the bulk liquid, ρ_∞ is the bulk equilibrium density, and d is the film thickness. When $e_0 < \mu_\infty$, then $\mu(n)$ can be a monotonically increasing function of the surface coverage n, but we shall see that this is practically never the case. If, on the other hand, $e_0 > \mu_\infty$, then there must be one or more regimes where $d\mu/dn < 0$. Depending on the precise nature of these regimes, this means "pre-wetting" and "non-wetting".

4.2. *Atomic Monolayers*

Let us start our quantitative discussions with an attractive plane substrate that contains no helium or one to two inert layers of solid helium as described by the above model. With the *caveat* that not every substrate supports monolayer films, the physisorption scenario of ^{4}He atoms on a plane substrate is as follows:

As atoms are adsorbed to such a substrate, a quasi–two–dimensional atomic liquid is formed. We know from the theory of the bulk two–dimensional liquid [7,1] that such liquid is stable against infinitesimal perturbations only above the *spinodal density,* which is about 0.031 Å^{-2}. Below that density, the system consists of two–dimensional liquid ^{4}He "clusters" coexisting with the vacuum or, at finite temperatures, with a quasi–two–dimensional gas. These "clusters" may, in fact, be one or more layers thick; we will encounter a case of multilayer clusters further below.

Fig. 2 shows the coverage dependence of the energetics of an atomic monolayer of ^{4}He atoms on the graphite/solid helium substrate (4.2), the relatively strongly attractive Mg substrate, and the equation of state of the two–dimensional liquid. Expectedly, one does not find translationally invariant solutions below the surface coverage $n_{\rm min} = 0.032\,\text{Å}^{-2}$, in good agreement with the estimates for the spinodal point in 2D. [39,40] Above that coverage, the ^{4}He film has the full planar symmetry forced on it by the substrate. The physical monolayers are slightly more strongly bound than the purely 2D system; this is a consequence of the additional degree of freedom perpendicular to the symmetry plane.

For low coverages, a two-dimensional system appears to give a reasonable approximation for the energetics, but the quasi–two–dimensional atomic monolayer becomes unstable at a surface coverage of about 0.068 Å^{-2}. The physical situation can be understood as follows: At the saturation density, the average particle distance is considerably smaller in the atomic monolayer than in three–dimensional ^{4}He. Therefore, by adding more atoms to the monolayer, the two–dimensional system must become highly compressed in order to approach the bulk equilibrium density. A surface coverage will eventually be reached where it is energetically favorable to elevate particles to a next layer instead of further compressing the previous one.

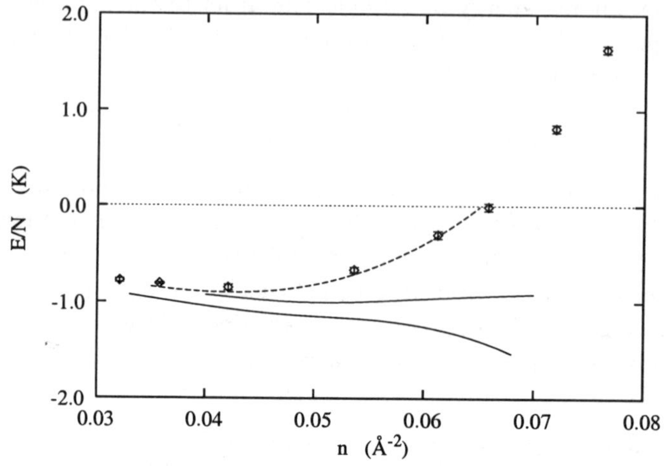

Fig. 2. The ground state energy of an atomic monolayer in the graphite/solid helium potential (4.2) (lower solid line) and on a Mg substrate (upper solid line). Also shown is the two-dimensional equation of state from the HNC-EL calculation (long dashed line) and from the Monte Carlo calculations of Ref. 41, (diamonds with error bars).

Precursors of this phenomenon are seen in Fig. 2 in the *drop* of the correlation energy with increasing coverage. At the same coverage, the incompressibility

$$mc_3^2 = n\frac{d\mu}{dn} \tag{4.7}$$

starts to drop, instead of further increasing as in a two–dimensional system. When the monolayer becomes unstable against the elevation of particles into the second layer, "patches" of ^{4}He can form *on top* of the previous layer, and the translational symmetry parallel to the helium surface is again spontaneously broken. The maximum coverage of the submonolayer two-phase coexistence region predicted using the static substrate is in very good agreement with the experimental value. [33]

4.3. *Multilayer Films*

One can learn most about many-body physics of the inhomogeneous system in the intermediate coverage regime where interaction effects between the ^{4}He adatoms become visible, but the energetics is not yet dominated by bulk features. The qualitative physical scenario described above predicts a sequence of transitions between a translationally-invariant, quasi–two–dimensional layered liquid and a system where "islands" of two-dimensional atomic clusters float upon it.

We shall first discuss the graphite/solid helium model. Fig. 3 shows the density profiles of the stable films up to approximately six layers. One finds three stable coverage regimes, corresponding to mono–, double–, and multi–layer films. At coverages larger than the stability limit of the atomic monolayer, the film consists of one liquid monolayer, and coexisting clusters and vacuum above it. The mechanism that

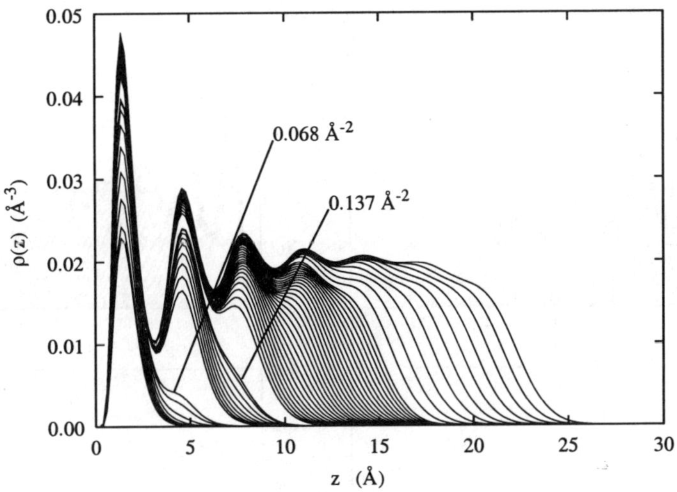

Fig. 3. The density profiles of stable configurations of helium films on a graphite/solid helium substrate. The profiles correspond to surface coverages of 0.033, 0.35, 0.40, ... , 0.065 and 0.068 $Å^{-2}$ for the monolayer, 0.100, 0.105, ... , 0.135 and 0.136 $Å^{-2}$ for the double–layer, and 0.165 0.170, ... , 0.450 $Å^{-2}$ for the multilayer films.

makes the ^{4}He films highly layered emerges from the realization that the instability region of the *second* layer once again has a width in coverages of 0.032 $Å^{-2}$. Beyond a coverage of 0.10 $Å^{-2}$, the second layer uniformly covers the surface. The same pattern is repeated for the third layer: An instability region with an approximate width of 0.035 $Å^{-2}$ exists between coverages of 0.13 $Å^{-2}$ and 0.165 $Å^{-2}$. At higher coverages a possible instability is indicated by a negative eigenvalue of Eq. (2.32).

A second family of substrates are alkali metals . Typically, alkali metals such as magnesium, lithium, and sodium are described by substrate potentials of the 3–9 form (4.1). These substrates differ, as seen in Fig. 1, from the graphite/solid helium substrate potential (4.2) because they tend to have longer ranges and, with the exception of Mg, substantially shallower well depths. Na and Cs have the weakest substrate potential studied here. These substrates have recently received much attention after the discovery of *non-wetting* of ^{4}He on cesium. [42–44] As pointed out above, the wetting behavior *per se* does not say much about the quality of the many–body treatment; more interesting and critical observations are made in the regime of mono–, double–, and triple–layer films.

Alkali metal substrates show a behavior that differs qualitatively from the graphite/solid helium potential. Representative profiles are shown in Fig. 4. For Na substrates, we did *not* find a solution of the HNC–EL equations that corresponds to a uniform monolayer. This observation implies that, in the early stages of the growth of the helium film on a Na substrate, the helium atoms form bulk-like clusters, that are several atoms thick in the dimension perpendicular to the substrate. A certain coverage (near that of a double layer) must be reached before

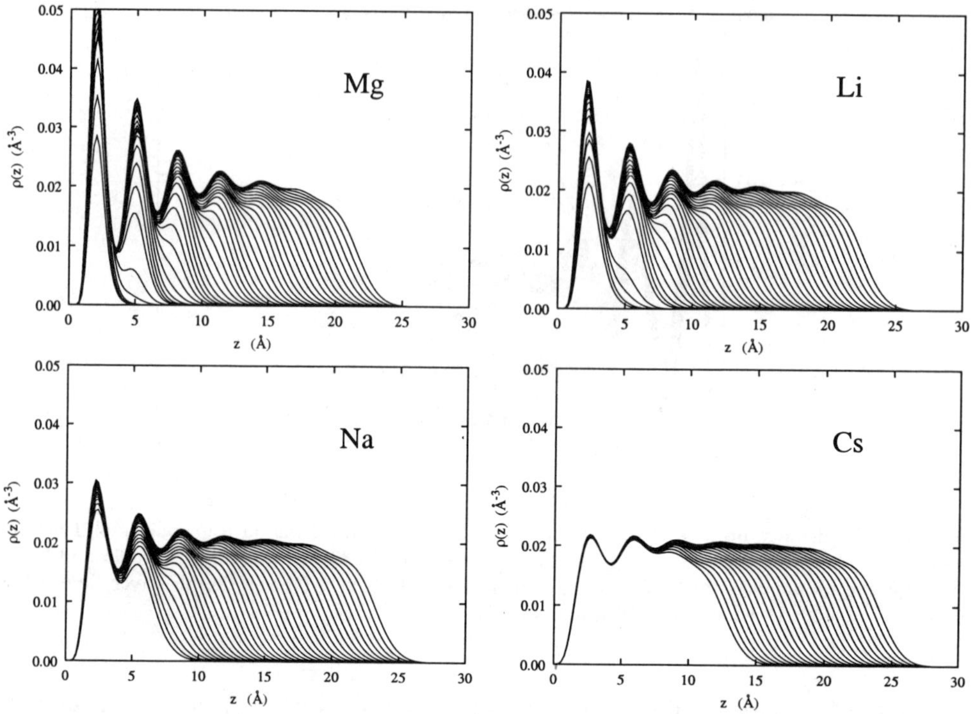

Fig. 4. The density profiles of ^{4}He films adsorbed to the alkali metal substrates as obtained from the HNC–EL theory. The curves correspond to coverages of $n = 0.04$ Å^{-2} … 0.45 Å^{-2} for Mg and Li, to $n = 0.10$ Å^{-2} … 0.45 Å^{-2} for Na, and to $n = 0.10$ Å^{-2} … … 0.45 Å^{-2} for Cs.

the helium clusters can fully connect to uniformly cover the surface. An extreme case is the cesium substrate, which has a very shallow bound state and, hence, will, similar to a hard wall, not support a film of any thickness. Of course, the interface between a Cs substrate and liquid ^{4}He could, in a realistic situation, be stabilized by gravity. The only way one can simulate a helium film on Cs is by adding a small attractive potential that acts at large distances from the film, or by artificially making the particle–hole interaction (2.33) repulsive at long wavelenghths. We have preferred the second route because in our experience such a modification does not change the nature of the film in any noticeable way. Using a very conservative long–wavelength modification of the effective interaction for $q_{\|} < 0.4$ Å^{-1}, we have been able to obtain stable films with coverages of $n \geq 0.24$ Å^{-2}. To test the validity of this procedure, we have calculated the dispersion relation of the ripplon at long wavelengths and high coverages and found that it is indeed indistinguishable from the familiar hydrodynamic ripplon dispersion relation (5.12) when the experimental surface energy is used.

Finally, we have also looked at helium on SiO_2 substrates. These describe aerogels and the hectorite gaps to be described below. The first layer of ^{4}He atoms on SiO_2 is solid, we use therefore the same method that we employed for graphite

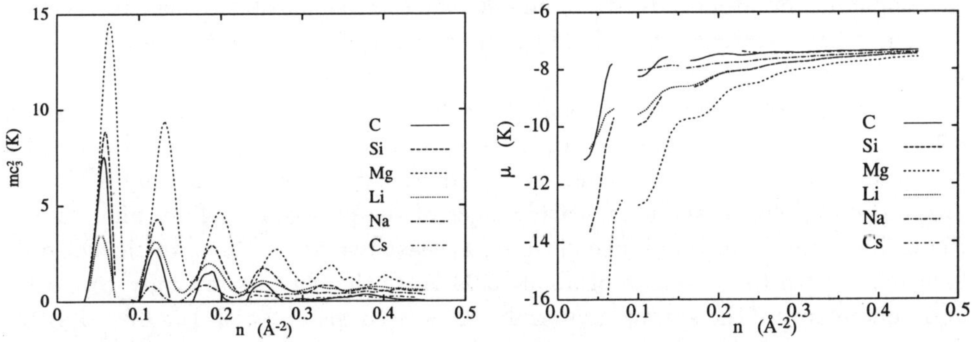

Fig. 5. The figure shows the coverage dependence of the incompressibility mc_3^2 (left figure) and the chemical potential (right figure) of ^{4}He films as a function of coverage and potential strength for graphite and Si substrates as well as a sequence of Alkali metals as indicated in the figures.

and modeled this solid layer with a potential of the form (4.3). To determine the parameters of the 3-9 potential (4.1), we have used the silicon-helium van der Waals interaction [35] with $C_3 = 2000$ KÅ^{-3}, and adjusted the short–ranged z^{-9} repulsion to reproduce the binding energy of single ^{4}He atoms [26] on SiO_2.

The alternation between translationally invariant and "patched" film configurations is reflected in the chemical potential, and in the speed of surface sound of the film. Plots of mc_3^2 and the chemical potential are shown in Fig. 5.

The growth scenario described above is most easily discussed by considering the coverage dependence of the chemical potential and the long-wavelength limit of the (third) sound velocity, which is conveniently expressed as mc_3^2. As a unique feature of the HNC–EL theory, we have found in Ref. 7 that the chemical potential of a ^{4}He atom on the graphite/solid helium substrate is *not* a monotonic function, and that the sound velocity of the homogeneous phase can become imaginary. In that situation, the HNC–EL equations have no solutions unless one projects out the very long-range excitations. A further prediction of our analysis is that, for low coverages, the chemical potential should be essentially equal to the binding energy e_0 of a single particle to the substrate, *plus* the chemical potential of the *two-dimensional* liquid,

The layering transitions on alkali metal substrate potentials are less pronounced than on the graphite/solid helium potential. We have calculated the structure of ^{4}He films on alkali metal substrates starting from the lowest stable coverage up to a coverage of $n = 0.45$ Å^{-2}, which corresponds, for a magnesium substrate, to slightly more than six layers. A comparison with the results of the non–local density–functional theory of Ref. 45 can be found in Ref. 46. In all cases, the HNC–EL calculations predict the expected oscillations and a minimum stable coverage.

Both the magnesium and the lithium substrates support a monolayer of sufficient coverage, but in both systems this monolayer is already rather "soft" in the sense that the chemical potential deviates, as a function of coverage, significantly from the two-dimensional estimate (4.6). This finding is consistent with our estimate that the single helium atom must be bound stronger to the substrate than to the bulk helium to have wetting. These binding energies are 19.35 K for Mg and 9.01 K for Li.

In Na, we *don't* find any monolayer solutions; the lowest stable coverage is 0.1 $Å^{-2}$, or about two layers. Only by projecting out the very long-wavelengths excitations were we able to extend the calculations to a coverage of 0.09 $Å^{-2}$. This observation implies that, in the early stages of the growth of helium films on Na, the helium atoms form bulk-like clusters, at least two atoms thick in the third dimension. Above the coverage of about 0.09 $Å^{-2}$, the clusters are fully connected and helium covers the surface uniformly. This is consistent with the fact that the binding energy of a single atom to the substrate is only -4.83 K, significantly less than the chemical potential of the bulk liquid. Indeed, Na is a borderline case where $\mu(n)$ is almost flat; in the coverage regime considered here it increases only by about 0.1 K. While the oscillations of the chemical potential and the speed of sound are still systematic functions of coverage, they are so small that we can not make a reliable estimate for their amplitude.

Globally, our theoretical results are in agreement with the observation by Nacher and Dupont-Roc, [43] who found that ^{4}He wets a Na surface. On the microscopic scale, one would, however, conclude that the minimum stable coverage is a double layer and not, as on stronger substrates, a monolayer. Such an effect might be detectable in torsion-oscillator experiments of the kind reported by Crowell and Reppy [34] since the minimum coverage for superfluidity appears to be more than one monolayer.

4.4. *Liquid Between Two Plane Substrates: Hectorite*

Hectorite is a material that consists of regular quasi–two–dimensional silicate layers of about 9.6 Å thickness and an open spacing of 17-20 Å between them. [26,47] In this environment one can study, among other properties, the interaction of *two* liquid surfaces. The layers will here be modelled by two attractive walls and the interaction of ^{4}He particles with the walls is described by the usual 3-9 potential. This assumes that the hectorite walls are flat and, hence, $U_{\rm ext}(\mathbf{r})$ is a function of one coordinate only, so that the lateral structure by the hectorite will be smoothed out by the first solid layer of helium atoms. We have used the long–ranged z^{-3} tail of the silicon-helium interaction [35] and have adjusted the short–ranged z^{-9} repulsion to reproduce the binding energy of single ^{4}He atoms. [26] The substrate attraction is strong enough to cause the first atomic layer of ^{4}He to solidify. We have taken this into account by adding, as described earlier, an inert solid monolayer of ^{4}He with a thickness of 3.3 Å to both hectorite walls. Since the distance between the hectorite walls is about 20 Å, we are considering a liquid phase of 14 Å width.[27,28]

Let us first discuss the scenario of filling the hectorite gaps with liquid helium, depicted in Fig. 6. Much of the above discussion of mono– and multi– layer films applies here as long as the the helium on the two walls does not interact. When the volume between the gaps is filled with ^{4}He, first one of the two walls will be filled

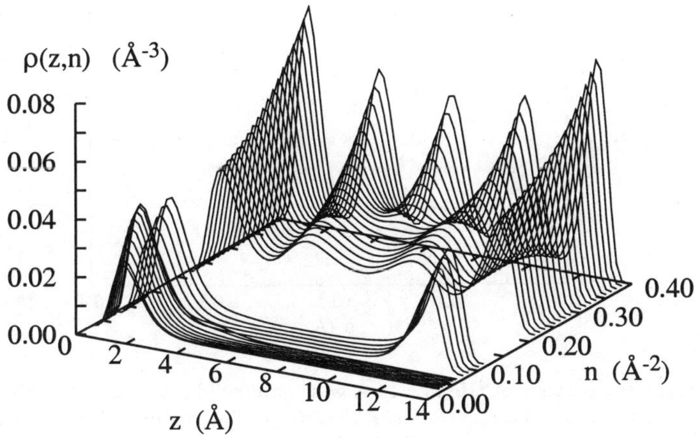

Fig. 6. Density profiles of ^{4}He in a 14 Å wide gap are shown as function of areal density n. The free spaces correspond to densities where no translationally invariant configurations exist.

with an atomic monolayer, or both walls with patches of ^{4}He. A complete coverage of *both* walls can be expected for areal densities above at least *twice* the spinodal density of an atomic monolayer, which is in this case 0.033 Å^{-2}. Incidentally, this value is close to the areal density where a single atomic monolayer becomes unstable against the promotion of particles to a second layer.

As the areal density is further increased, one finds that the interaction between the two liquid films attached to the two walls is strong enough to make a double–layer undergo capillary condensation. Consequently, there are no stable configurations between areal densities of $n = 0.122$ Å^{-2} and $n = 0.22$ Å^{-2}. Above that areal density, the gap is completely filled with a four–layer configuration.

We found that the lowest areal density for a stable double–layer configurations is $n = 0.068$ Å^{-2}, which is, expectedly, almost exactly twice the value of a monolayer. A very interesting *confined layering transition* is found between coverages of 0.28 Å^{-2} and 0.32 Å^{-2}: In this density regime, the film changes abruptly from four to five layers.

The structure of the system is determined by the energetics which is shown in Fig. 7. All three configurations discussed above —the monolayer, the symmetric double layer, and the filled gap— have an individual energy minimum; the slight mismatch between the chemical potential and the energy at its minimum is due to numerics. Most notable is again the gap between $n = 0.12$ Å^{-2} and $n = 0.22$ Å^{-2} where no stable systems exist that are translationally invariant in the direction parallel to the hectorite walls. The physical configuration of the film in this coverage region is to be found by executing the usual Maxwell construction.

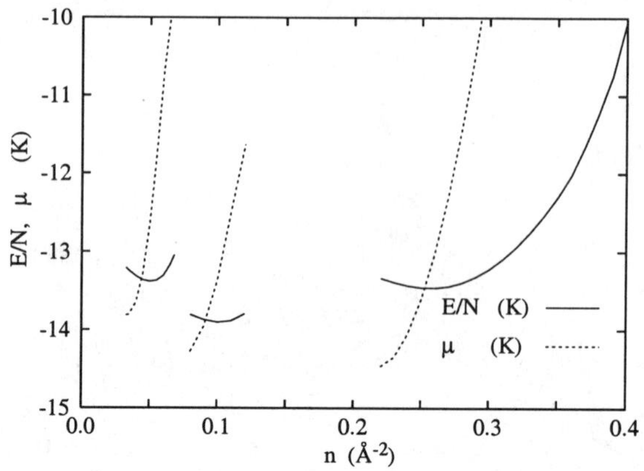

Fig. 7. The energy per particle of ^{4}He in a 14 Å wide gap (solid lines) and the chemical potential (dashed lines) are shown as a function of areal density n. The distinct regions correspond, from left to right, to a monolayer, a symmetric double-layer, and filling of the gap.

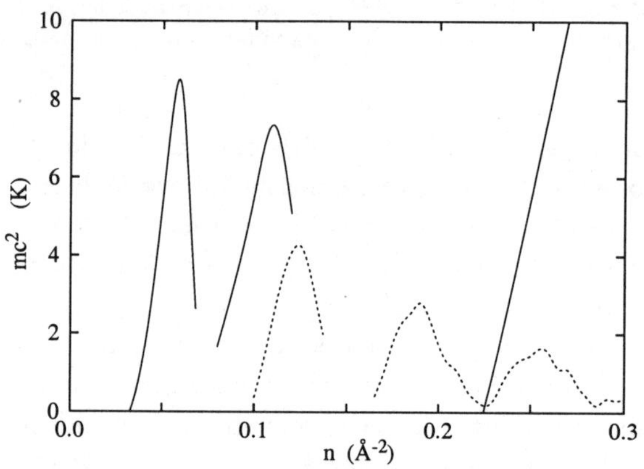

Fig. 8. The figure shows the speed of longitudinal sound as a function of areal density n (solid lines). Also shown is the speed of third sound on a one–sided substrate with the same potential (dashed line); the left–most curve is common to both cases

Fig. 8 shows the longitudinal incompressibility mc_s^2 as a function of coverage; we have calculated this quantity from the long–wavelength limit of the excitation spectrum. The figure also shows mc_s^2 for the case that ^{4}He is adsorbed on a one–sided substrate with the same potential. The results are identical for a single–monolayer filling, but start to deviate for the case of two layers: The fact that the double–layer configuration in the hectorite is *not* identical to that of two monolayers is the first sign of interaction between the two liquid monolayers. The speed of sound for configurations in the high–density regime, where the lowest excitations are longitudinal phonons is expectedly very different from that in a multi–layer adsorbed film where the lowest excitations are ripplons.

We conclude the discussion of hectorite by pointing out few exciting possibilities offered by this system. The environment provides the interesting possibility of examining ^{4}He at three-dimensional densities that would, in the bulk liquid, correspond to either significant negative pressures, or even to densities below the three–dimensional spinodal point: At the saturation areal density $n = 0.26$ Å^{-2}, the central density is $\rho = 0.015$ Å^{-3}, well below the spinodal density of bulk ^{4}He. At an areal density of $n = 0.29$ Å^{-2}, where the transition between the four-layer to the five-layer configuration occurs, the three-dimensional density in the center has an almost constant value of $\rho = 0.018$ Å^{-3}, which is also well inside the negative pressure regime of the 3D liquid.

The system itself may also be suitable to generate quasi–two–dimensional ^{4}He close to the spinodal density ρ_s where the speed of sound vanishes, and potentially verify its critical behavior which is *not* characterized by a critical exponent, but rather by the form [39]

$$\rho - \rho_s \propto c^4 \ln c \quad \text{as} \quad c \to 0+ \, . \tag{4.8}$$

A further interesting possibility is that one could examine quasi–two–dimensional ^{3}He-^{4}He mixtures. The absence of a free surface suppresses the population of Andreév states and, as a consequence, the mixture would be truly two–dimensional. We have recently shown [48] that the phase diagram of such a mixture is very different from that of its three–dimensional counterpart: At *low* concentrations, the ^{3}He atoms dimerize; the effect has been predicted some time ago by Bashkin. [49] Two-dimensional atomic mixtures can be found only at *finite concentrations* of ^{3}He in ^{4}He.

5. Film–Excitations

Quantum films have two structural characteristics not met in bulk: the free surface and the layer structure. In a complex situation like this the discussion of dispersion relations alone is insufficient for a complete characterization of the excitation types. We also need to determine *where* the excitation propagates. For that purpose we calculate *transition densities*, which is the time–dependent part of the fluctuating density. In the film geometry it has the form $\delta\rho_1(\mathbf{r};t) = \delta\rho_1(q_\parallel, z)e^{i(\mathbf{q}_\parallel \cdot \mathbf{r}_\parallel - \omega t)}$. Only the spatial shape $\delta\rho_1(q_\parallel, z)$ is of physical interest. Within the time–dependent Hartree-theory transition densities are given by Eq. (3.6); in the more general case they are the null eigenvectors of the density–density response function, represented by linear combinations of these functions. Occasionally one also needs to know *how* particles are moving. For that purpose, one calculates the *one–body probability current*, [12] which in the Feynman approximation reads

$$-i\mathbf{j}_1(\mathbf{r};t) = \frac{\hbar\rho_1(\mathbf{r})}{2m}\nabla\delta u_1(\mathbf{r};t) \quad . \tag{5.1}$$

The CBF form of the current is more involved, but it is seldom needed, because the most interesting, low–lying excitations have a rather well–defined Feynman character.

We start the discussion of quantum film excitations from surface excitations, whose phenomenology is introduced in the hydrodynamic limit.

5.1. *Surface Excitations*

The restoring force of a surface excitation can be either the van der Waals attraction of the substrate or the surface tension of the liquid itself. In the first case the surface sound mode of a superfluid is known as "third sound". The latter are capillary waves, whose quanta are called ripplons.

A simple view of surface excitations is provided by hydrodynamics. Waves on shallow water propagate with velocity $\sqrt{gd}$, where g is the gravitational acceleration and d is the water depth. For superfluids, the speed c_3 of third sound is

$$c_3 = \sqrt{\frac{\langle \rho_s \rangle}{\rho} a_{\rm vdW} d} \ , \tag{5.2}$$

where $a_{\rm vdW} = 3C_3/(md^4)$ is the acceleration due to the van der Waals force, d is the film thickness and $\langle \rho_s \rangle$ is the average superfluid density of the film within the two–fluid model. The density factor appears because the third sound involves only the superfluid component: The normal–fluid component is clamped to the substrate if the film thickness is less than the viscous penetration depth.

The speed of third sound of a superfluid placed in an external potential is given by [50,51]

$$mc_3^2 = n\frac{d\mu}{dn} \ . \tag{5.3}$$

We can calculate the microscopic expression for $d\mu/dn$ in the thick film limit from Eq. (4.1), for films that have not yet reached the asymptotic regime we can get $d\mu/dn$ directly from the Hartree–equation (2.29), realizing that the correlation energy E_c, and hence the Hartree potential $V_H(\mathbf{r}_1)$, depend on the particle number through the density $\rho_1(\mathbf{r})$ only. It is convenient to Fourier transform all quantities in the coordinate parallel to the surface and write

$$\begin{aligned} H_1(q_\parallel) &= \frac{\hbar^2}{2m}\left[-\frac{1}{\sqrt{\rho_1(z)}}\frac{d}{dz}\rho_1(z)\frac{d}{dz}\frac{1}{\sqrt{\rho_1(z)}} + q_\parallel^2\right] \\ &= -\frac{\hbar^2}{2m}\frac{d^2}{dz^2} + \frac{\hbar^2 q_\parallel^2}{2m} + U_{\rm ext}(z) + V_H(z) - \mu \ . \end{aligned} \tag{5.4}$$

where the last line follows from (2.23) and the Hartree–equation (2.29). Likewise, the particle–hole interaction is considered a function of z, z', and $q_\parallel$. Differentiating the Hartree–equation with respect to n gives

$$\begin{aligned} &\left[-\frac{\hbar^2}{2m}\frac{d^2}{dz^2} + U_{\rm ext}(z) + V_H(z) - \mu\right]\frac{d\sqrt{\rho_1(z)}}{dn} \\ &+ 2\sqrt{\rho_1(z)}\int dz' \frac{\delta V_H(z)}{\delta\rho_1(z')}\sqrt{\rho_1(z')}\,\frac{d\sqrt{\rho_1(z')}}{dn} = \frac{d\mu}{dn}\sqrt{\rho_1(z)} \ . \end{aligned} \tag{5.5}$$

Identifying

$$\tilde{V}_{\rm p-h}(z,z',0+) = \sqrt{\rho_1(z)}\,\frac{\delta V_H(z)}{\delta\rho_1(z')}\sqrt{\rho_1(z')} \tag{5.6}$$

and recalling (5.4), we can solve for $d\mu/dn$, which leads us to the microscopic result

$$mc_3^2 = \frac{1}{2}\frac{n}{\left(\sqrt{\rho_1}\middle|\left[H_1(0+)+2\tilde{V}_{\rm p-h}(0+)\right]^{-1}\middle|\sqrt{\rho_1}\right)}\ . \tag{5.7}$$

The positivity of this expression is also a condition for the existence of solutions of Eq. (2.25). As a *caveat* we stress that the long–wavelength limit of the collective excitations (5.7) and the hydrodynamic definition (5.3) are equal only in an exact theory.

Finally, we show that the third sound velocity (5.7) is indeed the slope of the small–$q_\parallel$ excitation mode. It is obvious that the lowest eigenvalue of (2.33) goes to zero in the limit $q_\parallel \to 0^+$. The corresponding eigenfunction is, to leading order in $q_\parallel$, proportional to $\sqrt{\rho_1(z)}$. Because the particle–hole potential has a short range we can write

$$\psi_0(z,q_\parallel) = K\left[\sqrt{\rho_1(z)} + t(q_\parallel)\delta\Psi(z)\right], \tag{5.8}$$

and require that $\left(\sqrt{\rho_1(z)}\mid\delta\Psi(z)\right) = 0$. We can now expand the lowest eigenfrequency $\epsilon_0(q_\parallel)$ for small $q_\parallel$ (note that $\epsilon_0(q_\parallel = 0) = 0$) and obtain

$$\left[H_1(0) + 2\tilde{V}(z,z',0+)\right]t(q_\parallel)\left[\sqrt{\rho_1(z)} + H_1(0)\delta\Psi(z)\right] = \epsilon_0^2(q_\parallel)\sqrt{\rho_1(z)}\ . \tag{5.9}$$

Inverting the operator on the left and integrating with $\sqrt{\rho_1(z)}$ gives

$$nt(q_\parallel) = \epsilon_0^2(q_\parallel)\left(\sqrt{\rho_1}\middle|\left[H_1(0+)+2\tilde{V}_{\rm p-h}(0+)\right]^{-1}\middle|\sqrt{\rho_1}\right)\ , \tag{5.10}$$

where n is the area density (2.28). We obtain therefore the dispersion relation

$$\epsilon_0(q_\parallel) = \hbar c_3 q_\parallel \tag{5.11}$$

where the third sound velocity is again given by Eq. (5.7).

Ripplons are typical to free surfaces far away from any substrate. Their long–wavelength dispersion is not linear, but hydrodynamics gives instead

$$\omega^2(q_\parallel) = \frac{\sigma}{m\rho_\infty}q_\parallel^3\ , \tag{5.12}$$

where σ is the surface energy, ρ_∞ is the asymptotic density, and $q_\parallel$ is the momentum parallel to the surface. Ripplons occur on the free surfaces of thick helium films and clusters. A weak substrate such as Cs will only slightly modify the helium surface next to it, and something known as "interfacial ripplon" survives close to the substrate. [52] A normal sound mode is longitudinal. In a ripplon excitation, on the other hand, the excitation function has the form

$$\psi_0(z,\mathbf{q}_\parallel) = C\sqrt{\rho_1(z)}e^{(q_\parallel z + i\mathbf{q}_\parallel\cdot\mathbf{r}_\parallel)} \tag{5.13}$$

and we obtain

$$\mathbf{j}_1(\mathbf{r}_1) \sim \rho_1(z)\left(\cos(\mathbf{q}_\parallel \cdot \mathbf{r}_\parallel)\mathbf{e}_z \;-\; \sin(\mathbf{q}_\parallel \cdot \mathbf{r}_\parallel)\hat{\mathbf{r}}_\parallel\right), \tag{5.14}$$

i.e., the current is *circular*: Ripplons are are like waves on deep water.

5.2. *Monolayer and Multilayer Excitations*

The excitations of the systems considered here are most naturally discussed in terms of the dynamic structure function $S(\mathbf{k},\omega)$. Neutron scattering probe the dynamic response of ^{4}He at medium and short wavelengths. In these experiments, different types of modes are excited, and their strength and energetics gives information about the structure of the system. The physical scenario can be quite complex: One has volume excitations, surface excitations, layer excitations, and combinations as well as transitions between them. In order to have a high event count, experiments are usually carried out at grazing angles, in which case one measures the parallel structure function

$$S(q_\parallel;\omega) = \int dz dz' d^2 r_\parallel e^{i\mathbf{q}_\parallel \cdot \mathbf{r}_\parallel} \sqrt{\rho_1(\mathbf{r}_1)} S(\mathbf{r}_1,\mathbf{r}_2;\omega)\sqrt{\rho_1(\mathbf{r}_2)}\ , \tag{5.15}$$

we therefore discuss the case of parallel momentum transfer first. Experimental difficulties, like instrumental broadening, substrate corrugation, and background scattering, tend to obscure the physics the more, the more complicated the spectrum is. Theoretically, one can include some experimental broadening, but many uncertainties prevail.

The situation is simplest for monolayers. In our discussion of the energetics, we were able to distinguish two characteristic types of such films: The films with *low* surface coverage can be approximated reasonably well by a two–dimensional model. When the coverage is increased, it becomes eventually energetically favorable for the atoms to populate the third dimension rather than further compressing the atomic monolayer. The population of states in the third dimension is accompanied by an abrupt drop of c_3, the third–sound velocity. The scenario is quite visible in Fig. 5. In our case the crossover coverage is approximately $n_{\rm co} = 0.05$ Å^{-2}.

The dispersion relations can be read off the dynamic structure functions shown in Figs. 9. For low–coverage films, $n < 0.05$ Å^{-2}, the dispersion curve of the lowest mode is in close agreement with the one obtained for a two–dimensional model system. Transition densities corresponding to the lowest–lying excitation are shown in Fig. 10. Excitations in the lower stable coverage regime are essentially two–dimensional longitudinal phonons propagating in the film; their transition densities are proportional to the ground–state density $\rho_1(\mathbf{r})$. For wavevectors shorter than 0.2 Å^{-1} these excitations are well described by the linear dispersion relation $\hbar\omega = \hbar c q_\parallel$.

The situation changes dramatically at the coverage $n = 0.055$ Å^{-2}. As the coverage increases, the $S(k,\omega)$ clearly shows clear evidence of *two* modes and a level crossing in the momentum regime $q_\parallel$= 1.2-1.3 Å^{-1} (*cf.* the case $n = 0.065$

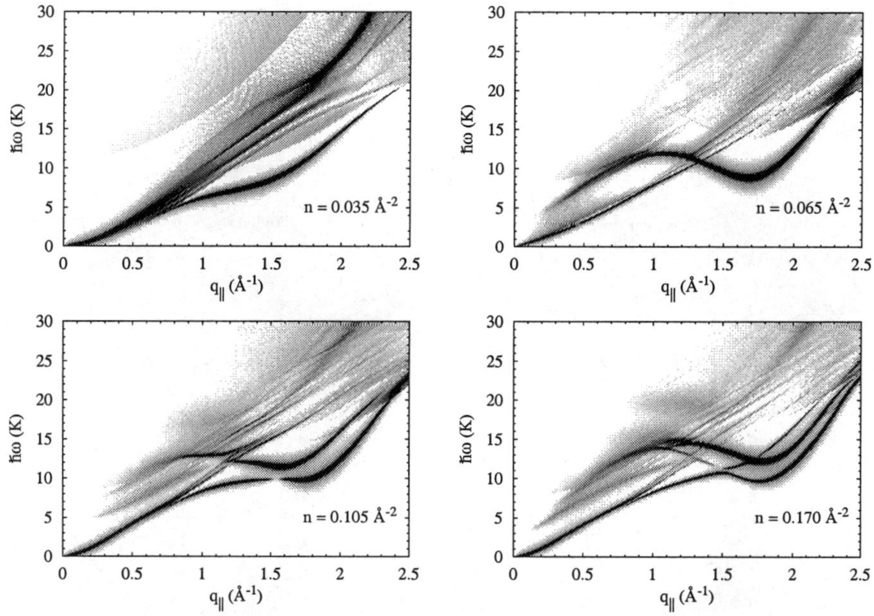

Fig. 9. The figures show the dynamic structure function $S(\mathbf{k},\omega)$, for momentum transfer in the parallel direction, for films with coverages of $n = 0.035$ Å^{-2} (upper left), $n = 0.065$ Å^{-2} (upper right), a double–layer film with $n = 0.105$ Å^{-2} (lower left) and a triple–layer film with $n = 0.170$ Å^{-2} (lower right). Dark areas correspond to high values of $S(q_{\|},\omega)$.

Å^{-2} in Fig. 9). We can identify this feature as a level-crossing between phonons and ripplons by looking at the transition density as a function of momentum transfer. At layer completion, the *long–wavelength* excitation evidently propagates in the surface, whereas the lowest mode for $q_{\|} > 1.2$-1.3 Å^{-1} is again proportional to the density profile.

Let us now turn to the excitations in multilayer films. For the purpose of discussion, we have chosen two representative cases: a double-layer film at 0.105 Å^{-2} and a triple–layer film at 0.170 Å^{-2}. Fig. 9 show the dynamic stricture function of these examples.

While the low–coverage double–layer case $n = 0.105$ Å^{-2} is still ambiguous, the triple–layer film with $n = 0.170$ Å^{-2} shows *two* phonons–roton modes and a ripplon. Again, an interpretation of the effect is obtained by looking at the transition density as a function of parallel momentum. In both cases, the lowest mode turns from a surface excitation at $q_{\|} \leq 1.3$ Å^{-1} to an excitation that *propagates in the first liquid layer close to the substrate.* This mode is evidently a two–dimensional "layer" phonon. This leaves for the third mode the interpretation of a bulk phonon.

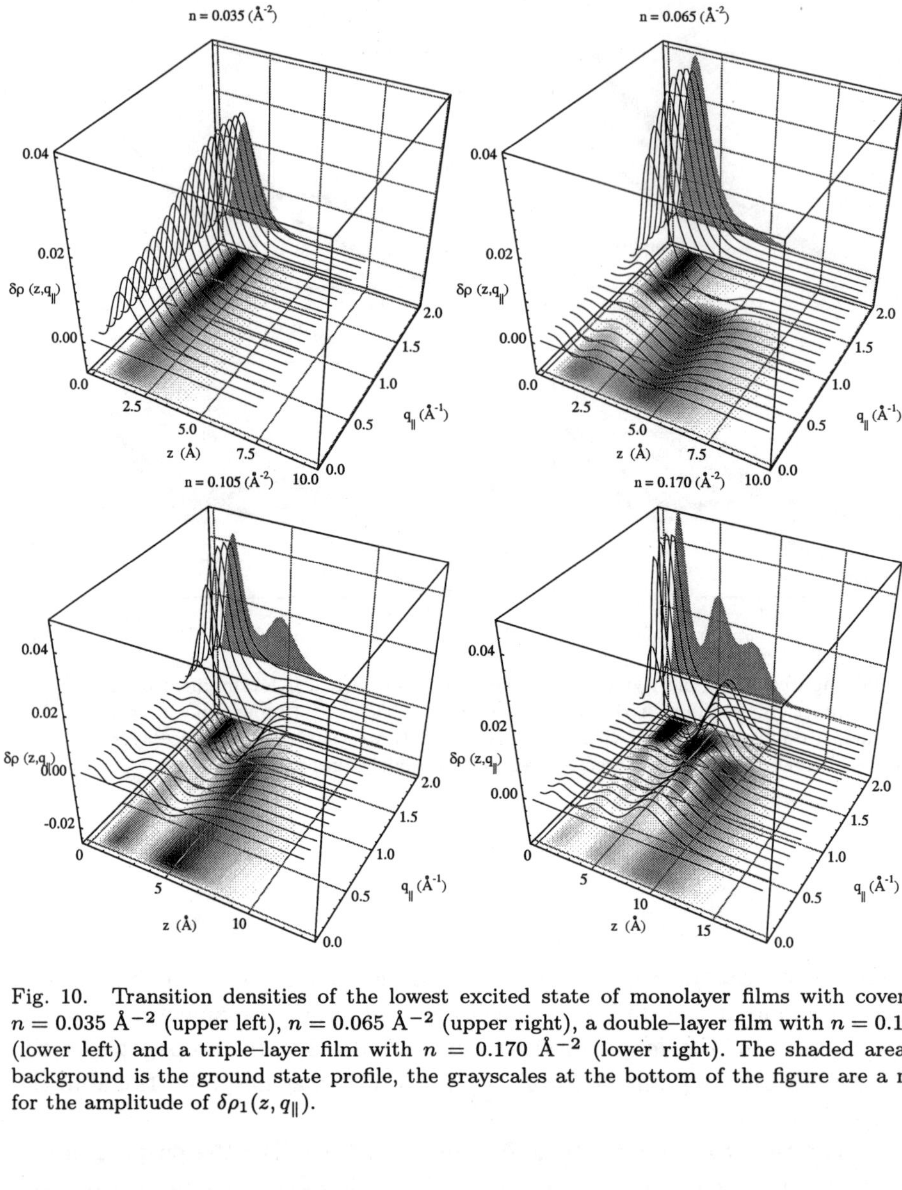

Fig. 10. Transition densities of the lowest excited state of monolayer films with coverages of $n = 0.035$ Å^{-2} (upper left), $n = 0.065$ Å^{-2} (upper right), a double–layer film with $n = 0.105$ Å^{-2} (lower left) and a triple–layer film with $n = 0.170$ Å^{-2} (lower right). The shaded area in the background is the ground state profile, the grayscales at the bottom of the figure are a measure for the amplitude of $\delta\rho_1(z, q_\parallel)$.

5.3. *Perpendicular Scattering*

As explained above, most experiments are done, for technical reasons, on parallel momentum transfer. Materials like graphite powder or *aerogel* surfaces are approximately randomly oriented, and one must take into account the effect of perpendicular scattering. Fig. 11 shows the dynamic structure function for ^{4}He on a graphite substrate for scattering perpendicular to the symmetry plane. In the perpendicular direction one can see a series of *dispersionless modes.* The excitations perpendicular to the surface are discrete below the evaporation threshold $\hbar\omega = -\mu$, where μ is the chemical potential of a single helium atom (dashed line in the right figure 11). The

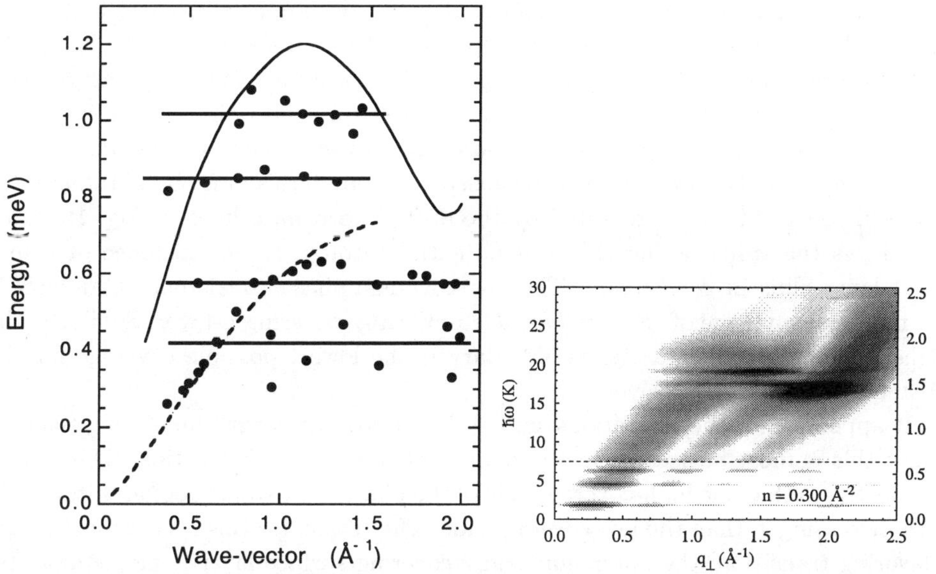

Fig. 11. The left figures shows neutron scattering data off a helium film of approximately five liquid layers adsorbed to graphite powder (from Ref. 53). The solid curve, shown for comparison, corresponds to the dispersion relation of bulk liquid ^{4}He. The dashed line indicates the ripplon mode (from Ref. 32). The right figure show the theoretical dynamic structure factor for neutron scattering off a ^{4}He film on a graphite substrate with the coverage 0.900 Å^{-2}. Here darker areas correspond to higher values; the theoretical results were broadened by 0.5 K. Apart from the low–energy dispersionless modes in $S(q_\perp, \omega)$, there are dispersionless resonances *above* the evaporation threshold (horizontal dashed line).

corresponding excitation functions are standing waves perpendicular to the film and confined to the thickness of the film. For a given energy, one observes in Fig. 11 a main response at a finite value of $q_\perp$, followed by a smaller response at higher $q_\perp$ values. This has a simple interpretation: the neutron plane wave interacts with a finite size system. The lack of translational invariance of the latter implies that the wave-vector is not a good quantum number for its excitations. As a consequence, the response is seen as a well defined excitation, which is generated by the neutron in the system and which has Fourier components at several wave-vectors with an intense first peak and smaller intensity in the following ones. These peaks are separated by a momentum difference $\Delta q_\perp \approx 2\pi/D$, where D is the thickness of the film. As the thickness of the film increases, the intensity distribution will obviously collapse onto the well-defined dispersion curve of an infinite system. The right part of Fig. 11 shows neutron scattering data of Ref. 32 where such dispersionless modes have actually been seen.

6. Quantum Film Thermodynamics

Let us now turn to thermodynamic properties of quantum films. [54] The implementation of the finite–temperature theory described in section 3.5 is relatively straightforward at low enough temperatures that do not need a consistent treatment of both the vapor and the liquid phases. Most interesting is, of course, the temperature–dependence of the layering transitions. At a fixed temperature, each layering transition has a low and a high coverage spinodal point, and a low and a high–coverage phase–coexistence boundary. The two spinodal points are determined directly from the theory. The phase coexistence boundaries must be obtained by a Maxwell construction, represented by the dotted horizontal lines in Fig. 12. Each line shows the value of the chemical potential through the coexistence of n and $n+1$-layer films in *equilibrium*. The low–coverage phase–coexistence boundary is the maximal coverage of the *locally stable* uniformly-covering n-layer. Similarly, the high–coverage phase–coexistence boundary is the lowest possible coverage for the existence of stable $n+1$ layers.

It appears that the transitions shift to lower coverages with increasing temperature; this is most noticeable in the double–to–triple–layer transition. The proper perspective is that the high-coverage side of the phase coexistence region shifts faster with increasing T than the low-coverage side. The reason for this is clear; just below a layering transition, the outer, uniformly-covering surface layer is metastable. By heating the system, atoms in this portion of the film will be promoted from the uniform to the cluster phase sooner than at lower temperatures. Consequently, the metastable film is driven into the unstable regime by the addition of thermal energy.

The temperature–dependent chemical potential $\mu(n,T)$ can be extracted directly from vapor-pressure measurements, [55–57] and it is also a key quantity of the theory. For our discussion, we will again consider the graphite substrate. Isotherms of the chemical potential $\mu(n,T)$ are shown in Fig. 12. The structured nature of the film is again visible in the non-monotonic behavior of $\mu(n,T)$; for each temperature there are several coverage regions with negative $d\mu/dn$; these indicate the above layering transitions.

Next we should estimate the accuracy of the theory. In bulk liquid ^{4}He the finite-temperature variational theory produces [16] for the liquid-gas phase transition a critical temperature T_c within 10% of the experimental value, and an accurate critical density ρ_c. In the films, the uncertainty in the theory is overshadowed by that in the substrate potential. Fig. 12 shows a comparison of the chemical potentials for the graphite/solid helium and magnesium substrates at $T = 0.6$ K (see also Fig. 1). We can compare with volumetric adsorption isotherm data at a nearby temperature of $T = 0.639$ K for a graphite foam substrate. [57] From the vapor pressure measurements by Chan *et al.* [57] we can extract the chemical potential by treating – quite appropriately – vapor as an ideal gas using

$$P = \frac{T}{\lambda_t^3} e^{\mu_{\rm vapor}/T} , \tag{6.1}$$

where $\mu_{\rm film} = \mu_{\rm vapor}$, and λ_t is the thermal wavelength. Removing solid layers by a coverage offset, our second layering transition coincides well with the plateau in the experimental μ. The plateau in the data suggests that a layering transition occurs

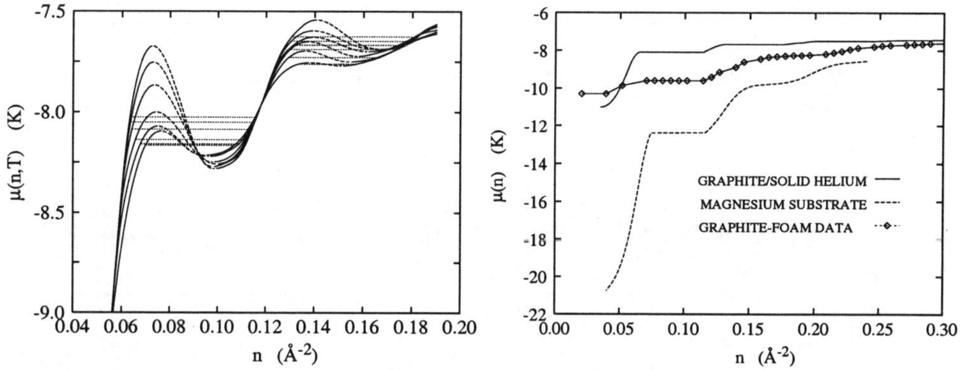

Fig. 12. *Left figure*: Chemical potential as a function of coverage for temperatures $T = 0.0$, 0.4, 0.6, 0.8, 1.0, and 1.2 K, for the graphite/solid helium substrate. The chemical potentials for higher temperatures correspond to curves with less intense oscillations. The dashed lines are cubic spline interpolations connecting the low and high coverage spinodal points for each transition. The horizontal dotted lines are the Maxwell tie-lines that give the equilibrium value of the chemical potential through the two-phase regions. *Right figure*: Theoretical chemical potentials for the graphite/solid helium and the magnesium substrates at T = 0.60 K compared with the experimental data of Ref. 57 on a graphite-foam substrate at T = 0.639 K.

there. The completion of three layers indicates that this would be the fourth layering transition in the experiments (and our second, since we have omitted the first two frozen layers). The final observation is that the experimental chemical potential falls between the two theoretical ones, indicating the sensitivity of the results to the substrate and that the graphite-foam substrate probably has a well–depth and range lying somewhere between the two potential models used here.

6.1. *Heat Capacity*

Another quantity of interest is the heat capacity C, which is the temperature derivative of the entropy (3.64) at fixed coverage n. For coverages away from the layering transitions the heat capacity is dominated by the explicit temperature dependence of the momentum distribution n_m:

$$C = T\left(\frac{\partial S}{\partial T}\right)_n \tag{6.2}$$

$$= \frac{A}{T^2}\sum_m \int \frac{d^2k_\parallel}{(2\pi)^2}(\epsilon_m(k_\parallel))^2 n_m(k_\parallel)[n_m(k_\parallel)+1] \tag{6.3}$$

$$\approx \frac{3}{\pi}A\zeta(3)\left(\frac{T}{\hbar c_3}\right)^2 . \tag{6.4}$$

Here the last form is obtained using a linear dispersion for the lowest excitation branch and ignoring higher branches (termed "linear spectrum" in the figures).

In Fig. 13 we compare the coverage and temperature dependence of the theoretical and the experimental heat capacities for helium on graphite. To accommodate

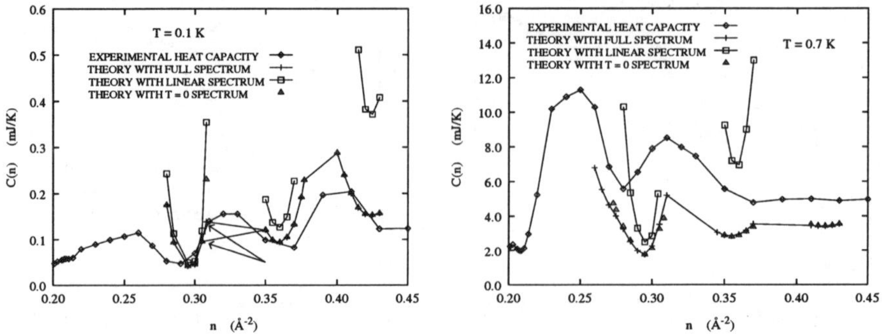

Fig. 13. Coverage dependence of the theoretical and experimental heat capacities of helium adsorbed to the graphite/solid helium substrate for T=0.1 K (left figure) and T=0.7 K (right figure). The experimental data are from Ref. 33. Straight lines connect the data points to guide the eye, only coverages corresponding to the liquid layers are shown. In the theoretical heat capacities, the linear segments extend over the entire two–phase regions determined by Maxwell constructions. The two arrows indicate the sensitivity of the heat capacity near the layering transitions. Note that the heat capacity calculated using the "FULL" and "T=0" spectra coincide for the majority of points shown.

solid layers, the theoretical coverage scale was shifted by 0.24 $Å^{-2}$ thus making it consistent with liquid layers of helium in the experiment of Greywall and Bush. [33]

Using our calculated mc_3^2, we found that the linear approximation gives reliable answers only for low temperatures, and only for coverages where the spectrum is linear over a substantial momentum regime. Most noticeably, it gives suitable results for our first stable island of coverages (0.28 $Å^{-2}$ to 0.32 $Å^{-2}$, in the shifted coverage scale). The important point is that c_3 is a hydrodynamic quantity; for even moderate temperatures the upward curvature of the spectrum plays a non-negligible role in the determination of the heat capacity.

Since the phase–coexistence boundaries, and not the spinodal points, determine the heat capacity, we again disregard the lowest coverage layering transition. The oscillations in C are easily correlated with the layered growth of the film and are an obvious consequence of the oscillatory nature of c_3 (Fig. 5) as a function of coverage, which has an important influence on the momentum integrals in (6.4). The parabolic regions in Fig. 13 are the coverages of stable uniformly covering films; the linear segments span the phase–coexistence boundaries of the layering transitions.

The qualitative agreement between our heat capacity by using our full variational theory (called "full spectrum" in the figures) and the experimental heat capacity is quite satisfactory. The heat capacities for the stable coverages (the parabolic minima) agree well at lower temperatures, considering that the heat capacity increases rapidly with temperature. This is an important observation since the heat capacity depends rather sensitively on c_3, especially at low T, and in turn, the sensitivity of c_3 to the substrate potential is well documented, see Fig. 5. The theoretical and experimental heat capacities differ more noticeably at $T = 0.7$ K. This is pre-

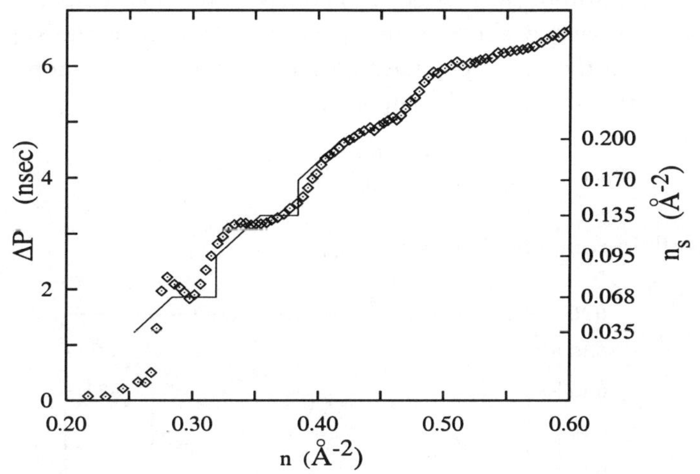

Fig. 14. The torsional oscillator periodicity ΔP, taken from Fig. 4 of Ref. 34 (diamonds and left scale) and the connected superfluid surface density from our theory (solid line and right scale) as a function of the total surface density, including the two solid ^{4}He layers.

dictable, since at this temperature higher momentum states are being populated and the differences between the generalized Feynman excitations and those of a more sophisticated theory begin to be significant.

6.2. *Superfluid Density*

Returning to the graphite/solid helium model, we notice that the chemical potential for the highest–coverage monolayer is notably *above* the one for the lowest coverage double layer. This is the well-known phenomena of *supersaturation* and *overexpansion* in a classical liquid-gas phase transition. These transitions provide an interpretation of the "re-entrant superfluidity" found in torsional oscillator experiments: [34] In the transition regions between two uniform configurations, the two-dimensional clusters on top of the "highest" uniform layer are disconnected from the superfluid and can couple to the substrate through, for example, hydrodynamic backflow. In that sense, these clusters behave very similar to single impurity atoms like ^{3}He, whose effective mass can also be determined by torsional oscillator experiments. [58] The numbers obtained above for the range of the uniform phase(s) determine our theoretical estimate of the *connected superfluid density* as a function of the total surface density. It is apparent from Fig. 14 that there is reasonable quantitative agreement between our results and those of Crowell and Reppy. It is known that there is a small surface density dependence in the relation between ΔP and the superfluid density. We believe this accounts for the slight downward bend of the experimental curve.

At all non-zero temperatures the total liquid ^{4}He present is divided into a normal and a superfluid component. At 1.2 K, which is the highest temperature considered here, the system is still below Kosterlitz–Thouless transition temperatures for most film thicknesses discussed above. Consequently, reduction in the superfluid component comes from thermally populating excited states and we apply the Landau theory [59] to get the normal fluid component

$$n_{\rm norm} = \frac{\hbar^2}{2mT} \sum_m \int \frac{dk_\parallel}{2\pi} k_\parallel^3 n_m^2(k_\parallel) e^{\beta\epsilon_m(k_\parallel)} \ , \tag{6.5}$$

and the superfluid component $n_{\rm s} = n - n_{\rm norm}$.

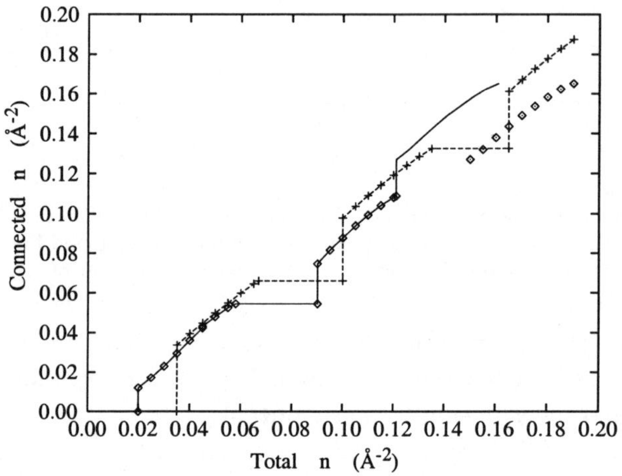

Fig. 15. Connected superfluid density as a function of the total ^{4}He loaded on a graphite/solid helium substrate. Two isotherms are shown: $T = 0.0$ K (plusses), and 1.2 K (squares).

At low temperatures, $n_{\rm norm}$ is a very small fraction of the total coverage. Nevertheless, the layering transitions influence the nature of the superfluidity with interesting consequences. Fig. 15 shows the *connected* superfluid density as a function of the total coverage. By connected superfluid we refer to those macroscopic regions of the film, where there is at least one superfluid percolating path. The layering transitions lead to plateaus, where the amount of connected superfluid material does not change with increasing total coverage until a percolation threshold is reached. In these regions the two–dimensional clusters on top of the outer–most uniform layer are disconnected from the superfluid underlayers and can couple to the substrate, *e.g.*, via excitations. Since there is no compelling argument for doing otherwise, the plateaus in the figure extend (approximately) between the spinodal points for each transition. To illustrate the temperature dependence of $n_{\rm s}$, we show representative isotherms of 0 K and 1.2 K.

The layering transitions provide a framework in which one can interpret the measurements by Crowell and Reppy [34] for ^{4}He films adsorbed to graphite substrates. A period shift ΔP of the torsional oscillator should be sensitive to the amount of *connected* superfluid; it detects deviations in the amount of connected superfluid

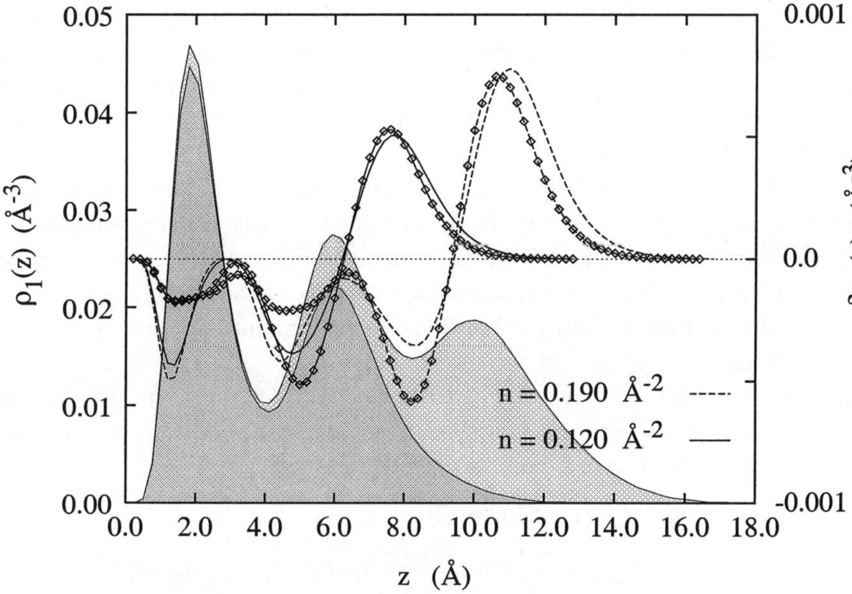

Fig. 16. The surface broadening $\delta\rho_1(z)$ is shown, for T=1.2 K, for two representative films of coverages $n = 0.12$ Å^{-2} (solid lines) and $n = 0.19$ Å^{-2} (dashed lines). The lines marked with diamonds are those obtained from the full calculation, the unmarked lines show the results from the low–temperature expansion. The shaded areas show the corresponding background density profiles. The expanded density scale on the right margin of the figure refers to the density change.

coverage from the total ^{4}He coverage on the film. Our interpretation of the Crowell–Reppy experiment is that the connected superfluid density, n_s, deviates from n at intervals equated to liquid layer completion. If n were equal to n_s for all coverages, ΔP would be approximately a linear function of n. The plateaus in ΔP occur near layer completion.

6.3. *Surface Broadening*

The broadening of the liquid surface due to thermal excitations can be computed directly by comparing the density profiles at different temperatures. It is nevertheless instructive to see how closed–form expressions for the surface broadening can be obtained within linear–response theory as the response of the system to its changed energetics at finite temperature.

The Hartree potential (3.68) depends on the temperature in two ways. One is the *explicit* temperature dependence due to the occupation numbers $n_m(T)$, and the second temperature dependence is the *implicit* dependence due to the temperature dependence of the density and the distribution functions. We can therefore write

$$V_H\left[\rho_1(T), g(T)\right](\mathbf{r}, T) = V_H\left[\rho_1(T), g(T), T=0\right](\mathbf{r}) + \Delta V_H(\mathbf{r}, T), \tag{6.6}$$

where the first term on the right hand side is the *zero–temperature* Hartree potential, but evaluated with finite–temperature densities, and the second term contains the explicit temperature dependence through the occupation numbers n_m. Since the n_m are considered independent variational parameters, they can be kept fixed in what follows. To reduce the problem to one that contains the temperature–dependent density as the only internal variable, one must further use the fact that $g(\mathbf{r},\mathbf{r}';T)$ has been optimized, in other words one has, in principle, a representation $g\left[\rho_1(\mathbf{r});T,n_m(T)\right](\mathbf{r},\mathbf{r}')$. It is important to note that the Euler equation for the pair distribution function also contains an explicit dependence on the occupation number $n_m(T)$; this temperature dependence also contributes to $\Delta V_H(\mathbf{r},T)$. The reader is reminded that we had to take this dependency into account to recover the correct low–temperature limit of the internal energy.

With these remarks, we are ready to linearize (3.68) in terms of the density change $\delta\rho_T(\mathbf{r},T)$. Allowing for a change in the chemical potential, one obtains for the finite–temperature Hartree potential the expression

$$
\begin{aligned}
V_H(\mathbf{r},T) &\equiv V_H(\mathbf{r}) + \delta V_H(\mathbf{r},T) \qquad (6.7)\\
&= V_H(\mathbf{r}) + \int d^3r' \frac{\delta V_H(\mathbf{r})}{\delta\rho_1(\mathbf{r}')}\delta\rho_1(\mathbf{r}',T) + \Delta V_H(\mathbf{r},T)\ .
\end{aligned}
$$

The functional derivative in the integral is just the particle–hole potential, so the linearized Hartree equation reads

$$
\begin{aligned}
&\left[-\frac{\hbar^2}{2m}\nabla^2 + U_{\rm ext}(\mathbf{r}) + V_H(\mathbf{r}) - \mu(0)\right]\delta\sqrt{\rho_1(\mathbf{r},T)}\\
&+2\int d^3r'\tilde V_{\rm p-h}(\mathbf{r},\mathbf{r}')\delta\sqrt{\rho_1(\mathbf{r}',T)} + \Delta V_H(\mathbf{r},T)\sqrt{\rho_1(\mathbf{r})} = \delta\mu(T)\sqrt{\rho_1(\mathbf{r})}
\end{aligned}
$$

or,

$$
\left[\left[H_1 + 2\tilde V_{\rm p-h}\right] * \delta\sqrt{\rho_1}\right](\mathbf{r},T) = \left[-\Delta V_H(\mathbf{r},T) + \delta\mu(T)\right]\sqrt{\rho_1(\mathbf{r})}. \qquad (6.8)
$$

The operator on the left hand side is related to the inverse static response function $\chi^{-1}(\mathbf{r},\mathbf{r}';\omega)$. Solving for $\delta\rho_1(\mathbf{r},T)$ we find

$$
\delta\rho_1(\mathbf{r},T) = \int d^3r'\chi(\mathbf{r},\mathbf{r}';0)\left[\Delta V_H(\mathbf{r}',T) - \delta\mu(T)\right] \qquad (6.9)
$$

where $\delta\mu(T)$ is determined by particle number conservation. At low temperatures we can identify

$$
\Delta V_H(\mathbf{r},T) = \sum_m n_m \frac{\delta\hbar\omega_m}{\delta\rho_1(\mathbf{r})}. \qquad (6.10)
$$

Thermal broadening arises from two effects. First, the film expands when the atomic zero–point motion is thermally enhanced. Thus, the *location* of the surface will be shifted away from the substrate. Second, the liquid-gas interface becomes more diffuse when the number of thermally populated surface states increases. Past phenomenological work [60] has treated these contributions separately and made the assumptions that the relevant states have either third sound ($\epsilon(k) = \hbar c_3 k$) or ripplon

($\omega^2(k) = (\sigma/m\rho_\infty)k^3$) dispersion relations. Within a microscopic theory there is no need to separate various sources of thermal broadening, or to make assumptions about the spectrum.

We have restricted ourselves to the most stable films, since it is difficult to properly assess the degree of broadening for coverages very near the transition regions. The profiles of the stable monolayers are quite resilient; below 1.2 K broadening is negligible. The double–layer and triple–layer films do however, show noticeable amounts of broadening. Fig. 16 shows two representative comparisons of the zero–temperature profiles with the 1.2 K one. Based on a 90% - 10% criterion, we find at $n = 0.19$ Å^{-2} a broadening of only 0.3 Å.

We can also determine the regions of broadening. The majority of the observable broadening takes place in the low–density tails of the outer–most layer. The increase in the local density of atoms there is accompanied by a depletion of atoms towards the inner region of the outer layer. Layers deeper inside the film are essentially unaffected by the increase in temperature.

It is interesting to compare the thermal broadening obtained from the full calculation with that obtained in the low–temperature limit, (6.9) and (6.10). Evidently, the agreement is quite good; the low–temperature expansion overestimates somewhat the density change deep inside the film. It appears that the quasiparticle approximation is accurate enough for most practical purposes.

7. Atomic Impurities

7.1. *Graphite Substrate*

We have calculated the binding energies of the ^{3}He impurity in a family of ^{4}He films in the *static approximation, i.e.*, neglecting the time dependence of the excited–state wave functions. The theory has been outlined in section 2.3 and we are using the background solutions presented in Figs. 3 and 4.

The energetics of the static ^{3}He states is shown in Fig. 17. For comparison, the figure shows also the results obtained using the *dynamic theory*, that will be described later. Also shown is the background chemical potential as a function of coverage. It appears that the results have approached an asymptotic limit at $n = 0.30$ Å^{-2}, and no surprises are expected for higher coverages. However, the background chemical potential appears to have converged to an asymptotic value of -7.45 K, but it has not completely settled down yet. At higher coverages we found that the background chemical potential increases to -7.35 K at a coverage of $n = 0.45$ Å^{-2}, confirming our assessment that the ground–state theory is accurate within about 0.1 K. A similarly slow approach of background quantities towards an asymptotic limit was found in the case of the much weaker Na and Cs substrates and in ^{4}He droplets.

Turning to the impurity energetics, we see at high coverages exactly the features anticipated by our analysis. The ground state results agree quite well with most recent experimental data. [61] Recall that our coverage scale involves liquid layers

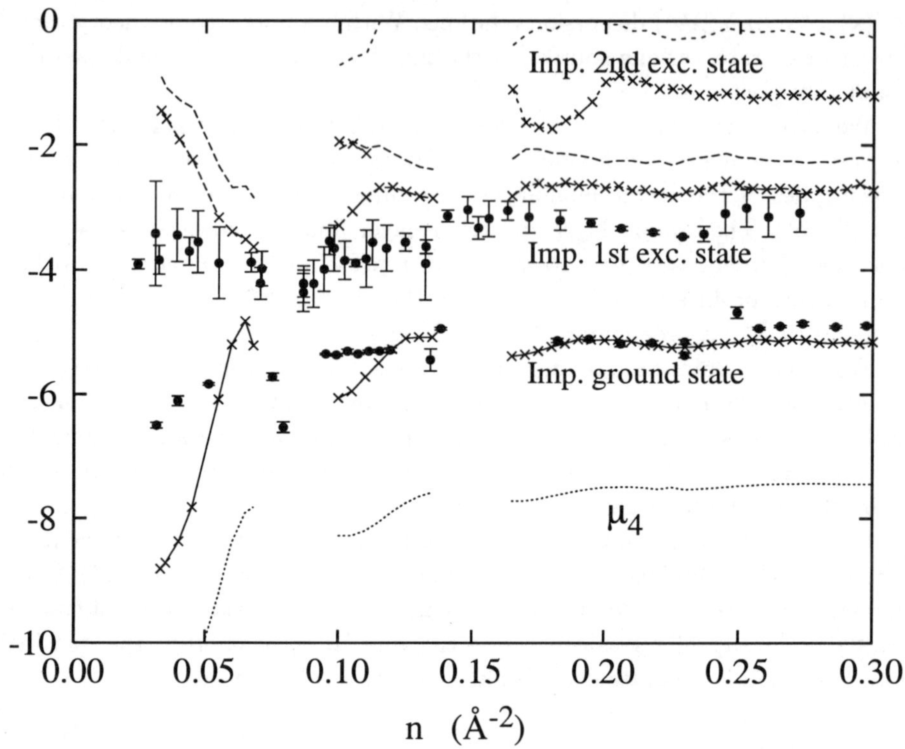

Fig. 17. The theoretical ground–state energy of ^{3}He impurities in ^{4}He films (solid line with markers) and the first two excited states in both the dynamic (long–dashed lines with markers) and the static theory (short-dashed lines without markers) are compared with the data of Ref. 61 for the ground state and the first excited state (the two sets of filled circles with error bars). Also shown is the chemical potential of the background (dotted line). The experimental data were shifted horizontally by 0.25 Å^{-2} to account for two solid layers.

only, whereas experimentally the first one or two atomic layers are solid. Therefore, in order to accommodate the first solid layer(s), we have shifted the coverage scale of the experimental data by 0.25 Å^{-2}. A similar coverage shift applies to other measurements as well, like, for example, to the volumetric adsorption isotherm data by Chan *et al.* [57] at T = 0.639 K for a graphite foam substrate. From their vapor pressure measurements we are able to extract the corresponding chemical potential by treating the vapor as an ideal gas. When we compare our chemical potential to the one obtained by this procedure, we find that our coverage scale is shifted by approximately 0.28 Å^{-2}. Moreover, the experiments of Ref. 61 were carried out on a nuclepore substrate and not on graphite as assumed here. Thus, one expects theory and experiment to agree *quantitatively* in the asymptotic regime, whereas the low coverage regime depends on the details of the substrate potential and can vary significantly. We attribute deviations between theory and experiment at lower coverages to the fact that our substrate potential is apparently somewhat steeper than the nuclepore potential. Instead of trying to model one specific substrate, we found it more instructive to study several typical cases.

The first excited state within the static approximation (2.44) is obviously too high when compared with experiment. This is expected, and its energy will, as we shall see, be lowered by the inclusion of dynamic effects. Moreover, the energy of the state is only -2.25 K, hence it is inconclusive to interpret it as a second surface–bound state, only consideration of the density profile of this state permits this interpretation. However, the impurity states of the films have still far too low level densities to make conclusive statements concerning whether this state will evolve, with increasing coverage, into the state of a ^{3}He atom dissolved in the bulk liquid. We will return to this question further below.

Evidently, *all* ^{3}He states are to some extent localized in the surface. As the energy increases, the states simply get broader and develop more nodes. Typically, the width of each oscillation in the wave function of the excited states is comparable to the width of an atomic monolayer, thus the first excited state has a width of about two layers, and the second excited state three layers. This causes these states to have, for thin films, an overlap with the first atomic monolayer of the background. Notice, that the second excited state is, in this approximation, only very weakly bound (*cf.* Fig. 17) and one might have expected that this state is spread out over a much larger area. As the film thickness increases further, this state should broaden and occupy most of the half–space available. Moreover, our calculation indicates that the impurity particle tends to *avoid* the area close to the substrate: It is energetically more favorable to keep the heavier ^{4}He particles in a confined and highly corrugated environment and the lighter ^{3}He atoms in the lower–density, less corrugated area.

The impurity location makes distinct jumps close to layer completion. The effect is most pronounced at the transition from a monolayer to a double layer. According to our theory of ^{4}He films, [7] at a coverage of 0.050 – 0.055 Å^{-2} the film is in a highly compressed, two–dimensional state. The density plot shows that the ^{3}He atom remains caged within the two–dimensional film up to that coverage; the film behaves like a two–dimensional mixture. The ground–state energy of the ^{3}He in the ^{4}He increases monotonically up to the coverage of 0.055 Å^{-2}. Above that coverage, it becomes energetically too expensive to compress the film further. First, the nature of the excitations changes from a longitudinal, two–dimensional phonon to a surface wave, [12,17] and the film eventually undergoes a first–order phase transition. Within the layering transition we expect that ^{4}He will form two–dimensional islands on top of the nearly completed underlayer.

As the ^{4}He coverage is increased, it is energetically unfavorable for the lighter particle to stay in the confined environment due to its larger zero–point energy. In fact it is promoted to the next layer *before* the ^{4}He background becomes unstable. The same effect is seen, somewhat less vigorously, for higher coverages. It is also quite clearly visible in the excited states: As soon as there is enough space for the impurity particle, it promptly uses the opportunity to avoid the compressed inner layers of the ^{4}He background. In the case of the first layer this leads even to a drop in the impurity chemical potential.

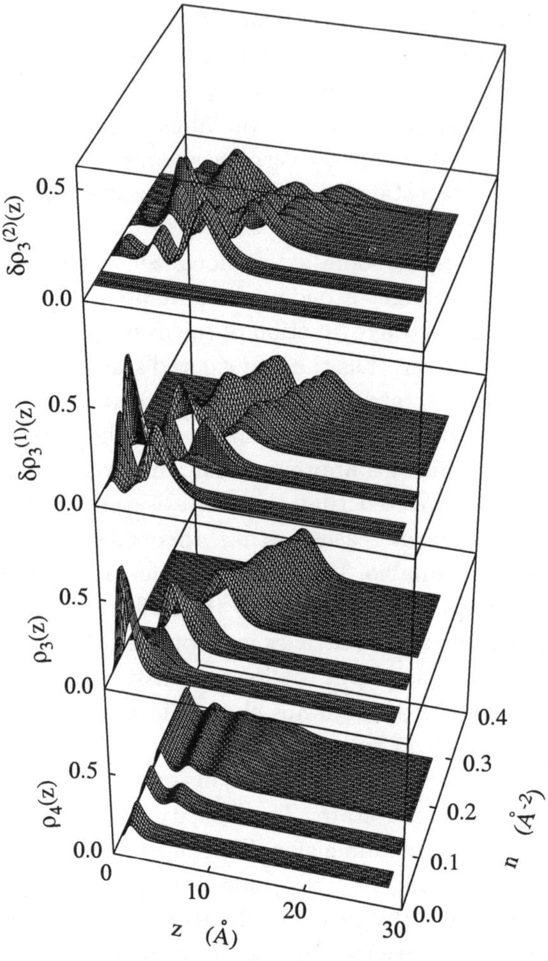

Fig. 18. The densities of ^{3}He impurities in a ^{4}He film on graphite for the *dynamic* theory. We show results in the ground state (second figure from below) and in the first two excited states (top two figures). For comparison, we show also the density profiles of the background (bottom figure). The wave functions of the excitations are normalized to one, the normalization of the background is arbitrary.

The sudden drop in the impurity ground–state energy is even more pronounced in the experimental data (*cf.* Fig. 17) at approximately 0.07 Å^{-2}. This might have some very interesting consequences since, while our calculations cannot be carried out in the coexistence region of a monolayer and a double–layer film, the experiments can. According to the growth scenario described earlier, this coverage point should be close to the first layering transition or even in the inhomogeneous phase between

Table 1. Ground–state and excited–state energies of ^{4}He and ^{3}He on various substrates. All energies are given in K, the row labeled "graphite" refers to graphite *plus* two solid layers of ^{4}He.

	^{4}He			^{3}He		
Substrate	ground	1st exc.	2nd exc.	ground	1st exc.	2nd exc.
graphite	-9.53	-0.49		-7.88	-0.15	
Mg	-19.38	-6.06	-1.43	-17.74	-4.41	-0.72
Na	-2.44	-0.23		-2.05	-0.10	
Cs	-1.49	-0.08		-1.21	-0.02	

transitions. Since the outer shoulder of the film is now highly compressible —it could even consists of "patches"— the larger zero–point motion of the ^{3}He atom becomes much less significant and the ^{3}He and a ^{4}He atom should have roughly the same energetics. As a consequence, the impurity ground–state energy approaches the ^{4}He chemical potential. Evidence that the impurity mass plays a secondary role for weakly–bound states is presented in Table 1, where we compare the binding energies of single ^{4}He and ^{3}He atoms to selected bare substrates.

It is energetically plausible, that the ^{3}He atom binds itself to the side of one of the two–dimensional ^{4}He clusters rather than jumps directly to the top of the cluster. Again, this side–bound configuration will maintain a low binding energy that will approach (approximately, as before) that of the equilibrium ^{4}He chemical potential. Eventually, one expects once again a sharp rise in the ground–state energy, when the local ^{4}He density becomes too large and the ^{3}He atom is promoted to the outer region of the ^{4}He film. We believe that this system offers another possibility for experimental verification of the layering transitions. One could, for instance, use the magnetic properties of the ^{3}He to promote the atom to the top of a cluster, and then watch it drop back to the lower energy side–bound configuration. For large clusters, there could be a notable time delay due to the diffusion time for ^{3}He to move across the cluster.

We hope that this observation inspires future experimental work, similar to that of Ref. 61. These studies should include dense sets of data, because the layering transitions are expected to span only a limited set of coverages. They should be also done with a variety of substrates, including the different morphologies of graphite.

For completeness and consistency we wish to discuss next the very low–coverage regime. One should expect the impurity energies to converge, in the low–coverage regime, towards the binding energy of a ^{3}He-particle on the bare substrate, *plus* the binding energy of the ^{3}He particle to the quasi–two dimensional monolayer. Indeed, we find that the ground state binding energy is -9.53 K, and that the excited state energy is -0.49 K for (*cf.* Table 1). Our results for finite films are consistent with the expected low–coverage limits. The weak binding of the first excited state on the graphite substrate also explains the abrupt increase of the first excited impurity state in the films as the coverage is decreased.

The inclusion of lifetime effects in the *dynamic theory* leads for strong substrate potentials to a qualitatively similar scenario as the static one. To calculate the excitation energies, one must solve the implicit equations (3.50) and (3.51) iteratively. The computation is somewhat complicated since the energy denominator in the self energy (3.51) can be zero, and thus the self energy can be complex. Physically, this describes the fact that the excitations are actually *resonances*, that have a finite lifetime and can decay into phonons, ripplons and impurity particles of lower energy. These resonances can be stationary excited states only if their decay is Pauli-blocked by other ^{3}He atoms; this was the case in the experiments of the Amherst group. [61]

Results for the energies are also shown in Fig. 17, and the corresponding transition densities are shown in Fig. 18. The excitation energies of the dynamical theory are lower than those of the static theory. Our first excitation energy is now only slightly above the experiment. This small deviation was expected: In our work on atomic impurities [62] in bulk ^{4}He a more complete solution of the equations of motion leads to an increase of the effective mass, in other words to a compression of the excitation energies. The effect can also be made plausible by considering the expression (3.51) for the self energy. The energy denominator contains bare impurity energies and Feynman phonon energies, which are known to be too high. A more complete – and more complicated – theory would contain proper self energies and lead to lower excitations energy. A possible way to simulate the proper self–energy corrections is to simply scale the background spectrum in Eq. (3.51) such that the roton energy roughly coincides with the experimental one. We have recently applied this scaling procedure quite successfully to the dynamic structure function of helium films, [17] but we refrain from such modifications here to keep the theory as clean as possible. Note that such self–energy improvements have no effect on the impurity ground–state energy.

In changing from the static to the dynamic theory, the energy of the second excited state is lowered remarkably, by about 1 K. Nevertheless this state remains energetically well above the bulk solvation energy of the ^{3}He impurity. The state still resides mostly in the outer layers of the film, but some enhanced strength towards the interior of the film is also found.

7.2. *Alkali Metal Substrates*

To supplement our analysis of impurity states we have studied three examples of alkali metal substrates, namely magnesium, sodium, and cesium. These three substrates are, in different aspects, distinctly different from the graphite substrate discussed above: The magnesium substrate has a somewhat deeper attractive well, but it is longer ranged. As a consequence, the ground state is more strongly corrugated and the maximum density of the first monolayer is somewhat higher, but the film exhibits only one weak layering transition. It is therefore possible to do meaningful calculations for a continuous sequence of coverages. The sodium potential, on the other hand, is overall quite weak. The minimum coverage at which a translationally invariant connected film can form is $n = 0.09$ Å^{-2} or *two* liquid layers. A discussion of the ground–state properties and excitation spectra of films on alkali metal substrates may be be found in Section 4.3 and in Ref. 46.

The behavior of the ^{3}He impurity in ^{4}He on the Mg substrate is expected to be qualitatively similar to that on graphite; quantitative differences can be anticipated at low coverages due to the deeper potential well. We have therefore restricted our calculations to the regime where differences are expected either due to the deeper potential well, or due the the absence of strong layering transitions, and have considered only the regime from a low–density monolayer ($n = 0.04$ Å^{-2}) up to triple–layer films with $n = 0.24$ Å^{-2}. Since one can obtain background configurations through a continuous series of coverages, the behavior of the impurity atom as a function of layer completion becomes more visible. Similar to the graphite substrate we see, for the monolayer, a cross-over where first the ground state is a two–dimensional mixture, and the excited state a surface state, to the reverse situation. As the coverage increases, the impurity is expelled from the most highly compressed layers as soon as there is enough space in the outer layers to accommodate the state. A second reason that we have examined this substrate is to examine the possible existence of "substrate states" on strongly attractive substrates as suggested, for example, in Refs. 45 and 63. No such substrate states were found. It is also noted that the *low* coverage limit of the ground state and both excited states is consistent with the binding energies of the ^{3}He atom on the bare substrate.

Compared to the results with the graphite and Mg substrates discussed above, a sodium substrate presents a rather different physical situation. According to our ground–state calculations of Ref. 46, an atomic *monolayer* on an Na substrate is unstable. Corresponding to the weak substrate attraction (the binding energy of a single ^{4}He atom is only -4.8 K) it tends to contract to clusters of at least two layers, and the minimum stable coverage is $n = 0.09$ Å^{-2}. The layers closest to the substrate are much less compressed than in graphite and Mg substrates. This has some consequences on the nature of impurity states: Trivially, if there were no substrate potential at all, a second "free" surface would exist at the "interface" and an Andreev state would be present at that surface. This does not abruptly change when a weak potential is turned on and therefore, in principle, a ^{3}He state can also form at the substrate-helium interface if the substrate potential is sufficiently weak. These observations are identical to those spelled out by Bashkin *et al.*, [63] and we note that they do not depend on any knowledge about the many–particle processes in this situation.

The ground state and the first excited state of ^{3}He impurities in ^{4}He films on a Na substrate are quite similar to those found on high–coverage films on graphite. Corresponding to the weakly layered ground–state structure, the energies are almost independent of coverage. Asymptotically the energies appear to converge to slightly higher values than on graphite; this is due to the discretization in a slightly smaller box which was done for computational economy. The *second* state is considerably

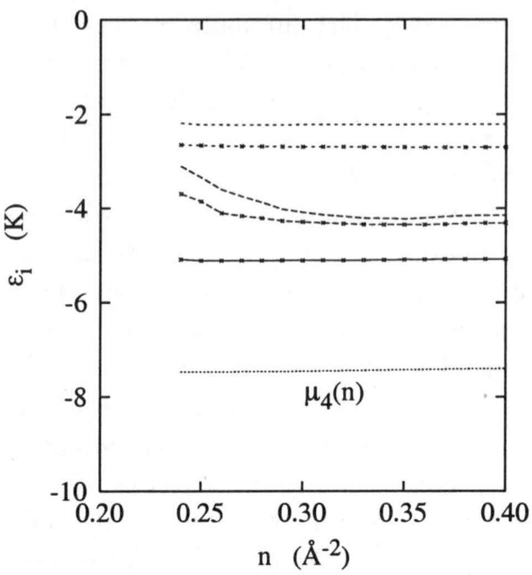

Fig. 19. Same as Fig. 17 for a ^{4}He film on a Cs substrate.

lower than the second state on the graphite substrate; it appears to be located close to the substrate, but also has some strength in the bulk of the film. Its energy is, with an asymptotic value of about -2.1K, conclusively above the solvation energy of a ^{3}He atom in the bulk liquid. In other words, there is no "substrate state" on a Na substrate, but there exists the possibility of a resonance that is located closely to the substrate. It appears that the potential used for Na is a borderline case; a substrate state *may* exist for slightly weaker potentials. This is consistent with the fact that our transition densities for this resonance are somewhat erratic; this simply reflects the feature that the substrate state is, or is close to being, unstable against solvation in the bulk and its actual shape depends sensitively on both the boundary conditions and the background corrugation.

We finally turn to our results on a cesium substrate. Fig. 19 shows our ground–state energies for that case. All energies as well as the background chemical potential are practically flat, consistent with the fact that the background structure shows only very weak layering. As discussed above, an impurity state at the Cs-^{4}He interface can develop in this situation because the ^{4}He density profile at the interface is similar to the one at the free surface, and since the Na substrate was found to be a borderline case. After self-energy corrections are applied, this state has an energy of -4.3 K. This is somewhat above the result of Ref. 64 who report approximately -4.8 K. However, note that our energy is an *upper bound* only, even within the variational theory: The theory also contains effects caused by the rearrangement of the background due to the presence of the impurity. These rearrangements are applied for the *impurity ground state,* the wave function is not re-optimized for the excited states. On the other hand, self-energy corrections are quite small and reduce the energy of the substrate state by 0.2 K only. This is understandable since there

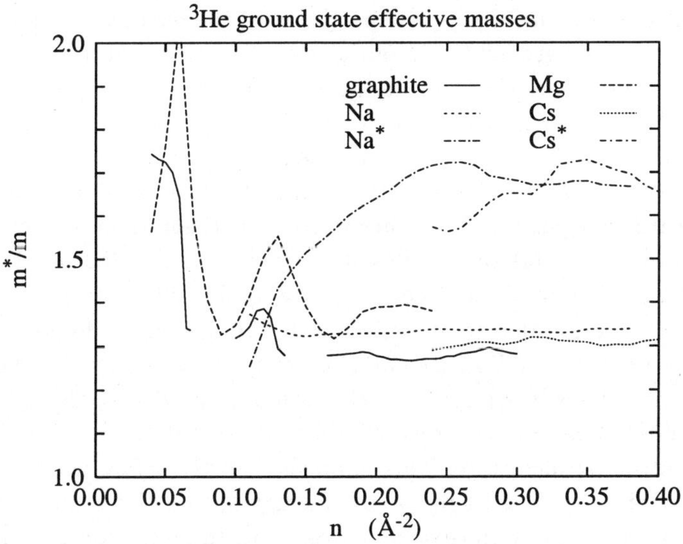

Fig. 20. The hydrodynamic effective masses of the ground state of ^{3}He impurities in a ^{4}He film on various substrates as labeled in the figure. Also shown is the effective mass of the ^{3}He impurity in the *substrate state* on Na and Cs (labeled by Na* and Cs*).

is virtually no overlap between the substrate–state wave function and that of any other state.

7.3. *Effective Masses and Lifetimes*

A *new* aspect of the dynamic theory is that the impurity particle now also has an effective mass. To be precise, we call the quantity calculated here the "hydrodynamic effective mass" since it is caused exclusively by the hydrodynamic backflow of the ^{4}He background around the impurity. The concept of an effective mass is, of course, rigorously valid for the ground–state particle moving with the momentum parallel to the surface; note also that Eq. (3.59) is not applicable for excited states.

To obtain the effective mass of the excited states, we have solved the self-energy equation (3.50) for the higher-lying bound impurity states for a sequence of momenta $0 < q_{\|} < 0.5$ Å^{-1} and fitted the resulting $\omega(q_{\|})$ by an effective mass formula. A useful consistency test for the procedure is that the effective mass of the ground state obtained by this procedure should agree with the one calculated directly from Eq. (3.59). Indeed, while the fitting procedure is occasionally affected by larger numerical uncertainties, we normally found good agreement between the effective masses of the impurity ground state obtained in these two ways. Only in the vicinity of the layering transitions, one expects and obtains visible deviations from a quadratic momentum dependence of the impurity spectrum. This is because one has, in that regime, low-lying excitations to couple to, which cause strong non-analytic contributions to the self-energy (3.51).

Fig. 20 shows our theoretical results for the effective mass of the Andreev state on all four substrates. The coverage dependence of the effective mass is a direct reflection of the overlap between the impurity wave function and the ^{4}He background. For large coverages, where the impurity ground state is a clear surface state, we obtain a theoretical value $m^*_H/m_I \approx 1.35$. Our results agree well with the value of 1.38 given by Higley *at al.* [65], are somewhat larger than the value of $m^*_H/m_I \approx 1.26$, reported by Valles *et al.* [66] and are at the lower end of the value $m^*_H/m_I = 1.45 \pm 0.1$ given by Edwards and Saam. [60] In other words, our theoretical prediction is within the spread of experimental values. We also observe, as pointed out by Wang and Gasparini, [58] a maximum of the effective mass shortly before completion of the first layer. This maximum is due to the effect that the system undergoes a transition between an essentially two-dimensional mixture to a ^{4}He film with a surface state. The effect is indeed easily seen in the ^{3}He ground-state density $\rho^I_1(z_0) = |\eta_0(z_0)|^2$; *cf.* Fig. 18. With increasing coverage (slightly above 0.05 Å^{-2}), the ^{3}He impurity is promoted into a surface state. This happens, shortly *before* the layering transition because of the significant energy cost for having the ^{3}He atom, with its larger zero-point motion, in the compressed environment. In fact, it occurs practically simultaneously with the change of the character of the low-lying excitations from a two-dimensional phonon to a surface mode. In a sense, the drop of the impurity effective mass is another precursor phenomenon of the layering transition where the whole film becomes unstable against population of the second ^{4}He layer. Due to the choice of our substrate potential, the peak in the effective mass is somewhat sharper than seen experimentally, and also obtained by us in Ref. 22. This is consistent with our finding that the ground-state energy drops, for low coverages, significantly faster than that observed in Ref. 61.

Fig. 20 also show the effective masses of those states that might be interpreted as "substrate states". This is the *second* state on the cesium substrate, and the third state (better, resonance) on the Na substrate. The calculation of the effective masses of these states is numerically somewhat delicate, the results are therefore somewhat noisy. In the bulk liquid one obtains, within the same theoretical approximation, $m^*/m \approx 1.95$. [62] We attribute the reduced effective mass of these states to incomplete backflow patterns due to the presence of the substrate. However, we found no way to reconcile these numbers with the data of Ross *et al.* [64] who estimate an experimental effective mass ratio $m^*/m \approx 0.75$. Further work is needed to explain this discrepancy.

Fig. 21 show the effective masses of the second surface state. As seen in Fig. 18, this state is more spread out in the surface region, the overlap with the background is larger and, hence, results in the shown effective mass.

Fig. 22 finally shows the imaginary part of the impurity self-energy for the two surface states. The transition from the monolayer to the double layer shows some essential features that are worth pointing out: As the coverage increases, the imaginary part of the self-energy goes through a sharp maximum and the excitation is very short-lived. Comparison of Figs. 22 and 20 shows that, at the same time, the effective mass of the ground state drops sharply. Fig. 18 reveals the explanation for

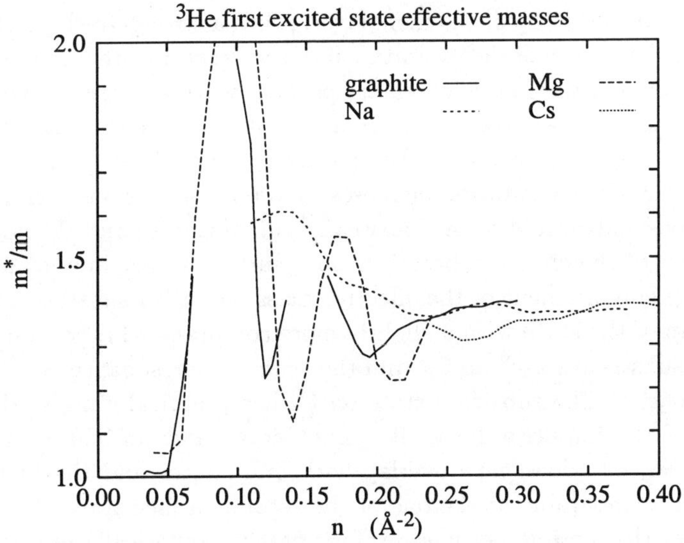

Fig. 21. The hydrodynamic effective masses of ^{3}He impurities in the first excited surface state in a ^{4}He film on various substrates as labeled in the figure.

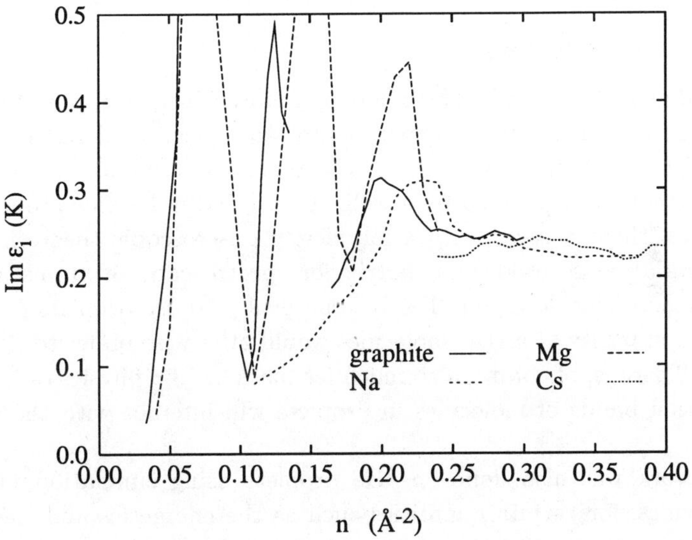

Fig. 22. The imaginary part of the impurity self-energy for the first excited state of a ^{3}He atom in a ^{4}He film on various substrates as labeled in the figure.

this behavior. The imaginary part of the excited state becomes large whenever it has an appreciable overlap with the ground state. For example, ground state is, for low coverage monolayers $n < 0.05$ Å^{-2}, essentially a two-dimensional mixture. The first excited surface state, whose energy is more than 6 K above the ^{3}He ground state, is essentially a free ^{3}He atom on top of a monolayer. At the point where the

^{3}He is promoted to the top of the monolayer, it has a strong overlap with the lowest excitation and the decay of the excitation into the ground state becomes more likely. Hence the sharp increase in the imaginary part of the self-energy shortly before layer completion. Shortly after layer completion, the films becomes broad enough for the excited state to also occupy partly the first layer. Hence the overlap decreases again and the lifetime of the excitation increases. A similar scenario is seen in the second layer, for the second excited state. The weaker substrates Na and Cs show essentially a flat behavior of the effective masses of both surface states; only at low coverages one sees some enhancement of the effective mass on an Na substrate, which is due to the overlap of the state with a slightly more compressed first atomic monolayer.

For the "substrate state" on Cs, and the "substrate resonance" on Na, we found different scenarios. The substrate state on Cs has practically no overlap with any other state of the impurity, hence it cannot decay and its lifetime is practically infinite. On Na, one does have overlap with bulk states and the lifetime of that state becomes comparable to the one of the second surface state. An accurate determination of the lifetime becomes, unfortunately, numerically quite cumbersome since the state itself is rather poorly defined. We have therefore deferred such an investigation until there is experimental need for the information.

8. Structure of Clusters

The second geometry that has been looked at intensively are spherically symmetric ^{4}He clusters (droplets); much experimental and theoretical research is presently being carried out on these systems. [67] Clusters share a number features with finite nuclei, but they are described by a well–defined and well–understood Hamiltonian which makes them accessible to a manifestly miscroscopic theoretical treatment. ^{4}He clusters also are used as a matrix for spectroscopy of atoms and molecules "under pressure" inside liquid ^{4}He. It is necessary to use droplets for this purpose because the impurity atoms or molecules would otherwise prefer to the walls of any container. There is, of course, also much feedback on the physics of helium because the excitation bands of molecules in droplets will interact with the excitations of the liquid.

Much work was also done on ^{4}He droplets using simulation methods [68–70] which provides, for certain quantities such as the energetics and the density profiles, higher accuracy than the semi–analytic methods discussed here and provide excellent means for testing and comparison.

The technical difference between the film– and the droplet geometry is that the relevant quantity in a finite system is not the total energy, but the "internal energy"

$$E_{\rm int} = E - T_{\rm cm} \; , \tag{8.1}$$

where $T_{\rm cm}$ is the expectation value of the kinetic energy connected with the center-of-mass motion:

$$T_{\rm cm} = \frac{\left\langle \Psi_0 \mid \hat{T}_{\rm cm} \mid \Psi_0 \right\rangle}{\langle \Psi_0 \mid \Psi_0 \rangle}, \qquad \hat{T}_{\rm cm} \equiv -\frac{\hbar^2}{2M}\left|\sum_i \nabla_i\right|^2 . \tag{8.2}$$

$M = Nm$ is the total mass of the droplet. The details of the treatment of center–of–mass corrections may be found in Ref. 71, we give here only the entry point to the calculations. Using the usual Jackson-Feenberg integration by parts, one obtains for the center–of–mass energy (8.2)

$$\begin{aligned} T_{\rm cm} &= -\frac{\hbar^2}{2M}\sum_i \frac{\langle \Psi_0 \mid \nabla_i^2 \mid \Psi_0\rangle}{\langle \Psi_0 \mid \Psi_0\rangle} - \frac{\hbar^2}{2M}\sum_{i\neq j} \frac{\langle \Psi_0 \mid \nabla_i \cdot \nabla_j \mid \Psi_0\rangle}{\langle \Psi_0 \mid \Psi_0\rangle} \\ &= -\frac{\hbar^2}{4M}\sum_i \frac{\langle \Psi_0 \mid [\nabla_i^2 U] \mid \Psi_0\rangle}{\langle \Psi_0 \mid \Psi_0\rangle} - \frac{\hbar^2}{4M}\sum_{i\neq j} \frac{\langle \Psi_0 \mid [\nabla_i \cdot \nabla_j U] \mid \Psi_0\rangle}{\langle \Psi_0 \mid \Psi_0\rangle} . \end{aligned} \tag{8.3}$$

The first term in Eq. (8.3) is structurally identical to the kinetic energy expression for non-uniform geometries; it causes a reduction of the ordinary Jackson-Feenberg kinetic energy expression (2.14) by a factor $1 - m/M$. Only pair (and possibly triplet) correlations contribute to the second term, which can be simplified by

$$\sum_{i\neq j} \langle \Psi_0 \mid [\nabla_i \cdot \nabla_j U] \mid \Psi_0\rangle = \int d^3r_1 d^3r_2 \rho_2(\mathbf{r}_1,\mathbf{r}_2)\nabla_1 \cdot \nabla_2 u_2(\mathbf{r}_1,\mathbf{r}_2) . \tag{8.4}$$

The relevant Euler equation is again conveniently derived using the prime-equation technique, but does not leads to much new information except that one has a null-mode corresponding to the rigid translation of the droplet. [71]

In nuclear physics, the *liquid–drop model* is a time–honored model in which finite nuclei are describing as droplets of nuclear matter, characterized by its energetics, compressibility and surface tension. Accurate calculations on simpler systems can give valuable information on the validity of such models. Interesting structural quantities are, besides the energetics and the density profiles, the *rms*–radius

$$r_{rms} \equiv \sqrt{\langle r^2 \rangle} \tag{8.5}$$

and the "hard sphere equivalent" radius

$$r_0 \equiv (5/3)^{1/2} \langle r^2 \rangle^{1/2} N^{-1/3} . \tag{8.6}$$

which is the radius that the droplet would have if it were a sphere of uniform density, and *rms*–radius r_{rms}. This quantity is a useful measure of the actual size of the droplet.

If the energetics of ^{4}He droplets were the sole interest, one could consider the problem solved by simulations. However, recent experiments are much more concerned with excitations, atomic and molecular impurities, and atom scattering, where microscopic many–body methods have many advantages. We will return to these issues, at the present level of the development of our theoretical tools we restrict ourselves ground state properties and their comparison with simulation data.

The left panel of Fig. 23 shows our results for the energy along with the DMC results of Ref. 70 and a fit of the energy of the form

$$\frac{E(N)}{N} = -\epsilon_\infty + \left(\frac{36\pi}{\rho_\infty^2}\right)^{1/3} \sigma N^{-1/3} + cN^{-2/3}\ K. \tag{8.7}$$

where ϵ_∞ is the binding energy of the bulk liquid, $\rho_\infty = 0.02185$ Å^{-3} the saturation density, and σ the surface energy. Using the HNC energies for particle numbers $20 \le N \le 1000$ leads to the results $\epsilon_\infty = 7.110 \pm 0.037$ and $\sigma = 0.2738 \pm 0.0068$. Our values compare favorably with the experimental value of $\sigma = 0.256$ KÅ^{-2} reported by Iino, [72] and even better with the older value $\sigma = 0.274$ KÅ^{-2} of Edwards and Saam [60] and the most recent measurement [73] of $\sigma = 0.272$ KÅ^{-2}.

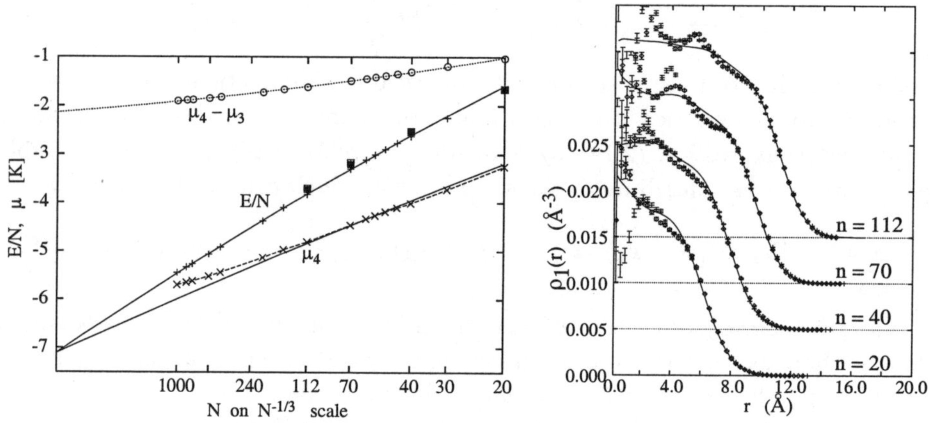

Fig. 23. The *left figure* shows the HNC–EL energies of helium clusters (+-symbols) for a particle number range of $N = 20$ to $N = 1000$. and the DMC results of Ref. 70 (boxes). Also shown is the fit (8.7) (solid line) to the energy, the chemical potentials as calculated from the generalized Hartree-equation (2.29) (long–dashed line with crosses) and by differentiating the mass formula (8.7). Also shown is the difference of chemical potentials $\mu_4 - \mu_3$ between ^{4}He and ^{3}He atoms as a function of particle number (circles) as well as a quadratic fit to these results (dotted line). The *right figure* shows the ground state density profiles of ^{4}He droplets from the HNC–EL theory (solid lines) compared with the DMC results of Ref. 70 (crosses and diamonds with error bars). The curves for higher particles numbers have been shifted upwards by 0.005 each to improve legibility.

The mass formula fit (8.7) should be considered with some caution because the results depend somewhat on how many powers of $N^{-1/3}$ are retained in Eq. (8.7) and which particle numbers are used for the fit. By changing these parameters, we estimated that the prediction for the asymptotic energy is accurate within a percent, whereas the surface energy can be extracted with a confidence of about five percent.

Another piece of information that follows directly from the ground state calculations are the density profiles. A comparison with the DMC calculations of Ref. 70 is shown in right panel of Fig. 23. The agreement between our HNC–EL calculations and the DMC results for the one–body densities is quite satisfactory. The HNC-EL data show somewhat weaker density–oscillations than the DMC data; these oscillations are extremely sensitive to the calculational procedure. As long as there are no directly observable facts connected with these oscillations, we are not prepared to spend much effort in determining their precise strength.

The asymptotic central density extrapolated from the the calculated cluster sized between 20 and 1000 atom is just above 0.021 $Å^{-2}$. This is slightly too low, but the result is still well within the expected accuracy of the HNC–EL theory. Some small amplitude density oscillations can be seen in the surface. Overall, we feel that our ground–state calculation has led to a satisfactory description of the structural properties which we can now use, as an input, for other purposes.

9. Summary and Conclusions

We have given in this review an extensive discussion of our methods and results. There is no need for repetition, but we shall highlight the most important conclusions.

Before attempting physical predictions, it is good practice to scrutinize any theory by applying it to situations that are well known and relevant to the physical situation under consideration. Equally important, these test situations, themselves, should not appear in the choosing process (for example, the function parameterization) for the approximations to be made in the theory. The main point where one might suspect that our approach has deficiencies is the treatment of "elementary diagrams" and triplet correlations, and we have carried out comparisons [7,62,71] with every piece of information that is known, either from experiment [74] or from simulation methods, [41,40] to verify that our approach is reliable. We found [7] that the present implementation of the HNC theory provides indeed excellent agreement with experiments and/or Monte Carlo data.

At this point, a brief comparison of our results with those of a non-local density functional theory [45] is appropriate. By construction, that method reproduces those quantities that were used for the parametrization or the choice of the analytic form of the functional. These are properties of the bulk, three-dimensional liquid and liquid mixtures. The most visible discrepancies between the results of the theories for the background structure appear in the compressed atomic layer close to the substrate. The most important known [41,40] feature to compare with is the behavior of the two-dimensional liquid. For properties of atomic monolayers, the density functional approach is somewhat less accurate [46] than our earlier HNC calculations. [75]

Acknowledgments

The work was supported, in part, by the Austrian Science fund under grants No. P12832–TPH and the Austrian Academic Exchange Service. Discussions with numerous senior and junior colleagues are gratefully acknowledged.

References

1. M. Saarela, V. Apaja, and J. Halinen, *Structure and dynamics of the bulk liquid and bulk mixtures,* 2001, this volume.
2. T. Morita and K. Hiroike, *Progress of Theor. Physics,* **25**, 537 (1960).
3. E. Krotscheck, in *Microscopic Quantum Many-Body Theories and their Applications*, Vol. 510 of *Lecture Notes in Physics*, edited by J. Navarro and A. Polls (Springer, Heidelberg, 1998), pp. 187–250.
4. E. Feenberg, *Theory of Quantum Fluids* (Academic, New York, 1969).
5. E. Krotscheck, Q.-X. Qian, and W. Kohn, *Phys. Rev.* **B 31**, 4245 (1985).
6. B. E. Clements, E. Krotscheck, and M. Saarela, *Phys. Rev.* **B 55**, 5959 (1997).
7. B. E. Clements, J. L. Epstein, E. Krotscheck, and M. Saarela, *Phys. Rev.* **B 48**, 7450 (1993).
8. D. Pines and P. Nozieres, *The Theory of Quantum Liquids* (Benjamin, New York, 1966), Vol. I.
9. A. K. Kerman and S. E. Koonin, *Ann. Phys. (NY)* **100**, 332 (1976).
10. R. P. Feynman, *Phys. Rev.* **94**, 262 (1954).
11. E. Krotscheck, *Phys. Rev.* **B 31**, 4258 (1985).
12. B. E. Clements *et al.*, *Phys. Rev.* **B 50**, 6958 (1994).
13. C. C. Chang and C. E. Campbell, *Phys. Rev.* **B 13**, 3779 (1976).
14. M. Saarela, *Phys. Rev.* **B 33**, 4596 (1986).
15. M. Saarela and J. Suominen, in *Condensed Matter Theories*, edited by J. S. Arponen, R. F. Bishop, and M. Manninen (Plenum, New York, 1988), Vol. 3, pp. 157–165.
16. B. E. Clements, E. Krotscheck, J. A. Smith, and C. E. Campbell, *Phys. Rev.* **B 47**, 5239 (1993).
17. B. E. Clements, E. Krotscheck, and C. J. Tymczak, *Phys. Rev.* **B 53**, 12253 (1996).
18. H. W. Jackson, *Phys. Rev.* **A 8**, 1529 (1973).
19. C. E. Campbell, in *Progress in Liquid Physics*, edited by C. A. Croxton (Wiley, London, 1977), Chap. 6, pp. 213–308.
20. J. W. Clark, in *Progress in Particle and Nuclear Physics*, edited by D. H. Wilkinson (Pergamon Press Ltd., Oxford, 1979), Vol. 2, pp. 89–199.
21. B. E. Clements *et al.*, *Phys. Rev.* **B 53**, 12242 (1996).
22. E. Krotscheck, M. Saarela, and J. L. Epstein, *Phys. Rev. Lett.* **61**, 1728 (1988).
23. C. E. Campbell, K. E. Kürten, M. L. Ristig, and G. Senger, *Phys. Rev.* **B 30**, 3728 (1984).
24. G. Senger, M. L. Ristig, K. E. Kürten, and C. E. Campbell, *Phys. Rev.* **B 33**, 7562 (1986).
25. C. E. Campbell, in *Condensed Matter Theories*, edited by L. Blum and F. B. Malik (Plenum, New York, 1993), Vol. 8, pp. 149–158.
26. N. Wada, A. Inoue, H. Yano, and K. Torii, *Phys. Rev.* **B 52**, 1167 (1996).
27. V. Apaja and E. Krotscheck, *J. Low Temp. Phys.* **123**, 241 (2001).
28. V. Apaja and E. Krotscheck, *Phys. Rev.* **B64**, 134503 (2001).
29. M. J. McKenna, T. B. Brosius, and J. D. Maynard, *Phys. Rev. Lett.* **69**, 3346 (1992).
30. B. E. Clements, E. Krotscheck, and H. J. Lauter, *Phys. Rev. Lett.* **70**, 1287 (1993).
31. H. J. Lauter, H. Godfrin, V. L. P. Frank, and P. Leiderer, in *Excitations in Two-Dimensional and Three-Dimensional Quantum Fluids*, Vol. 257 of *NATO Advanced Study Institute, Series B: Physics*, edited by A. F. G. Wyatt and H. J. Lauter (Plenum, New York, 1991), pp. 419–427.
32. H. J. Lauter, H. Godfrin, V. L. P. Frank, and P. Leiderer, *Phys. Rev. Lett.* **68**, 2484 (1992).
33. D. S. Greywall and P. A. Busch, *Phys. Rev. Lett.* **67**, 3535 (1991).

34. P. A. Crowell and J. D. Reppy, *Phys. Rev. Lett.* **70**, 3291 (1993).
35. G. Vidali, G. Ihm, H.-Y. Kim, and M. W. Cole, *Surface Science Reports,* **12**, 133 (1991).
36. M. W. Cole, D. R. Frankl, and D. L. Goodstein, *Rev. Mod. Phys.* **53**, 199 (1981).
37. E. Zaremba and W. Kohn, *Phys. Rev.* **B 15**, 1769 (1977).
38. R. A. Guyer, M. D. Miller, and J. Yaple, *Phys. Rev.* **B 25**, 4570 (1982).
39. C. E. Campbell, R. Folk, and E. Krotscheck, *J. Low Temp. Phys.* **105**, 13 (1996).
40. S. Giorgini, J. Boronat, and J. Casulleras, *Phys. Rev.* **B 54**, 6099 (1996).
41. P. A. Whitlock, G. V. Chester, and M. H. Kalos, *Phys. Rev.* **B 38**, 2418 (1988).
42. K. S. Ketola, S. Wang, and R. B. Hallock, *Phys. Rev. Lett.* **68**, 201 (1992).
43. P. J. Nacher and J. Dupont-Roc, *Phys. Rev. Lett.* **67**, 2966 (1991).
44. P. Taborek and J. E. Rutledge, *Phys. Rev. Lett.* **68**, 2184 (1992).
45. N. Pavloff and J. Treiner, *J. Low Temp. Phys.* **83**, 331 (1991).
46. B. E. Clements, H. Forbert, E. Krotscheck, and M. Saarela, *J. Low Temp. Phys.* **95**, 849 (1994).
47. M. Hieda *et al.*, *Physica,* **B263-B264**, 370 (1999).
48. E. Krotscheck, J. Paaso, M. Saarela, and K. Schörkhuber, *Phys. Rev. Lett.* **85**, 2344 (2000).
49. E. P. Bashkin, *Sov. Phys. JETP* **51**, 181 (1980).
50. K. R. Atkins, *Physica* **23**, 1143 (1957).
51. D. Goodstein, *Phys. Rev.* **183**, 327 (1969).
52. B. E. Clements, E. Krotscheck, and C. J. Tymczak, *J. Low Temp. Phys.* **107**, 387 (1997).
53. H. J. Lauter, H. Godfrin, and H. Wiechert, in *Proceedings of the Second International Conference on Phonon Physics*, edited by J. Kollàr, N. Kroo, M. Meynhard, and T. Siklos (World Scientific, Singapore, 1985), p. 842.
54. C. E. Campbell, B. E. Clements, E. Krotscheck, and M. Saarela, *Phys. Rev.* **B 55**, 3769 (1997).
55. W. E. Keller, *Helium-3 and Helium-4* (Plenum, New York, 1969).
56. J. Wilks, *The Properties of Liquid and Solid Helium* (Clarendon, Oxford, 1967).
57. G. Zimmerli, G. Mistura, and M. H. W. Chan, *Phys. Rev. Lett.* **68**, 60 (1992).
58. X. Wang and F. M. Gasparini, *Phys. Rev.* **B 38**, 11245 (1988).
59. I. M. Khalatnikov, *An Introduction To The Theory of Superfluidity* (Benjamin, New York, 1965), Chap. 1, p. 13.
60. D. O. Edwards and W. F. Saam, in *Progress in Low Temperature Physics*, edited by D. F. Brewer (North Holland, New York, 1978), Vol. **7A**, pp. 282–369.
61. D. T. Sprague, N. Alikacem, P. A. Sheldon, and R. B. Hallock, *Phys. Rev. Lett.* **72**, 384 (1994).
62. M. Saarela and E. Krotscheck, *J. Low Temp. Phys.* **90**, 415 (1993).
63. E. Bashkin, N. Pavloff, and J. Treiner, *J. Low Temp. Phys.* **99**, 659 (1995).
64. D. Ross, P. Taborek, and J. E. Rutledge, *Phys. Rev. Lett.* **74**, 4483 (1995).
65. R. H. Higley, D. T. Sprague, and R. B. Hallock, *Phys. Rev. Lett.* **63**, 2570 (1989).
66. J. M. Valles, Jr., R. H. Higley, B. R. Johnson, and R. B. Hallock, *Phys. Rev. Lett.* **60**, 428 (1988).
67. The Physics of Helium Clusters, *Journal of Chemical Physics,* **115**, (2001).
68. M. V. R. Krishna and K. B. Whaley, *J. Chem. Phys.* **93**, 6738 (1990).
69. M. V. R. Krishna and K. B. Whaley, *Phys. Rev. Lett.* **64**, 1126 (1990).
70. S. A. Chin and E. Krotscheck, *Phys. Rev.* **B 45**, 852 (1992).
71. S. A. Chin and E. Krotscheck, *Phys. Rev.* **B 52**, 10405 (1995).
72. M. Iino, M. Suzuki, and A. J. Ikhushima, *J. Low Temp. Phys.* **61**, 155 (1985).

73. G. Deville, P. Roche, N. J. Appleyard, and F. I. B. Williams, *Czekoslowak Journal of Physics Suppl.* **46**, 89 (1996).
74. R. de Bruyn Ouboter and C. N. Yang, *Physica* **144B**, 127 (1986).
75. E. Krotscheck, M. Saarela, and J. L. Epstein, *Phys. Rev.* **B 38**, 111 (1988).

CHAPTER 6

DENSITY FUNCTIONAL DESCRIPTIONS OF LIQUID ^{3}He IN RESTRICTED GEOMETRIES

E. S. Hernández

Depto. de Física, Facultad de Ciencias Exactas y Naturales,
Universidad de Buenos Aires, 1428 Buenos Aires, Argentina,
E-mail: shernand@df.uba.ar

J. Navarro

IFIC (Centro Mixto CSIC Universidad de Valencia),
Apdo. 22085, E-46071 Valencia, Spain
E-mail: navarro@ific.uv.es

In this chapter we review the applications of density functional theory to configurations of ^{3}He atoms in restricted geometries. The confinement may be imposed by an external field, like the adsorbing wells provided by planar substrates and porous environments, or can originate in the saturation properties of the many body system, as in the case of droplets containing a sufficiently large number of ^{3}He particles. For these arrangements of fermions theoretical investigations of the ground state properties and excitations are inhibited by the restriction to a fully antisymmetric subspace of the many body Hilbert space. Density functional techniques become then advantageous instruments, which for several purposes may override the practical limitations of more fundamental methods of many-body theory. We discuss the density functional forms and parametrizations proposed by different authors, starting from the basic descriptions of the homogeneous liquids ^{4}He, ^{3}He and their mixture, and show their extensions to treat various cases of symmetry-breaking. We summarize the present experimental scenarios and density functional descriptions of helium atoms adsorbed on substrates of diverse geometries, emphasizing on the abilities of this approach to anticipate the excitation spectrum of these systems. We review the current developments concerning the energy systematics, the structural and the excitation properties of pure and doped ^{3}He drops within density functional theory, including the case of mixed ^{3}He-^{4}He clusters with and without impurities.

1. Introduction

In the past three decades, a large variety of new and interesting phenomena involving liquid ^{3}He in confined geometries became experimentally accessible. The two-dimensional character of small amounts of ^{3}He dissolved in bulk ^{4}He or adsorbed on exfoliated graphite was already recognized in the early seventies. Films of mixtures of liquid helium isotopes were extensively investigated both from the theoretical and experimental viewpoints, and more recently, the production of liquid droplets entered a promising channel, which among various possibilities, permits to investigate the evolution of finite systems to bulk liquids in a systematic fashion. In addition, these clusters can furnish laboratories for investigation of the interaction of helium atoms with external fields, whose sources are foreign atoms or molecules trapped inside the drop. In the last few years, adsorption properties of helium on differently shaped substrates, such as alkali and alkaline earth flat surfaces, carbon nanotubes and aerogels, became a fertile topic for research. In fact, increasing interest in graphite nanotubes as potential containers for gases and liquids opened a new field, where experimental data are now starting to accumulate.

One important feature of these systems is the fact that their internal structure becomes more easily observable than in bulk liquids, due to the restricted motion of the particles in the confining potential. Consequently, it is important to possess appropriate theoretical instruments to describe these objects. In this context, it is worthwhile noticing that although most experiments involve ^{3}He, ^{4}He or their mixture, accurate microscopic calculations of properties of confined systems are only available for the bosonic isotope, and not for the fermionic counterpart, which is the main topic of the present chapter.

The theoretical understanding of liquid ^{3}He started with the celebrated Landau theory of Fermi liquids. [1] In this theory, dynamics and thermodynamics of a normal Fermi liquid are derived from the properties of a gas of quasiparticles, which are the elementary excitations of the liquid. The merit of Landau theory is to connect the relevant properties of the many body system to the interaction between quasiparticles, which is described in terms of a limited set of coefficients, known as Landau parameters. Their values, which can be derived from experiment, [2] are the basic ingredients for other phenomenological theories of liquid ^{3}He (for a general discussion see e.g. Ref. 3). Landau theory applies to the description of macroscopic properties and low energy - low momentum excitations of the system starting from the change in total energy induced by variations in the fermionic populations, and cannot provide keys about the ground state itself. Consequently, specific applications of Landau's Fermi liquid theory demand an ansatz for the total energy, which in its simplest version can be accounted for by a mean field description, whose one body level is the Hartree-Fock (HF) approximation.

The HF technique has been widely applied to atomic nuclei, starting from an effective zero-range two-body interaction. In particular, the so-called Skyrme forces — a type of density-dependent effective interactions — became largely successful in the 70's, bringing into evidence the high power of the HF-Skyrme approach to yield the energetic and structural systematics of finite nuclei. [4] One characteristic of the HF procedure is that it provides a density-functional form for the total energy; in fact, the well-known Hohenberg-Kohn's theorem [5] guarantees that the energy per particle of a many body system can be expressed in terms of a functional of the density, out of which not only the ground state properties, but also the elementary excitation spectrum, can be derived solving the so-called Kohn-Sham (KS) equation. [6] However, this theorem does not provide any clue for the construction of such a functional, which has to be written down resorting to physical insight and/or phenomenological approaches. Electron systems and classical liquids soon became the subject of many applications of density functional theory, and in the early 80's Stringari [7] inaugurated the description of ground state properties of liquid ^{3}He within a specific Skyrme-HF scheme.

In the frame of Landau's theory of Fermi liquids many aspects of the atom-atom interaction — such as the presence of the core — and their consequences on the dynamical behavior of helium are accurately taken care of by Pines' polarization potentials. [8,9] Although this is a powerful tool, very carefully designed to describe both helium liquids, further use of these forces is restricted to homogeneous systems. Moreover, polarization potentials demand density-dependent coefficients in order to extend their range of applicability to the whole density interval where the bulk liquids exist. As shown in the nextcoming sections, this limitation can be removed to some extent by complementing the polarization core with conveniently parametrized interaction terms which contain either velocities or density gradients, with force strenghts that depend on the local density according to a preestablished power form. The flexibility achieved in this fashion permits the application of density-functional techniques to nonhomogeneous helium systems such as the liquid free surface, drops, adsorbed films, and adsorbed gases and liquids in aerogels and nanotubes. This is an important achievement, in view of the fact that elaborated many body techniques — such as variational descriptions which include high order correlations beyond the one-body level, correlated clusters and diffusion or Green's function Monte Carlo methods — cannot be applied to Fermi systems within reasonable computational costs. In fact, it is only recently that diffusion Monte Carlo calculations of the equation of state of liquid ^{3}He have been reported. [10] In this context, DF methods provide a useful tool which can explore territories not yet accesible by sophisticated many body physics.

The aim of this chapter is to review the past and current developments in DF theory as applied to finite configurations of ^{3}He atoms. For this sake, in Section 2 we present and discuss the various DF parametrizations, starting from the basic descriptions of the bulk liquids ^{4}He and ^{3}He as well as their mixture. These functional forms are the source of the DF's later employed to study the inhomogeneous systems. For the sake of completeness, we also briefly comment on recent DF calculations for weakly magnetized, homogeneous ^{3}He, as a special case of symmetry-breaking in a degree of freedom other than a spatial one. In Section 3, we summarize the relevant experimental achievements and DF-based descriptions of adsorbed helium atoms on flat substrates and in carbon nanotubes, with special emphasis on the abilities of this technique to anticipate the excitation spectrum of these systems. In Section 4 we review the state-of-the-art concerning the energy systematics and structural as well as excitation properties of pure and doped ^{3}He drops within DF theory. This Section also addresses the case of mixed ^{3}He-^{4}He clusters and the inclusion of impurities into these drops. Section 5 summarizes this chapter.

2. Density functionals for liquid Helium

Phenomenological models to investigate surface properties of superfluid ^{4}He were developed during the seventies. At zero temperature and in the absence of currents, the superfluid order parameter is simply the square root of the particle number density ρ, and it becomes natural to think of a density functional to describe superfluid helium. [11,12] To illustrate this point, we shall briefly refer to the work of Ebner and Saam, [12] who start from a density functional of the form

$$E[\rho] = \int d^3r \left\{ \varepsilon[\rho(\mathbf{r})] + \int d^3r' W(\mathbf{r},\mathbf{r}') \left(\rho(\mathbf{r}) - \rho(\mathbf{r}')\right)^2 \right\} \tag{2.1}$$

where a local and a nonlocal interaction contribution have been separately written, $\varepsilon[\rho(\mathbf{r})]$ being the local energy density. The nonlocal kernel $W(\mathbf{r},\mathbf{r}')$ reflects the interaction between different regions of the nonuniform fluid. For small perturbations of the density with respect to its constant equilibrium value, this kernel can be related to the density-density response χ_q of the system. Explicit separation of the noninteracting terms gives

$$\begin{aligned} E[\rho] &= \int d^3r \left\{ \frac{\hbar^2}{2m}\tau[\rho(\mathbf{r})] + \varepsilon[\rho(\mathbf{r})] \right\} \\ &- \int\int d^3r d^3r' \left[\int \frac{d^3q}{(2\pi)^3} e^{i\mathbf{q}\cdot(\mathbf{r}-\mathbf{r}')} \left(\frac{1}{\chi_q(\hat{\rho})} - \frac{1}{\chi_q^0(\hat{\rho})} \right) \right] \left(\rho(\mathbf{r}) - \rho(\mathbf{r}')\right)^2 \end{aligned} \tag{2.2}$$

where

$$\tau = \frac{|\nabla\rho(\mathbf{r})|^2}{4\rho(\mathbf{r})} \tag{2.3}$$

is the kinetic energy density of the non-interacting inhomogeneous boson system, χ_q^0 is the free density-density response, and $\hat{\rho} = (\rho(\mathbf{r}) + \rho(\mathbf{r}'))/2$. The response function χ_q is related to the static structure factor S_q, and Ebner and Saam obtained a specific form for both S_q and ε starting from an effective atom-atom interaction. Indeed, one of the appealing features of this functional is that it makes room to an explicit connection with the atom-atom interaction. However, although it was applied with some success to the plane surface of ^{4}He it seems to be unpractical for other geometries. Moreover, it requires much computational effort, losing thus one of the interesting aspects of a density funcional.

2.1. *Zero-range functionals*

In the last term of functional (2.2) we can visualize an approach to the surface energy as an expansion in powers of $\rho(\mathbf{r}) - \rho(\mathbf{r}')$. Thus, a crude approximation consists in the replacement of the integrals upon $\mathbf{r}'$ and $\mathbf{q}$ by a phenomenological term proportional to the square of the density gradient, *i.e.*

$$E[\rho] = \int d^3r \left\{ \frac{\hbar^2}{2m}\tau[\rho(\mathbf{r})] + \varepsilon[\rho(\mathbf{r})] + d|\nabla\rho(\mathbf{r})|^2 \right\} \tag{2.4}$$

where d is a parameter to be determined, and the density of local energy ε has not been defined yet. From this viewpoint, we cannot overlook an analogy with other self-saturating systems, like e.g. atomic nuclei. In particular, computation of the HF energy of a configuration of nucleons subjected to a Skyrme interaction naturally generates the term with density gradient, in addition to the kinetic energy. [4]

This was the philosophy adopted by Stringari, [7] who pioneered the application of DF theory to liquid ^{3}He inspired in the analogies between this system and atomic nuclei. In fact, under this description, ^{3}He droplets display shell effects and magic numbers. [13] The idea is to write down a density functional, resigning any connection to an underlying microscopic atom-atom interaction. Instead, one assumes an effective interaction in charge of yielding a DF in a HF scheme. The interaction, or the functional, depends upon parameters to be determined so as to fit experimental values of given physical magnitudes.

To fix ideas, let us consider the simplest case, namely homogeneous ^{4}He, where only the term ε enters the density functional. This local energy density is parametrized as

$$\varepsilon = \frac{1}{2}b\rho^2 + \frac{1}{2}c\rho^{2+\gamma} \tag{2.5}$$

in terms of three parameters b, c and γ, which may be determined by fitting the values of saturation density ρ_0, energy per particle ϵ_0 and isothermal compressibility κ_0 at bulk saturation, being $1/\kappa\rho$ the bulk modulus. The values of the parameters are given in Table 1. The coefficient d entering Eq. (2.4) is fixed so as to reproduce the experimental surface tension of the liquid.

In Fig. 1 are displayed the calculated equation of state (EOS) and the compressibility of liquid ^{4}He as functions of pressure, as well as the experimental points, extracted from Ref. 14. It may be seen that just the fit of three quantities at saturation enables this functional to achieve a good agreement with experiment. It has been shown [15] that the DF results agree with those obtained from a diffusion Monte Carlo (DMC) calculation from the solidification down to the spinodal density. See also Chapter 2 of this volume. [16]

The same type of functional is immediately generalized to liquid ^{3}He as

$$E[\rho] = \int d^3r \left\{ \frac{\hbar^2}{2m^*}\tau + \varepsilon[\rho] + d|\nabla\rho|^2 \right\} \tag{2.6}$$

In the case of ^{4}He, the effective mass is identical to the bare one, $m^* = m_4$, and the kinetic energy density is the one of an homogeneous system of bosons given in Eq. (2.3). In the liquid ^{3}He case, the experimental analysis of the specific heat indicates that an effective mass has to be considered. The parametrization

$$\frac{1}{m^*} = \frac{1}{m_3}\left(1 - \frac{\rho}{\rho_c}\right)^2 \tag{2.7}$$

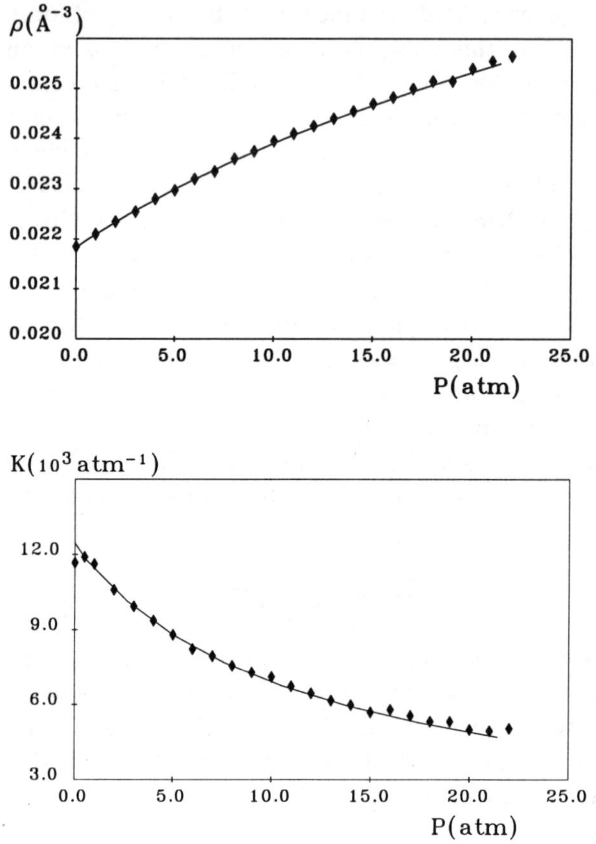

Fig. 1. Equation of state and compressibility of liquid ^{4}He, as calculated with the ST functional. Experimental points are taken from Ref. 14.

reproduces very well the pressure dependence of the effective mass. A Thomas Fermi approximation for the kinetic energy density is usually employed, which adds gradient terms to Eq. (2.6). The coefficient d is adjusted to fit the surface tension of the free surface. The parameters are in Table 1.

In Fig. 2 are displayed the EOS and the incompressibility of ^{3}He as functions of pressure, compared with the experimental values extracted from Ref. 17. Again, one can see that the empirical DF results are in excellent agreement with experiment.

Zero-range DF's have also been applied to homogeneous mixtures of ^{3}He -^{4}He [18] to study the properties of the solution as the relative concentration varies continuously, seeking also to describe the behavior of nonhomogeneous systems such as the interface between both liquids and the mixed films. In this case, one considers that the total energy of a liquid-helium mixture can be expressed as a density functional of their particle densities ρ_3, ρ_4, and of the kinetic energy density τ_3; the total density is $\rho = \rho_3 + \rho_4$ with $x = \rho_3/\rho$ being the concentration of ^{3}He atoms.

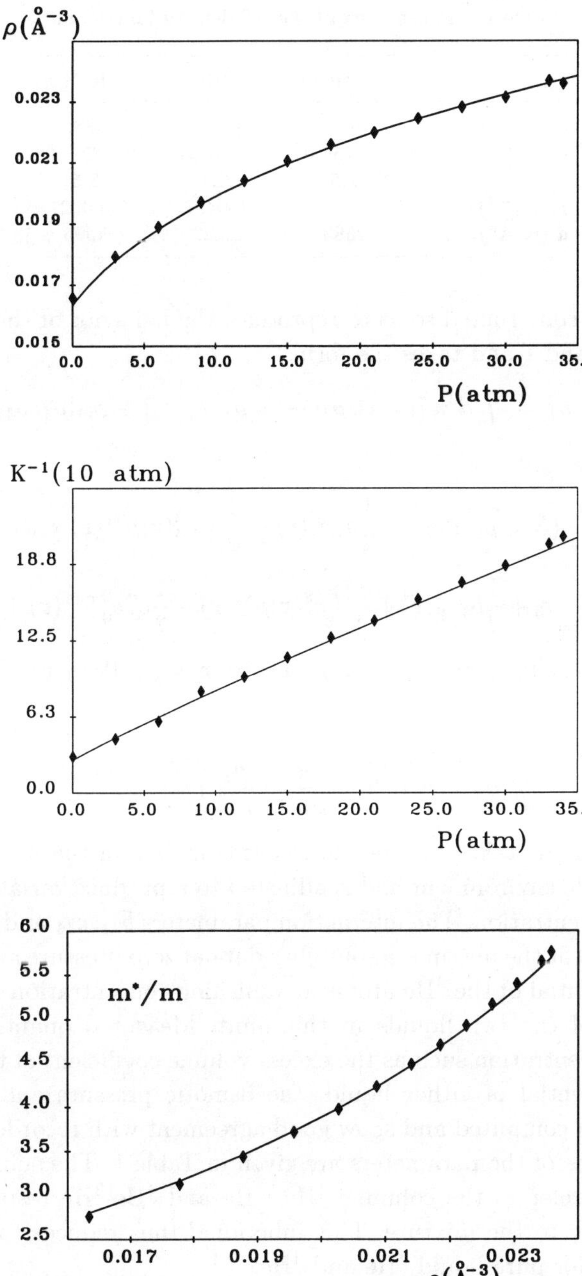

Fig. 2. Equation of state, incompressibility and effective mass of liquid ^{3}He, as calculated with the ST functional. Experimental points are taken from Ref. 17.

Table 1. Parameters of the ST density functional.

	^{4}He	^{3}He	^{4}He-^{3}He
b (K Å^3)	-444.5	-341.5	-388.5
c (K Å$^{3(1+\gamma)}$)	5.2275·10^6	7.025·10^5	2.2825·10^6
γ	2.8	2.1	2.5
ρ_c (Å^{-3})		0.0406	0.062
d (K Å^5)	2383	2222	4505

The functional is constructed so as to reproduce the behavior of the pure liquids in the limits $x = 0$ and 1 and takes the form

$$E[\rho_3, \tau_3, \rho_4] = \int d^3r \, \{\varepsilon_4[\rho_3, \rho_4] + \varepsilon_3[\rho_3, \tau_3, \rho_4] + \varepsilon_{34}[\rho_3, \rho_4]\} \tag{2.8}$$

where

$$\varepsilon_4[\rho_3, \rho_4] = \frac{\hbar^2}{2m_4}\left(\nabla\sqrt{\rho_4(\mathbf{r})}\right)^2 + \frac{1}{2}b_4\rho_4^2(\mathbf{r}) + \frac{1}{2}c_4\rho_4^2(\mathbf{r})\rho^{\gamma_4}(\mathbf{r}) + d_4\left[\nabla\rho_4(\mathbf{r})\right]^2 \tag{2.9}$$

$$\varepsilon_3[\rho_3, \tau_3, \rho_4] = \frac{\hbar^2}{2m_3^*}\tau_3 + \frac{1}{2}b_3\rho_3^2(\mathbf{r}) + \frac{1}{2}c_3'\rho_3^2(\mathbf{r})\rho^{\gamma_3}(\mathbf{r}) + \frac{1}{2}c_3''\rho_3^{2+\gamma_3}(\mathbf{r}) + d_3\left(\nabla\rho_3\right)^2 \tag{2.10}$$

$$\varepsilon_{34}[\rho_3, \rho_4] = b_{34}\rho_3(\mathbf{r})\rho_4(\mathbf{r}) + c_{34}\rho_3(\mathbf{r})\rho_4(\mathbf{r})\rho^{\gamma_{34}}(\mathbf{r}] + d_{34}\left[\nabla\rho_3(\mathbf{r})\right]\left[\nabla\rho_4(\mathbf{r})\right] \tag{2.11}$$

being [19]

$$m_3^* = m_3\left(1 - \frac{\rho_3}{\rho_{3c}} - \frac{\rho_4}{\rho_{4c}}\right)^{-2} \tag{2.12}$$

The parameter ρ_{4c} takes care of the renormalization of the mass of ^{3}He atoms induced by the ^{4}He environment and is adjusted to reproduce variations of m_3^* with pressure and concentration. The interaction parameters b_{34}, c_{34} and the strength c_3'' are chosen so as to fit the maximum solubility data at zero pressure and temperature, the chemical potential of the ^{3}He atoms at vanishing concentration and the relative specific volume of the two liquids in this limit. Measured quantities at varying pressure and concentration such as the excess volume coefficient of the mixture, [113] the chemical potential of either liquid, the osmotic pressure and the maximum solubility are then computed and show good agreement with recorded experimental data. [19] The values of the parameters are given in Table 1. The subindices 4, 3 and 43 in b, c and γ refer to the columns ^{4}He, ^{3}He and ^{4}He-^{3}He respectively, except for ρ_{4c} which refer to the mixture. This functional thus reduces to the previously commented ones for pure liquid ^{4}He and ^{3}He.

Formally, the DF (2.5) can be constructed by a Hartree calculation of the ground state energy of a boson system with a contact Skyrme-like interaction. The term proportional to b corresponds to an effective attraction, the one in c to a density-dependent repulsion, which simulates a correlation repulsion. It is worth keeping in mind that while within a microscopic description, dynamical correlations are explicitly taken into account when constructing the many-body wave function, within the DF approach correlations are implicitly considered in the fit of the parameters to experimental quantities. As a consequence, their effects show up in an indirect fashion, or implicitly in quantities such as the energy per particle or the density profiles. In this philosophy, it is meaningless to strive for a microscopic interpretation of the correlations. Notwithstanding the lack of a full microscopic foundation, a DF based on the HF energy of ^{3}He atoms subjected to an effective two-body interaction permits to incorporate terms, according to physical insight to cover specific needs, as well as to present a unified description of both helium isotopes and their mixture. This type of functional has been employed for a systematic investigation of ^{3}He and ^{4}He droplets [20] and liquid free surface properties. [21,22] Apart from minor modifications in the values of the parameters, mostly oriented to make room to thermal effects, the zero-range DF including gradient terms was applied to a variety of subjects such as the gas-liquid transition, [23] liquid-gas equilibrium and phase transitions of spin-polarized ^{3}He, [24,25] temperature dependence of surface tension, [26,27] cavitation phenomena (see Chapter 7 of this volume [28]), collective excitations [29] and multipole response in droplets. [30]

2.2. *Finite-range functionals*

Zero-range DF proved inconvenient to reproduce dynamical characteristics as soon as spatial variations come into play. For instance, as was shown in Refs. 31–33 it is not possible to describe simultaneously surface properties and zero sound at large transferred momenta (typically beyond half the Fermi momentum), and finite range effects must be incorporated into the DF for this purpose. It is quite obvious that the previous zero-range DF misses two important characteristics of the atom-atom interaction, namely its asymptotic r^{-6} behavior and the strong repulsion at short distances. In Ref. 34 a way to include these effects into the ST functional for ^{4}He was suggested, and later on it was incorporated to a DF for ^{3}He. [31] The idea is to substitute the b and c terms by suitable finite range terms. The b term has been replaced by a two-body interaction $V(r)$ consisting of a Lennard-Jones potential, but screened at short distances

$$\begin{aligned} V(r) &= 4\varepsilon\left[\left(\frac{\sigma}{r}\right)^{12} - \left(\frac{\sigma}{r}\right)^{6}\right] \qquad && r \geq h \\ &= V_{\rm in}(r) && \text{otherwise} \end{aligned} \tag{2.13}$$

As this is an effective interaction, there is no fundamental reason for keeping the free-atom values of ϵ=10.22 K and σ=2.556 Å. Indeed, the polarization potentials differ appreciably from the bare LJ potential, and there are some DF's which take different values for these quantities. There exist in the literature several choices regarding the effective range h and the form of the softened potential $V_{\rm in}(r)$ at short distances. Of course, the simplest possibility is $V_{\rm in} = 0$. Another option is

$$V_{\rm in}(r) = V_0 \left(\frac{r}{h}\right)^{n} \tag{2.14}$$

which avoids the discontinuity at point $r = h$, and goes to zero at origin. A third choice, much in the spirit of the polarization potential is

$$V_{\rm in}(r) = V_0 \left[1 - \left(\frac{r}{h}\right)^n\right] \tag{2.15}$$

In any case, to guarantee that the finite range DF does not disrupt the description of the bulk properties of the zero-range DF, the volume integral of $V(r)$ is fixed to the previously determined value of b (to its half, in fact).

The c term is generalized by replacing the density with a coarse-grained density $\tilde{\rho}$, as introduced in the study of classical fluids, [35] defined as

$$\tilde{\rho}(\mathbf{r}) = \int d^3r' w(\mathbf{r} - \mathbf{r}')\rho(\mathbf{r}') \tag{2.16}$$

where the weight function $w(r)$ has been taken as

$$\begin{aligned} w(r) &= \frac{3}{4\pi h'^3} \qquad && r < h' \\ &= 0 && \text{otherwise} \end{aligned} \tag{2.17}$$

The use of a coarse-grained density has an interesting consequence. In a Skyrme interaction the density dependence ρ^γ of the repulsive term causes the density profile to stay smooth, as any deviation from an average density is energetically unfavourable. In contrast, such local deviations are so not punished by a coarse-grained density.

The values of h and h' are usually around σ. These parameters control the surface tension and the zero sound dispersion relation of liquid ^{3}He. For instance, in Ref. 31 the softened potential form (2.14) was employed with the values $n = 4$, $h = 2.3563$ Å, and $h' = 1.8h$, togheter with the ST parameters. The zero sound dispersion relation was in a good agreement with experiment up to a transferred momentum of around twice the Fermi momentum, *i.e.*, up to the maxon region.

It is worth mentioning that a closely related, semimicroscopic energy functional for KS calculations in the homogeneous liquid was presented in the literature. [32] The energy density is

$$\begin{aligned} \varepsilon[\rho] = {} & \frac{\hbar^2}{2m}\tau(\mathbf{r}) + \frac{1}{2}\int d^3r'\rho(\mathbf{r})\Phi_0(\mathbf{r},\mathbf{r}')\rho(\mathbf{r}') \\ & + \frac{1}{2}\int d^3r'\rho(\mathbf{r})\Phi_2(\mathbf{r},\mathbf{r}')[\tau(\mathbf{r}') - \frac{1}{4}\Delta\rho(\mathbf{r}')] - \frac{1}{4}\int d^3r'\mathbf{j}(\mathbf{r})\Phi_2(\mathbf{r},\mathbf{r}')\mathbf{j}(\mathbf{r}') \end{aligned} \tag{2.18}$$

where $\mathbf{j}$ is the current density. This functional has been justificated starting from a gradient expansion of a general vertex function, coming out from a T-matrix calculation. The zeroth and second integrals of the vertex function originate the quasi-potentials Φ_0 and Φ_2. The density-dependence of Φ_i has been parametrized in close analogy to the ST functional, and the spatial shape has been constructed to resemble as close as possible the polarization potentials. [8] This DF has been applied to investigate the zero sound dispersion relation of liquid ^{3}He and the shell structure of ^{3}He droplets. [33]

2.3. *Spin-density dependent functionals*

In the process of fixing the functional parameters only some properties of the spin-saturated liquid ^{3}He at zero temperature have been considered, namely the saturation density and energy per particle and the dependence of the effective mass with pressure. In terms of the Landau theory of Fermi liquids, this procedure guarantees that the dimensionless spin-symmetric Landau parameter F_0^s is reproduced at saturation density, and the parameter F_1^s is reproduced in the full range of densitites from saturation up to solidification. However, a careful analysis shows that this type of functional cannot simultaneously give the experimental surface tension and the Landau parameters as functions of the density. As an attempt to describe magnetic properties of liquid ^{3}He, in Refs. [36,37] a suggestion to include spin-density terms in the functional was put forward. An explicit dependence of the functional on the spin-current density was proposed in Ref. 38; this form was later employed to investigate the structure and pairing properties of helium drops. [39] To guarantee galilean invariance, a slightly modified DF was constructed in Ref. 39 starting from a density-dependent effective interaction of the form

$$V(1,2) = \mathcal{V}_0\delta(\mathbf{r}_{12}) + \frac{1}{2}\left[\mathbf{k}'^2\mathcal{V}_1\delta(\mathbf{r}_{12}) + \delta(\mathbf{r}_{12})\mathcal{V}_1\mathbf{k}^2\right] + \mathbf{k}'\mathcal{V}_2\delta(\mathbf{r}_{12})\mathbf{k} \tag{2.19}$$

where $\delta(\mathbf{r}_{12})$ is the delta function $\delta(\mathbf{r}_1 - \mathbf{r}_2)$, $\mathbf{k}$ is the operator $(\nabla_1 - \nabla_2)/2i$ acting on the right, and $\mathbf{k}'$ is the operator $-(\nabla_1 - \nabla_2)/2i$ acting on the left. The functions $\mathcal{V}_i$ depend on the particle $\rho(\mathbf{r})$, spin $\mathbf{S}(\mathbf{r})$, spin-kinetic $\mathbf{T}(\mathbf{r})$, and spin-current tensor $\mathbf{J}(\mathbf{r})$ densities. The explicit form is

$$\mathcal{V}_0 = t_0 + t_0'\tilde{\rho}^\gamma + u_0\tilde{\rho}^\beta\,\mathbf{S}\cdot\mathbf{S} + v_0\,(\mathbf{J}\cdot\mathbf{J} - \mathbf{S}\cdot\mathbf{T}) \tag{2.20}$$

$$\mathcal{V}_i = t_i + t_i'\tilde{\rho} + u_i\mathbf{S}\cdot\mathbf{S} + v_i\,(\mathbf{J}\cdot\mathbf{J} - \mathbf{S}\cdot\mathbf{T}) \quad ,i = 1,2 \tag{2.21}$$

The various densities appearing in this expression are related to the single-particle density matrix $\rho(\mathbf{r},\sigma;\mathbf{r}',\sigma')$, as follows

$$\rho(\mathbf{r}) = \sum_\sigma \rho(\mathbf{r},\sigma;\mathbf{r}',\sigma)|_{\mathbf{r}=\mathbf{r}'} \tag{2.22}$$

$$\mathbf{S}(\mathbf{r}) = \sum_{\sigma,\sigma'} \rho(\mathbf{r},\sigma;\mathbf{r}',\sigma')\langle\sigma'|\vec{\sigma}|\sigma\rangle|_{\mathbf{r}=\mathbf{r}'} \tag{2.23}$$

$$\mathbf{T}(\mathbf{r}) = \sum_{\sigma,\sigma'} \nabla\nabla'\rho(\mathbf{r},\sigma;\mathbf{r}',\sigma')\langle\sigma'|\vec{\sigma}|\sigma\rangle|_{\mathbf{r}=\mathbf{r}'} \tag{2.24}$$

$$\mathbf{J}(\mathbf{r}) = \frac{1}{2i}\sum_{\sigma,\sigma'}(\nabla - \nabla')\rho(\mathbf{r},\sigma;\mathbf{r}',\sigma')\langle\sigma'|\vec{\sigma}|\sigma\rangle|_{\mathbf{r}=\mathbf{r}'} \tag{2.25}$$

where $\vec{\sigma}$ is the Pauli matrices vector, with $\sigma, \sigma' = \pm 1/2$. Furthermore, a coarse-grained density $\tilde{\rho}$ as defined in Eq. (2.16) has been introduced in the funtions $\mathcal{V}_i$. In the case of a homogeneous system $\tilde{\rho} = \rho$, and the densities (2.23)-(2.25) vanish if the system is unpolarized. However, all these densities are important to determine the particle-hole interaction, out of which the Landau parameters are obtained.

Table 2. Parameters of the BHN density functional. $\rho_0 = 0.0163489$Å is the experimental saturation density of liquid ^{3}He.

t_0 (K Å^3)	$t'_0\rho_0^\gamma$ (K Å^3)	t_1 (K Å^5)	t_2 (K Å^5)
-1369.351	496.3449	-772.236	-772.236
$t'_1\rho_0$ (K Å^5)	$t'_2\rho_0$ (K Å^5)	γ	$u_0\rho_0^2$ (K Å$^{3+3\beta}$)
96.0680	160.004	2.1251	-0.2690275
$(u_1+3u_2)\rho_0^2$ (K Å$^{5+3\beta}$)	β	$v_0\rho_0^2$ (K Å^3)	$(v_1+3v_2)\rho_0^2$ (K Å^5)
2011.3517	-2.085	-399.5911	1628.351
ϵ (K)	σ (Å)	h (Å)	h' (Å)
6	2.91	2.91	4.3

For unpolarized liquid ^{3}He, the total energy is given by the expectation value of the Hamiltonian assuming the wave function to be a Slater determinant of plane waves saturated in the spin third component. The energy density is written as

$$\epsilon[\rho,\tau] = \frac{\hbar^2}{2m^*}\tau + \frac{1}{4}\rho^2[t_0 + t'_0\rho^\gamma] + \frac{3}{32}[t_1 - t_2]|\nabla\rho|^2 + \frac{1}{32}[5t'_1 - 3t'_2]\rho|\nabla\rho|^2 \quad (2.26)$$

where the effective mass is

$$\frac{\hbar^2}{2m^*} = \frac{\hbar^2}{2m} + \frac{1}{8}[t_1 + 3t_2]\rho + \frac{1}{8}[t'_1 + 3t'_2]\rho^2 \quad (2.27)$$

Note that apart from the last term entering Eq. (2.26) this energy density looks similar to the ST one.

After inclusion of finite-range effects, choosing the form (2.15) for the screened Lennard-Jones potential with $n = 8$, the above DF was redefined as the BHN functional [40]. Its parameters are given in Table 2. Note that the chosen parametrization implies that no gradient terms related to the interaction appear in Eq. (2.26), and only the finite-range terms determine the surface tension. With this functional a simultaneous description of the available spin symmetric and spin antisymmetric Landau parameters [2] as well as of the experimental surface tension, was achieved. It is also capable of reproducing the zero sound spectrum of liquid helium at various pressures and temperatures within an extended RPA description.[40] The dynamical structure factor in the spin-antisymmetric channel could be also visualized for various temperatures attained in experimental situations.

Available data for the induced magnetic field and susceptibility at low magnetization permitted to solve the remaining ambiguity in the determination of the BHN parameters u_i and v_i, $(i = 1, 2)$ (cf. Eqs. (2.20)-(2.21)), which appears in the total energy of the unpolarized liquid only through the combinations $u_1 + 3u_2$ and

$v_1 + 3v_2$i. [42] One is then allowed to extract the Landau parameters of polarized helium as functions of both particle and spin density, and examine the dependence of bulk compressibility, specific heat —*i.e.*, effective mass of the quasiparticles— and magnetic susceptibility on the imposed magnetization. In Ref. 42, it has been shown that this form of the finite range DF does not support magnetizations above 30%; in fact, extensions of the BHN parametrizations to describe the strongly or the fully polarized liquid pay a high price in extra dependences of the force strengths upon density and spin density (GHN functionals). In the earlier paper on this topic, Stringari [36] showed that a DF constructed to describe bulk properties of liquid helium could account for various low magnetization features, but higher powers of the polarization were necessary to consider several aspects of the fully polarized liquid. With this extension it was possible to take a close look at the liquid-gas phase transition in these systems. [24] Although the DF in Ref. 42 represents an improvement over the original approach in Ref. 36, the validity of DF descriptions of spin-polarized helium remains an open topic at the frontier of an active field of research. Recent measurements [41] of the specific heat of liquid ^{3}He with up to 70% polarization seem to shed light on an old standing controversy (see Ref. 42 and references therein) regarding the validity of the nearly localized picture *vs* the paramagnon model and provide accurate data to be considered in the determination of the DF parameters. In addition, DMC calculations of the fully polarized system [43] open a new perspective, since the predicted EOS is an important input for a reliable formulation of the DF. The new GHN parametrization was later applied to prediction of magnetic zero sound [44] and transverse Silin waves [45] in weakly polarized liquid helium. Regardless the ambiguities above discussed, the appearance of two gapless branches of longitudinal zero sound and one gapped dispersion curve for spinflip excitations, in addition to the paramagnon mode, is a robust prediction of finite range DF theory, as well as the disappearance of the damped roton minimum with increasing magnetization. [44]

The systematics of droplets, as well as the mass formula giving rise to the proper figure of binding energy per particle and surface tension, has been reproduced by the BHN parametrization, and an estimate of the superfluid transition temperature for BCS-like pairing has been furnished. [39] A similar DF, which gives identical bulk properties and differs only in the parametrization of effective mass and gradient terms, has been proposed in Ref. 46 and successfully applied to predicting the prewetting transition of liquid ^{3}He adsorbed on weak alkali substrates. We shall discuss this parametrization in more detail in Section 3.

2.4. *Finite-range functional for mixtures*

The density functional was modified in the mid 90's to keep track of increasing interest in pure and doped helium clusters, bearing in mind that zero range DF's cannot account for finite size effects induced by a strong density compression. In 1995 a new density functional for bulk and nonhomogeneous ^{4}He was reported, [47] which overcomes previous restrictions on the abilites of zero and finite range DF's to describe the static and dynamical response of the homogeneous liquid. This so-called Orsay-Trento (OT) functional improves by construction the description of relevant properties of the bulk liquid ^{4}He, such as the momentum dependence of the static response function and the phonon-roton dispersion. Inspired in this OT parametrization, a finite range alternative to the representations (2.9) to (2.11) has been proposed, [48] which reads

$$\begin{aligned}\varepsilon_4[\rho_3,\rho_4] &= \frac{\hbar^2}{2m_4}\left(\nabla\sqrt{\rho_4(\mathbf{r})}\right)^2 + \frac{1}{2}\int \mathrm{d}\mathbf{r}'\rho_4(\mathbf{r})V_4(|\mathbf{r}-\mathbf{r}'|)\rho_4(\mathbf{r}') \\ &+ \frac{1}{2}c_4'\rho_4(\mathbf{r})\left[\tilde{\rho}_3(\mathbf{r})+\tilde{\rho}_4(\mathbf{r})\right]^2 + \frac{1}{3}c_4''\rho_4(\mathbf{r})\left[\tilde{\rho}_3(\mathbf{r})+\tilde{\rho}_4(\mathbf{r})\right]^3 \\ &- \frac{\hbar^2}{4m_4}\alpha_s\int d^3r' F(|\mathbf{r}-\mathbf{r}'|)\left[1-\frac{\bar{\rho}_4(\mathbf{r})}{\rho_{0s}}\right]\nabla\rho_4(\mathbf{r})\cdot\nabla'\rho_4(\mathbf{r}')\left[1-\frac{\bar{\rho}_4(\mathbf{r}')}{\rho_{0s}}\right]\end{aligned} \tag{2.28}$$

$$\begin{aligned}\varepsilon_3[\rho_3,\tau_3,\rho_4] &= \frac{\hbar^2}{2m_3^*}\tau_3 + \frac{1}{2}\int d^3r'\rho_3(\mathbf{r})V_3(|\mathbf{r}-\mathbf{r}'|)\rho_3(\mathbf{r}') \\ &+ \frac{1}{2}c_3'\rho_3^2(\mathbf{r})\left[\tilde{\rho}_3(\mathbf{r})+\tilde{\rho}_4(\mathbf{r})\right]^{\gamma_3} + \frac{1}{2}c_3''\rho_3^2(\mathbf{r})\tilde{\rho}_3^{\gamma_3}(\mathbf{r})\end{aligned} \tag{2.29}$$

$$\varepsilon_{34}[\rho_3,\rho_4] = \frac{1}{2}\int d^3r'\rho_3(\mathbf{r})V_{34}(|\mathbf{r}-\mathbf{r}'|)\rho_4(\mathbf{r}') + c_{34}\rho_3(\mathbf{r})\rho_4(\mathbf{r})\left[\tilde{\rho}_3(\mathbf{r})+\tilde{\rho}_4(\mathbf{r})\right]^{\gamma_{34}} \tag{2.30}$$

In these expressions, $\tilde{\rho}_i(\mathbf{r})$ for $i = 3, 4$ is an averaged density given by (2.16) with coarse-graining radii h_i in the weight function (2.17).

In addition, $V_i(|\mathbf{r} - \mathbf{r}'|)$ (i =3, 4 or 34) is a finite range interaction of the form (2.13), with $V_{\rm in}(r) = 0$, and parameters σ_i, h_i. The function $F(r)$ is a Gaussian kernel with dispersion l equal to unity

$$F(r) = \frac{1}{\pi^{3/2}l^3}\mathrm{e}^{-r^2/l^2} \tag{2.31}$$

which is also used to define the other averaged density entering Eq. (2.28)

$$\bar{\rho}_4(\mathbf{r}) = \int d^3r'\rho_4(\mathbf{r}')F(|\mathbf{r}-\mathbf{r}'|) \tag{2.32}$$

ε_4 reduces to the Orsay-Trento (OT) density functional [47] setting ρ_3 to zero.

The effective mass of an ^{3}He atom in the mixture was chosen as

$$m_3^* = m_3\left(1-\frac{\tilde{\rho}_3}{\rho_{3c}}-\frac{\tilde{\rho}_4}{\rho_{4c}}\right)^{-2} \tag{2.33}$$

and the parameters corresponding to pure ^{3}He and to the mixture were adjusted; in particular, the LJ range σ_3 was changed from its standard value to exactly reproduce the experimental surface tension of liquid ^{3}He. The values of the LJ vanishing cores h_i were fixed as in Ref. 49 and the remaining set was chosen so as to achieve the best fit to the maximum solubility of ^{3}He in ^{4}He, the excess volume coefficient and the osmotic pressures, measured for varying pressures between 0 and 20 atm, and the surface tension of the 3-4 interface at zero temperature and at saturation pressure.

Table 3. Parameters of the density functional for mixtures

b_4 (KÅ^3)	c'_4 (KÅ^6)	c''_4 (KÅ^9)	σ_4 (Å)	α_s (K^{-1}Å^3)	ρ_{0s} (Å^{-3})	h_4 (Å)
-718.99	-2.41186 $\cdot 10^4$	1.8585 $\cdot 10^6$	2.556	54.31	0.04	2.190323
b_3 (KÅ^3)	c'_3 (KÅ$^{3+\gamma_3}$)	c''_3 (K Å$^{3+\gamma_3}$)	σ_3 (Å)	γ_3	ρ_{3c} (Å^{-3})	h_3 (Å)
-684.676	1.58879 $\cdot 10^6$	-3.5 $\cdot 10^4$	2.46	2.1251	0.0406	2.11311
b_{34} (KÅ^3)	c_{34} (KÅ$^{3+\gamma_{34}}$)	ϵ_{LJ} (K)	σ_{34} (Å)	γ_{34}	ρ_{4c} (Å^{-3})	h_{34} (Å)
-662.8	4.5 $\cdot 10^6$	10.22	2.5455	2.6565	0.062	2.176374

The final parameters are presented in Table 3, with b_i denoting the volume integral of the LJ potential V_i. The agreement with the experimentally determined solubility and surface tension shows sizable improvement with respect to preceding fits. [19] The application of the parametrization (2.28) to (2.32) to mixed helium clusters will be examined in detail in Sec. 4.

3. Adsorbed systems

3.1. *General theoretical aspects: the band spectrum*

Helium films adsorbed on substrates of various shapes exhibit an interesting behavior due to the restriction on quantum and interaction effects imposed by the confining field. Illustrations of such restricted geometries are films on flat surfaces, cylindrical shells of adsorbed atoms on the walls of carbon nanotubes, and axial phases in the interstitial sites of nanotube bundles or along the central axis of a single cylinder. A common feature of these systems is partial suppresion of translational invariance, so that the 3D wave function for the Landau quasiparticles adopts the form

$$\phi_{\mathbf{k}\nu}(\mathbf{r}) = \frac{1}{\sqrt{L^D}}\, e^{i\mathbf{k}\cdot\mathbf{r}_D}\, f_{k\nu}(\mathbf{r}_\perp) \tag{3.1}$$

with D the number of nonrestricted dimensions, each with characteristic length L, $\mathbf{r}_\perp = \mathbf{r} - \mathbf{r}_D$ the transverse vector along which the substrate potential unfolds, and $f_{k\nu}$ the transverse wave function displaying the features of the confinement. The single-particle spectrum is, in most cases of interest, of the form

$$\varepsilon_{k\nu} = \frac{\hbar^2 k^2}{2m^*} + \varepsilon_\nu \tag{3.2}$$

with m* an effective mass and with band excitation quanta $\varepsilon_{\nu+1} - \varepsilon_\nu$ much larger than the D-translational one $\Theta_D = h^2/(2mL^2)$.

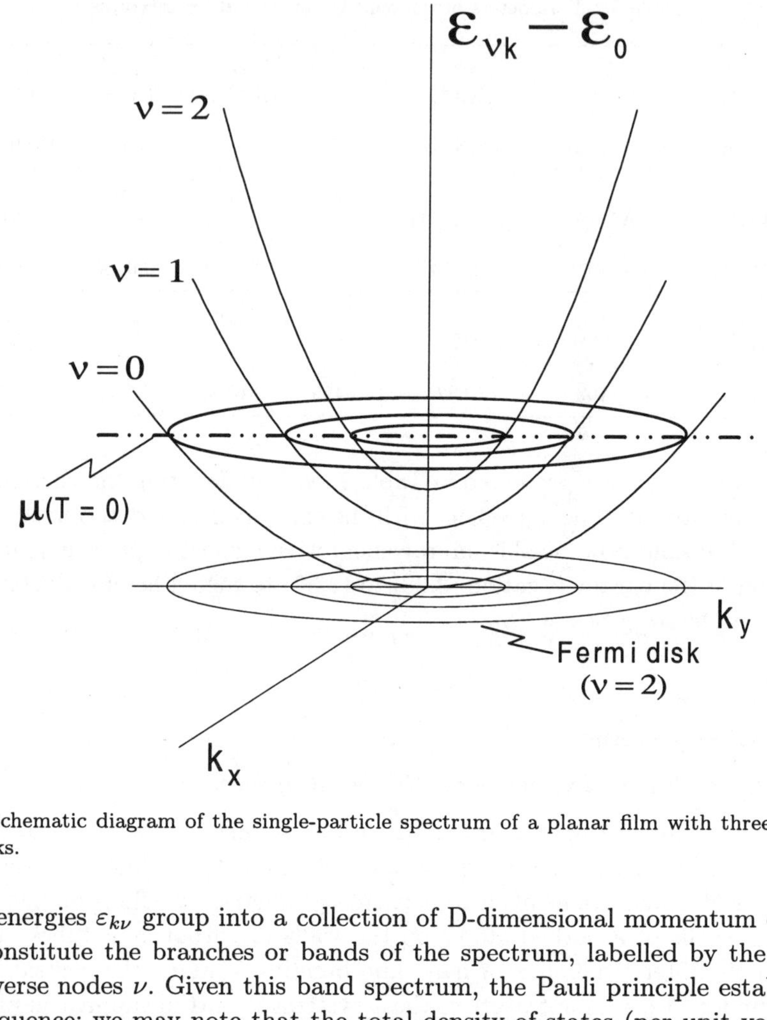

Fig. 3. Schematic diagram of the single-particle spectrum of a planar film with three occupied Fermi disks.

The energies $\varepsilon_{k\nu}$ group into a collection of D-dimensional momentum continua which constitute the branches or bands of the spectrum, labelled by the number of transverse nodes ν. Given this band spectrum, the Pauli principle establishes a filling sequence; we may note that the total density of states (per unit volume) of spin 1/2 qp's may be cast as a summation of D-dimensional contributions over the bands,

$$g(\varepsilon) = \frac{2}{\Gamma(D/2)} \left(\frac{m}{2\pi\hbar^2}\right)^{D/2} \sum_{\nu} (\varepsilon - \varepsilon_\nu)^{D/2-1} \tag{3.3}$$

where $\Gamma(x)$ is the Gamma function. The number of particles per unit D-volume of the Landau qp's, namely

$$\rho_D = \frac{N}{L^D} = \int_0^\infty d\varepsilon\, g(\varepsilon)\, n(\varepsilon) \tag{3.4}$$

with $n(\varepsilon) = \left[e^{(\varepsilon - \mu)/T} + 1\right]^{-1}$ the Fermi occupation number, defines the chemical potential μ. At zero temperature, the constraint which establishes the total Fermi energy takes the form

$$\rho_D = \frac{4}{D\Gamma(D/2)} \left(\frac{m}{2\pi\hbar^2}\right)^{D/2} \sum_\nu (\varepsilon_F - \varepsilon_\nu)^{D/2} \, \Theta\,(\varepsilon_F - \varepsilon_\nu) \tag{3.5}$$

with the usual step function $\Theta(x)$. The fluid of qp's described by Eqs. (3.3) to (3.5) is a quasi-D-dimensional (QD) system whose spectrum consists of a sequence of ν_M D-dimensional Fermi spheres, *i.e.*, Fermi disks (D=2) or Fermi segments (D=1), the number ν_M being determined by the coverage ρ_D according to $\varepsilon_{\nu_M - 1} < \varepsilon_F \le \varepsilon_{\nu_M}$. As the coverage increases, one may observe a dimensionality crossover between the pure D-regime ($\varepsilon_0 < \varepsilon_F \le \varepsilon_1$) to the (D+1) system, once a substantial number of bands has been incorporated to the spectrum to span the extra dimension. The $\nu - th$ Fermi D-sphere lies on a bandhead ε_ν and extends up to a Fermi momentum $k_{F\nu} = [2m^* \, (\varepsilon_F - \varepsilon_\nu)/\hbar^2]^{1/2}$. The nonhomogeneous density profile reads

$$\rho(\mathbf{r}_\perp) = \sum_\nu \rho_{D\nu} \, \rho_\nu(\mathbf{r}_\perp) \tag{3.6}$$

where

$$\rho_{D\nu} = \frac{4}{D\Gamma(D/2)\,(4\pi)^{D/2}} \, (k_{F\nu})^D \tag{3.7}$$

is the D-density of the ν-th D-sphere. Given a particular value of the coverage ρ_D

$$\rho_D = \int d\mathbf{r}_\perp \rho(\mathbf{r}_\perp) \tag{3.8}$$

the qp's occupy the $N/2$ states of lowest energy, filling each band ν to satisfy the relation (3.5). A typical situation is illustrated in Fig. 3.

It is customary to disregard thermal effects and consider the zero temperature spectrum for the evaluation of the chemical potential. According to the calculations in Refs. 46, 50 the chemical potential exhibits an oscillatory behavior as a function of coverage ρ_D due to sequential completion of the Fermi D-spheres. The presence of regions of coverage where $\partial\mu/\partial\rho_D$ is negative indicates that the fluid is unstable and cannot support continuous transverse growth.

3.2. *The response of a free quasiparticle gas in the Fermi D-spheres model*

Within the Fermi D-spheres model the exact, retarded ph propagator of the qp gas, whose general form is given, for example, in Ref. 51, reads

$$G_0(\mathbf{r}, \mathbf{r}', \omega) = \int\!\!\int \frac{d^Dk}{(2\pi)^D} \frac{d^Dk'}{(2\pi)^D} \, e^{i\,(\mathbf{k} - \mathbf{k}')\cdot(\mathbf{r_D} - \mathbf{r'_D})}$$

$$\sum_{\nu\nu'} f_\nu(\mathbf{r}_\perp)\, f^*_\nu(\mathbf{r}'_\perp)\, f^*_{\nu'}(\mathbf{r}_\perp)\, f_{\nu'}(\mathbf{r}'_\perp) \, \frac{n(\varepsilon_{k\nu}) - n(\varepsilon_{k'\nu'})}{\hbar\omega - (\varepsilon_{k'\nu'} - \varepsilon_{k\nu}) + i\,\eta} \tag{3.9}$$

The exact free response to a density fluctuation induced by an external field $V_{ext}(\mathbf{r}) = e^{i\mathbf{q}\cdot\mathbf{r}}$, per spin degree of freedom and per unit translationally invariant volume L^D, reads

$$\chi_0(\mathbf{q},\omega) = \int\int d\mathbf{r}\, d\mathbf{r}'\, e^{-i\mathbf{q}\cdot(\mathbf{r}-\mathbf{r}')}\, G_0(\mathbf{r},\mathbf{r}',\omega) \tag{3.10}$$

It is convenient to split the momentum of the perturbation into parallel and transverse components, $\mathbf{q} = \mathbf{q}_D + \mathbf{q}_\perp$. The Fourier transforms of overlapping wave functions,

$$F_{\nu\nu'}(q_\perp) = \int d\mathbf{r}_\perp\, e^{-i\mathbf{q}_\perp\cdot\mathbf{r}_\perp} f_\nu(\mathbf{r}_\perp)\, f_{\nu'}(\mathbf{r}_\perp) \tag{3.11}$$

provide the transition strengths of the transverse excitations, and integration of variables $\mathbf{r}_D, \mathbf{r}'_D$ and $\mathbf{k}'$ gives the free response

$$\chi_0(\mathbf{q},\omega) = \sum_{\nu,\nu'} |F_{\nu,\nu'}(q_\perp)|^2\, \chi_{\nu\nu'}(\omega, q_D) \tag{3.12}$$

where

$$\chi_{\nu\nu'}(q_D,\omega) = \int \frac{d^Dk}{(2\pi)^D} \frac{n_\nu(k) - n_{\nu'}(k')}{\hbar\omega - \dfrac{\hbar^2}{2}\left[\dfrac{(\mathbf{k}+\mathbf{q_D})^2}{m^*_{\nu'}} - \dfrac{k^2}{m^*_\nu}\right] - (\varepsilon_{\nu'} - \varepsilon_\nu) + i\eta} \tag{3.13}$$

being $k' = |\mathbf{k} + \mathbf{q}_D|$ and $n_\nu(k)$ the occupation for qp's with effective chemical potential $\mu_\nu = \mu - \varepsilon_\nu$. In Eq. (3.12) we recognize a) intraband or longitudinal ph transitions $\nu = \nu'$, determined by $\mathbf{q_D}$ which contributes the D-response of the $\nu - th$ band, [52] and b) interband or transverse ph transitions $\nu \neq \nu'$, which occur to an upper branch shifted by $\mathbf{q_D}$.

When the transferred momentum is $\mathbf{q} = \mathbf{q_D}$, one has $F_{\nu\nu'} = \delta_{\nu\nu'}$ and the momentum integration in (3.13) leads to the D-dimensional Linhard function of a single Fermi D-sphere with density $\rho_{D\nu}$ and effective mass m^*_ν. The corresponding expressions for D = 1 and 2 can be found i.e., in Refs. 53 and 52, respectively. The longitudinal response is then

$$\chi_0^L(q_D,\omega) = \sum_\nu \chi_0^D(q_D,\omega) \tag{3.14}$$

Although different bands respond independently to the external field, their contributions are related by global properties of the entire system such as coverage and temperature, which determine the D-densities $\rho_{D\nu}$ and the chemical potential. At zero temperature only the occupied bands contribute to the total response $\chi_0^L(q_D,\omega)$ while all branches participate at finite temperature.

If the incident momentum is $\mathbf{q} = \mathbf{q}_\perp$, we get instead the transverse response

$$\chi_0^t(q_\perp,\omega) = \int \frac{d^Dk}{(2\pi)^D} \sum_{\nu\neq\nu'} |F_{\nu\nu'}(q_\perp)|^2 \frac{n_\nu(k) - n_{\nu'}(k)}{\hbar\omega_{\nu\nu'} - \dfrac{\hbar^2 k^2}{2m^*_{\nu'\nu}} + i\eta} \tag{3.15}$$

where $\hbar\omega_{\nu\nu'} = \hbar\omega - (\varepsilon_{\nu'} - \varepsilon_\nu)$ and

$$\frac{1}{m^*_{\nu'\nu}} = \left(\frac{1}{m^*_{\nu'}} - \frac{1}{m^*_\nu}\right) \tag{3.16}$$

In (3.15) we observe that ph transitions take place between states belonging to different bands and each exciton involves energy levels with the same momentum k. The available energy $\hbar\omega$ is spent in the energy difference between bandheads and the jump in kinetic energy associated to the change in effective mass undergone by the fermion being excited. We can write

$$\chi^t_0(q_\perp,\omega) = \sum_{\nu\nu'} |F_{\nu\nu'}(\mathbf{q}_\perp)|^2 \, \mathcal{P} \int \frac{d^D k}{(2\pi)^D} \frac{n_\nu(k) - n_{\nu'}(k)}{\hbar\omega_{\nu\nu'} - \dfrac{\hbar^2 k^2}{2m^*_{\nu'\nu}}} - \tag{3.17}$$

$$-i\,\pi\, |g_D(\hbar\omega_{\nu\nu'})| \left\{ n_\nu \left[\frac{m^*_{\nu'\nu}}{m^*_\nu}\hbar\omega_{\nu\nu'}\right] - n_{\nu'}\left[\frac{m^*_{\nu'\nu}}{m^*_{\nu'}}\hbar\omega_{\nu\nu'}\right]\right\}$$

with $g_D(x)$ a D-dimensional density of states per unit spin, i.e., a term in Eq. (3.3) for qp's with effective mass $m^*_{\nu\nu'}$, provided that $m^*_\nu \neq m^*_{\nu'}$.

At zero temperature, expression (3.17) acquires the simplified form

$$\chi^t_0(q_\perp,\omega) =$$

$$-\sum_{\nu\nu'} |F_{\nu\nu'}(q_\perp)|^2 \, g_D(\hbar\omega_{\nu\nu'}) \ln \left[\frac{\hbar\omega_{\nu\nu'} + \dfrac{\hbar^2 k^2_{F\nu}}{2m^*_{\nu'\nu}}}{\hbar\omega_{\nu\nu'} + \dfrac{\hbar^2 k^2_{F\nu'}}{2m^*_{\nu'\nu}}}\right] \tag{3.18}$$

Each term $(\nu\nu')$ in the imaginary part of the response represents the strength of ph transitions from the $\nu - th$ into the $\nu' - th$ branch and is nonvanishing within the band

$$\varepsilon_{\nu'} - \varepsilon_\nu + \frac{\hbar^2 k^2_{F\nu'}}{2m^*_{\nu'\nu}} \leq \hbar\omega \leq \varepsilon_{\nu'} - \varepsilon_\nu + \frac{\hbar^2 k^2_{F\nu}}{2m^*_{\nu'\nu}} \tag{3.19}$$

if $m^*_{\nu'\nu} > 0$ (i.e. $\omega > 0$), and

$$-(\varepsilon_{\nu'} - \varepsilon_\nu) - \frac{\hbar^2 k^2_{F\nu}}{2m^*_{\nu'\nu}} \leq \hbar\omega \leq -(\varepsilon_{\nu'} - \varepsilon_\nu) - \frac{\hbar^2 k^2_{F\nu'}}{2m^*_{\nu'\nu}} \tag{3.20}$$

if $m^*_{\nu'\nu} < 0$ (i.e. $\omega < 0$).

A special situation of possible interest arises when the effective mass of the D-spheres is a constant. This is likely the case for strongly dilute fluids as those adsorbed on the cylindrical walls of nanotubes, where the dependence of the effective mass with coverage is negligible. The kinetic energy term in the energy denominator in Eq. (3.15) disappears and the transverse response reads

$$\chi^t_0(q_\perp,\omega) = \sum_{\nu\nu'} |F_{\nu\nu'}(q_\perp)|^2 \, \frac{N_\nu - N_{\nu'}}{\hbar\omega_{\nu\nu'} + i\eta} \tag{3.21}$$

where N_ν is the total occupation of the $\nu - th$ Fermi D-sphere. The strength of transverse ph excitations is just the population difference between the participating bands.

3.3. *The Random-Phase-Approximation in the Fermi D-spheres model*

Adsorbed helium systems combine characteristics of D-homogeneous systems, which in the free gas limit exhibit a continuous excitation spectrum, with the transverse confined geometry that gives rise to discrete elementary excitations, similar to those in the spectrum of spherical helium droplets. The Random Phase Approximation (RPA) in coordinate space provides the reference background to compute the dynamical response, and the ph effective interaction derived from the energy density supplies the restoring force for the coherent qp oscillations. The RPA equation for the response of an inhomogeneous system is expressed in configuration space as

$$G(\mathbf{r},\mathbf{r}',\omega) = G_0(\mathbf{r},\mathbf{r}',\omega) + \int\int d\mathbf{r}_1 d\mathbf{r}_2\, G_0(\mathbf{r},\mathbf{r}_1,\omega)\, V_{ph}(\mathbf{r}_1,\mathbf{r}_2)\, G(\mathbf{r}_2,\mathbf{r}',\omega) \quad (3.22)$$

where $G_0(\mathbf{r},\mathbf{r}',\omega)$ and $G(\mathbf{r},\mathbf{r}',\omega)$ respectively represent the free ph (3.9) and the RPA propagator, $V_{ph}(\mathbf{r},\mathbf{r}')$ is the ph interaction and ω the transferred energy.

For the external field under consideration, after expanding the RPA propagator as

$$G(\mathbf{r},\mathbf{r}',\omega) = \int \frac{d^D k}{(2\pi)^D}\, G_{\mathbf{k}}(\mathbf{r}_\perp,\mathbf{r}'_\perp,\omega)\, e^{i\mathbf{k}\cdot(\mathbf{r_D}-\mathbf{r'_D})} \quad (3.23)$$

one gets a transition density $\delta\rho(\mathbf{r}) = \delta\rho(\mathbf{r}_\perp)\, e^{i\,\mathbf{q}_D\cdot\mathbf{r}_D}$ with

$$\delta\rho(\mathbf{r}_\perp) = \int d\mathbf{r}'_\perp\, G_{\mathbf{q}_D}(\mathbf{r}_\perp,\mathbf{r}'_\perp,\omega)\, V_{ext}(\mathbf{r}'_\perp) \quad (3.24)$$

and a susceptibility

$$\chi(q_D,q_\perp,\omega) = \int d\mathbf{r}_\perp\, V^*_{ext}(\mathbf{r}_\perp)\, \delta\rho(\mathbf{r}_\perp) \quad (3.25)$$

Here $G_{\mathbf{q}_D}(\mathbf{r}_\perp,\mathbf{r}'_\perp,\omega)$ is the component $\mathbf{k}=\mathbf{q_D}$ in expansion (3.23), which satisfies the integral equation

$$\begin{aligned} G_{\mathbf{q}_D}(\mathbf{r}_\perp,\mathbf{r}'_\perp,\omega) &= G_{0\mathbf{q}_D}(\mathbf{r}_\perp,\mathbf{r}'_\perp,\omega) \\ &+ \int\int d\mathbf{x}_\perp\, d\mathbf{y}_\perp\, G_{0\mathbf{q}_D}(\mathbf{r}_\perp,\mathbf{x}_\perp,\omega)\, V_{\mathbf{q}_D ph}(\mathbf{x}_\perp,\mathbf{y}_\perp)\, G_{\mathbf{q}_D}(\mathbf{y}_\perp,\mathbf{r}'_\perp,\omega) \end{aligned} \quad (3.26)$$

where the free qp contribution is (cf. Eq. (3.9))

$$G_{0\mathbf{q}_D}(\mathbf{r}_\perp,\mathbf{r}'_\perp,\omega) = \sum_{\nu\nu'} f_\nu(\mathbf{r}_\perp)\, f_\nu(\mathbf{r}'_\perp)\, f_{\nu'}(\mathbf{r}_\perp)\, f_{\nu'}(\mathbf{r}'_\perp)\, \chi_{\nu\nu'}(q_D,\omega) \quad (3.27)$$

The effective interaction $V_{ph}(\mathbf{r},\mathbf{r}')$ is derived from the total energy as

$$V_{ph}(\mathbf{r},\mathbf{r}') = \frac{\delta^2 E}{\delta\rho(\mathbf{r})\,\delta\rho(\mathbf{r}')} \quad (3.28)$$

3.4. *Films on planar substrates*

Starting with bosonic ^{4}He, experimental and theoretical studies of helium films on different adsorbers have taken place since the early fifties (see, for example, Ref. 54 and therein). In addition, films of ^{3}He-^{4}He mixtures were investigated both experimentally [55–58] and theoretically, [59] and data on surface tension, surface sound, heat capacity, [55,57,58] propagation of third sound, [56] nuclear magnetic susceptibility, [60] phase separation and binding energy [61] became available along the years. Most of these earlier results have been summarized in Refs. 62-64

The early observation of adsorbed ^{3}He atoms on liquid ^{4}He, [65,66] and on exfoliated graphite as reviewed in Ref. 67 brought into evidence the twodimensional character of these degenerate thin films, and oriented the theorists' interest towards the many body physics of imperfect, low density Fermi gases in 2D. [68] The phase diagrams of films of ^{3}He of increasing thickness on graphite, with succesive layering as a function of temperature and areal coverage, was established in the last decade, [69–71] and substantial work was also devoted to examining the solid phases in the first and second layers of adsorbed helium. [72–75] The submonolayer regime received special attention and intense experimental research of the thermodynamic [76,77] and magnetic [78–82] properties was performed. A recent overview of the field has been presented in Ref. 83 From the theoretical viewpoint, understanding the character of a twodimensional helium system became a challenge and the Landau theory of Fermi liquids, adapted to the restricted dimensionality, proved adequate to describe the main trend of the magnetic susceptibility of ^{3}He films on Grafoil at various temperatures. [84] The properties of twodimensional liquid ^{3}He in the frame of Landau Fermi liquid theory were revisited in Ref. 85 and in a similar spirit, the dynamical susceptibility in the restricted dimensionality was investigated in Ref. 52 for both vanishing and finite temperatures. In this work, the Landau limit was explicitly analyzed in the collisional, as well as in the collisionless, regime, and density zero sound modes were examined within the Random-Phase Approximation for a purely monopolar particle-hole interaction.

Studies of the equation of state of twodimensional ^{3}He resorting to variational Monte Carlo calculations [86] showed that no selfbound state should be expected; however, inclusion of the transverse motion of the atoms seemed to indicate that the quasi-twodimensional system would exhibit a small, however nonvanishing, binding energy. [87] By contrast, density functional calculations of a fluid submonolayer of ^{3}He on graphite predict density profiles consistent with a 2D phase of the adsorbed system, whose energy per particle remains larger than the energy of a single adsorbed atom. [88] More recent Green's function Monte Carlo calculations [43] carried for the strictly twodimensional helium fluid interacting through a realistic Aziz potential indicate that the fluid remains gaseous throughout the meaningful density regime.

Structural properties like energetics and wetting behaviour of ^{4}He layers adsorbed on different substrates have been examined resorting to a wide range of theoretical methods, from phenomenological functionals to variational methods beyond the one-body level. See Chapter 5 in this volume. [89] Density functional theory proved adequate to describe adsorption of ^{4}He on alkali substrates, [90–94] in good agreement with available experimental evidence; consequently, the wetting behavior of liquid ^{3}He also became a topic of interest. Employing a finite range density functional which contains gradient terms [46] a prewetting transition for ^{3}He adsorbed on cesium (Cs) was predicted, and experimentally verified [95] within the correct order of magnitude for the prewetting temperature. This experiment demontrates that ^{3}He is a universal wetting agent, capable of uniformly covering the weakest adsorbers with films of finite thickness even at zero temperature.

The success of nonlocal DFT applied to structure and thermodynamics of confined, degenerate ^{3}He systems encouraged further theoretical efforts to predict unobserved dynamical quantities such as density fluctuation spectra. [96] Although neutron scattering experiments are the main source of information on the spectrum of bulk liquid helium, [97] and have been also used to study the excitations of ^{4}He films on graphite, [98] the dynamical susceptibility of films of ^{3}He has not been experimentally investigated yet. A theoretical approach that includes the computation of the free susceptibility and the collective spectrum has been presented within the frame of finite range Density Functional Theory (FRDFT), which permits to describe the inhomogeneous structure of the system in a HF or KS mean field approach. [46,88] The HF+RPA scheme gives rise to an exploratory outlook of both the ground state and the excitations of a Fermi liquid, overriding the difficulties of more fundamental approaches like the microscopic descriptions of liquid ^{4}He.

3.4.1. *The Fermi disks model*

The general form of the nonlocal, zero temperature density functionals for ^{3}He films reported in the literature is

$$\begin{aligned} E(\rho) = \int d\mathbf{r}\, \frac{\hbar^2}{2m^*(\mathbf{r})}\, \tau(\mathbf{r}) + \frac{1}{2} \int d\mathbf{r} \int d\mathbf{r}'\, V(\mathbf{r}-\mathbf{r}')\, \rho(\mathbf{r})\, \rho(\mathbf{r}') \\ + \int d\mathbf{r} \left\{ \frac{c}{2}\, \rho(\mathbf{r})\, \tilde{\rho}^{\gamma+1}(\mathbf{r}) + d_1\, [\nabla\rho(\mathbf{r})]^2 + d_2\, \rho(\mathbf{r})\, [\nabla\rho(\mathbf{r})]^2 \right. \\ \left. + V_s(\mathbf{r})\, \rho(\mathbf{r}) \right\} \end{aligned} \tag{3.29}$$

where V_s is the external field acting on the helium atoms.

The density functionals employed in the description of these nonhomogeneous systems differ mostly in the choice of the nonlocal effective mass, as well as in the gradient and finite range effects incorporated through the pair interaction and the averaged density. For instance, in Ref. 39 the effective mass is parametrized as

$$\frac{\hbar^2}{2m^*} = \frac{\hbar^2}{2m} + \frac{t_1 + 3t_2}{8}\, \rho(\mathbf{r}) + \frac{t_1' + 3t_2'}{8}\, \rho(\mathbf{r})\, \tilde{\rho}(\mathbf{r}) \tag{3.30}$$

and the gradient terms are disregarded. In contrast, in Ref. 46 density gradients are retained and the effective mass is described by the density-dependent parabola

$$\frac{\hbar^2}{2m^*(\mathbf{r})} = \frac{\hbar^2}{2m}\left(1 - \frac{\tilde{\rho}(\mathbf{r})}{\rho_c}\right)^2 \tag{3.31}$$

Although these density functionals have proven to yield identical thermodynamical properties of bulk ^{3}He, the perspectives for their application to finite systems are very different, in view of the fact that the effective mass (3.30) can become negative for large values of the local density, which can be reached in highly compressed systems such as films on strong adsorbers or doped droplets (cf. Sec. 4). On the other hand, under the parametrizations of Ref. 39 the gradient terms can be safely suppressed when describing smooth interfaces like the free surface of semiinfinite helium, where the finite range in the two body interaction $V(\mathbf{r}-\mathbf{r}')$ and the coarse-grained density $\tilde{\rho}$ correctly account for most inhomogeneity effects. This is no longer true in the presence of sharp surfaces, i.e., the substrate-helium interface or the edge of the dense shell surrounding an impurity near the center of a cluster. Consequently, density functionals close to the original Stringari ansatz have proven to be the most convenient for calculations of confined ^{3}He systems, as described below.

The basis of the theoretical approach within DFT is the Fermi disks model,[46,88,96] where the film is a system with translational symmetry on the (x, y) plane of total area A, and the single-particle wave functions are labelled by 2D momentum $\mathbf{k}$ and ν nodes in the transverse motion (cf. Eq. (3.1)) The transverse wave functions $f_{\nu k}(z)$ are obtained solving a onedimensional HF equation, and if they depend only weakly upon k, as it is usually the case except except in the vecinity of layering transitions in the film, one can consider $f_{\nu k}(z) \approx f_\nu(z)$ as the solutions of the simplified HF equation

$$\left[-\frac{\hbar^2}{2m^*_\nu}\left(\frac{d^2}{dz^2} - k^2\right) + V(z;\rho)\right] f_\nu(z) = \varepsilon_{\nu k}\, f_\nu(z) \tag{3.32}$$

with eigenvalues

$$\varepsilon_{\nu k} = \frac{\hbar^2}{2m^*_\nu} k^2 + \varepsilon_\nu \tag{3.33}$$

Here the averaged effective mass m^*_ν is defined as

$$\frac{\hbar^2}{2m^*_\nu} = \int dz \frac{\hbar^2}{2m^*(z)}\, \rho_\nu(z) \tag{3.34}$$

being $\rho_\nu(z) = |f_\nu(z)|^2$, and the mean field $V(z;\rho)$ is computed as the first functional derivative of the energy density functional $E(\rho)$

$$V(z;\rho) = \frac{\delta E}{\delta \rho(z)} \tag{3.35}$$

The total kinetic energy density can be written as

$$\tau(z) = \sum_\nu \rho_{2\nu}\left[\frac{1}{2}\, k^2_{F\nu}\, \rho_\nu(z) + \tau_\nu(z)\right] \tag{3.36}$$

where $\tau_\nu(z) = [f'_\nu(z)]^2$. Within the Fermi disks model, the energy density functional (per unit area) takes then the form

$$\frac{E}{A} = \sum_\nu \rho_{2\nu} \left[\frac{1}{2} \frac{\hbar^2}{2m^*_\nu} k^2_{F\nu} + \int dz \, \frac{\hbar^2}{2m^*(z)} \, \tau_\nu(z) + \int dz \, V_s(z) \, \rho_\nu(z) \right]$$
$$+ \frac{1}{2} \sum_{\nu\nu'} \rho_{2\nu} \, \rho_{2\nu'} \left\{ \int dz \int dz' \, V(z - z') \, \rho_\nu(z) \, \rho_{\nu'}(z') \right.$$
$$\left. + \frac{c}{2} \int dz \, \rho(z)^\gamma \, \rho_\nu(z) \, \rho_{\nu'}(z) + \frac{1}{2} \int dz [d_1 + d_2 \, \rho(z)] \, \rho'_\nu(z) \, \rho'_{\nu'}(z) \right\} \quad (3.37)$$

This expression describes the system as a collection of elementary films (ef's); the $\nu - th$ film is associated to a Fermi disk in momentum space, and in real space, to a quasi-2D object with constant areal density $\rho_{2\nu}$ in the plane of the substrate, which extends along the perpendicular direction with local density $\rho_\nu(z)$. In other words, each ef contributes a local term to the total density of the form (cf. Eq. (3.6)

$$\rho_{3\nu}(z) = \rho_{2\nu} \, \rho_\nu(z) \quad (3.38)$$

It is important to remark that the present description is model dependent, its validity being restricted to the case where all single-particle states with the same number of nodes ν in the transverse motion can be described by the unique wave function $f_\nu(z)$. Only in this case the film can be viewed as a set of interacting ef's. When the areal coverage is low and a single Fermi disk appears, the corresponding ef coincides with the real film; it is possible, and usually convenient, to introduce effective 2D variables, whose genesis however takes into account the transverse zero point motion of the particles.

The first quantum calculation of the growth of multilayer ^{3}He films on weak binding substrates was performed by Pricaupenko and Treiner, [46] who showed than in addition to the universal wetting behaviour of this liquid, prewetting transitions —in other words, first order transitions in film thickness, intimately related to the fermionic nature of the system— should be observed on weak adsorbers such as the alkali metals. In this work, a Fermi disks model was extracted from the following density functional

$$E = \int d\mathbf{r} \left[\frac{\hbar^2}{2m^*(\mathbf{r})} \, \tau(\mathbf{r}) + \left(\frac{\hbar^2}{2m} - \frac{\hbar^2}{2m^*(\mathbf{r})} \right) \frac{|\nabla \rho(\mathbf{r})|^2}{4\tilde{\rho}(\mathbf{r})} \right]$$
$$+ \frac{1}{2} \int d\mathbf{r} \, d\mathbf{r}' \, \rho(\mathbf{r}) \, V(\mathbf{r} - \mathbf{r}') \, \rho(\mathbf{r}') + \frac{c}{2} \int d\mathbf{r} \, \rho(\mathbf{r}) \, [\tilde{\rho}(\mathbf{r})]^{1+\gamma}$$
$$+ \int d\mathbf{r} \, V_s(\mathbf{r}) \, \rho(\mathbf{r}) \quad (3.39)$$

with $V(\mathbf{r} - \mathbf{r}')$ a standard Lennard-Jones potential with depth $V_0 = 10.22$ K, hard-core radius $\sigma = 2.556$ Å, and which vanishes for distances shorter than h_{LJ}=2.1866 Å. The coarse-grained density $\tilde{\rho}$ is averaged within a sphere with radius $h_c = 2.6289$ Å and the remaining parameters, determined together with h_{LJ} and h_c

so as to reproduce the energy per particle, saturation density and compressibility of the bulk liquid, are $c = 1.40505 \cdot 10^7$K, $\gamma = 2.1$. The effective mass fitted to the experimental pressure dependence of the specific heat is given by Eq. (3.31). with $\rho_c = 0.0406$ Å^{-3}. The substrate potential is a standard 3-9 one

$$V_s(z) = \frac{4}{27}\frac{C_3^3}{D^2}\frac{1}{z^9} - \frac{C_3}{z^3} \tag{3.40}$$

Recent *ab initio* calculations of various adsorption potentials which introduce corrections to the above standard form have been presented in Ref. 99.

Employing the above density functional, the HF equations are solved together with the number-of-particles constraint which gives rise to the chemical potential $\mu(T, \rho_2)$; it should be kept in mind that this is the 3D one determined by particle number conservation. This constraint can be expressed as

$$\rho_2 = \frac{T}{\pi}\sum_\nu \frac{2m_\nu^*}{\hbar^2} \ln\left[1 + e^{-(\varepsilon_\nu - \mu)/T}\right] \tag{3.41}$$

A essential feature of the Fermi disks model is the ν-dependent effective mass to be associated to each ef. On thermodynamic grounds, one can see that both the specific heat and the magnetization of the film present a step-like behavior as functions of coverage, which reflects the sequential filling of consecutive Fermi disks; however, as shown in Ref. 46 these steps are not horizontal, due to the dependence of the effective mass —and thus, of the 2D density of states— with coverage. The discrete nature of the transverse spectrum can be also tested with dynamical probes. This model is a very reliable instrument to explore thermodynamics and dynamics of adsorbed helium films, the accuracy of the approach being limited by the density functional itself, the characteristics of the confining field and the coverage. For a strong adsorber such as graphite, the twodimensional character of the thin submonolayer film is very clearly established; however, the predictive power of the method is only qualitative, in view of the large discrepancy between experimental and calculated values of the effective mass parameters. [88] This apparent drawback can be anyway attributed to an inadequate substrate potential and could be removed with an updated choice of its form and parameters. [100]

3.4.2. *The response of adsorbed ^3He in the Fermi disks model*

In Ref. 96 explicit expressions for the free dynamical susceptibility have been derived and illustrated with numerical computations for a helium film adsorbed on a Cs substrate with a coverage $\rho_2 = 0.14$ Å^{-2}. The description is valid for any substrate insofar as the transverse wave functions are weakly dependent upon the 2D momentum of the particles. Specific numerical calculations [96] indicate that for a typical configuration representing an adsorbed film of medium thickness, the lowest band $\nu = 0$ provides the most important contribution to the total longitudinal response, with decreasing weigth of the excited bands $\nu > 0$ governed by the smaller effective masses. Furthermore, the thicker the film, the more closely $\chi_0(\mathbf{q},\omega)$ resembles the susceptibility of a free 3D Fermi gas; [148] however, thermal excitation of the film suppresses the layered structure of momentum space, regardless the number of occupied disks.

The effective interaction $V_{ph}(\mathbf{r},\mathbf{r}')$ is derived from the same energy density functional that gives rise to the Fermi disks [46,88] according to Eq. (3.28), and the general expression for $V_{ph}(z,z')$ has been presented in Ref. 96

To extract the collective spectrum of the film, one performs a onedimensional Hartree-Fock (HF) calculation of the single-particle band head spectrum ε_ν, wave functions $f_\nu(z)$, effective masses m^*_ν and Fermi momenta $k_{F\nu}$, out of which the free ph band-to-band responses $\chi_{\nu\nu'}(\mathbf{q}_{//},\omega)$ and the propagator $G_{0\mathbf{q}_{//}}(z,z',\omega)$ are constructed. One then computes the effective ph interaction and inverts the matrix equation (3.26) to obtain the RPA propagator $G(z,z',\omega)$. The transition density $\delta\rho(z)$ and the susceptibility χ are then investigated to establish the collective spectrum.

In Ref. 96 the submonolayer regime ($N/A = 0.04$ Å^{-2}) of a ^{3}He film on two substrates of very different adsorbing power like cesium and graphite has been analyzed. In both cases, the coverage permits the filling of a single Fermi disk; however the spatial characteristics are substantially dissimilar. While the film spreads over about 10 Å upon Cs and roughly 2 Å on graphite, the maximum density of helium on the latter is about four times higher than on Cs. These features are responsible of very different collective spectra, which however share the important feature that every dispersion relation can be adjusted by a polynomial of the form $\omega(q) = a_0 + a_1 q + a_2 q^2$. Moreover, the broadening of the ph continua with transferred momentum is a characteristic of the free susceptibility; due to the different efective masses of the disks, the transverse continua posess finite extensions even for vanishing longitudinal momentum and their growth with $q_{//}$ permits damping of the high energy modes at zero temperature.

Since collective oscillations are expected to appear at low energies, the long wavelength - low energy limit deserves special attention. The calculations and an explicit analysis indicate that two types of modes appear: (i) modes with energy ω close to zero and finite phase velocity $v = \omega/q_{//}$ (phonons in the Landau limit) and (ii) interband modes with finite frequency for vanishing momentum. The first situation mainly consists of interband, longitudinal ph excitations, the second one exhibits mostly transverse interband transitions. For the Cs substrate, the numerical procedure does not predict the lowest energy branch of excitations within the occupied disk; instead, a series of high energy branches between two transverse continuum bands appears, which merge into a transverse ph band as the transferred momentum increases as seen in Fig. 4.

For finite values of transferred momentum, the longitudinal mode damps into the 2D ph continuum, remaining very close to its edge. This mode must be always

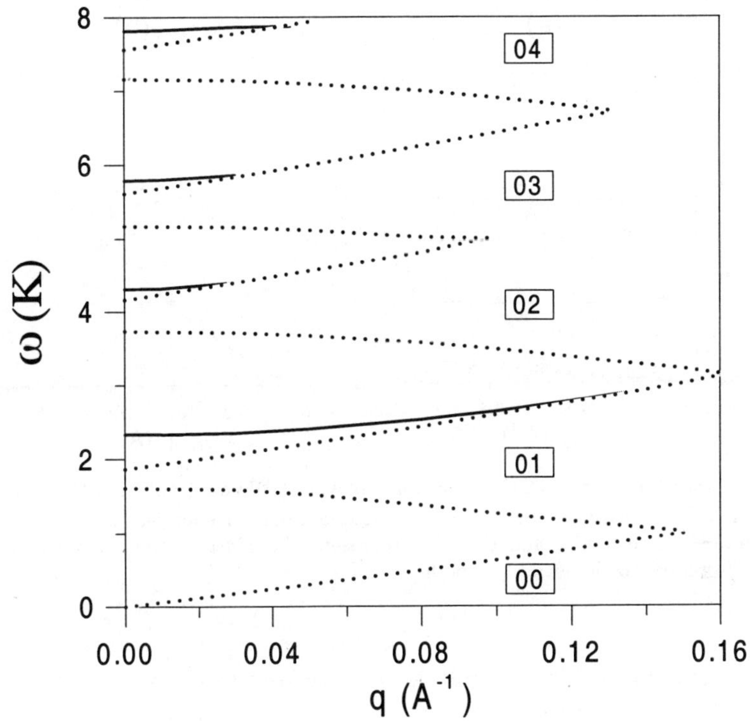

Fig. 4. The dispersion relations $\omega(q_{//})$ for a monolayer of ^{3}He on Cs with coverage $N/A = 0.04\text{Å}^{-2}$, as functions of transferred momentum. The dashed lines indicate the edges of the ph continuum for the transverse ph excitations $\nu\nu'$ indicated in each square box.

present in the Landau limit, with a phase velocity $v = \omega/q_{//}$ given by the root of

$$\text{Re}\,\chi^L_{00}(v) = \frac{1}{V^{eff}_{00}} \tag{3.42}$$

for a repulsive renormalized effective interaction. [52] In this case, the smaller the value of V^{eff}_{00}, the closer to the Fermi velocity is v. For the helium monolayer on Cs, the effect of the transverse susceptibilities is to lower the 2D repulsion by two orders of magnitude; the collective phase velocity equal to $1.002\ v_F$ is unobservable in the current scale, the mode is damped for increasing transferred momentum and appears as a sharp resonance immediately below the ph edge.

The situation on a strong graphite adsorber displayed in Fig. 5 is remarkably different: due to the large energy differences between bandheads, [88,96] one expects that monolayer excitations involve only the occupied and first unoccupied band. For vanishing momentum two transverse modes appear, which rapidly damp away for $q_{//}$ larger than 0.05 Å^{-1}, and a clearly distinguishable longitudinal mode is present, whose dispersion relation can be fitted to $\omega(q)a_1q + a_2q^2$ with a_1=5.6 K Åand a_2=2.57 K Å^2. The value of the phase velocity at the origin ($a_1/\hbar =$

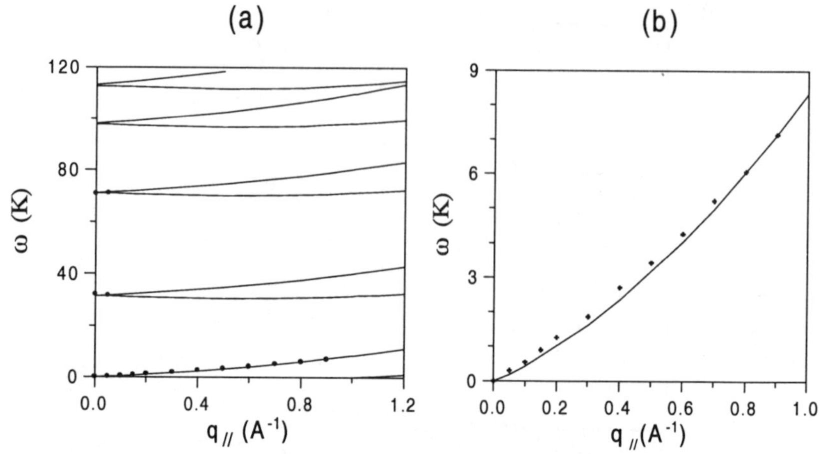

Fig. 5. The dispersion relations $\omega(q_{//})$ for a monolayer of ^{3}He on graphite with coverage $N/A = 0.04$ Å^{-2}. (a) Complete spectrum in the investigated range of energies. The dots indicate the energy of the collective modes and the lines represent the edges of the ph continuum.(b) The dispersion relation for the longitudinal mode.

$73.8m/s$) is close to the third sound velocity derived from the equation of state of the film, [88] 120 m/s for the given coverage. It should, however, be kept in mind that these excitations are density zero sound modes arising from distorsions of the Fermi sphere, whose physical origin and structure is different from hydrodynamic sound. [102] One can verify that the dispersion relation $\omega(q_{//})$ coincides with the purely 2D one computed as in Ref. 52 In the Landau limit, the transverse bands renormalize the effective interaction by a factor 1.08 with respect to the 2D value.

The transition densities $\delta\rho(z)$ associated with each mode for a transferred momentum $q_{//} = 0$ involve the whole film, extending beyond the equilibrium thickness and displaying finite amplitudes where the unperturbed density is practically vanishing. This effect is more noticeable the higher the collective energy, suggesting increasing surface localization. In each density fluctuation profile one can clearly distinguish the z-dependence introduced by the ν-th wave function; even though the relative amplitudes change, the shapes of these profiles are not modified with increasing transferred momentum.

Denser films on a Cs substrate display a multilayer structure which has been analyzed in Ref. 96 and it has been shown that since the number of ph bands that build up the continuum is greater than in the monolayer situation, the collective modes are quickly damped as either the transferred momentum or the temperature increases. The overall pattern is similar to the monolayer one, with no longitudinal mode and various transverse-like excitations that due to the relative proximity between transverse ph bands, cannot be associated with a particular interband transition. The spatial shapes of the density fluctuations exhibit the joint participation of several wave functions; this is not surprising, in view of the fact that the equilibrium density already contains contributions of three different wave functions.

In the case of a helium monolayer upon graphite, the collective excitations bring into clear evidence the manifestly 2D character of the film. The effective interaction is able to generate a collective longitudinal mode, whose dispersion relation coincides with the prediction of a strictly 2D model. For the same value of the coverage, a monolayer adsorbed on Cs reveals the importance of transverse atomic motion. In the Landau limit, the relatively low cost accessibility to higher branches of the single-particle spectrum is the origin of noticeable weakening of the effective ph interaction in charge of providing a longitudinal excitation. Consequently, this mode appears practically on the border of the ph continuum and rapidly damps with increasing transferred momentum. On the other hand, a series of collective branches with higher energies shows up, which can be interpreted as coherent interband excitations from the low occupied disk to the empty bands. As the coverage increases, the richness of the collective spectrum also enlarges; however, the available energy between adjacent bands becomes small and excitations cannot develop. In this case, Landau damping becomes a very efficient agent for disappearance of low energy collectivity. The overall behavior differs substantially from the pattern of ^{4}He films; examination of the monolayer transition densities indicates that the localization of the perturbation is stable against increase of the transferred momentum, in contrast with the changes reported in the case of bosonic systems [103–105].

3.5. *One and quasi-one dimensional helium fluids*

In addition to the wetting and prewetting transitions undergone by adsorbed liquid ^{3}He, as well as features of layered growth of films on flat surfaces in general, like the step structure of the specific heat and magnetization of ^{3}He, other interesting phenomena taking place in confined geometries are capillary condensation [100,106] and appearance of 1D and 2D phases in nanotube ropes. [107] Investigation of these shapes became popular in the last years, motivated by active research on physical properties of carbon nanotubes and their fluid adsorbates. [108] In particular, capillary condensation of liquid ^{3}He has been analyzed within DFT for a simplified geometry, [100] where the walls of the pore are simulated by two adjacent flat substrate surfaces a distance L apart. In this way, the single-particle wave functions retain the form (3.1), while the HF wave equation for the transverse eigenfunctions contains an external potential of the double-well form

$$V_{slab}(z) = V_s(z) + V_s(L - z) \tag{3.43}$$

In this frame, specific calculations have been performed for slabs of ^{3}He within Cs, Li and Au walls employing the same density functional as in Ref. 88. The general trend is similar for the three adsorbers, with the chemical potential of the confined atoms displaying the nonmonotonic pattern indicative of transitions between unstable and stable regions in the thermodynamic (μ, ρ_2) plane. The DFT results show various regimes, which in the case of the Cs surfaces can be summarized as follows.

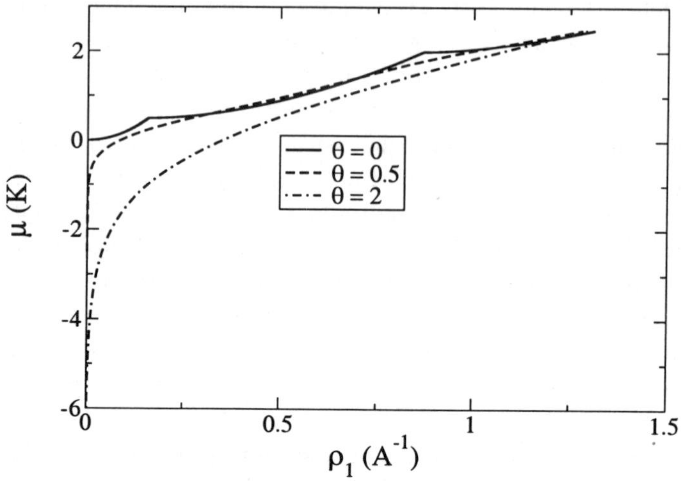

Fig. 6. Chemical potential (in K) of a gas of ^{3}He atoms confined to a cylinder of radius R = 4 Å as a function of linear density, for zero temperature, and for $\theta = T\Theta_\varphi = 0.5$, 2.

For very small separation ($L \leq = 10$ Å) a compressed quasimonolayer is visible for any number of particles per unit area. For L between 15 and 40 Å, unstable density profiles appear at low real density, with large particle concentration in the neighbourhood of the walls. The pattern evolves into a stable multilayer structure as the coverage increases, in correspondence with the appearance of a jump in the T = 0 adsorption isotherm. Wall separations larger than 40 Å display a clear difference between unstable density patterns with essentially two disjoint films at the lowest coverages, which give rise to almost uniform filling of the inner space —*i.e.* capillary condensation— as bulk saturation is approached. In this case the overall thermodynamic behavior resembles closely that of the prewetting transition of helium films; [46] moreover, the general trend predicted by these explicit DFT calculations can be adequately formulated by a simpler thermodynamical model which yields a universal description of wetting and capillary condensation transitions, in terms of only two parameters, such as wall separation and adsorber strength.

In spite of the lack of experimental evidence, it is likely that ^{3}He may be confined in the channels present in carbon nanotubes and bundles. This possibility has been confirmed for ^{4}He in nanotubes [109] and other porous materials. [110,111] Moreover, theoretical adsorption isotherms of gaseous H and ^{4}He indicate a sequence of phase transitions as the linear density of atoms increases. [112] At the lowest densities, the atoms are adsorbed next to the nanotube wall, and subsequently several 1D phases appear filling the interstitial channels, the outer grooves of the bundle and the axis

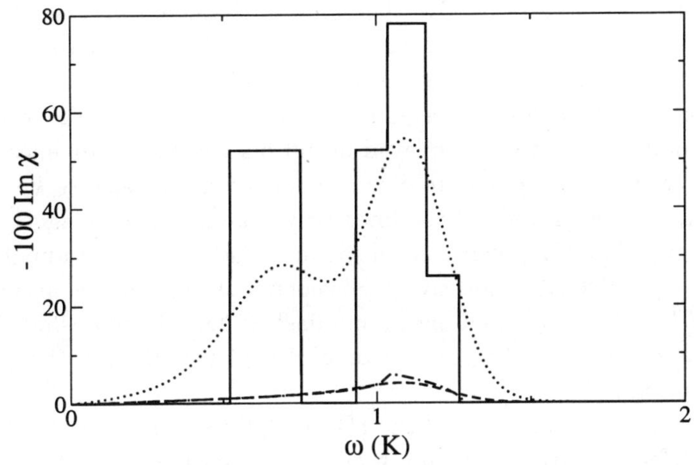

Fig. 7. The longitudinal response for ^{3}He atoms adsorbed on a cylindrical wall so as to fill three Fermi segments. See text for details.

of each tube. In addition to the crossover between 1D and higher dimensionalities which may take place due to coupling among neighbouring channels, the shell phase itself contains a band spectrum, since for large dilution the elementary excitation spectrum of the qp's adsorbed on a cylinder of several tens of microns length L, and radius R of a few Å, is essentially

$$\varepsilon_{kl} = \frac{\hbar^2}{2m}\left(k^2 + \frac{l^2}{R^2}\right) + \varepsilon_0 \tag{3.44}$$

for integer angular momentum l and angular quantum $\Theta_\varphi = \hbar^2/2mR^2$, which amounts to a few tenths of degree K and is much larger than the 1D translational one. Here ε_0 represents the ground-state energy for radial motion, which remains frozen since the corresponding quantum is around 100 K. The wave function (3.1) reads

$$\phi_{kl}(\mathbf{r}) = \frac{1}{\sqrt{2\pi L}}\, e^{i(kz + l\varphi)}\, f_0(r) \tag{3.45}$$

with a ground-state factor $f_0(r)$ narrowly peaked around the cylinder radius R.

In this case, the band structure of 1D momentum continua on bandheads $\varepsilon_l = \Theta_\varphi\, l^2$ arises from the decoupling between angular and axial motion. As an illustration of band filling and dimensional crossover from one to two dimensions through a Q1D regime, in Fig. 6 the chemical potential of free ^{3}He atoms on a cylinder of radius R = 4 Å is displayed as a function of linear density, for T = 0, 0.5 and 2 Θ_φ. The angular points in the zero temperature isotherm, which disappear at nonvanishing temperature, indicate completion of a Fermi segment. At the lowest densities the chemical potential is negative, as it corresponds to an ideal classical gas. Stronger compression drive the gas into the quantum degenerate regime; we appreciate that for any temperature, at sufficiently high density the chemical potential exhibits the linear density dependence of 2D Fermi gases. A dynamical manifestation of this crossover appears in the free susceptibility; the imaginary part of the longitudinal response is displayed in full lines in Fig. 7 for linear density $\rho_1 = 0.94$ Å^{-1}, which corresponds to Fermi momentum and energy $k_F = 0.6$ Å^{-1}, $\varepsilon_F = 2.88$ K. The magnitude of the linear coverage has been exaggerated so that for this configuration, three Fermi segments are filled. The longitudinal transferred momentum is $q = 0.2\ k_F$, respectively. For completeness, Im χ is also shown for a temperature T $= 0.1\ \varepsilon_F$ (dotted lines), and dashed-dotted and dashed lines respectively illustrate the responses per unit area of a planar 2D gas [52] with the same Fermi momentum.

The low momentum behavior of the 1D susceptibility indicates that a 1D phase of adsorbed ^{3}He is a Luttinger liquid, with no quasiparticle excitations. Preliminar RPA calculations for a dilute 1D gas of ^{3}He atoms results indicate that the phase velocity v_ρ of density fluctuations is about one order of magnitude higher than the Fermi velocity, for linear densities compatible with gaseous liquid helium. When the temperature increases, these excitations are expected to decay into 2D ones, or into the bulk spectrum; moreover, interchannel coupling among neighbouring interstitial sites or grooves may also destroy the Luttinger liquid behavior. These are topics of current research and will be presented shortly, a preliminary overview may be found in Ref. 50.

4. Self-saturating systems

Helium systems are dominated by quantum effects and remain liquid at zero temperatures. Because of the absence of a triple point, helium clusters remain liquid under all conditions of formation, thus differing significantly from clusters of heavier rare gases. For instance, the structure of argon and xenon clusters is due to the packing of spheres, which leads to some "magic number" effects. These effects cannot appear in helium clusters as no localization of atoms may be assumed. Moreover, statistical effects are expected to introduce an important difference between ^{4}He and ^{3}He drops. Indeed, the Pauli principle will manifest in the latter through shell effects, as it happens with electrons in atoms and nucleons in atomic nuclei. In this respect clusters formed by both helium isotopes are of particular interest. Since the atom-atom interaction does not distinguish between isotopes, these mixed systems offer the possibility of studying features due entirely to quantum effects, related both to the zero-point motion and to statistics.

Helium droplets provide ideal inert matrices to perform high resolution spectroscopy of embedded atoms and molecules, [114] due to their low temperatures: 0.4 K for ^{4}He and 0.15 K for ^{3}He. [115–118] In turn, the doping atoms or molecules can be used as probes to obtain very useful information about the properties of the clusters. In fact, the experimental research on superfluidity in ^{4}He clusters has used weakly interacting probes attached to the droplet; the excitation spectrum of the guest molecule could bring some insight into the fluid properties of the host. A large variety of impurities can be captured by helium clusters during their formation in the supersonic beam expansion, as rare gases, alkali and alkaline earth atoms, molecules as HF, SF_6, OCS, HCN and even bigger molecules as aminoacids. All these impurities are massive enough to support a theoretical semiclassical treatment, consisting in the addition of a helium-impurity potential to the helium-helium mean field within the DF approach. The interaction of helium atoms with impurities such as rare gases and SF_6 molecules is rather well established. [120,121]

The structure and dynamics of helium drops have been the subject of many experimental and theoretical studies. For a general overview see for instance Refs. 122, 123 and Chapter 9 of this volume, [124] where other aspects not considered here are also addressed. In this section review the density functional approaches employed to study the structural and energetic properties of pure droplets of ^{3}He, mixed droplets of ^{4}He-^{3}He, and the effects of introducing molecular impurities in these systems.

4.1. *Pure drops*

For a long time, a severe experimental limitation has been the impossibility of selecting and identifying the clusters quantitatively. Beams containing Helium drops are easily produced in a low temperature nozzle expansion of the gas. [124] Although variations in the source conditions of pressure and temperature change in principle the size of the drops, in practice this method gives only very rough estimates of the average sizes. [125] Another attempt to determine size distributions relied upon ionization of clusters by electron scattering, [126] but this method is not very reliable for neutral systems. Deflection of clusters by a secondary beam has proven to be more useful to establish the size distribution of large drops (between $10^3 - 10^4$ atoms [127,128]). More recently, a new method based on diffracting a molecular beam from a transmission gratting has been developed [129] to detect small droplets. Experimental indication on the dimer $^4He_2$ was reported in Ref. 130 using electron impact ionization techniques. The diffraction technique of Ref. 129 has provided conclusive evidence of its existence, [131] and has also lead to the determination of its binding energy [132] 1.1+0.3-0.2 mK, very close to the limit of stability, with a bond length of 52 ± 4 Å. Indeed, the weak helium-helium interaction makes the dimer $^4He_2$ an extremely fragile system, with the zero-point energy almost identical to the potential depth. This circumstance requires calculations to be performed with a very high accuracy. It appears that any number of ^{4}He atoms form a self-bound system; [133] in contrast, a substantially large number of ^{3}He atoms is necessary for self-binding. The precise determination of the minimum number of ^{3}He atoms is still an open question, as we shall show below.

The first systematic calculations of the ground state properties of ^{4}He and ^{3}He

Table 4. Energies per particle E/N and unit radius r_0 for ^{4}He droplets

N	VMC	DMC			DF	
	E/N(K)	E/N(K)	r_0(Å)	Ref.	E/N(K)	r_0(Å)
8	-0.5919	-0.6417		133	-0.556	3.26
10	-0.7916	-0.8654		133	-0.78	3.06
20	-1.510	-1.688	2.29	123	-1.60	2.70
40	-2.430	-2.575	2.53	123	-2.48	2.51
70	-3.043	-3.253	2.42	123	-3.16	2.42
112	-3.476	-3.780	2.39	123	-3.68	2.37
240	-4.192			123	-4.41	2.31

droplets were carried out using a variational microscopic wave function [134] and a local, zero-range density functional, [24] exhibiting good overall agreement both in ground state energies and density profiles. At present, better variational and DF results as well as DMC calculations are available for ^{4}He drops. In the density functional approach the ground state of a ^{4}He droplet at $T = 0$ is simply described as a Bose condensate where the N atoms occupy the lowest energy level, characterized by a single-particle wave function ϕ_0. In such a case, the particle number density is given by $\rho = N|\phi_0|^2$, and since the kinetic energy density is given by Eq. (2.3), establishing the density profile of a drop is equivalent to finding the single-particle wave function of a particle in the condensate. This is carried out by minimizing the total energy given by the density functional with the constraint of a fixed number of particles. It is usually assumed that the ground state of the droplets is spherically symmetric.

In Table 4 are collected some results for the energies per particle. Columns labelled VMC and DMC contain microscopic results for the Aziz atom-atom interaction HFD-B(HE), [135] taken from Ref. 133 for $N = 8, 10$, and from Ref. 123 otherwise. The DF results correspond to the OT functional. [47] The overall agreement obtained between microscopic and density functional calculations is striking. Parametrization OT includes finite range terms, and has been fitted so as to reproduce some selected properties of the homogeneous liquid at saturation, as well as the static structure factor of bulk ^{4}He. Consequently, the predictions for finite droplets do not rely on free parameters. The outcoming energies per particle lie in between the VMC and the DMC ones for a number of atoms $N \geq 20$.

In the case of a system with N atoms of ^{3}He the density functional $E[\rho, \tau]$ depends on the particle and kinetic energy densities, defined in terms of the single-particle wave functions ϕ_α in the following way

$$\rho(\mathbf{r}) = \sum_\alpha n_\alpha |\phi_\alpha(\mathbf{r})|^2 \tag{4.1}$$

$$\tau(\mathbf{r}) = \sum_\alpha n_\alpha |\nabla\phi_\alpha(\mathbf{r})|^2 \tag{4.2}$$

Table 5. Energies per particle (in K) for ^{3}He droplets

N	VMC			DF			
	PPW	GN1	GN2	ST	WR	SkV-LJ	BHN
20	0.21	0.17	0.15	0.08	0.049	0.103	0.135
40	-0.04	-0.063	-0.098	-0.19	-0.245	-0.187	-0.107
70	-0.28			-0.50	-0.533	-0.481	-0.387
112	-0.46			-0.75	-0.772	-0.726	-0.624
168	-0.62			-0.95	-0.967	-0.923	-0.832

where n_α refer to the single-particle occupation numbers of the state characterized by quantum numbers $\{\alpha\}$. The single-particle wave functions are optimally determined by solving the KS variational equations

$$\frac{\delta}{\delta\phi^*_\alpha}\left\{E[\rho,\tau]-\sum_\alpha \epsilon_\alpha \int d^3r|\phi_\alpha(\mathbf{r}|^2\right\}=0 \tag{4.3}$$

This lead to mean field equations which have to be solved self-consistently. For instance, starting from the functional (2.6) one gets

$$\left\{-\nabla\frac{\hbar^2}{2m^*(\rho)}\nabla+U(\rho,\tau)\right\}\phi_\alpha=\epsilon_\alpha\phi_\alpha \tag{4.4}$$

where the self-consistent field is given by

$$U(\rho,\tau)=\tau\frac{d}{d\rho}\frac{\hbar^2}{2m^*(\rho)}+\frac{d}{d\rho}\epsilon(\rho)-2d\Delta\rho \tag{4.5}$$

One assumes spin saturated systems so that the ground state of the droplet is a Slater determinant built up from the $N/2$ single-particle states of lowest energy. Spherical symmetry is also assumed, which means that only completely occupied states with quantum numbers (n,l) i.e. with degeneracy $2(2l+1)$ need to be considered. This guarantees spherical symmetry in both the density ρ and the mean field U.

In Table 5 are displayed the ground state energies obtained either by means of VMC calculation or a DF approach. Columns PPW and GN1 displays respectively the results of Ref. 134 and [138], for the HFDHE-2 Aziz interaction [136], whereas column GN2 contains those of Ref. 138 for the HFD-B(HE) Aziz interaction [135]. Note that the latter interactions produces slightly more binding than the former. In the DF columns are displayed the ground state energies for the following functionals: ST, which is the original zero-range functional proposed by Stringari and Treiner [24]; WR, constructed by Weisberger and Reinhard [33], inspired in the philosophy of both the ST functional and the Pines polarization potentials; finally, SkV-LJ [39] and BHN [40] which contain finite range terms as well as an explicit spin dependence, as they include the Landau parameters in the determination of their force strenghts. Note that, similarly to the ^{4}He case, the DF produces more bound systems as

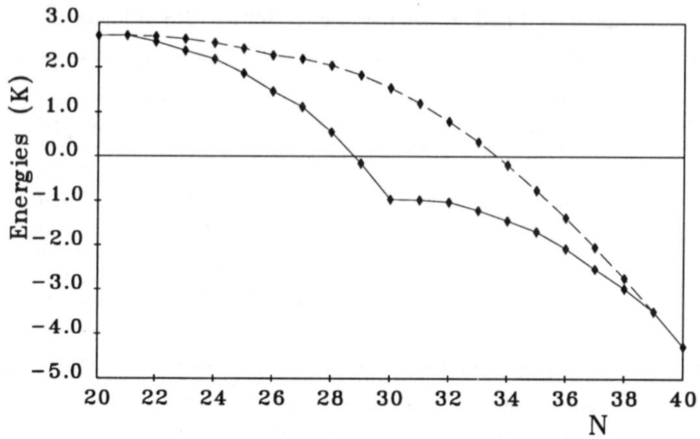

Fig. 8. Mean field DF energies (dashed line) and total energies including correlation between open $1f$-$2p$ subshells (solid line) as a function of the number N of atoms in the droplet.

compared to VMC based on Aziz interactions. It is worth noticing that the VMC approach of [134] underbinds the homogeneous liquid by 0.13 K/atom. [139] Let us stress again that all parameters of the DF's have been determined so as to reproduce selected bulk liquid properties plus, for some functionals, the surface tension of the free surface, so that the results for finite systems are predictions independent of free parameters. The ground state energies depend on the parametrization to a considerable extent, indicating that bulk plus surface properties of the homogeneous liquid do not suffice to determine the ground state of a finite system unambiguously.

As expected, for a given number of atoms, ^{3}He drops are less bound than ^{4}He, due to the larger zero point motion and the Pauli principle. Calculations have been done for a number of ^{3}He atoms which correspond to shell closure in an harmonic oscillator scheme. Indeed, up to $\simeq 240$ atoms the self-consistent equations give rise to such a shell scheme: the magic numbers are given by the sequence $(p+1)(p+2)(p+3)/3$, with p a nonnegative integer. The WR functional predicts instead shell closure at 198 and 274 atoms. For larger systems the magic numbers do not rely on the harmonic oscillator pattern (see Table 9).

Finally, we also note that the minimum number of atoms to form a self-bound system lies between 20 and 40 atoms. An estimate of this minimum number $N_{\rm min}$ has been obtained [137] using the uniform filling approximation for open shell systems, which consists in allowing states (n, l) to have fractional occupation numbers, in order to uniformly distribute the valence atoms between the magnetic substates. This yields $\mathrm{N_{min}} = 34$, as can be seen in Fig. 8.

Since correlations not included in the mean field are known to play an important role in open shell drops, such estimates may be too crude. It is then necessary to perform a configuration interaction calculation with the $1f$ and $2p$ active subshells, and this was considered in Ref. 137 for the density functional BHN. [40] The basic ingredients are the two-body matrix elements of the residual atom-atom interaction, which have been taken as the effective ^{3}He-^{3}He interaction derived from the functional. For each specific drop, the mean field calculation in the uniform filling approximation provides both the single-particle wave functions to evaluate the two-body matrix elements, and the single-particle energies. The variational space reaches a dimension of 184,756 different Slater determinants for N=30. The calculated correlation energies depend smoothly on the number of particles in the shell and display a pronounced minimum at midshell. The total energy, also shown in Fig. 8, is obtained by adding the correlation energy to the ground state energy arising from the uniform filling or mean field calculation. It is seen that correlations move the limit of stability down by about 5 atomic units, so that the value of $N_{\rm min}$ may be established in 29 atoms. The very pronounced E_C minimum at midshell, allows us to state that the droplet with 30 atoms is bound, even taking in consideration the uncertainties of the DF energies as shown in Table 5.

An interesting outcome from these DF plus shell-model calculation is the structure of droplets with N=20 to 40. A common feature of these droplets is that their ground states have the maximum spin value allowed by Pauli's principle, i.e., $S_{\rm max} = \tilde{n}/2$, where $\tilde{n}$ is the number of particles (holes) in the valence space below (above) midshell. This can be understood by looking at the properties of the residual interaction. Table 6 displays the most important matrix elements obtained for the droplets with N=30 and N=40. This Table corrects an error in the values quoted in Ref. 137 which does not change quantitatively the results. The matrix elements in a single shell are repulsive for L=0, essentially zero for even L-values, and attractive for odd L-values. The diagonal matrix elements involving two l-shells are much more attractive in the S=1 channel than in the S=0 one. Moreover, the S=1 matrix elements are nearly L-independent. As a consequence, the energy minimum within one l-shell is obtained for configurations maximally antisymmetric in orbital space, and thus maximally symmetric in spin space. In addition the aligned states of two l-shells couple to maximum total spin due to the dominance of the spin vector channel in the cross shell interaction. Almost exact degeneracy with orbital angular momentum is also found.

These results have been qualitatively confirmed by the VMC calculations of Refs. 138, 140. In this case, the trial wave function combines additive two- and three-body correlations, coming from a configuration interaction description, and multiplicative two-body Jastrow-type correlations, the whole acting on Slater determinantal states containing backflow correlations. In Ref. 138 the determinants have been constructed in a cartesian basis, and the occupied levels are selected to be invariant under 90° rotations. These trial wave functions have not well-defined orbital angular momentum. In Ref. 140 the trial wave functions do have good orbital angular momentum, but the occupied states are restricted to a single valence subshell (either $1f$ or $2p$) in the $p = 3$ major shell. In either case no configuration mixing between different subshells has been considered. The atom-atom interaction

Table 6. Matrix elements $\langle l_1, l_2(L)|V|l_3, l_4(L)\rangle$ (K) for N=30 and 40 drops

l_1, l_2, l_3, l_4	LS	N=30	N=40
3,3,3,3	00	0.207	0.280
	11	–0.210	–0.232
	20	–0.015	–0.007
	31	–0.103	–0.116
	40	0.014	0.009
	51	–0.074	–0.084
	60	0.075	–0.074
1,1,1,1	00	0.141	0.263
	11	–0.120	–0.135
	20	–0.003	0.004
3,1,3,1	20	–0.046	–0.048
	21	–0.141	–0.161
	30	–0.052	–0.065
	31	–0.060	–0.073
	40	–0.066	–0.068
	41	–0.109	–0.132

is the HFD-B(HE) Aziz interaction. The binding threshold has been found to be 34 atoms in both calculations. Moreover, the ground state is characterized by the maximum spin, and for a given spin it is degenerate with respect to the orbital angular momentum. Finally, the matrix elements are in a qualitative agreement with those in Table 6.

4.2. *Mixed drops*

The properties of the free surface of liquid ^{4}He change dramatically when a small amount of ^{3}He atoms are added. For instance, its surface tension is substantially lowered even for a relatively small ^{3}He concentration. This effect is explained in terms of the difference in kinetic energy between one ^{3}He and one ^{4}He atom in the bulk. The excess in kinetic energy pushes the ^{3}He atom to the surface, where it acquires a quasi two-dimensional wave function known as the Andreev state [141]. The system formed by addition of one ^{3}He atom impurity to liquid ^{4}He has been largely studied from the microscopic point of view. This has naturally led to the study of mixed systems of ^{3}He and ^{4}He. The first investigation of ^{3}He-^{4}He drops in the frame of finite range DF theory was presented in Ref. 48 where the DF (2.28)-(2.32) was introduced, and its parameters were determined as explained in Section 2. The results discussed here refer to this DF whose parameters are given in Table 3.

The structure and energetics of clusters with N_4 and N_3 atoms have been investigated solving the coupled Hartree and KS equations which respectively describe the ^{4}He and the ^{3}He isotope, obtained as usual from functional differentiation of the total energy with respect to the corresponding density. The particle and kinetic energy densities of the ^{3}He atoms are computed from the single-particle wave functions $\phi_j(\vec{r})$ as

$$\rho_3(\vec{r}) = \sum_{j=1}^{N_3} \mid \phi_j(\vec{r}) \mid^2 = \sum_{nlm} \mid \frac{R_{nl}(r)}{r} Y_{lm}(\hat{r}) \mid^2 \tag{4.6}$$

and

$$\tau_3(\vec{r}) = \sum_{j=1}^{N_3} \mid \nabla \phi_j(\vec{r}) \mid^2 \tag{4.7}$$

The kinetic energy terms are further renormalized as

$$\frac{\hbar^2}{2m_3} \rightarrow \frac{\hbar^2}{2m_3}\left(1 - \frac{1}{N_3 + \frac{m_4}{m_3} N_4}\right) \tag{4.8}$$

$$\frac{\hbar^2}{2m_4} \rightarrow \frac{\hbar^2}{2m_4}\left(1 - \frac{1}{N_4 + \frac{m_3}{m_4} N_3}\right) \tag{4.9}$$

to take into account center-of-mas corrections.

The simplest mixed system consists in adding only one ^{3}He atom to a given ^{4}He drop. Several groups [48,142,143] have investigated in a systematic way the structural and energetic features of theses systems as functions of the number of particles in the drop. In a DF scheme, the calculation first determines the energy per particle E_4/N_4 and the chemical potential μ_4, and then solves for the spectrum of the ^{3}He atom in the external field furnished by the ^{4}He density. The DF calculations of Ref. [48] indicate that the energy of the $1s$ state can be very well adjusted to a mass formula of the type

$$\varepsilon_{1s} = \varepsilon_0 + \frac{C}{N_4^{1/3}} \tag{4.10}$$

with $\varepsilon_0 = -4.81$ K and C = 8.44 K. Drop sizes as large as $N_4 = 10^4$ have been considered for this fit. This value of ε_0 is in reasonably agreement with the energy of the Andreev state (-5.00 $\pm$ 0.03) K, [144] whereas for the zero-range DF in Ref. [142] the fit gives the values –5.44 K and 9.8 K, respectively. The present finite range DF provides more repulsion on the ^{3}He atoms than the zero-range one, and ε_0 compares favorably with the VMC calculation in Ref. [143] which yields $\varepsilon_0 \sim -4.90$ K.

Moreover, examination of the radial probability density as a function of the distance to the center of the droplet, for different N_4, indicates that the ^{3}He atom always sits at the surface of the cluster, as shown in Fig. 9. It is also found that as N_4 increases, the spectrum of the ^{3}He atom becomes rather independent of the orbital quantum number, reflecting the fact that in the limit of a large drop, the surface ^{3}He states would no longer be adequately characterized by an angular momentum

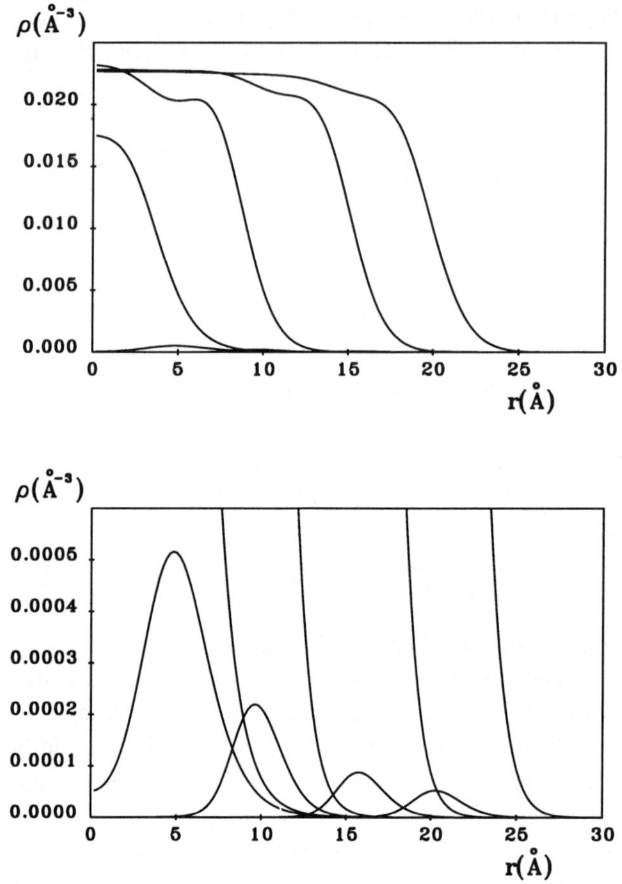

Fig. 9. Density profile ρ_4 (top figure) and radial probability density of one ^{3}He atom in $^4\text{He}_{N_4}$ (bottom figure), as a function of the distance to the center of the droplet. From left to right, the number of ^{4}He atoms in the drop is N_4=8, 70, 330, and 728. The density profile ρ_4 is also seen in the bottom figure.

quantum number, but by a linear momentum parallel to the free surface, whose multipolar decomposition demands a large superposition of partial waves.

The more general case of a mixed drop with variable numbers of ^{4}He and ^{3}He atoms is analyzed below for $N_4 = 70$ and 728, combined with N_3 between 8 and 288, chosen so as to fill a shell of the ^{3}He single-particle spectrum.

In Fig. 10 are displayed the density profiles $\rho_4(r)$ and $\rho_3(r)$, as functions of the radial distance, for drops with N_4=70, 728, and varying N_3. In both cases, the dashed line displays the density of a pure drop of ^{4}He. We may notice in these two pictures that whichever the value of N_3, the density ρ_3 remains peaked at the surface of the $^4\text{He}_\text{N}$ drop; we can observe as well a slight inward compression of this surface as the ^{3}He bubble grows larger. It is also clear that the larger the hosting

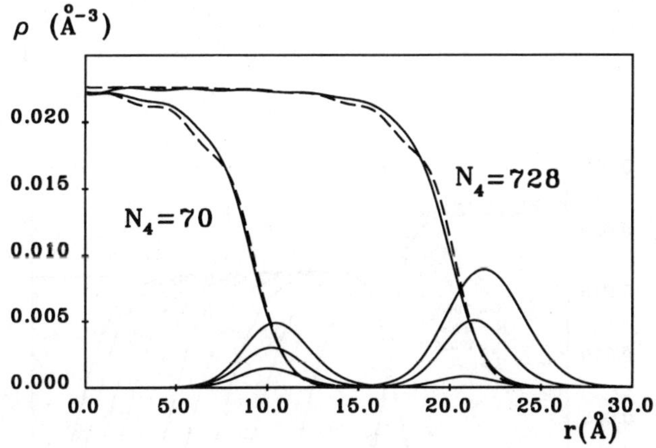

Fig. 10. The densities $\rho_4(r)$ and $\rho_3(r)$ for N_4=70 and 728, and varying numbers of ^{3}He atoms: 8, 18 and 32 for N_4=70, and 18, 32 and 70 for N_4=728. Dashed lines are the density profiles of pure ^{4}He drops.

Table 7. Ground state energies (in K) and chemical potentials (in K) of mixed drops ($N=N_4+N_3$).

	N_4=70			N_4=728		
N_3	E/N	μ_4	μ_3	E/N	μ_4	μ_3
0	-2.97	-4.17		-5.16	-5.80	
8	-3.06	-4.35	-2.68			
18	-3.00	-4.45	-2.59	-5.14	-5.80	-3.86
32	-2.92	-4.60	-2.47			
50	-2.81	-4.80	-2.29			
70	-2.68	-5.05	-2.08			
128				1.04	-5.80	-3.40
288				-4.63	-6.08	-2.71

^{4}He drop, the larger its surface and consequently, the higher the number of ^{3}He atoms it may accommodate. The above mentioned compressional effect induced by the latter on the bosonic cluster is also present.

Concerning the energy systematics of these systems, the major characteristics reported are that increasing amounts of ^{3}He atoms introduce important attractive contributions into the chemical potential of the ^{4}He atoms, and repulsive ones in the chemical potential of ^{3}He atoms. Also, the drops become less bound as N_3 grows at fixed N_4. This is due to the energy per particle difference between liquid ^{3}He and ^{4}He, -2.49 and -7.15 K, respectively. These facts can be visualized in Table 7, where the total energy per particle and the chemical potentials of ^{4}He and ^{3}He for pure and doped $^4He_{70}$ and $^4He_{270}$ clusters are displayed as functions of N_3.

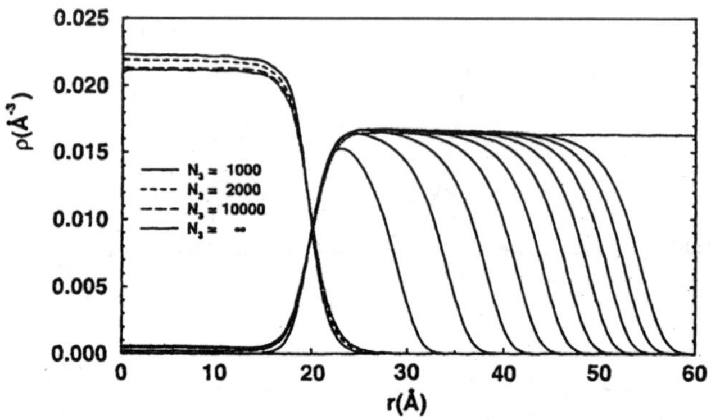

Fig. 11. Density profiles of $^4He_{728}$-$^3He_{N_3}$ droplets, for values of N_3 from 1000 to 10000, in steps of 1000.

According to the above discussion, a single ^{3}He atom attached to a ^{4}He cluster has a single-particle spectrum whose lowest energy state smoothly approaches the Andreev state in the liquid free surface. Increasing the number of ^{3}He atoms gives rise to interesting structural features, one of which is that they locate on the surface of the ^{4}He cluster. The experimental findings of Refs. 118, 119 confirm this structure of mixed droplets. They consist of an inner core of nearly pure ^{4}He atoms and an outer shell of ^{3}He atoms. The temperature of the mixed droplet, which is of only 0.15 K, [124] is determined by the outside ^{3}He atoms. Taking into account that the structure of liquid mixtures, for concentrations above the maximum solubility of ^{3}He in ^{4}He, corresponds to an homogeneous ^{3}He-^{4}He solution plus a segregated phase consisting of pure ^{3}He, it is apparent that much larger amounts of atoms are needed to visualize the onset of ^{3}He dilution. This problem has been addressed in Ref. 145, where large mixed drops $^4He_{728}$-$^3He_{N_3}$ with $N_3 \geq 1000$ have been analyzed using the DF (2.8). As the single-particle energies become closer when large values of N_3 are considered it is more convenient to use a Thomas-Fermi approximation. The kinetic energy density τ_3 is thus written in terms of the particle density ρ_3 and its gradients

$$\tau_3 = \alpha\rho^{5/3} + \beta\frac{|\nabla\rho|^2}{\rho} + \gamma\Delta\rho$$

where α, β, and γ are known coefficients. In that way one deals with an energy functional depending only on ρ_4 and ρ_3, and one has to solve two coupled Euler Lagrange equations.

Table 8. Bulk ^{3}He concentrations as a function of N_3.

$N_3/1000$	1	2	3	4	5	6	7	8	9	10
$x\%$	$6.25 \cdot 10^{-5}$	0.996	2.01	2.41	2.57	2.65	2.71	2.70	2-71	2.71

Fig. 11 displays the situation in which a $^4He_{728}$ drop, large enough to clearly distinguish a surface and bulk region, is coated with an increasing number of ^{3}He atoms, and the limiting situation of the same drop immersed into liquid ^{3}He. The evolution with N_3 of the ^{3}He concentration inside the ^{4}He drop, defined as $x_3 = \rho_3/\rho|_{bulk}$, with $\rho = \rho_4 + \rho_3$, is shown in Table 8. It is apparent that a fairly large amount of ^{3}He is needed before it is appreciably dissolved in the interior of the ^{4}He drop. For $N_3 = 1000$, ρ_3 near the origin is $\simeq 1.4 \cdot 10^{-8}$Å^{-3}. The solubility is appreciably reduced by finite size effects. Indeed, one can see from the Table that the limiting solubility in the N_4=728 drop is $\simeq$ 2.5%, as compared to the 6.6% value in the liquid mixture. It is also worth noting that for high N_3 the bulk solubility is slightly above the limiting one, indicating that finite size effects still appear in rather large drops. Another manifestation of finite size is that the average ^{3}He density exceeds the saturation value even for the larger drops, showing that the existence of the outer ^{3}He surface still causes a visible density compression. Because of the high incompressibility of helium, the bulk density of ^{4}He decreases when ^{3}He is dissolved, and the rms radius of the ^{4}He drop manifests a peculiar N_3 behavior. It decreases when N_3 increases up to a few hundreds due to the initial compression of the outermost ^{4}He surface, and then steadily increases as ^{4}He is pushed off the center by intruder ^{3}He atoms. This is a very tiny effect anyway.

4.3. *Doped drops*

Doping a helium cluster with atomic or molecular impurities constitutes a useful probe of the structural and energetic properties of the drop. The best studied systems are ^{4}He clusters doped with a large variety of dopants, [116,148,151,153,154,156] as rare gases, alkali metals, alkaline earths, SF_6, OCS, HCN and larger molecules. It has been established that rare gases and closed-shell molecules as HF, OCS or SF_6 are located in the bulk of the drop. In contrast, alkali and alkaline earth atoms, and presumably open-shell molecules, remain attached to the surface of the drop. If the foreign atom or molecule is heavy enough as compared to the helium atom, it can be regarded as an infinite mass object at the coordinate origin, which provides an external field to the helium atoms. In the cases where the helium-impurity potential $V_{\rm imp}$ is well known, the doped helium cluster can be described by simply adding the interaction term $\rho(r)V_{\rm imp}(r)$ to the density functional of the pure drop.

Doped ^{3}He droplets have been studied in Ref. 158, using the following density

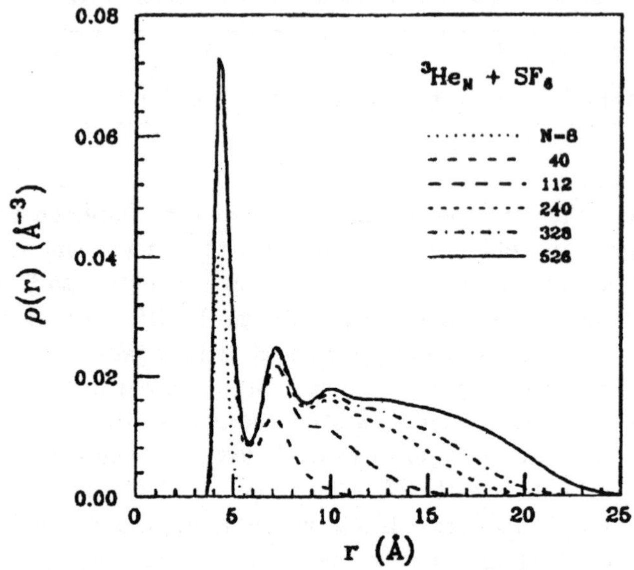

Fig. 12. Density profile of ^{3}He clusters doped with SF_6 molecule.

functional

$$E[\rho,\tau] = \int d^3r \ \epsilon[\rho,\tau] \tag{4.11}$$

with

$$\epsilon[\rho,\tau] = \frac{\hbar^2}{2m^*}\tau + \frac{1}{2}\int d^3r'\rho(r)V(|\mathbf{r}-\mathbf{r}'|)\rho(\mathbf{r}') + \frac{1}{2}c\rho^2(r)\bar{\rho}^\gamma(r) + \rho(r)V_{\rm imp}(r) \tag{4.12}$$

The impurities considered are rare gases and SF_6 molecules, because their interaction potential with the helium atom has been well determined [121,157] and they are always located inside the drop. In contrast, alkali atoms are expelled from the bulk of the cluster [153,155] and the study of such systems require a different methodology. A center of mass correction has been considered multiplying the kinetic energy by the factor $1 - 1/N$, where N is the number of ^{3}He atoms.

As a typical example, in Fig. 12 are displayed the helium density profiles of clusters doped with SF_6. As the size of the drop increases the number of ^{3}He atoms contained in the first solvation shell quickly reaches a limiting value. Typical values are $\simeq$ 19 atoms for SF_6, 15 for Xe, and 9 for Ne. The more attractive the helium-impurity potential, the more atoms are contained in the first solvation shell. The isotopic effects are compared in Fig. 13, where the density profiles of the drops with 528 Helium atoms doped with SF_6 are plotted. In general, as compared with the analogous ^{4}He cluster, the first ^{3}He shell contains around 5 atoms less. This is a clear

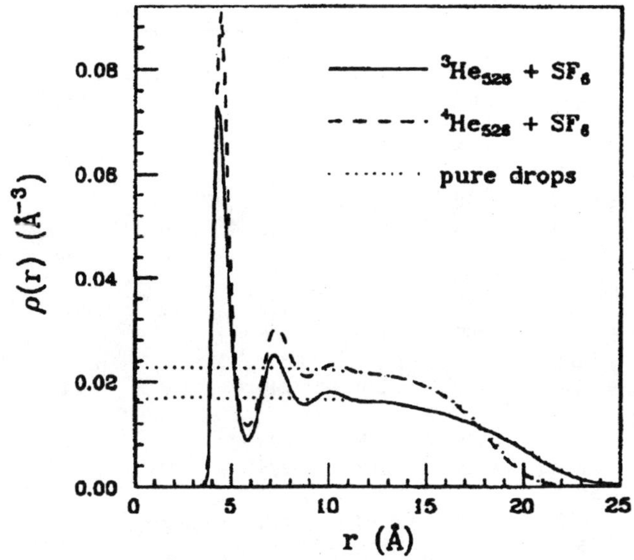

Fig. 13. Comparison of the effect of a SF_6 impurity in the density profiles of ^{3}He and ^{4}He drops.

effect of statistics, which can be attributed to the increase in kinetic energy of the fermionic atoms in this first shell. For all cases studied, the solvation energy of the impurity is negative but smaller in absolute value than in a ^{4}He cluster of the same size. As already observed in the ^{4}He case [146] the helium chemical potential quickly goes towards that of the pure drop as N increases, indicating that the impurity essentially affects the first coating shell.

The helium-impurity potential provides an additional attractive well to the ^{3}He atoms, able to bind at least 8 atoms. The precise determination of the minimum number of ^{3}He atoms bound by a given impurity should require a microscopic approach beyond DF. An interesting question which can be addressed within DF theory is the modification of the magic numbers by the presence of a specific dopant. In Table 9 are shown for several impurities, the numbers of atoms that give a stable closed shell structure with spherical symmetry for the doped cluster in its minimum energy configuration. Apart from minor differences due to the filling of ns orbitals, the shell structure is robust up to N$\simeq$ 328, irrespectively of the dopant. In some cases, however, a rather weak structure appears, indicated by an asterisk in the Table.

Mixed clusters of ^{3}He and ^{4}He atoms doped with a single OCS molecule have been employed [159] in the first experimental observation of superfluidity in a finite system. The analysis of the infrared spectrum of the dopant molecule led unambo-

Table 9. Magic numbers of pure and doped ^{3}He clusters.

	Doped				
Pure	Ne	Ar	Kr	Xe	SF_6
	8	8	8	8	8
		18	18	18	18
	20	20			
40	40	40	40	40	40
		68	68	68	68
70	70				
112	112	112	112	112	112*
	166	166	166	166	166
168					
240	240	240	240	240	240
274*					
	328	328	328	328	328
368	368	368	368		
398*					
482	482	482	482	482	
516					
					526*

gously to the conclusion that the OCS molecule rotates freely when the number of ^{4}He atoms is $\simeq 60 - 70$, preventing the ^{3}He atoms from getting too close to the molecule. Indeed, the ^{3}He atoms are at temperatures of $\simeq 0.15$K, and are thus in the normal state. Interestingly enough, a path integral calculation [161] of pure ^{4}He clusters predicted that superfluidity should manifest at around 70 atoms.

Within the DF formalism, general systems of mixed drops with variable numbers of ^{3}He and ^{4}He atoms, doped with a single SF_6 impurity, have been first considered in Ref. 48. As compared with the non-doped cluster with N_3=1, the trend of the ^{3}He s-states indicates that the distortion of the mean potential created by the SF_6 molecule is important only for the smallest drops. For $N_4 \leq 300$ the effect of the molecular attraction is to strongly bind the ^{3}He atom, producing a significative lowering and compression of the whole spectrum. The effect of the impurity disappears for N_4 above $\simeq 300$; for instance, the energy ε_{1s} is insensitive to the presence of the SF_6 impurity. This is a consequence of the fact that the ^{3}He always remains located at the surface of the drop, similarly to the case without impurity. For a large drop, the molecular potential felt by the ^{3}He atom is practically zero in the surface region, where the ^{3}He single-particle wave functions concentrate.

In Fig. 14 are displayed the radial probability densities $|R_{1s}(r)|^2$ of the ^{3}He atom in the doped drop. They can be compared with the densities of Fig. 9 corresponding to the non-doped cluster. The density profiles indicate that for small drops, the molecule not only prevents that the ^{3}He atom reaches the central region, but compresses the whole density pattern; in fact, for any value of N_4, the ^{3}He atom is

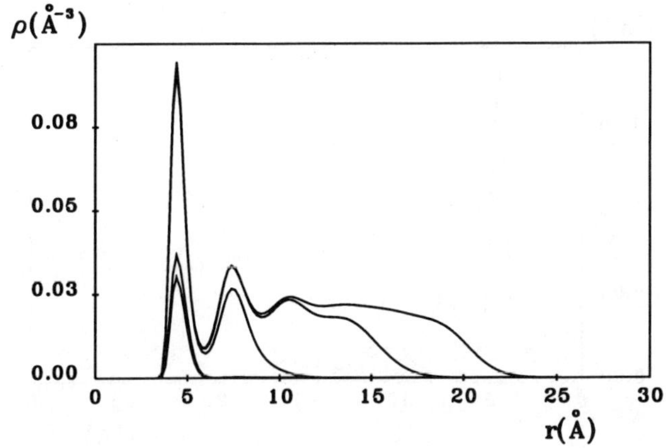

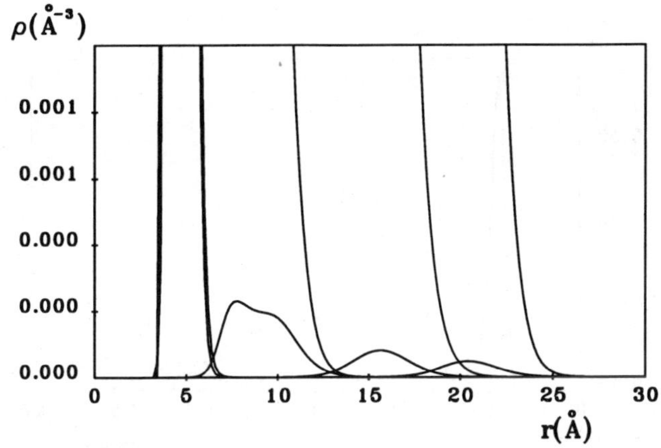

Fig. 14. Density profile ρ_4 (top figure) and radial probability density of one ^{3}He atom in $^4\text{He}_{N_4}$ (bottom figure), as a function of the distance to the center of the droplet. The drops are doped with SF_6. From left to right, the number of ^{4}He atoms in the drop is N_4=8, 70, 330, and 728. The density profile ρ_4 is also seen in the bottom figure.

pushed towards smaller radii. One can observe that for $N_4 = 70$, the fermion probability density already presents two peaks; the outer one becomes the most important as the drop grows. For $N_4 = 112$ the peak is moderately sharper than in the pure cluster and lies slightly inwards. For N_4 above 300, the probability densities of the ^{3}He atom in either pure or doped clusters are very similar, reflecting the fact that the impurity has little influence on the ^{3}He surface states, as mentioned before. It is worthwhile noticing that the accuracy of this type of calculations is comparable [146] to that of variational descriptions of small clusters of liquid ^{4}He [151].

When the number of ^{3}He atoms is increased a slight tendency to penetrate the host cluster appears. In Fig. 15 are shown the densitites ρ_3 and ρ_4 of a doped

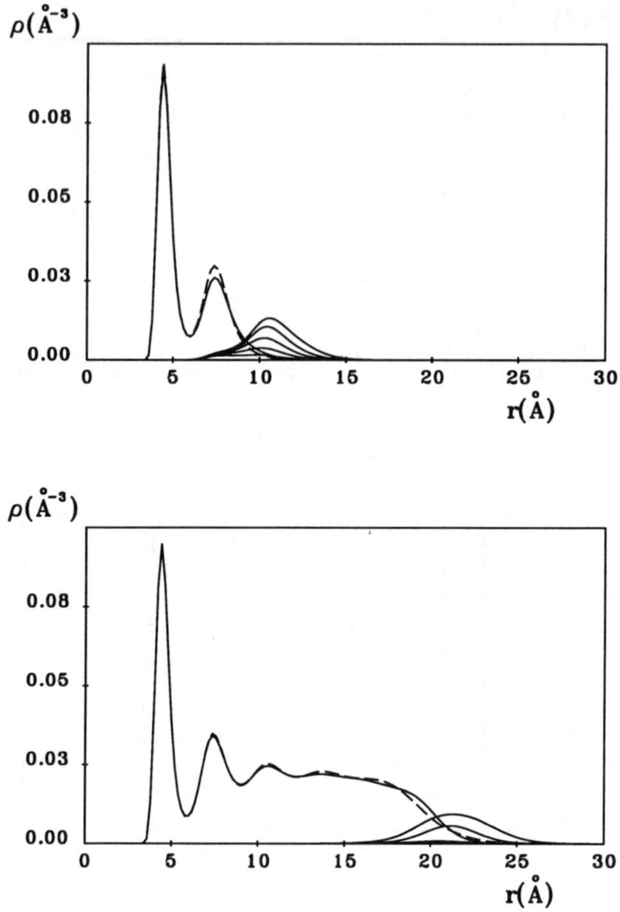

Fig. 15. Density profiles of ^{4}He and ^{3}He in a mixed drop with an SF_6 impurity. The top panel corresponds to N_4=70, and N_3= 8, 18, 32, and 72 from bottom to top. Full and dashed lines are the ^{4}He densities for drops with N_3=8 and 72 respectively. The bottom panel corresponds to N_4=728 and N_3= 18, 128 and 288. Full and dashed lines are the ^{4}He densities for drops with N_3=18 and 288.

drop with N_4=72 (upper panel) and 728 (lower panel) with varying magic values of N_3. This figure is to be compared with the non-doped case represented in Fig. 10. Whichever the value of N_3, the density ρ_3 remains peaked at the surface of the ^{4}He drop. An inner peak in ρ_3 insinuates for N_3 above 18 (not clearly seen in the N_4=728 case due to the scale) which may indicate the attraction of the low ^{4}He density in the dip of the drop profile, intending to build a second coating shell.

In Table 10 are shown the total energy per particle and the chemical potentials of ^{4}He and ^{3}He atoms. Comparing these values with those of Table 7 corresponding to the undoped case, one realizes that, according to the previous discussions, the

Table 10. Ground state energies (in K) and chemical potentials (in K) of mixed drops ($N=N_4+N_3$) doped with a single SF_6 molecule.

	N_4=70			N_4=270		
N_3	E/N	μ_4	μ_3	E/N	μ_4	μ_3
0	-12.95	-5.46		-6.30	-5.80	
8	-12.72	-5.38	-3.28			
18	-11.62	-5.39	-3.13	-6.24	-5.81	-3.87
32	-10.42	-5.49	-2.96			
50	-9.25	-5.63	-2.68			
70	-8.16	-5.80	-2.31			
128				-5.90	-5.89	-3.40
288				-5.44	-6.08	-2.71

impurity modifies the energetics to an important amount in the smaller clusters, and its effects are less significant for larger number of ^4He atoms.

Fig. 16 shows the density profiles of several SF_6-doped clusters with N_4=728 and N_3 varying from 4000 to 10 000. Note the appearance of two high density solvation shells with a density depleted region. The small peak of ρ_3 in this inner ^4He surface contains about one ^3He atom.

The most recent application of DF theory to doped mixed drops has been an

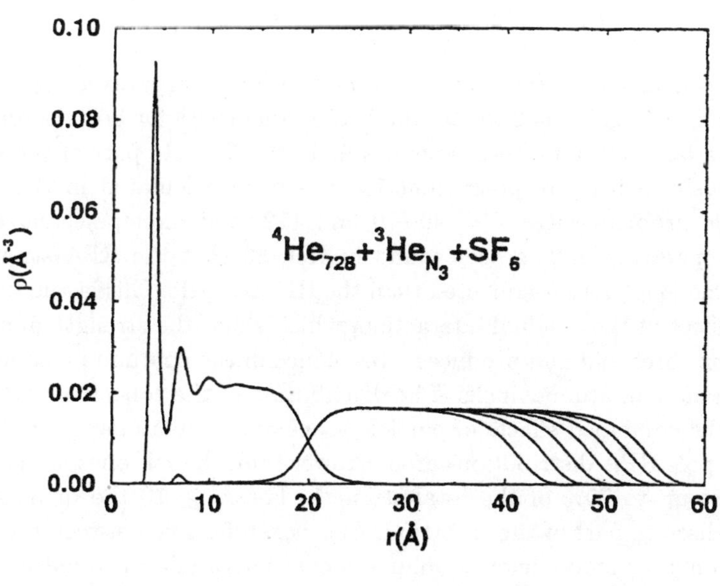

Fig. 16. Effect of a SF_6 impurity in mixed large drops with 728 atoms of ^4He and N_3 values from 4000 to 10 000 in steps of 1000.

investigation of their structure with addition of a quantized vortex, pinned to the impurity.[167] It is found that in the absence of a dopant, the ^{3}He atoms are capable of filling the vortex core, thus developing a central density peak in addition to that on the outer surface. The presence of the impurity pushes ^{3}He towards the pinning points on the vortex axis, modulating its density along the vortex line.

4.4. *Response in pure and doped helium clusters.*

Collective states of helium droplets have also been studied. In the case of ^{4}He droplets, both microscopic [161–163] and DF [164,165] calculations have been performed, showing that these clusters can sustain collective oscillations of different multipolarities.

The first systematic description of some global properties of these excitations in ^{3}He systems was presented in Ref. 29, employing RPA M_k sum rules, defined as the $k-$th energy weighted integrals of the strength function with respect to energy. The strength function has been calculated for surface excitations generated by external multipole fields $r^L Y_{L0}$ and by the monopole operator r^2. In this study, the zero range DF of Ref. [13] was selected. The average collective energy $E_3 = \sqrt{M_3/M_1}$ shows a smooth, well-defined dependence with both the number of atoms N and the multipolarity L. For small sizes, E_3 increases, and for $N \geq 150$, it decreases following an $N^{-1/3}$ law. Such a dependence can be explicitly obtained for $L \neq 0$ in the limit of a large drop within the Thomas-Fermi approximation, which gives the following analytical formula:

$$E_3(L) = \frac{1.21}{N^{1/3}} \frac{\hbar^2}{m r_0^2} \left[\frac{m}{m^*} (L-1)(2L+1) \right]^{1/2} \tag{4.13}$$

The main conclusion of this paper is that the low energy excitations of helium clusters for $L = 0$ and $L = 2$ to 10 exhibit the same trend for both isotopes, i.e., for small L and large clusters, these energies lie in the discrete part of the spectrum.

The density-density response function has been calculated in Ref. 30 for the three magic droplets with $N = 40$, 70 and 112, and monopole and quadrupole excitation operators. It has been found in Ref. 30 that the RPA strength of the monopole mode is more fragmented than the HF strength. This is due to the strong repulsive effect of the residual interaction, which shifts the strength above the particle emision threshold and produces a resonance in the continuum, similarly to the giant resonances in atomic nuclei. The distribution of the collective states is broad, reflecting the combined effects of particle space and Landau damping. In contrast, the quadrupole RPA distributions are narrower than the HF ones, and a single line can exhaust up to 70% of the total strength. For $N \geq 70$ the quadrupole mode lies in the discrete part of the spectrum. Another difference between these modes is that the monopole mode involves volume oscillations while the quadrupole mode is a surface vibration.

A preliminary calculation of the density-density response of ^{3}He droplets doped with an SF_6 molecule, within the RPA formalism and using finite range DF's, was carried in Ref. 158. In this work, both the mean field and the effective interaction are derived from the same DF, selected as the one previously adjusted in Ref. 149, where the energetics and structure of the pure and doped helium clusters was investigated. Plots of the dipole oscillator strength distributions for closed shell ^3He_N + SF_6 clusters demonstrate that some fragmentation takes place even for large drops; moreover, the collective energy, which in this case always lies above the continuum threshold, decreases with increasing size. This disappearance of the dipole energy indicates that the impurity rapidly delocalizes inside the drop, as already observed in the case of doped ^4He clusters [146,148].

The distortions on the RPA excitation spectrum of helium drops, caused by solvation of an impurity, have been recently investigated in Ref. 150 for the same DF. In this work, the comparison among the monopole spectra of the ^3He, $^3He + SF_6$ and ^3He + Xe drops shows that the presence of the impurity enlarges fragmentation in the high energy region, the effect becoming more important for the smallest clusters and for the most attractive foreign potential. As in Ref. 149, the mean energy M_1/M_0 lies above the atom emission threshold and decreases for growing N_3, in contrast with the reported trend for large ^4He clusters [147]. However, in all cases the monopole oscillation localizes in the bulk of the drop, as indicated by the shape of the transition densities. The dipole mode is also heavily fragmented, with a mean energy below the continuum threshold, whose trend as a function of drop size reinforces the previous finding [149] concerning delocalization of the impurity.

The surface quadrupole oscillation offers an interesting behavior; although the collective energies of the doped clusters are pushed downwards in energy with respect to the pure drops, it is also seen that when the mean quadrupole energy moves into the discrete part of the spectrum, the strength becomes more concentrated. Although most quantitative remarks in this study are model dependent and sensitive to the parametrization of the DF, it can be safely assumed that the predictions concerning stability of dipole fluctuations and evolution of the collective energy centroids with particle number are robust.

5. Summary

This chapter summarizes the current state-of the-art concerning applications of DF theory to confined ^3He fluids. As anticipated in the Introduction and developed along the chapter, after a presentation of the background of DF descriptions of liquid helium isotopes, we have focused upon two representative illustrations of restricted geometries. An important collection of finite helium systems are those subjected to a confining field which limits the motion of the ^3He atoms. In these geometries, *i.e.*, films adsorbed on planar substrates or fluids in porous environments, one may observe dimensionality crossover from D to (D+1)-dimensional homogeneous structures, as the number of atoms per unit area or length increases. The transition takes place through intermediate configurations consisting of inhomogeneous fluids, whose structure cannot be described at a low cost employing elaborated methods of many-body theory. In these cases, DF techniques have proven powerful to envisage thermodynamic phase transitions and excitation spectra of the zero sound type. Another subset consists of the selfsaturated systems which occur for sufficiently large numbers of particles, where the helium-helium interaction is able to counteract the fermionic kinetic energy and Pauli repulsion so as to produce a spherical drop. These ^{3}He clusters, together with their ^{4}He and mixed ^{3}He-^{4}He counterparts, are currently a subject of active experimental and further theoretical research in view of various open problems, such as phase separation in confined mixtures, the onset of superfluidity in finite systems, and distorsions of the spherical shape due to the presence of vortices in the bulk of the clusters, [166,167] or the vicinity of adsorbing walls.

This overview contains enough illustrations to prove that DF theory is a serviceable and rewarding instrument, if handled with care and aiming at very specific targets. Its predictive abilities are large, since once the parametrization has been chosen so as to reproduce measurable magnitudes of the bulk liquid, ground-state properties, thermodynamics and general trends of excitation spectra of finite numbers of helium atoms in restricted geometries, can be forecasted. A possible weakness of the approach could be the fact that the amount and strength of the many-body correlations, which have been incorporated in a phenomenological fashion regarding exclusively the 3D homogenous system, may depend on the dimensionality and degree of confinement. In such a case, one might expect that DF descriptions provide a first, qualitative idea of relative figures and tendencies of measurable entities, as functions of size parameters like the number of atoms in a drop or the coverage of an adsorbed film. In such a spirit, and as a final remark, we wish to state that although DF theory does not intend to substitute the well-founded and microscopically sound machinery of many-body physics, in the present stage of development of the latter, it endows researchers with substantial landmarks where to focus more fundamental efforts.

Acknowledgments

Discussions with M. Barranco, M.M. Calbi, M.W. Cole, S.M. Gatica, E. Krotscheck, and M. Pi are gratefully acknowledged. This work was supported by UBA, Argentina (grant TW81), and DGI, Spain (grant BFM2001-0262).

References

1. L.D. Landau, *JETP*, **3**, 920 (1957); *ibid.* **5**, 101 (1957).
2. D. Greywall, *Phys. Rev.* **B27**, 2747 (1983).
3. K. Levin and O.T. Valls, *Phys. Rep.* **98**, 1 (1983).
4. D. Vautherin and D.M. Brink, *Phys. Rev.* **C5**, 626 (1972).
5. P. Hohenberg and W. Kohn, *Phys. Rev.* **B136**, 864 (1964).
6. W. Kohn and L.J. Sham, *Phys. Rev.* **140**, A1133 (1965).
7. S. Stringari, *Phys. Lett.* **A106**, 267 (1984).
8. C.H. Aldrich III and D.M. Pines, *J. Low Temp. Phys.* **25**, 677 (1976); C.H. Aldrich III, C.J. Pethick and D.M. Pines, *J. Low Temp. Phys.* **25**, 691 (1976); C.H. Aldrich

III, C.J. Pethick and D.M. Pines, *Phys. Rev. Lett.* **37**, 845 (1976); C.H. Aldrich III and D.M. Pines, *J. Low Temp. Phys.* **32**, 689 (1978).
9. D.M. Pines, Proc. Int. School of Physics "E. Fermi", LXXXIX Course, Varenna, 1985.
10. J. Casulleras and J. Boronat, *Phys. Rev. Lett.* **84**, 3121 (2000).
11. T.C. Padmore and M.W. Cole, *Phys. Rev.* **A9**, 802 (1974).
12. C. Ebner and W.F. Saam, *Phys. Rev.* **B12**, 923 (1975).
13. S. Stringari, *Phys. Lett.* **A107**, 36 (1985).
14. A.E. Watson, J.D. Reppy and R. Richardson, *Phys. Rev.* **188**, 384 (1969).
15. J. Boronat, J. Casulleras and J. Navarro, *Phys. Rev.* **B50**, 3427 (1994).
16. J. Boronat, Monte Carlo simulations at zero temperature: helium in one, two and three dimensions, 2001, this volume.
17. J.C. Wheatley, *Rev. Mod. Phys.* **47**, 467 (1975).
18. F. Dalfovo and S. Stringari, *Phys. Lett.* **112**, 171 (1985).
19. F. Dalfovo. Ph. D. Thesis, University of Trento, 1989.
20. S. Stringari and J. Treiner, *J. Chem. Phys.* **87**, 5021 (1987).
21. S. Stringari and J. Treiner, *Phys. Rev.* **B36**, 8369 (1987).
22. E. Krotscheck, S. Stringari and J. Treiner, *Phys. Rev.* **B35**, 4754 (1987).
23. M. Barranco, A. Polls and S. Stringari, *J. Phys.* (Paris) **48**, 911 (1987).
24. S. Stringari, M. Barranco, A. Polls, P.J. Nacher, and F. Lalöe, *J. Phys.* (Paris) **48**, 1337 (1987).
25. M. Barranco, A. Polls, S. Stringari, P.J. Nacher, and F. Laloë, *J. Phys.* (Paris) **48**, 101 (1987).
26. M. Barranco, M. Pi, A. Polls, and X. Viñas, *J. Low Temp. Phys.* **80**, 77 (1990).
27. A. Guirao, M. Centelles, M. Barranco, M. Pi, A. Polls, and X. Viñas, *J. Phys. : Cond. Matt.* **4**, 667 (1992).
28. M. Barranco, M. Guilleumas, M. Pi, and D.M. Jezek, Cavitation in liquid helium, 2001, this volume.
29. Ll. Serra, F. Garcias, M. Barranco, J. Navarro and Nguyen Van Giai, *Z. Phys.* **D20**, 277 (1991).
30. Ll. Serra, J. Navarro, M. Barranco and Nguyen Van Giai, *Phys. Rev. Lett.* **67**, 2311 (1991).
31. C. García-Recio, J. Navarro, Nguyen Van Giai, and L.L. Salcedo, *Ann. Phys.* (NY) **214**, 293 (1992).
32. S. Weisgerber and P.-G. Reinhard, *Phys. Lett.* **A158**, 407 (1991).
33. S. Weisgerber and P.-G. Reinhard, *Z. Phys.* **D23**, 275 (1992).
34. J. Dupont-Roc, M. Himbert, N. Pavloff, and J. Treiner, *J. Low Temp. Phys.* **81**, 31 (1990).
35. P. Tarazona, *Phys. Rev* **A31**, 2672 (1983).
36. S. Stringari, *Europhys. Lett.* **2**, 639 (1986).
37. S. Weisberger, P.G. Reinhard and C. Toepffer, in *Spin-polarized Quantum Systems*, Ed. S. Stringari, World Scientific (1988).
38. E.S. Hernández, M. Barranco, and A. Polls, *Phys. Lett.* **A171**, 119 (1992).
39. M. Barranco, D.M. Jezek, E.S. Hernández, J. Navarro, and Ll. Serra, *Z. Phys.* **D28**, 257 (1993).
40. M. Barranco, E.S. Hernández and J. Navarro, *Phys. Rev.* **B54** 7394 (1996).
41. O. Buu, L Puech and P.E. Wolf, *Phys. Rev. Lett.* **85**, 1278 (2000).
42. S.M. Gatica, E.S. Hernández and J. Navarro, *Phys. Rev.* **B58** 12300 (1998).
43. J. Boronat, J. Casulleras and V. Grau, *Cond. Matt. Theor.* **16**, in press (2001).
44. S.M. Gatica, E.S. Hernández and J. Navarro, *Phys. Rev.* **B60** 15302 (1999).
45. E.S. Hernández, *J. Low Temp. Phys.* **121**, 755 (2000).

46. L. Pricaupenko and J. Treiner, *Phys. Rev. Lett.* **72**, 2215 (1994).
47. F. Dalfovo, A. Lastri, L. Pricaupenko, S. Stringari and J. Treiner, *Phys. Rev.* **B52**, 1193 (1995).
48. M. Barranco, M. Pi, S. Gatica, E.S. Hernández and J. Navarro, *Phys. Rev.* **B56**, 8997 (1997).
49. N. Pavloff and L. Treiner, *J. Low Temp. Phys.* **83**, 15 (1991).
50. E. S. Hernández, *Recent Progress in Many Body Theories 11*, ed. R. F. Bishop, World Scientific, in press.
51. A.L. Fetter and J.D. Walecka, *Quantum Theory of Many-Particle Physics*, McGraw-Hill, New York, 1971.
52. M.M. Calbi, S.M. Gatica and E.S. Hernández, *Phys. Rev.* **B54**, 13097 (1996).
53. S. Das Sarma and E. H. Hwang, *Phys. Rev.* **B54**, 1936 (1996).
54. F. D. Manchester, *Rev. Mod. Phys.* **39**, 383 (1967).
55. M. J. DiPirro and F. M. Gasparini, *Phys. Rev. Lett.* **44**, 269 (1980).
56. F.M. Ellis, R.B. Hallock, M.D. Miller and R.A. Guyer, *Phys. Rev. Lett.* **46**, 1461 (1981).
57. B.K. Bhattacharya and F.M. Gasparini, *Phys. Rev. Lett.* **49**, 919 (1982).
58. B.K. Bhattacharya and F.M. Gasparini, *Phys. Rev.* **B31**, 2719 (1982).
59. R.A. Guyer, *Phys. Rev. Lett.* **53**, 795 (1984).
60. J.M. Vallies, Jr., R.H. Higley, R.B. Johnson and R.B.Hallock, *Phys. Rev. Lett.* **60**, 428 (1988).
61. D.T. Sprague, N. Alikacem, P.A. Sheldon and R.B. Hallock, *Phys. Rev. Lett.* **72**, 384 (1994).
62. N. Alikacem, R.B. Hallock, R.H. Higley and D.T. Sprague, *J. Low Temp. Phys.* **87**, 279 (1992).
63. R.B. Hallock, *J. Low Temp. Phys.* **121**, 441 (2000).
64. P.C. Ho and R.B. Hallock, *J. Low Temp. Phys.* **121**, 501 (2000).
65. H. M. Guo, D. O. Edwards, R. E. Sarwinski and J. T. Tough, *Phys. Rev. Lett.* **27**, 1259 (1971).
66. J.R. Eckardt, D.O. Edwards, P.P. Fatours, F.M. Gasparini and S.Y. Shen, *Phys. Rev. Lett.* **32**, 706 (1974).
67. M. Bretz, J.G. Dash, D.C. Hiuckernall, E.O. McLean and O.E. Vilches, *Phys. Rev.* **A3**, 15889 (1973).
68. P. Bloom, *Phys. Rev.* **B12**, 125 (1975).
69. D. Greywall, *Phys. Rev.* **B41**, 182 (1990).
70. D. Greywall and P.A. Busch, *Phys. Rev. Lett.* **65**, 64 (1990).
71. D. Greywall and P.A. Busch, *Phys. Rev. Lett.* **65**, 2788 (1990).
72. C. Bauerle, Yu. M. Bunkov, A. S. Chen, S. N. Fisher and H. Godfrin, *J. Low Temp. Phys.* **110**, 333 (1998).
73. C. Bauerle, J. Bossy, Yu. M. Bunkov, A. S. Chen, H. Godfrin and M. Roger, *J. Low Temp. Phys.* **110**, 345 (1998).
74. C. Bauerle, A. S. Chen, S. Triqueneaux, Yu. M. Bunkov, H. Godfrin and M. Roger, *J. Low Temp. Phys.* **113** 259 (1998).
75. M. Roger, C. Bäuerle, Yu. M. Bunkov, A.S. Chen and H. Godfrin, *Phys. Rev. Lett.* **80**, 1308 (1998).
76. D. Greywall, *Physica* **B197**, 1 (1994).
77. K. Ishida, M. Morishita, K. Yawata and H. Fukuyama, *Phys. Rev. Lett.* **79**, 3451 (1997).
78. H. Godfrin, *CJP* **65**, 1430 (1987).
79. H. Godfrin, R. Ruel and D. D. Osheroff, *Phys. Rev. Lett.* **60**, 305 (1988).

80. K.D. Morhard, J. Bossy and H. Godfrin, *Phys. Rev.* **B51**, 446 (1995).
81. K.D. Morhard, C. Bäuerle, J. Bossy, Yu. Bunkov, S.N. Fisher and H. Godfrin, *Phys. Rev.* **B53**, 2658 (1996).
82. E.V.L. de Mello and H. Godfrin, *J. Low Temp. Phys.* **108**, 407 (1997).
83. H. Godfrin and R. E. Rapp, *Adv. Phys.* **44**, 113 (1995).
84. S. M. Havens-Sacco and A. Widom, *J. Low Temp. Phys.* **40**, 357 (1990).
85. J.R. Engelbrecht, M. Randeria and L. Zhang, *Phys. Rev.* **B45**, 10135 (1992).
86. M.D. Miller and L.H. Nosanow, *J. Low Temp. Phys.* **32**, 145 (1990).
87. B. Brami, F. Joly and C. Lhuillier, *J. Low Temp. Phys.* **94**, 63 (1994).
88. M.M. Calbi and E.S. Hernández, *Phys. Rev.* **B57**, 13258 (1998).
89. V. Apaja and E. Krotscheck, A microscopic view of confined quantum liquids, 2001, this volume.
90. E. Cheng, M.W. Cole, W.F. Saam, and J. Treiner, *Phys. Rev. Lett.* **67**, 1007 (1991).
91. E. Cheng, M.W. Cole, W.F. Saam, and J. Treiner, *Phys. Rev.* **B46**, 13967 (1992).
92. E. Cheng, M.W. Cole, J. Dupont-Roc, W.F. Saam, and J. Treiner, *Rev. Mod. Phys.* **65**, 557 (1993).
93. F. Ancilotto, A. M. Sartori and F. Toigo, *Phys. Rev.* **B58**, 5085 (1998).
94. F. Ancilotto, F. Faccio and F. Toigo, cond-mat 0007126 (2000).
95. D. Ross, J.A. Phillips, J.E. Rutledge, and P. Taborek, *J. Low Temp. Phys.* **106**, 81 (1997).
96. M.M. Calbi and E.S. Hernández. *J. Low Temp. Phys.* **120**, 1 (2000).
97. H. R. Glyde, *Excitations in Liquid and Solid Helium*, Clarendon Press, Oxford, 1994.
98. B. E. Clements, H. Godfrin, E. Krotscheck, H. J. Lauter, P. Leiderer and C. J. Tymczak, *J. Low Temp. Phys.* **102**, 1 (1996).
99. A. Chizmeshya, M. W. Cole and F. Zaremba, *J. Low Temp. Phys.* **110**, 677 (1998).
100. M. M. Calbi, F. Toigo, S. M. Gatica and M. W. Cole, *Phys. Rev.* **B60**, 14935 (1999).
101. E. S. Hernández, J. Navarro, A. Polls and J. Ventura, *Nucl. Phys.* **A596**, 1 (1996).
102. D.M. Pines and P. Nozières, *The Theory of Quantum Liquids*, Addison Wesley, Mass. 1994.
103. B. E. Clements, H. Forbert, E. Krotscheck, H. J. Lauter, M. Saarela and C. J. Tymczak, *Phys. Rev.* **B50**, 6954 (1994).
104. B. E. Clements, E. Krotscheck and C. J. Tymczak, *Phys. Rev.* **B53**, 12253 (1996).
105. L. Pricaupenko and J. Treiner, *J. Low Temp. Phys.* **96**, 19 (1994).
106. S.M. Gatica, M.M. Calbi and M.W. Cole, *Phys. Rev.* **E59**, 4484 (1999).
107. S.M. Gatica, G. Stan, M.M. Calbi, J.K. Johnson, and M.W. Cole, *J. Low Temp. Phys.* **120**, 337 (2000).
108. T.W. Ebbesen, *Physics Today*, June 1996, p. 26.
109. W. Teizer, R.B. Hallock, E. Dujardin and T.W. Ebbesen, *Phys. Rev. Lett.* **82**, 5305 (1999).
110. H. Cho and G.A. Williams, *Phys. Rev.* **B**, in press (2001).
111. N. Wada, J. Taniguchi., H. Ikegami, S. Inagali, and Y. Fukushima, *Phys. Rev. Lett.* **86**, 4322 (2001).
112. M.M. Calbi, M.W. Cole, S.M. Gatica, M.J. Bojan, and G. Stan, *Rev. Mod. Phys.* **73**, 857 (2001).
113. G. Baym and C.J. Pethick, *Landau Fermi Liquid Theory: Concepts and Applications*, Wiley, New York, 1991.
114. K.K. Lehmann and G. Scoles, *Science* **279**, 2065 (1998).
115. D.M. Brink and S. Stringari, *Z. Phys.* **D15**, 257 (1990).
116. M. Hartmann, R.B. Miller, J.P. Toennies, and A.F. Vilesov, *Phys. Rev. Lett.* **75**, 1566 (1995).

117. A. Guirao, M. Pi and M. Barranco, *Z. Phys.* **D21**, 185 (1991).
118. J. Harms, M. Hartmann, J.P. Toennies, A.F. Vilesov, and B. Sartakov, *J. Mol. Spect.* **185**, 204 (1997)
119. J. Harms, M. Hartmann, B. Sartakov, J.P. Toennies, and A.F. Vilesov, *J. Chem. Phys.* bf 110, 5124 (1999).
120. S.H. Patil, *J. Chem. Phys.* **94**, 8089 (1991).
121. K.T. Tang and J.P. Toennies, *Z. Phys.* **D1**, 91 (1986).
122. J.P. Toennies, in *The Chemical Physics of Atomic and Molecular Clusters*, (North Holland, Amsterdam, 1990), p. 597.
J.P. Toennies and V.F. Vilesov, *Annu. Rev. Phys. Chem.*, **49**, 1 (1998).
123. K.B. Whaley, *Int. Rev. Phys. Chem.* **13**, 41 (1994).
K.B. Whaley, in Advances in Molecular Vibrations and Collision Dynamics, edited by J.M. Bowman and Z. Bacic (JAI Press, Greenwich, Conn., 1998), Vol. III, p. 397.
124. J.P. Toennies, Microscopic superfluidity of small ^{4}He and para-H_2 clusters inside Helium clsuters, 2001, this volume
125. E.W. Becker, *Z. Phys.* **D3**, 101 (1986).
126. A.P.J. van Deursen and J. Reuss, *J. Chem. Phys.* **63**, 4559 (1975); P.W. Stephen and J.G. King, *Phys. Rev. Lett.* **51**, 1538 (1983).
127. H. Buchenau, E.L. Knuth, J. Northby, J.P. Toennies, and C. Winkler, *J. Chem. Pjys.* **92**, 6875 (1990).
128. M. Lewerenz, B. Schilling, and J.P. Toennies, *Chem. Phys. Lett.* **206**, 381 (1993).
129. W. Schoellkopf and J.P. Toennies, *Science* **256**, 1345 (1993).
130. F. Luo, G.C. McBane, G. Kim, C.F. Giese, and W.R. Gentry, *J. Chem. Phys.* **98**, 3564 (2000).
131. W. Schoellkopf and J.P. Toennies, *J. Chem. Phys.* **104**, 1155 (1996).
132. R.E. Grisenti, W. Schöllkopf, J.P. Toennies, T. Köhler, and M. Stoll, *Phys. Rev. Lett.* **85**, 2284 (2000).
133. M. Lewerenz, *J. Chem. Phys.* **106**, 4596 (1997).
134. V.R. Pandharipande, S.C. Pieper and R.B. Wiringa, *Phys. Rev.* **B34**, 4571 (1986).
135. R.A. Aziz, F.R. McCourt and C.C.K. Wong, *Mol. Phys.* **61**, 1487 (1987).
136. R.A. Aziz, V.P.S. Nain, J.S. Carley, W.L. Taylor, and G.T. McConville, *J. Chem. Phys.* **70**, 4330 (1979).
137. M. Barranco, J. Navarro and A. Poves, *Phys. Rev. Lett.* **78**, 4729 (1997).
138. R.Guardiola and J. Navarro, *Phys. Rev. Lett.* **84**, 1144 (2000).
139. E. Manousakis, S. Fantoni, V.R. Pandharipande, and Q.N. Usmani, *Phys. Rev.* **B28**, 3770 (1983).
140. R. Guardiola, *Phys. Rev.* **B62**, 3416 (2000).
141. A.F. Andreev, *Sov. Phys. JETP* **23**, 939 (1966).
142. F. Dalfovo, *Z. Phys.* **D14**, 263 (1989).
143. A. Belić, F. Dalfovo, S. Fantoni and S. Stringari, *Phys. Rev.* **B49**, 15253 (1994).
144. D.O. Edwards and W.F. Saam, *Progress in Low Temperature Physics*, edited by D.F. Brewer (North Holland, Amsterdam, 1978), Vol. IIA, p. 283.
145. M. Pi, R. Mayol and M. Barranco, *Phys. Rev. Lett.* **82**, 3093 (1999).
146. S.M. Gatica, E.S. Hernández and M. Barranco, *J. Chem. Phys.* **107**, 927 (1997).
147. M. Barranco and E.S. Hernández, *Phys. Rev.* **B49**, 12078 (1994).
148. E.S. Hernández and M. Barranco, *Phys. Rev.* **B51**, 9364 (1995).
149. M. Barranco, F. Garcias, Ll. Serra, and M. Casas, *J. Low Temp. Phys.* **113**, 381 (1998).
150. F. Garcias, Ll. Serra, M. Casas, and M. Barranco, *J. Chem. Phys.* **115**, in press (2001).

151. M. A. McMahon, R. N. Barnett and K. B. Whaley, *J. Chem. Phys.* **104**, 5080 (1996).
152. M. Barranco, E.S. Hernández and C.E. Vignolo, *Z. Phys.* **D25**, 233 (1993).
153. F. Dalfovo, *Z. Phys.* **D29**, 61 (1994).
154. S. Goyal, D.L. Schutt and G. Scoles, *J. Phys. Rev.* **97**, 2236 (1993).
155. S.A. Chin and E. Krotschek, *Phys. Rev.* **B52**, 10405 (1995).
156. G. DeToffel, F. Ancilotto and F. Toigo, *J. Low Temp. Phys.* **102**, 381 (1996).
157. T. Pack, E. Piper, G.A. Pfeffer, and J.P. Toennies, *J. Chem. Phys. 80*, 4940 (1984).
158. F. Garcias, Ll. Serra, M. Casas, and M. Barranco, *J. Chem. Phys.* **108**, 9102 (1998).
159. S. Grebenev, J. P. Toennies and A. F. Vilesov, *Science* **279**, 2083 (1998).
160. S.A. Chin and E. Krotscheck, *Phys. Rev.* **B45**, 852 (1992).
161. P. Sindzingre, M.J. Klein, and D.M. Ceperley, *Phys. Rev. Lett.* **63**, 1601 (1986).
162. M.V. Krishna and K.B. Whaley, *J. Chem. Phys.* **93**, 746 (1990).
163. R.Guardiola, J. Navarro, and M. Portesi, *Phys. Rev.* **B60**, 6288 (1999).
164. M. Casas and S. Stringari, *J. Low Temp. Phys.* **79**, 135 (1990).
165. M.V. Krishna and K.B. Whaley, *Phys. Rev. Lett.* **64**, 1126 (1990).
166. F. Dalfovo, R. Mayol, M. Pi, and M. Barranco, *Phys. Rev. Lett.* **85**, 1028 (2000).
167. R. Mayol, M. Pi,M. Barranco, and F. Dalfovo, *Phys. Rev. Lett.* **87**, 145301 (2001).

CHAPTER 7

CAVITATION IN LIQUID HELIUM

M. Barranco, M. Guilleumas, M. Pi

Departament ECM, Facultat de Física. Universitat de Barcelona. 08028 Barcelona, Spain

D.M. Jezek

Departamento de Física, Facultad de Ciencias Exactas y Naturales. Universidad de Buenos Aires and CONICET. 1428 Buenos Aires, Argentina

The status of homogeneous cavitation in liquid helium is reviewed. Thermal and quantum regimes in pure and isotopic mixtures are described, and special attention is paid to experimental and theoretical findings together with questions that still remain open. The effect of vortices and electrons as seeds of heterogeneous nucleation is also discussed.

1. Introduction

Phase transitions under equilibrium conditions are experimentally well determined and take place in the coexistence regime, for example the normal boiling point of a liquid or the equilibrium freezing point. However, they do not always occur under equilibrium conditions. As the new phase forms, the free energy of the system is lowered, but the original phase can be held in a metastable state close to the equilibrium transition point. Superheated liquids and supercooled vapors are examples of metastable systems. Although they are internally stable, in each case there exists another configuration that has a lower thermodynamical potential. The metastable state is separated from the stable state by a thermodynamic barrier. Due to statistical fluctuations in density or concentration, this barrier can be overcome as the result of the formation and growth of small clusters of the new phase in the metastable state (bubbles in the liquid or droplets in the vapor; bubbles and droplets will be generically referred to as clusters). Nucleation is the process of the first localized appearance of a new stable phase in a metastable state which potentially plays a role in the dynamics of all first-order phase transition. The nucleation process can proceed by thermal activation or by quantum tunnelling, and depending on the energy barrier the nucleation rate will be very slow or very fast. When the rate is slow, large deviations from equilibrium may be required before the first localized cluster of the stable phase appears.

Within the classical theory of nucleation, [1] the grand potential of the growing cluster is evaluated in the capillarity approximation. In the case of nucleation for example, this means that the droplet is treated as a piece of bulk liquid limited by a sharp surface. Such a macroscopic approximation is intuitively appealing and allows for simple and sometimes analytical estimates, but it has at least two obvious shortcomings. Firstly, it cannot describe a surface region having a finite, and sometimes quite large thickness. And secondly, this theory is unable to take into account the presence of vapor inside the bubble that may appear as the temperature increases. Moreover, the capillarity approximation also neglects any compression effect on the central density of the droplet due to surface tension. This is what is often called a finite-size effect, as it is also the change in the energy of the cluster due to curvature corrections. These limitations, certainly important for small size clusters, can be overcome using the density functional theory (DFT) to describe the thermodynamical properties of the system. [2–6] Density functional theory has been the most successful approach in addressing nucleation in liquid helium so far.

Liquid helium is especially appealing for nucleation studies due to its particular features at low temperatures. Since it does not wet the walls of the experimental cells and the samples can be prepared free of impurities, this avoids undesired heterogeneous nucleation and allows the study of *homogeneous nucleation*, a true property of the bulk liquid. Moreover, both helium isotopes remain liquid at zero temperature. This fact can be potentially exploited to study the transition from thermal to quantum nucleation regimes. As ^{4}He is superfluid below the lambda temperature T_λ= 2.17 K and ^{3}He is in the normal phase down to 3 mK, the comparative study of both isotopes is expected to shed light on the role played by superfluidity in the manifestation of quantum tunnelling.

In recent years, theoretical and experimental interest has been focused on the study of the negative pressure region of the phase diagram of pure ^{3}He and ^{4}He liquids. [7] Obviously, liquid helium in this region can only be in a metastable state which is obtained by focusing sound waves generated by a hemispherical ultrasonic transducer into the bulk of the liquid. Negative pressures are produced during the negative part of the pressure swing at the acoustic focus. This technique was used in the early 80's by the Portland State University group, [8–10] and further developed by Balibar, Maris and co-workers. [4,11–13] The analysis of these experiments is complicated by the fact that only the static pressure and temperature of the experimental cell are known. Neither the pressure (P) nor the temperature (T) at the focus, where cavitation takes place, can be directly measured. This has caused gross errors in the past when trying to calibrate the potential voltage applied to the transducer in terms of the pressure generated at the focus.

Strictly speaking, cavitation is the formation of bubbles in a liquid held at a pressure below the saturation pressure at given temperature, and consequently it may occur at positive pressures and relatively high temperatures. Experimental results were obtained [8,9] for ^{4}He around 4 K, and for ^{3}He around 2.5 K which agreed with the classical theory of nucleation. Later on, these experiments were also succesfully described [6] within DFT. In this case cavitation is a thermally driven process. It is crucial for the application of density functional theory to the quantal regime, to reproduce this experimental situation because one aims at a unified description of the cavitation process in all regimes, and also because some of the ingredients entering the quantum description, such as critical configurations and barrier heights, are obtained as in the thermal regime.

This review is organized as follows. In Sections 2 and 3 we describe the nucleation process in the thermal and quantum regime, respectively. Nucleation in liquid mixtures is presented in Section 4, and some attention is paid to heterogeneous nucleation caused by the presence of quantized vortices.

2. Thermal nucleation

The formation of a new phase from a metastable phase proceeds through the formation of a 'critical cluster' of the novel phase. The nucleation rate J, *i.e.*, the number of critical clusters formed in the homogeneous system per unit time and volume, is given in the original Becker-Döring theory by the expression [1]

$$J = J_{0T} \exp(-\Delta\Omega_{max}/kT) \ , \tag{2.1}$$

where $\Delta\Omega_{max}$ is the free energy required to form the critical nucleus and therefore is the difference between the grand canonical potential of the critical cluster and that of the homogeneous metastable system, and k is the Boltzmann constant. The prefactor J_{0T} depends on the dynamics of the nucleation process, and there are many proposals of different degrees of complexity in the literature. [1,4,9,14,15] We shall see that J_{0T} can be varied by several orders of magnitude without appreciably changing the physical results. A simple estimate of J_{0T} consists in writing it as an 'attempting frequency per unit volume', $J_{0T} = kT/(hV_{cl})$, where V_{cl} is the volume of the critical cluster roughly represented by a sphere of 10 Å radius, [4,6] and h is the Planck constant.

It is quite instructive to use the capillarity model to obtain $\Delta\Omega_{max}$ in spite of the limitations mentioned above. In this approach, the nucleation barrier to create a cluster of radius R is written as a balance between surface and volume terms [14]

$$U(R) = \mathcal{S}R^2 - \mathcal{V}R^3 \ , \tag{2.2}$$

where in the case of bubbles one has

$$\mathcal{S} = 4\pi\sigma \ , \quad \mathcal{V} = \frac{4\pi}{3}\Delta P \tag{2.3}$$

where σ is the surface tension of the liquid free surface, and $\Delta P \geq 0$ is the pressure difference between the bubble and the bulk. The barrier has a maximum

$$U_{max} = \frac{4}{27}\frac{\mathcal{S}^3}{\mathcal{V}^2} \tag{2.4}$$

at a radius $R_c = 2\mathcal{S}/3\mathcal{V}$ which corresponds to the critical cluster configuration. The physical meaning of this barrier is clear, it prevents the metastable system from decaying into the stable phase. Thus, if a thermal fluctuation produces a cluster smaller than the critical cluster, it will spontaneously disappear, whereas if it is larger it will continue to grow into a macroscopic domain of the novel phase thus triggering the phase transition. The capillarity estimate of $\Delta\Omega_{max}$ is U_{max}. It is worth noting that U_{max} diverges at the saturation curve since $\Delta P = 0$.

Fig. 1 shows a schematic picture of the phase equilibrium diagram in the pressure-density plane (P, ρ), which may represent either helium isotope. Strictly speaking, the $T = 0$ isotherm corresponds to the ^{4}He case, since for ^{3}He at low densities one would have the pressure of the free Fermi gas, and the pressure should increase with density even at zero temperature. From now on, zero temperature for ^{3}He will always mean a very small temperature above 3 mK so that it is in the normal phase.

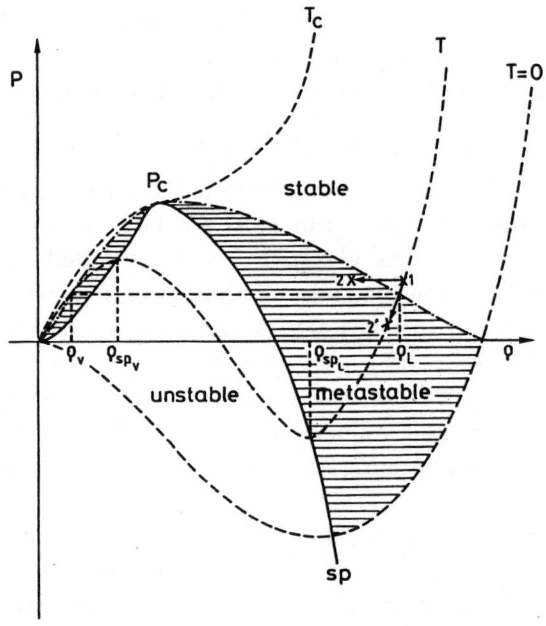

Fig. 1. Schematic representation of the liquid-gas phase equilibrium. The dash-dotted line is the two-phase equilibrium (saturation) line, and the solid line labelled *sp* is the spinodal line.

The region below the dashed-dotted line is the two-phase coexistence region. At given T, the densities of liquid and vapor in equilibrium are found imposing that the pressure and chemical potential μ of both phases are the same:

$$\mu(\rho_L, T) = \mu(\rho_V, T) \tag{2.5}$$
$$P(\rho_L, T) = P(\rho_V, T) \ .$$

These equations have a nontrivial solution only for T below a critical value T_c. For $T < T_c$ the two-phase equilibrium region splits into two domains. One is the unstable region in which the system cannot exist as a uniform phase. The other is a metastable region where the system can remain homogeneous until a small perturbation produces the two-phase equilibrium state. These domains are separated by the classical spinodal curve defined as

$$\left(\frac{\partial P}{\partial \rho}\right)_T = 0 \ . \tag{2.6}$$

The spinodal line is represented in Fig. 1 by the solid line labelled *sp*, and the metastability region corresponds to the hatched zone limited by the two-phase coexistence and the spinodal lines. Both curves are tangent at the critical point (P_c, T_c); three isotherms have also been drawn on the figure.

The system may be driven into a metastable state, for instance, by superheating it at constant P (going from point 1 to point 2; see Fig. 1) or decreasing P at constant T (going from point 1 to point 2'). These processes cause the system to cross the liquid-gas equilibrium line, penetrating into the metastable zone. How deeply it can be driven defines the degree of metastability attained in the process. It has a limit fixed by the spinodal line: if the system reaches this line, the homogeneous phase becomes macroscopically unstable and the liquid undergoes a phase separation.

The application of DFT to the nucleation problem starts with the determination of the liquid-gas equilibrium phase diagram. Within DFT one has a rather simple, analytical expression for the free energy density which in most applications has been written for either helium isotope as (or similar to) an expression of the kind: [4,16,17,18]

$$f(\rho, T) = f_{vol}(\rho, T) + \beta \frac{(\nabla \rho)^2}{\rho} + \xi (\nabla \rho)^2 \ , \tag{2.7}$$

where $f_{vol}(\rho, T)$ consists of a non-interacting part consisting of the well-known free energy density of a Bose or Fermi gas plus phenomenological density dependent terms that take into account the effective interaction of helium atoms in the bulk liquid. The parameters of both these terms and those of the density gradient terms in Eq. (2.7) are adjusted so as to reproduce physical quantities such as the equation of state (EOS) of the homogeneous liquid, and the surface tension of the liquid free surface. Knowledge of $f(\rho, T)$ enables solution of the phase equilibrium equations and to determine the spinodal line. It is worthwhile noticing that it also supplies an equation of state in the negative pressure regime, inaccessible to the experimental determination, through the thermodynamical relationship $f(\rho, T) = -P + \mu\rho$ with $\mu = \partial f(\rho, T)/\partial \rho|_T$. DFT has predicted spinodal lines (or spinodal points, if one restricts the analysis to $T = 0$) in excellent agreement with fully microscopic approaches. [19–22] For instance, the zero temperature ^{4}He spinodal point of Guirao *et al.* [17] is $P = -9.08$ atm, $\rho = 0.0159$ Å^{-3} as compared to that of Boronat *et al.*, [19] $P = -9.30$ atm, $\rho = 0.0158 \pm 0.0001$ Å^{-3}. The corresponding values for ^{3}He are $P = -3.06$ atm, $\rho = 0.0119$ Å^{-3} (Barranco *et al.* [16]), and $P = -3.08 \pm 0.10$ atm,

$\rho = 0.0121$ Å^{-3} (Casulleras and Boronat. [22]) The agreement is seen to be good, especially if one bears in mind that in these calculations the density functional has been used in a region which is far from the region where the parameters have been adjusted. It has been recently proposed [23] that the spinodal line of ^{3}He reaches a minimum value at 0.4 K. This behaviour has been related to the change in sign of the expansion coefficient of liquid ^{3}He along the saturation line at low temperatures. However, this is a rather small effect, and at $T = 0$ the spinodal points of Ref. 16 and 23 are both within the error bar of the microscopic value obtained by Casulleras and Boronat.

Once the phase diagram has been established, the application of DFT to the nucleation problem proceeds in two steps. One first determines the critical cluster for P and T values that correspond to metastable states. This allows calculation of the nucleation barrier $\Delta\Omega_{max}(P,T)$. Next, for a given T, the pressure is determined at which the produce of the nucleation rate Eq. (2.1) times the experimental volume times nucleation time $(V\tau)_e$ equals to a number which is conventionally chosen to be one, $[(V\tau)_e \cdot J = 1]$. This indicates that there is an appreciable probability of cluster formation which causes the onset of phase separation. This pressure is called homogeneous cavitation pressure (P_h), and it is one of the magnitudes (or equivalently the tensile strength defined as -P_h) that the experiments aim to determine.

At negative pressures (see Fig. 1) $|P_{sp}| \geq |P_h|$, and the importance of establishing the spinodal line becomes apparent: it fixes a limit to the degree of metastability the system may sustain, represented by P_h. It is quite obvious that P_{sp} can only be obtained theoretically; experimentally one only has access to P_h. The first experiments to determine P_h in ^{4}He seemed to yield values ~ -15 bar, [10] thus violating the above-mentioned inequality. This was first realized by Xiong and Maris, [24] and has been confirmed later on Ref. 5, 25 using liquid ^{4}He EOS's at negative pressures obtained from rather different approaches.

The density profile of the critical cluster is obtained by solving the Euler-Lagrange (EL) equation in spherical coordinates for the grand potential density $\omega(\rho, T) = f(\rho, T) - \mu\rho$:

$$\frac{\delta\omega}{\delta\rho} = 0 \tag{2.8}$$

imposing the physical conditions that $\rho'(0) = 0$, and $\rho(r \to \infty) = \rho_m$, where ρ_m is the density of the metastable homogeneous liquid. This supposes that the critical cluster is spherically symmetric, which is a plausible approximation in the case of homogeneous nucleation. The nucleation barrier $\Delta\Omega_{max}$ is obtained by subtracting the critical cluster grand potential $\omega(\rho, T)$ from that of the homogeneous metastable liquid:

$$\Delta\Omega_{max} = \int d\mathbf{r}[\omega(\rho,T) - \omega(\rho_m,T)] = \int d\mathbf{r}\,[f(\rho,T) - f_{vol}(\rho_m,T) - \mu(\rho - \rho_m)] \,. \tag{2.9}$$

Since $P = -f_{vol}(\rho_m, T) + \mu\rho_m$, Eq. (2.9) gives $\Delta\Omega_{max}$ as a function of P and T. Nucleation barriers $\Delta\Omega_{max}(P,T)$ are shown in Fig. 2. As indicated, the barriers diverge near the saturation curve, and become negligible when the system approaches the spinodal line. The latter fact is missed in the capillarity approximation, which is completely unphysical in this region. This is crucial for liquid helium: we will see in the next Section that cavitation occurs near the spinodal line for both isotopes. [26]

Eq. (2.1) has been solved to obtain P_h for two selected values of $(V\tau)_e$, namely 1 cm^3 sec, and 2.5×10^{-13} cm^3 sec that roughly correspond to the experimental conditions of Xiong and Maris. [27] The results are displayed in Fig. 3. Note, on the one hand, that agreement between theory and experiment is good, especially in the case of ^{4}He, and, on the other hand, that changing the prefactor by 13 orders of magnitude only produces a moderate shift in P_h. This is of relevance, as usually either J_{0T} and/or $(V\tau)_e$ are poorly known. This reflects the fact that the decay

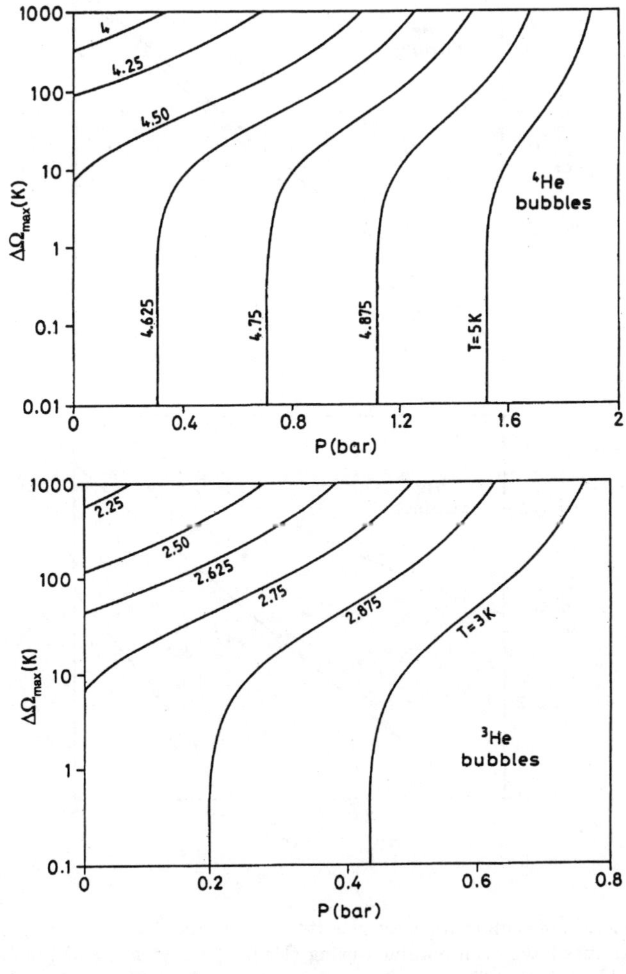

Fig. 2. Nucleation barriers as a function of pressure for several temperatures. Top panel corresponds to ^{4}He, and bottom panel to ^{3}He.

rate of the metastable phase depends drastically on the degree of metastability, which varies by several orders of magnitude within a narrow range of metastability (critical metastability). This allows for a sensible determination of P_h even if some aspects of the process are not very precisely known.

Fig. 4 shows the critical bubbles that correspond to the $P_h(T)$ configurations. Their size grows with T due to the decrease of the surface tension. Notice the filling of the bubble as T increases, and the increase of the surface diffuseness, effects also found in the description of the liquid free-surface.[16,17] The density of bulk liquid ^{4}He has a maximum at $T \sim 2.2$ K, and this is reflected in the curves shown in the

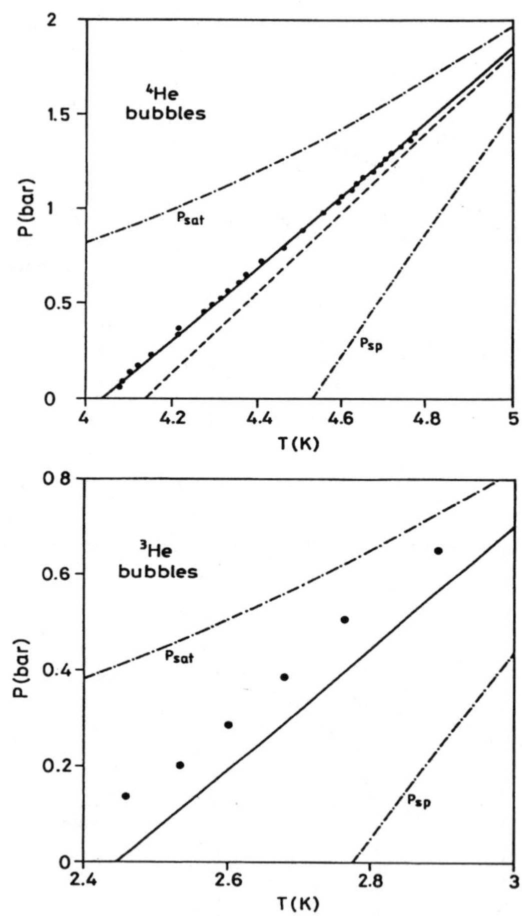

Fig. 3. P_h as a function of temperature for ^{4}He (top panel) and ^{3}He (bottom panel) and positive pressures. The solid lines have been obtained using $(V\tau)_e = 1$ cm^3 sec, and the dashed ones using $(V\tau)_e = 2.5 \times 10^{-13}$ cm^3 sec. The experimental points are from Sinha *et al.* [8] (^{4}He) and from Lezak *et al.* [9] (^{3}He). The saturation vapor pressure P_{sat} and spinodal P_{sp} lines are also indicated.

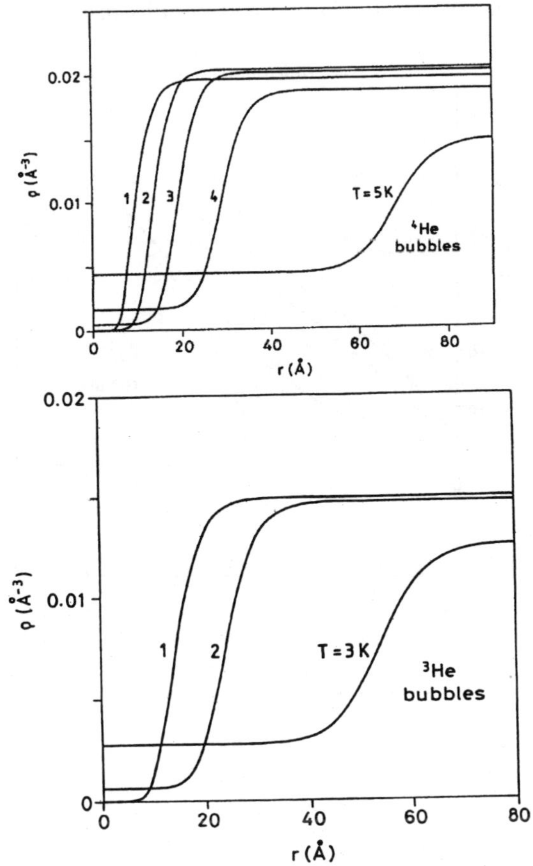

Fig. 4. Density profiles of the critical bubble configurations at the indicated temperatures. Top panel corresponds to ^{4}He, and bottom panel to ^{3}He.

top panel of this figure.

Fig. 5 displays P_h from $T = 0$ to near the critical point (cross). Below ~ 0.2 K the results are indicative, since we have not considered quantum tunnelling through the barrier. The lines have the same meaning as in Fig. 3, and the dots are the experimental points. [27] At first sight, there is a serious discrepancy between theory and experiment (see also Xiong and Maris [27]). It now seems well established that the disagreement is due to difficulties in relating the voltage applied to the transducer to the pressure it causes at the focus. [11,27,28] Although the experiments were subsequently continued, no attempt has been made to produce an experimental $P_h(T)$ curve. The unified treatment of the cavitation process that DFT offers from a region near the critical point region, where ^{4}He is in the normal phase (Fig. 3), to below the lambda point where it is superfuid, as well as the quantum-to-thermal transition which will be discussed in the next Section, leads us to believe that its results are likely to be reliable in the whole temperature range.

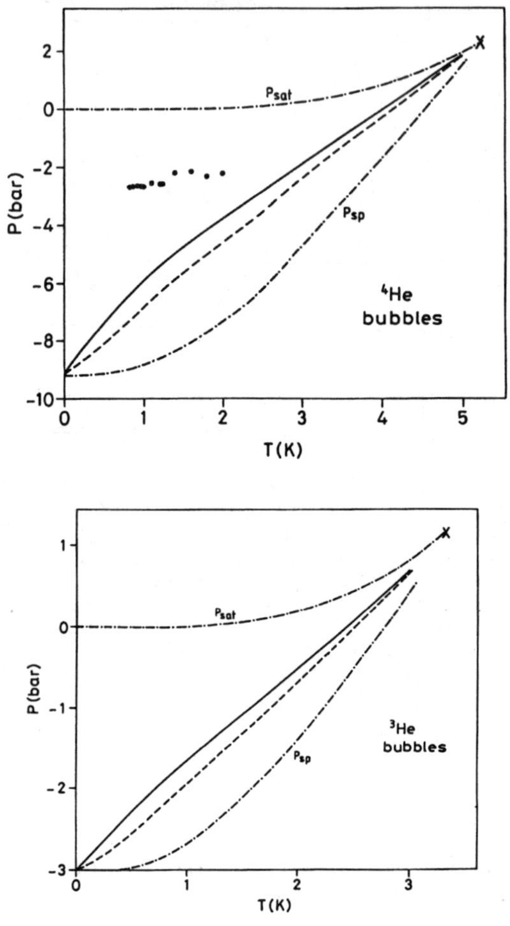

Fig. 5. Homogeneous cavitation pressure (solid line) for ^{4}He (top panel) and ^{3}He (bottom panel) as a function of temperature.

DFT has also been succesfully used to study heterogeneous cavitation produced by electrons introduced in the liquid by means of a β-rays source. [29,30,31] They produce large electron bubbles, of $\sim$ 20 Å radius, which act as cavitation seeds. As in most heterogeneous processes, the critical metastability is drastically reduced and cavitation then proceeds nearer the saturation line. In this case, the capillarity approximation works fairly well and encompasses the essential physics. [15,29] This can be seen by writing the energy of an electron bubble in liquid helium as the sum of the electron zero-point energy plus a surface and a volume term as in Eq. (2.2):

$$E(R) = \frac{\pi^2 \hbar^2}{2m_e R^2} + 4\pi R^2 \sigma - \frac{4\pi}{3} R^3 \Delta P \,, \tag{2.10}$$

where m_e is the electron mass, σ is the surface tension of the liquid, and ΔP is the pressure difference between the empty bubble and the bulk liquid. Minimizing with respect to the bubble radius, the equilibrium condition is obtained

$$-\frac{\pi \hbar^2}{4m_e R^5} + \frac{2\sigma}{R} - \Delta P = 0 \,. \tag{2.11}$$

This equation has a solution at any positive pressure, i.e., $\Delta P \leq 0$. In particular, at $\Delta P = 0$ one gets $R_{eq} = [\pi\hbar^2/(8m_e\sigma)]^{1/4} \sim 19$ Å for ^{4}He, and ~ 23 Å for ^{3}He. On the other hand, when the liquid is submitted to a tensile strength ($\Delta P > 0$) the electron bubble is metastable, and thus there exists a barrier between this state and the more stable state. When the pressure is decreased, the barrier also decreases and it vanishes at a pressure that we shall call P_{cr}. This value constitutes a lower pressure limit that can be attained by the system. The metastable electron bubble becomes unstable at a critical tensile strength that can be obtained from the equations

$$\begin{aligned} \frac{dE}{dR} &= -\frac{\pi^2\hbar^2}{m_e R^3} + 8\pi\sigma R - 4\pi R^2 \Delta P = 0 \\ \frac{d^2E}{dR^2} &= \frac{3\pi^2\hbar^2}{m_e R^4} + 8\pi\sigma - 8\pi R \Delta P = 0 \,. \end{aligned} \tag{2.12}$$

This yields $R_{cr} = [5\pi\hbar^2/(8m_e\sigma)]^{1/4}$ and $\Delta P = (8/5) \times [8m_e/(5\pi\hbar^2)]^{1/4}\sigma^{5/4}$. Consequently, liquid ^{4}He becomes macroscopically unstable at $P_{cr} \sim -2$ bars at $T \sim 0$ K, and liquid ^{3}He at $P_{cr} \sim -0.71$ bars. These pressures are much closer to zero than the spinodal pressures, and are an example of how the presence of impurities affects the cavitation process.

A fully quantitative description of this so-called electron bubble explosion phenomenon is only achieved if one considers that helium partially fills the bubble, and the electron is described quantum mechanically, allowing it to spread beyond the sharp bubble surface. These effects have been nicely incorporated in the DFT description. [30,31] It is worth mentioning that in the case of electron bubble explosions in ^{4}He, a calibration procedure was applied to convert the transducer voltage into a local pressure at the focus, which seems to work quite well. [30]

It has been recently argued [32] that electron bubbles might fission into two smaller daughter bubbles, each containing half of the original electron's wave function and each allowing the detection of fragments of the original electron. This interpretation has been questioned, [33] arguing that electrons in bubbles are not fractional but they are entangled.

Another interesting case of heterogeneous cavitation is that caused by the existence of quantized vortices in liquid ^{4}He below T_λ. The fact that the presence of quantized vortices results in a decrease of the tensile strength has been known experimentally for quite a long time. [34] Detailed DFT calculations [35,36] have indeed found this effect. Some of the experimental results on heterogeneous cavitation caused by electron bubbles have been interpreted as arising from electron bubbles attached to vortex rings previously created by the electrons moving in the liquid. [30]

Finally, we would like to point out that thermal nucleation of droplets in a supersaturated helium vapor has also been theoretically studied, [6] but to our knowledge no experimental data are available.

3. Quantum nucleation

The fact that nucleation is a thermally activated process at high enough temperatures and the energy barrier is overcome by the energy provided to the system by a thermal bath has been discussed in the previous Section. At low T this is no longer possible. However, nucleation may proceed by quantum tunnelling: the metastable state 'tunnels' through the energy barrier. The transition from one regime to the other is very abrupt, so that a thermal-to-quantum crossover temperature T^* may be defined by indicating whether nucleation takes place thermally ($T > T^*$) or quantically ($T < T^*$). In the limit of zero temperature, the transition is purely quantal, but for $T^* > T > 0$ thermally assisted quantum nucleation is the physical process. This is one of the many thermally assisted quantum tunnelling phenomena occuring in physics (see for example Chudnovsky [37] and references therein).

Quantum cavitation in superfluid ^{4}He has been observed by Balibar, Maris and co-workers, [12,38] and T^* has been determined. We want to stress that this determination relies on the theoretical EOS and cavitation barrier near the spinodal point, and, consequently, it is not a model-independent quantity. In this section we present a determination of T^* based on the application of the functional-integral approach (FIA) in conjunction with a density functional (DF) description of liquid helium. [39] This method gives T^* in the whole pressure range and overcomes the conceptual limitations of zero-temperature multidimensional WKB methods, [40] although, in practice, it yields quite similar results. The first detailed description of quantum cavitation in liquid helium was provided by Lifshitz and Kagan [14] who used the capillarity model near the saturation line and a kind of density functional-like description near the saturation line. In this Section we closely follow Guilleumas *et al.* [39] and Barranco *et al.* [41]

For $T < T^*$ the tunnelling rate is

$$J_Q = J_{0Q} \exp(-S^Q) , \tag{3.1}$$

where $\mathcal{P} = \exp(-S^Q)$ is the tunnelling probability, and the prefactor J_{0Q} is of the order of the number of nucleation sites per unit volume times an attempting frequency.

To obtain the tunnelling probability one formally starts from the statistical average of the transition probability over a time $t = t_f - t_i$:

$$\mathcal{P} = \frac{\sum_{i,f} |\langle\Psi_f| \exp\left[-\frac{i}{\hbar}\int_{t_i}^{t_f} dt\, \hat{\mathcal{H}}\right] |\Psi_i\rangle|^2 \exp(-E_i/kT)}{\sum_i \exp(-E_i/kT)} , \tag{3.2}$$

where $\hat{\mathcal{H}}$ is the Hamiltonian of the system, E_i are the energy levels, and Ψ_i and Ψ_f are the wave functions of the initial and final states. This expression can be written in a more workable form by using the path integral formulation of quantum mechanics, and its connection with statistical mechanics. According to Feynman [42] (see also Refs. 43, 44, 45), $\mathcal{P}$ can be written as the functional integral

$$\int D[q(\tau)] \exp\left[-\frac{1}{\hbar}\oint d\tau \mathcal{L}[q(\tau)]\right] , \tag{3.3}$$

where $\mathcal{L}[q(\tau)]$ is the imaginary-time ($\tau = \imath t$) classical Lagrangian of the system and $D[q(\tau)]$ denotes integration over all periodic trajectories $q(\tau)$ with period $\tau_p = \hbar/kT$ in the potential well that results from inverting the energy barrier. The integral in the exponent of Eq. (3.3) is the imaginary-time action $S(T) = \oint d\tau \mathcal{L}$ evaluated over the period τ_p. In the semiclassical limit $S(T) \gg \hbar$, the trajectory that contributes the most from all possible periodic orbits is that which minimizes the action. It leads to Eq. (3.1) with $S^Q = S_{min}(T)/\hbar$, where S_{min} is the minimum action.

The practical usefulness of Eq. (3.3) for the problem at hand is that physical insight allows one to guess $\mathcal{L}$ as a functional of densities and collective velocities instead of dealing with the impracticable Eq. (3.2). To implement this scheme, it is quite clear that a sound approximation is needed for the imaginary-time action. This means a realistic energy barrier and a simple yet reliable choice of the integration path $q(\tau)$.

Let us now work out in some detail the case in which only one collective coordinate δ is considered. This collective coordinate may represent, for instance, the displacement of the surface of the critical nucleus from its equilibrium position, or the density fluctuation from its metastable value. It is then a simple task to minimize the imaginary-time effective action:

$$S(T) = \oint d\tau \mathcal{L}[\delta(\tau)] = \int_{-\tau_p/2}^{\tau_p/2} d\tau \left[\frac{1}{2}M(\delta)\dot{\delta}^2 + \Delta\Omega(\delta)\right] . \tag{3.4}$$

As indicated, the effect of continuing the action to imaginary time is to invert the 'potential', i.e., $\Delta\Omega \to -\Delta\Omega$ in the equation of motion, and the identification of kT with $\hbar/\tau_p$; the path $\delta(\tau)$ defined in imaginary time τ has to fulfill the periodic boundary condition $\delta(-\tau_p/2) = \delta(\tau_p/2)$. This is illustrated in the schematic Fig. 6. We have supposed that the collective mass M depends on δ, which will be the practical case. [46] Imposing the extremum condition on the action yields the following equation of motion for $\delta(\tau)$:

$$M(\delta)\ddot{\delta} + \frac{1}{2}\frac{dM}{d\delta}\dot{\delta}^2 = \frac{d\Delta\Omega}{d\delta} . \tag{3.5}$$

Multiplying Eq. (3.5) by $\dot{\delta}$ we have

$$\frac{d}{d\tau}\left[\frac{1}{2}M(\delta)\dot{\delta}^2 - \Delta\Omega(\delta)\right] = 0 . \tag{3.6}$$

Thus

$$\frac{1}{2}M(\delta)\dot{\delta}^2 - \Delta\Omega(\delta) = constant \equiv -E \tag{3.7}$$

with $\Delta\Omega_{max} \geq E \geq 0$.

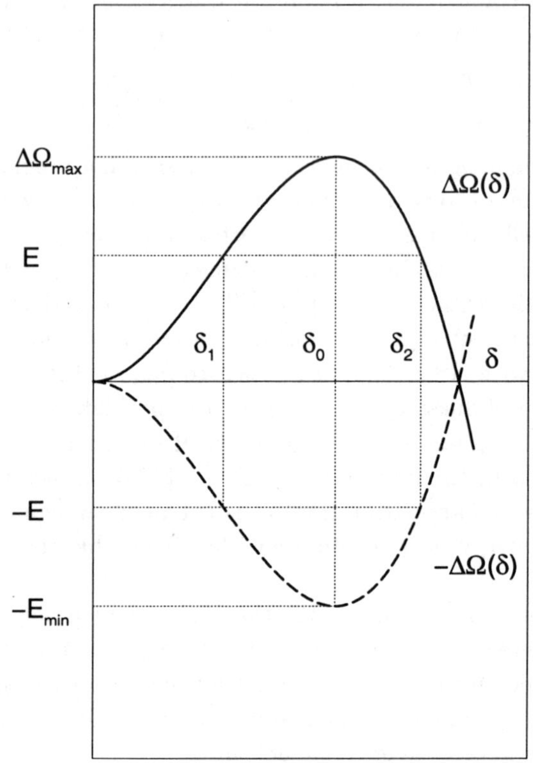

Fig. 6. Schematic barrier $\Delta\Omega(\delta)$ and inverted barrier well.

Eq. (3.7) has the trivial solution $\delta = \delta_0$ corresponding to the minimum of $-\Delta\Omega$: the system is 'at rest' at the bottom of the inverted barrier potential well. In this case, $E = \Delta\Omega_{max}$, and the integration of Eq. (3.4) yields

$$S^Q = \frac{1}{\hbar} S_{min}(T) = \frac{\Delta\Omega_{max}}{\hbar} \int_{-\tau_p/2}^{\tau_p/2} d\tau = \frac{\Delta\Omega_{max}}{kT} \, . \tag{3.8}$$

Thus, the trivial solution yields the exponent for classical thermal activation, Eq. (2.1). It means that within FIA, the transition between the thermal and quantal regime is smooth. For $E < \Delta\Omega_{max}$, one has to seek periodic solutions $\delta(\tau)$ whose turning points δ_1 and δ_2 are such that $\Delta\Omega(\delta_1) = \Delta\Omega(\delta_2) = E$ (see Fig. 6). Integrating Eq. (3.7) we get the period τ_p:

$$\tau_p(E) = 2 \int_{\delta_1(E)}^{\delta_2(E)} d\delta \sqrt{\frac{M(\delta)}{2[\Delta\Omega(\delta) - E]}} \, . \tag{3.9}$$

Using

$$d\tau = \sqrt{\frac{M(\delta)}{2[\Delta\Omega(\delta) - E]}}\, d\delta \tag{3.10}$$

the action Eq. (3.4) becomes

$$S^Q(E) = \frac{2}{\hbar}\int_{\delta_1(E)}^{\delta_2(E)} d\delta\, [2\Delta\Omega(\delta) - E]\sqrt{\frac{M(\delta)}{2[\Delta\Omega(\delta) - E]}}\,. \tag{3.11}$$

At $T = 0$, $\tau_p = \infty$ and $E = 0$. In this case, the solution to Eq. (3.7) is the usual instanton, [44] and $S^Q(E = 0)$ coincides with the WKB approximation at zero energy: [47]

$$S^{WKB} \equiv \frac{2}{\hbar}\int_{\delta_1(E=0)}^{\delta_2(E=0)} d\delta\, \sqrt{2M(\delta)\Delta\Omega(\delta)}\,. \tag{3.12}$$

The crossover temperature is obtained by equating Eqs. (2.1) and (3.1)

$$S^Q(E = \Delta\Omega_{max}) = \frac{\Delta\Omega_{max}}{kT^*}\,. \tag{3.13}$$

Taking a trajectory δ corresponding to $E \approx \Delta\Omega_{max}$ and using Eq. (3.11), we can write

$$S^Q(E \approx \Delta\Omega_{max}) \approx \frac{2}{\hbar}\Delta\Omega_{max}\int_{\delta_1(E\approx\Delta\Omega_{max})}^{\delta_2(E\approx\Delta\Omega_{max})} d\delta\, \sqrt{\frac{M(\delta)}{2\,[\Delta\Omega(\delta) - E]}}\,. \tag{3.14}$$

Comparison with Eq. (3.1) yields $\hbar/(kT^*) = \tau_p(E \approx \Delta\Omega_{max})$. An analytical expression for T^* is obtained by expanding Eq. (3.9) around the maximum of $\Delta\Omega$ located at δ_0:

$$kT^* = \frac{\hbar}{2\pi}\sqrt{-\frac{1}{M(\delta_0)}\frac{d^2\Delta\Omega}{d\delta^2}\bigg|_{\delta_0}}\,. \tag{3.15}$$

Eq. (3.15) shows that the value of T^* is determined by *small variations around* δ_0. This is a well-known result that was suggested long ago by Goldanskii, [48] the validity of which goes beyond the simple model we have used to derive it. Generally speaking, the crossover temperature is determined from the frequency of the small amplitude oscillations around the minimum of the inverted barrier potential well.

We note that S^{WKB} can be used to estimate T^* through the expression [4,49]

$$kT^*_{WKB} = \frac{\Delta\Omega_{max}}{S^{WKB}}\,. \tag{3.16}$$

When more realistic methods are employed to generate the critical nucleation configurations, such as for example DFT, the problem becomes an infinite-dimensional one whose solution is quite a formidable task, but essentially it follows the steps of the one-dimensional case; since T^* is expected to be small, one resorts to a zero temperature DF and first determines the critical cluster ρ_0 by solving Eq. (2.8) and the barrier height $\Delta\Omega_{max}$ from Eq. (2.9). The next step consists in describing the dynamics of the cavitation process in the inverted barrier well, whose equilibrium configuration corresponds to $\rho_0(r)$ and has an energy $-\Delta\Omega_{max}$. The imaginary-time Lagrangian $\mathcal{L}$ can be easily written supposing that the collective velocity $\vec{u}(\vec{r},t)$ associated with the bubble growth is irrotational, which is not a severe restriction as one expects only radial displacements (spherically symmetric bubbles). Introducing a velocity potential $s(\vec{r},t)$ such that $\vec{u}(\vec{r},t) = \nabla s(\vec{r},t)$, one obtains

$$\mathcal{L} = m\dot{\rho}s - \mathcal{H}(\rho, s) \,, \tag{3.17}$$

where $\mathcal{H}(\rho, s)$ is the imaginary-time Hamiltonian density:

$$\mathcal{H} = \frac{1}{2}m\rho\vec{u}^2 - [\omega(\rho) - \omega(\rho_m)] \,. \tag{3.18}$$

Hamilton's equations yield

$$m\dot{\rho} = \frac{\delta\mathcal{H}}{\delta s} = -m\nabla(\rho\vec{u}) \tag{3.19}$$

$$m\dot{s} = -\frac{\delta\mathcal{H}}{\delta\rho} \,. \tag{3.20}$$

Eq. (3.19) is the continuity equation. Taking the gradient of Eq. (3.20) one gets the equation of motion

$$m\frac{d\vec{u}}{dt} = -\nabla\left[\frac{1}{2}m\vec{u}^2 - \frac{\delta\omega}{\delta\rho}\right] \,. \tag{3.21}$$

Periodic solutions of Eqs. (3.19)-(3.21) are called thermons. [37] From Eq. (3.3), using Eqs. (3.17) and (3.19) one can write

$$S_{min} = \oint d\tau \int d\vec{r}\left[\frac{1}{2}m\rho\vec{u}^2 + \omega(\rho) - \omega(\rho_m)\right] . \tag{3.22}$$

To determine T^* we proceed as in the one-dimensional case: only the linearized version of Eqs. (3.19) and (3.21) is needed. Writing the T^* thermon as

$$\rho(r,t) = \rho_0(r) + \rho_1(r)\cos(\omega_p t) \,, \tag{3.23}$$

where $\rho_1(r) \ll \rho_0(r)$ and $\omega_p = 2\pi/\tau_p$, and keeping only first-order terms in $\vec{u}(r,t)$ and $\rho_1(r)$ we get

$$m\,\omega_p^2\rho_1 = \nabla\left[\rho_0(r)\,\nabla\left(\frac{\delta^2\omega}{\delta\rho^2}\bullet\rho_1(r)\right)\right] \,. \tag{3.24}$$

In the above equation $(\delta^2\omega/\delta\rho^2)\bullet\rho_1$ means that $(\delta\omega/\delta\rho)$ has been linearized keeping only terms in $\rho_1(r)$ and its r-derivatives.

Eq. (3.24) is a fourth-order linear differential, eigenvalue equation whose physical solutions have to fulfill $\rho_1'(0) = \rho_1'''(0) = 0$ and fall off exponentially to zero. The linearized continuity equation $\rho_1(r) \propto \nabla(\rho_0\vec{u})$ shows that the integral of $\rho_1(r)$ is zero when it is taken over the whole space, and guarantees that the thermon Eq. (3.23) is properly normalized.

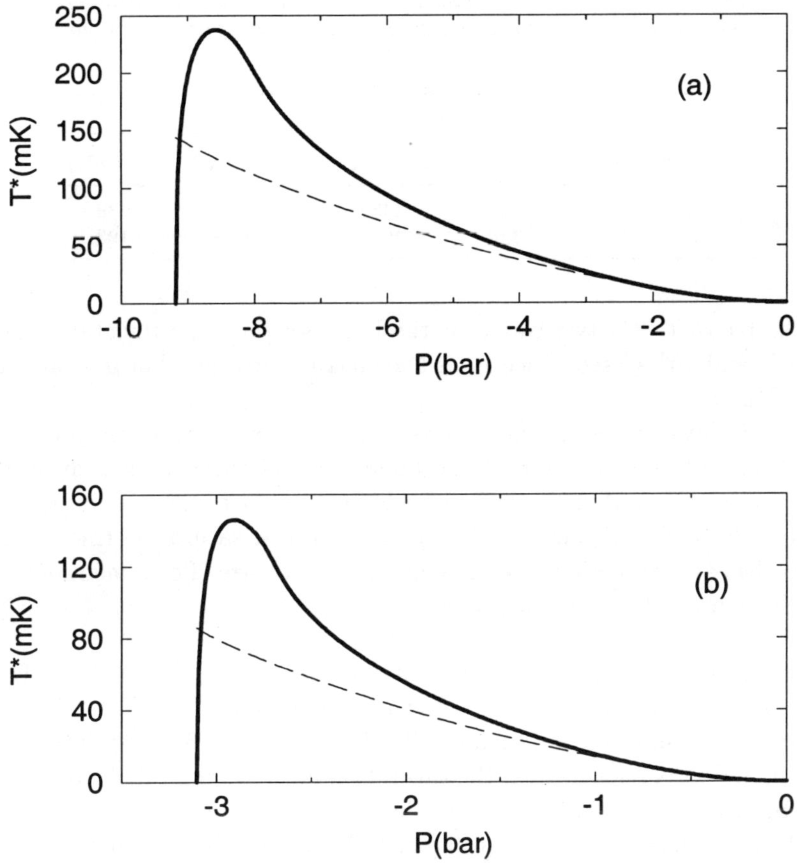

Fig. 7. T^* (mK) as a function of pressure (bar) (solid line) for ^{4}He (top panel) and ^{3}He (bottom panel). Also shown is the prediction of the capillarity model (dashed lines).

Fig. 7 shows $T^*(P)$ for $P \leq 0$. For both isotopes it presents a maximum rather near the spinodal point, falling to zero at the saturation and spinodal $T = 0$ points (recall we have been using a zero temperature DF). Note that T^* strongly depends on P in the spinodal region. In the case of ^{4}He the maximum is 238 mK at -8.58 bar, and for ^{3}He it is 146 mK at -2.91 bar. These maxima set a lower bound to thermal cavitation: irrespective of the negative pressure attained in the process, it is thermal-type cavitation if the temperature at the focus of the transducer is above $\sim$ 240 mK in the case of ^{4}He, and $\sim$ 150 mK in the case of ^{3}He. This constitutes one of the main theoretical results.

To determine which of the $T^*(P)$ displayed in Fig. 7 corresponds to the actual experimental conditions we have to calculate P_h either from Eq. (2.1) if $T > T^*$, or from Eq. (3.1) if $T < T^*$. In the latter case, we have taken $J_{0Q} = J_{0T}(T = T^*)$,

Table 1. Crossover temperatures T^* and homogeneous cavitation pressures P_h for two different values of the experimental conditions $(V\tau)_e$.

	^{3}He		^{4}He	
$(V\tau)_e$ (Å^3 sec)	P_h (bar)	T^* (mK)	P_h (bar)	T^* (mK)
10^{14}	-2.97	143	-8.57	238
10^4	-3.06	106	-8.99	198

and for the factor $(V\tau)_e$ two values at the limits of the experimental range, [38,40] namely 10^{14} and 10^4 Å^3 sec. Once P_h is known, the corresponding T^* value can be read from Fig. 7.

Table 1 displays the results and shows that the thermal-to-quantum cavitation transition takes place near the spinodal point. This is another fundamental result of the theory, which invalidates the use made in the past of the capillarity approximation near the spinodal point, [10,50] only justified near saturation (more precisely, only when the cavitation seeds are large and the finite size effects mentioned in the previous Section can be safely neglected).

A look at the density profiles of the critical cluster $\rho_0(r)$ and of the $\rho_1(r)$ component of the thermon discloses the power and flexibility of the FIA+DF method. As an example, we show several such profiles for ^{4}He in Fig. 8. Away from the spinodal region, and in particular near the saturation line, the critical cluster is a large, empty bubble. For large bubbles, $\rho_1(r)$ is localized at the surface, and the thermon is a well-defined 'surface vibration' (top panel). In the simple one-dimensional model described at the beginning of this Section, one might use the surface displacement as collective coordinate δ; the use of the capillarity approximation is justified in this region. This is apparent from Fig. 7, where we have also drawn (dashed line) the capillarity

$$T^* = \frac{1}{4\pi\sigma}\sqrt{\frac{\hbar^2}{m}\frac{|P|^3}{\rho_m}} \, ,$$

which can be analytically obtained from Eq. (3.15). Another case of interest in which the capillarity approximation works well is that of quantum cavitation from electron bubbles; [51] as we have seen in the previous Section, metastable electron bubbles are rather large, and the associated thermons are well-defined surface modes.

When the density inside the bubble becomes sizeable, the thermon is a mixed 'surface-volume vibration' (middle panel), which eventually becomes a pure 'volume vibration' in the spinodal region, where one might use a density fluctuation around ρ_m as collective coordinate. Our FIA+DF method handles all these situations on the same footing, producing a continuous $T^*(P)$ curve instead of delivering it in disconnected pieces of loose limits. [4,14]

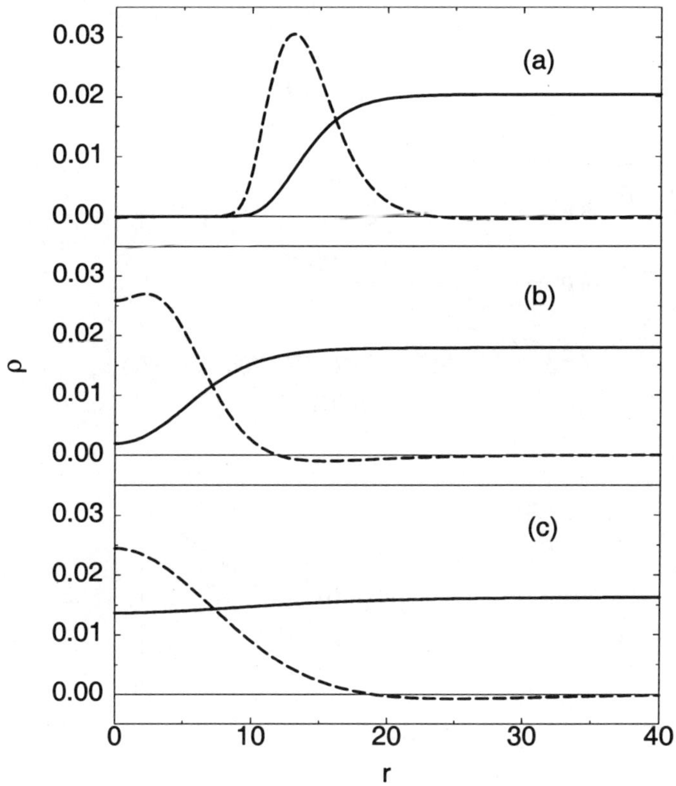

Fig. 8. For ^{4}He we show the critical bubble density profile $\rho_0(r)$ (solid line) and the density $\rho_1(r)$ of the T^* thermon (dashed line). The top panel corresponds to $P = -4.59$ bar, the middle panel to $P = -8.35$ bar, and the bottom panel to $P = -9.16$ bar. Units are Å^{-3} for ρ_0 and Å for r The scale for ρ_1 is arbitrary.

P_h as a function of T is displayed in Fig. 9 for both isotopes. The dashed line below T^* is the result obtained by continuing the thermal cavitation approach down to the spinodal point, which is reached at $T = 0$. It turns out that for ^{4}He agreement between theory and experiment is excellent. [12,39,40] We recall that the analysis of the experimental data to extract P_h and T^* requires the theoretical EOS at negative pressures as a crucial input. Given the agreement between the rather different theoretical approaches regarding this magnitude, we may consider that, at present, the problem of thermal and quantum cavitation in liquid ^{4}He is fairly well solved.

On the other hand, for ^{3}He the problem is still open. It has been found [13] that bubble nucleation in ^{3}He proceeds stochastically down to 40 mK. However, the process is thermal, in clear disagreement with the calculations [39,40] that predict $T^* \sim 120$ mK for ^{3}He. It is worth recalling that above $T \sim 3$ mK ^{3}He is a Fermi liquid in the normal phase, whereas ^{4}He is superfluid up to 2.17 K. Since the viscosity

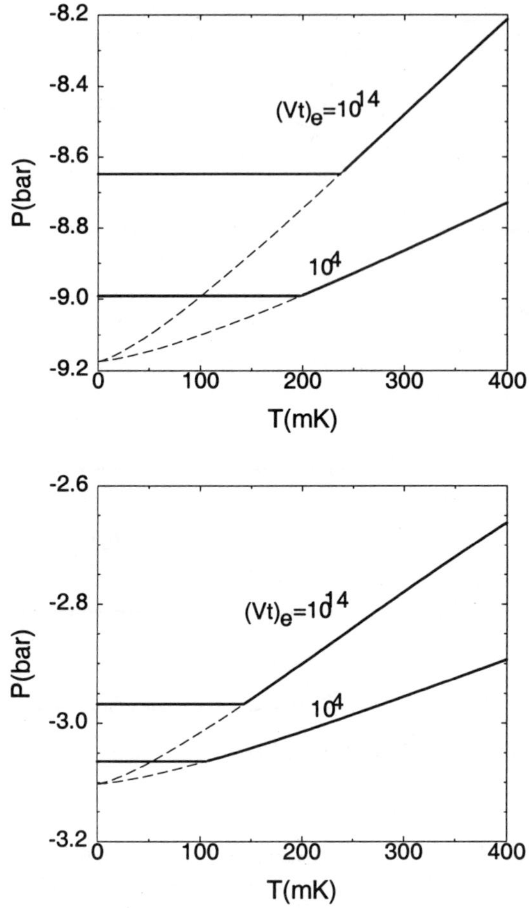

Fig. 9. Homogeneous cavitation pressure (solid line) as a function of temperature for ^{4}He (top panel) and ^{3}He (bottom panel).

of ^{3}He (which diverges at low T) has to be taken into account in the propagation of the sound waves in the transducer, this makes the experimental determination of the cavitation threshold voltage versus T curve especially complex. A plateau in this curve at low T would be the experimental signature of the existence of a crossover temperature. At present, an experiment is planned to extend that of Caupin *et al.* [13] to lower temperatures and to refine the interpretation of the raw data. [52]

In the previous Section we have seen that, on the one hand, thermal cavitation in ^{3}He is fairly well reproduced by DF calculations (Fig. 3), and, on the other hand, DF also yields an EOS in the negative pressure region in very good agreement with current microscopic calculations. [22] Consequently, if the absence of a crossover temperature is confirmed, it is very likely that the theoretical ingredient that has to be reexamined is the imaginary time dynamics. Two different ways of improvement have been proposed so far. It has been argued [53,54] that one should use zero-sound instead of first, ordinary-sound dynamics to describe cavitation in low-T liquid ^{3}He. While this seems quite adequate since ^{3}He is in the collisionless regime, some arguments have been presented [41] indicating that T^* is likely not to be very sensitive to whatever description one uses. The second way is to take into account dissipation effects while keeping an ordinary-sound dynamics description. In this case, it has been quantitatively shown [55] that a rather small dissipation decreases T^* sizeably. A general formalism that incorporates dissipative effects in the phase-separation kinetics of ^{3}He-^{4}He solutions has been developed by Burmistrov and Dubovskii [56] (see also Jezek *et al.* [57]).

The approach followed by Jezek *et al.* [55] to describe the dynamics in the dissipative regime while still being able to deal with inhomogeneous ^{3}He, which we have seen is crucial for a proper description of cavitation in liquid helium, consists in introducing a phenomenological Rayleigh dissipation function [58,59]

$$\mathcal{F} = \frac{1}{2}\,\xi\,\frac{\dot{\rho}^2}{\rho^2}\;. \tag{3.25}$$

¿From Lagrange's equations

$$\frac{\partial}{\partial t}\left(\frac{\delta \mathcal{L}}{\delta \dot{x}}\right) - \frac{\delta \mathcal{L}}{\delta x} = -\frac{\partial \mathcal{F}}{\partial \dot{x}}\;, \tag{3.26}$$

where x is either s or ρ, one gets the continuity and motion equation, respectively:

$$\dot{\rho} + \nabla(\rho\,\vec{u}) = 0 \tag{3.27}$$

$$m\left\{\frac{\partial\,u_k}{\partial\,t} + u_i \nabla_k u_i\right\} = -\nabla_k\left(\frac{\delta\omega}{\delta\rho}\right) + \nabla_k\left[\xi\frac{1}{\rho^2}\nabla(\rho\,\vec{u})\right]\;, \tag{3.28}$$

which for a homogeneous fluid ressembles the Navier-Stokes equation. [60] For liquid ^{3}He at low T, dissipation depends on the quasiparticle mean free path, and a precise estimation of the magnitude of this effect in the tunnelling process is difficult. If the scope is just to explore the effect of a small viscosity on T^*, one may adopt the pragmatic viewpoint of obtaining results for different ξ's close to the macroscopic viscosity coefficient value. Using this approach, one should bear in mind that dissipation effects are probably overestimated.

Upon linearization, we end up with the following equation for ω_p and $\rho^1(r)$:

$$\left[m\,\omega_p^2 - \mathcal{M}_1 - \xi\omega_p\mathcal{M}_2\right]\rho^1(r) = 0\;. \tag{3.29}$$

The differential operators $\mathcal{M}_1$ and $\mathcal{M}_2$ in Eq. (3.29) are, respectively, the linearization of

$$\nabla\left\{\rho\nabla\left(\frac{\delta\omega}{\delta\rho}\right)\right\} \text{ and } \nabla\left\{\rho\nabla\left(\frac{1}{\rho^2}\right)\right\}. \tag{3.30}$$

Since ξ depends on the density [61] as $\rho^{5/3}$, in actual calculations use has been made of a local density approximation, taking the expression $1/(\rho_{sat}^{5/3}\rho^{1/3}(r))$ as form factor in Eq. (3.25), where ρ_{sat} is the density of the liquid at $T = 0$ and $P = 0$, and ξ is then density-independent.

We show T^* in Fig. 10 as a function of pressure for different ξ values. In particular, $\xi = 100$ μP roughly corresponds to the experimental value [62] of the viscosity coefficient at $P = 0$ and $T = 100$ mK. The results displayed in this figure indicate that for liquid ^{3}He even a moderate dissipation may reduce the crossover temperature by a non-negligible amount, by displacing the homogeneous cavitation pressure towards the spinodal value. Viscosity may then be the reason for the inconclusive results for quantum cavitation reported by Caupin *et al.* [13] which, if confirmed, would indicate that dissipation plays a crucial role in quantum cavitation in liquid helium. The experimental study of cavitation in undersaturated ^{3}He-^{4}He mixtures might then uncover a very rich structure, since ^{4}He is still superfluid and ^{3}He is in the normal phase. This would open the possibility of studying the influence of dissipation in the cavitation process by varying the ^{3}He concentration. Some aspects of this problem are discussed in the next Section.

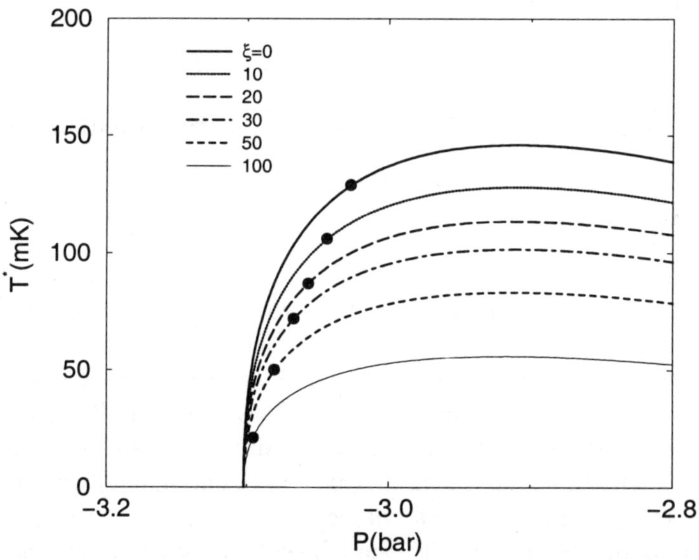

Fig. 10. T^* as a function of pressure for different ξ values (in μP). $P_h(T^*)$ is shown as circles for $(V\tau)_e = 10^8$ Å^3 sec.

4. Nucleation in ^{3}He-^{4}He liquid mixtures

^{3}He-^{4}He liquid mixtures have also been recognized as another ideal system to study nucleation phenomena. On the one hand, these mixtures can be made extremely pure, and on the other, ^{4}He covers the walls of the experimental cells, preventing undesired heterogeneous nucleation that is difficult to avoid in other substances. We will see, however, that the possible presence of quantized vortices in supersaturated mixtures, acting as nucleation seeds, may have a dramatic effect on the nucleation process.

Experiments on supersaturated helium mixtures were carried out in the sixties, [63,64] but one had to wait more than twenty years for a systematic study of this phenomenon. [65–69] These experiments are carried out under quite different conditions than those aiming to study cavitation in pure helium. In particular, the time scale of the supersaturation process is considerably large, of the order of 1-3 hours, [65,67] whereas the ultrasound pulses used for cavitation last between 30 and 70 μsec. [12] The kinetics of nucleation in dilute helium mixtures was first addressed by Lifshitz *et al.* [70]

Before studying nucleation, it is again convenient to delimit the metastable region where it may occur. The phase diagram is now richer than for the pure substance, and we will only discuss the $T = 0$ case. Let ρ_i be the particle density of the ^{i}He isotope and ρ the total density $\rho = \rho_3 + \rho_4$. The boundaries of the different regions in the (P, x) plane, where x is the ^{3}He concentration ($x = \rho_3/\rho$), can be determined as follows. Necessary and sufficient stability conditions for a binary system are given by the inequalities on the compressibility

$$K = \left(\frac{\partial P}{\partial \rho}\right)_x \geq 0 \tag{4.1}$$

as in the pure case, Eq. (2.6), and the chemical potentials

$$\text{either} \quad \left(\frac{\partial \mu_4}{\partial x}\right)_P \leq 0 \quad \text{or} \quad \left(\frac{\partial \mu_3}{\partial x}\right)_P \geq 0 \ . \tag{4.2}$$

A positive compressibility guarantees mechanical stability, whereas the condition on the chemical potentials (one inequality implies the other) ensures diffusive stability. [71] Taken as equalities, the above equations determine two curves on the (P, x) plane which are shown in Fig. 11 as a dot-dashed line (Eq. (4.1)) and as a dashed line (Eq. (4.2)). This phase diagram was calculated by Guilleumas *et al.* [72] using the zero temperature DF proposed by Dalfovo [73] for the mixture. It reads:

$$f(\rho_3, \rho_4) = f_{vol}(\rho_3, \rho_4) + \frac{\hbar^2}{2m_4}\tau_4 + \frac{\hbar^2}{2m_3^*}\tau_{3s} + d_4(\nabla\rho_4)^2 + d_3(\nabla\rho_3)^2 + d_{34}\nabla\rho_4 \cdot \nabla\rho_3 \ , \tag{4.3}$$

where

$$f_{vol}(\rho_3, \rho_4) = \frac{1}{2}b_4\rho_4^2 + \frac{1}{2}c_4\rho_4^2\rho^{\gamma_4} + \frac{\hbar^2}{2m_3^*}\tau_{3v} + \frac{1}{2}b_3\rho_3^2 + \frac{1}{2}c_3'\rho_3^2\rho^{\gamma_3} + \frac{1}{2}c_3''\rho_3^{2+\gamma_3} + b_{34}\rho_3\rho_4 + c_{34}\rho_3\rho_4\rho^{\gamma_{34}} \ , \tag{4.4}$$

where τ_i are the corresponding kinetic energy densities and m_3^* is the ^{3}He effective mass. They are taken to be:

$$\tau_4 = \frac{1}{4}\frac{(\nabla\rho_4)^2}{\rho_4} \tag{4.5}$$

$$\tau_3 \equiv \tau_{3v} + \tau_{3s} = \left[\frac{3}{5}(3\pi^2)^{2/3}\rho_3^{5/3}\right] + \left[\frac{1}{18}\frac{(\nabla\rho_3)^2}{\rho_3} + \frac{1}{3}\Delta\rho_3\right] \tag{4.6}$$

$$m_3^* = m_3\left(1 - \frac{\rho_3}{\rho_{3c}} - \frac{\rho_4}{\rho_{4c}}\right)^{-2} . \tag{4.7}$$

The parameters of the functional can be found in the references. They have been fixed so as to reproduce the zero temperature EOS of both isotopes, the maximum solubility of ^{3}He in ^{4}He at zero pressure and the surface tensions of the different interfaces. Other quantities, such as osmotic pressure, P-dependence of the maximum solubility and excess volume of ^{3}He in ^{4}He are also well reproduced.

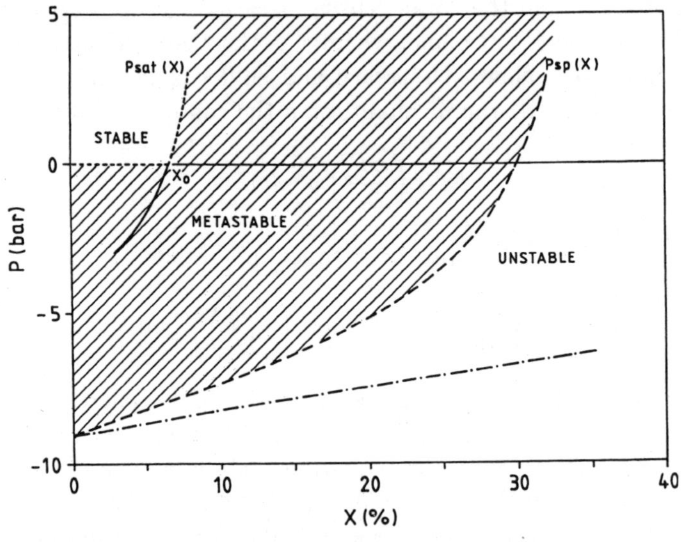

Fig. 11. DF phase diagram of the ^{3}He-^{4}He liquid mixture at $T = 0$.

It can be seen from Fig. 11 that condition Eq. (4.2) is violated first and thus defines the spinodal line $P_{sp}(x)$, where the mixture becomes macroscopically unstable. At $P = 0$, it cuts the x axis at $x_{sp} \sim 30\%$. To draw the border between metastable and stable regions one has to obtain the maximum concentration (or saturation) line by solving the two-phase equilibrium conditions

$$\begin{aligned} P(\rho, x) &= P(\rho_{3p}, x = 1) \\ \mu_3(\rho, x) &= \mu_3(\rho_{3p}, x = 1) \ , \end{aligned} \tag{4.8}$$

where ρ_{3p} is the density of segregated (pure) ^{3}He. These equations determine the curve denoted as $P_{sat}(x)$ in the (P, x) plane. It is interesting to see that Eqs. (4.8) have solutions at negative pressures down to the value corresponding to the spinodal point of pure ^{3}He.

Experimentally, [63,74] the ^{3}He concentration in the mixture at saturation is $x_s \sim$ 6.6% at $P = 0$, which reaches a maximum value of $\sim 9.4\,\%$ at $P \sim 10$ atm. However, supersaturated ^{3}He-^{4}He mixtures can be found in a metastable state for concentrations above the saturation value; at $P \geq 0$, the formation of critical ^{3}He drops is responsible for phase separation.

As in the pure case, another kind of metastability arises from the application of a tensile strength that drives the system into the negative pressure region. Although no experimental information for mixtures is available up to now in the $P < 0$ regime, its study constitutes a natural extension of that carried out in pure liquid helium. In the case of mixtures, the system either will develop a free surface for low x values, or will segregate ^{3}He, producing a mixture-pure ^{3}He interface as in the $P \geq 0$ regime. Phase separation will proceed in the first case by nucleating bubbles 'coated' with ^{3}He, and in the second case by nucleating ^{3}He-rich drops. These processes have been thoroughly studied [72,75] and will be briefly discussed below.

We would like to point out that as opposed to the case of pure helium liquids, no fully microscopic determination of $P_{sp}(x)$ exists. Even $P_{sat}(x)$ is extremely difficult to obtain microscopically. The best variational calculation carried out so far [76] yields $x_s(P = 0) \sim 1\%$. However, phenomenological DF calculations reproduce the phase diagram region around the saturation line once the experimental $x_s(P = 0)$ value is used for fixing some of the density functional free parameters. [72,73] Whether the predicted phase diagram in Fig. 11 is accurate or not in the spinodal region remains to be seen. This is of little practical relevance for nucleation, since at present the region of experimental interest is that near the $P \geq 0$ saturation line. The dynamics of phase separation in liquid ^{3}He-^{4}He mixtures near the tricritical point at $T \sim 0.867$ K has been experimentally investigated in the past. [77] We will only address the low temperature regime below $\sim 0.15 - 0.2$ K, as it is the only regime that can be sensibly studied by current density functionals for liquid helium mixtures. [72,73,78]

Let us now discuss the appearance of cavitation in undersaturated mixtures. [72] The supersaturated case has been addressed by Guilleumas *et al.* [75] using the same method. As in the pure case, the density profiles of the critical cluster ρ_3^0, ρ_4^0, are obtained by solving the coupled EL equations for the grand potential density $\omega(\rho_3, \rho_4) = f(\rho_3, \rho_4) - \mu_3\rho_3 - \mu_4\rho_4$:

$$\frac{\delta\omega}{\delta\rho_3} = 0 \qquad \frac{\delta\omega}{\delta\rho_4} = 0\,, \tag{4.9}$$

imposing, as before, the physical conditions that $\rho_i'(0) = 0$, and $\rho_i(r \to \infty) = \rho_{im}$. The cavitation barrier $\Delta\Omega_{max}$ is

$$\Delta\Omega_{max} = \int d\mathbf{r}\,[f(\rho_3, \rho_4) - f_{vol}(\rho_{3m}, \rho_{4m}) - \mu_3(\rho_3 - \rho_{3m}) - \mu_4(\rho_4 - \rho_{4m})]\ . \tag{4.10}$$

This equation gives $\Delta\Omega_{max}$ as a function of x and pressure $P = -f_{vol}(\rho_{3m}, \rho_{4m}) + \sum_i \mu_i \rho_{im}$. Several cavitation barriers are shown in Fig. 12 for $x = 0$ and $1\% - 6\%$.

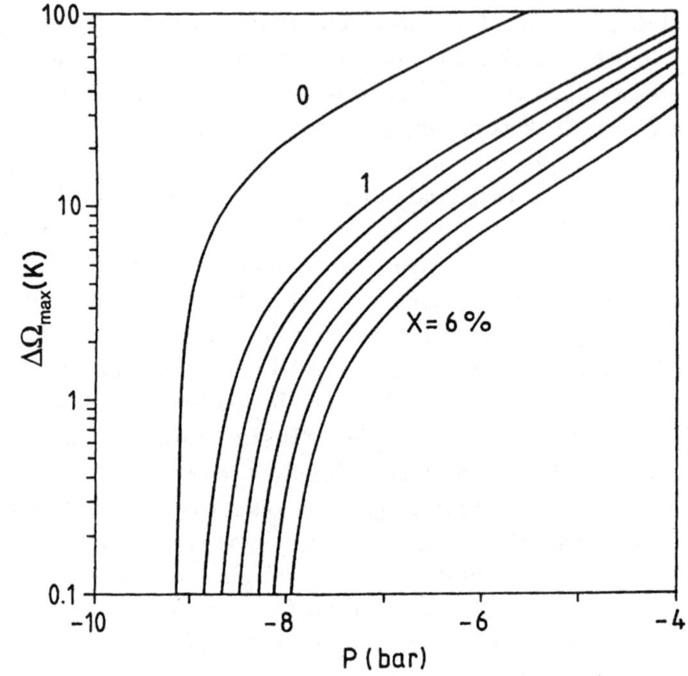

Fig. 12. From top to bottom, cavitation barriers as a function of P for pure ^{4}He and for ^{3}He concentrations $x = 1\% - 6\%$.

In Fig. 13 we display some density profiles corresponding to the critical bubbles for $x = 4\%$ and $P = -7, -5, -4,$ and -2 bar. As mentioned, the cavitation seed evolves from a ^{4}He bubble coated with ^{3}He to a ^{3}He-rich droplet. It is quite apparent from this figure that the flexibility of the DF approach is especially indicated for helium mixtures, where the limited miscibility of both isotopes at low T and the existence of surface Andreev states make the shape of the nucleation clusters hard to guess and mimic by means of simple-minded sharp-surface models.

Thermally assisted quantum cavitation in undersaturated solutions of ^{3}He in ^{4}He has also been studied within FIA+DF. [79] For mixtures, the method becomes extremely cumbersome to apply, but proceeds as in the pure case. Assuming that only spherical bubbles develop, the collective velocities $\vec{u}_q(\vec{r}, t), q = 3, 4$ of both helium fluids are irrotational and one can then define a velocity potential field $s_q(\vec{r}, t)$ for each isotope such that $\vec{u}_q(\vec{r}, t) = \nabla s_q(\vec{r}, t)$. It follows that

$$\mathcal{L} = \sum_q m_q \dot{\rho}_q s_q - \mathcal{H}(\rho_3, \rho_4, s_3, s_4) , \tag{4.11}$$

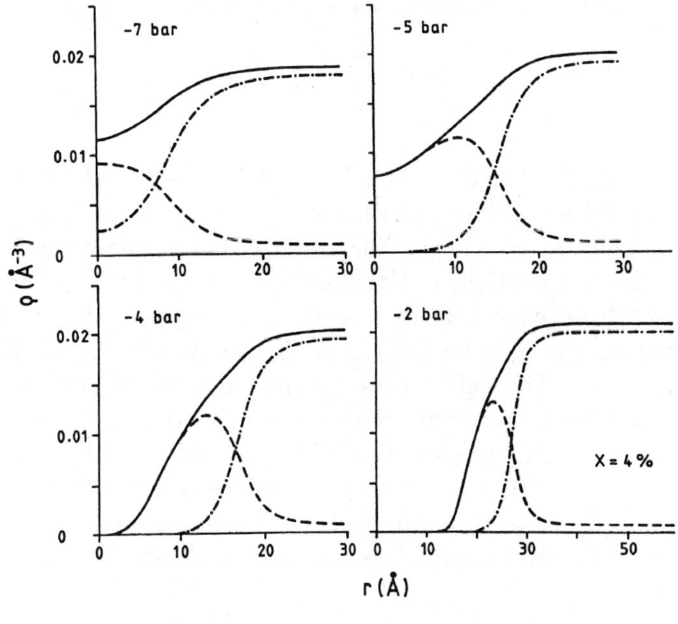

Fig. 13. Density profiles corresponding to the critical bubbles for $x = 4\%$ and several values of P.

where $\mathcal{H}(\rho_q, s_q)$ is the imaginary-time Hamiltonian density

$$\mathcal{H}(\rho_3, \rho_4, s_3, s_4) = \frac{1}{2} \sum_q m_q \rho_q \vec{u}_q{}^2 - [\omega(\rho_3, \rho_4) - \omega(\rho_{3m}, \rho_{4m})] \ . \tag{4.12}$$

Hamilton's equations yield the following four equations :

$$m_q \dot{\rho}_q = \frac{\delta \mathcal{H}}{\delta s_q} = -m_q \nabla(\rho_q \vec{u}_q) \tag{4.13}$$

$$m_q \dot{s}_q = -\frac{\delta \mathcal{H}}{\delta \rho_q} \ . \tag{4.14}$$

Eqs. (4.13) are the continuity equations. Taking the gradient of Eqs. (4.14) one gets the equations of motion

$$m_q \frac{\mathrm{d}\vec{u}_q}{\mathrm{dt}} = -\nabla \left\{ \frac{1}{2} m_q \vec{u}_q{}^2 - \frac{\delta \omega}{\delta \rho_q} \right\} \ . \tag{4.15}$$

To determine T^* one has to again find the small amplitude, periodic solutions of Eqs. (4.13) and (4.15) linearized around ρ_3^0 and ρ_4^0. Defining the 'transition densities' $\rho_q^1(r)$ as

$$\rho_q(\vec{r}, t) \equiv \rho_q^0(r) + \rho_q^1(r) \cos(\omega_p t) \ , \tag{4.16}$$

and keeping only first-order terms in $\vec{u}_q(r,t)$ and in $\rho^1_q(r)$, one gets:

$$\omega_p^2 \rho_q^1(r) = \frac{1}{m_q} \nabla \left[\rho_q^0(r) \nabla \left(\sum_{q'=3,4} \frac{\delta^2 \omega}{\delta\rho_q \delta\rho_{q'}} \bullet \rho^1_{q'}(r) \right) \right] \quad , \quad q = 3, 4. \tag{4.17}$$

In this equation, $\frac{\delta^2 \omega}{\delta\rho_q \delta\rho_{q'}} \bullet \rho^1_{q'}(r)$ means that $\delta\omega/\delta\rho_q$ has to be linearized, keeping only terms in ρ^1_3 and ρ^1_4, and their derivatives.

Eq. (4.17) is a fourth-order linear differential, eigenvalue equation for the 'vector' $(\rho^1_3(r), \rho^1_4(r))$. Physical solutions to Eqs. (4.17) have to fulfill $(\rho^1_q)'(0) = (\rho^1_q)'''(0) = 0$, and have to fall exponentially to zero at large distances. From the linearized continuity equation $\rho^1_q(r) \propto -\nabla(\rho^0_q \vec{u}_q)$, it is obvious that the integral of $\rho^1_q(r)$ over the whole space is zero. For a given pressure and ^{3}He-concentration, only a positive eigenvalue ω_p^2 has been found, from which we get $T^* = \hbar\omega_p/2\pi$.

Fig. 14 shows T^* (mK) as a function of P (bar) for $x = 0.1, 1, 2, 3, 4$ and 5 %. Compared to the pure ^{4}He case [39] (Fig. 7, top panel), $T^*(P)$ has now a more complex structure. It is worth noting that the maximum of the $T^*(P)$ curve has decreased from ~240 mK for pure ^{4}He down to ~140 mK for ^{3}He-concentrations as small as 1 %.

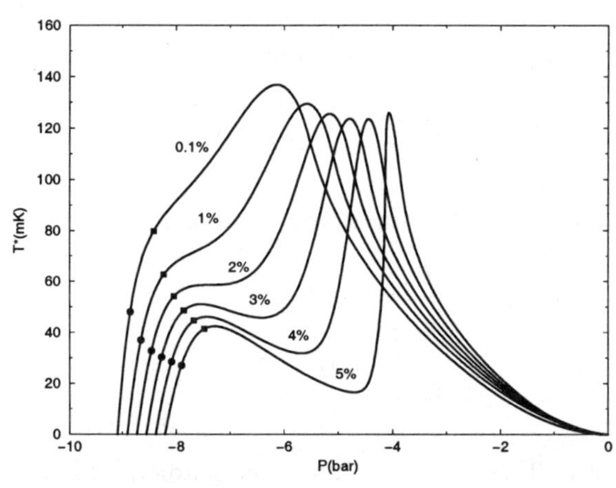

Fig. 14. T^* as a function of P for different ^{3}He concentrations. $P_h(T^*)$ is shown as circles(squares) for $(V\tau)_e = 10^4$ Å^3 sec (10^{14} Å^3 sec), respectively.

Fig. 15 shows two different bubble configurations for $x = 1\%$. Configuration (a) corresponds to $P = -8$ bar and $T^* = 67.6$ mK, and configuration (b) to $P = -5$ bar and $T^* = 102.1$ mK. The solid lines represent the ^{3}He and ^{4}He critical bubble densities in Å^{-3}, and the dashed (dash-dotted) lines represent $\rho^1_3(r)$ $(\rho^1_4(r))$

in arbitrary units. Near the spinodal region, the 'bubble' configuration is filled with ^{3}He: the surface tension that matters for bubble formation is that of the ^{3}He-^{4}He interface. Away from the spinodal region (configuration (b)), the critical bubble is a true bubble covered with ^{3}He: the surface tension that matters now is that of the ^{3}He-^{4}He liquid free-surface, which is about ten times larger than the previous surface tension.

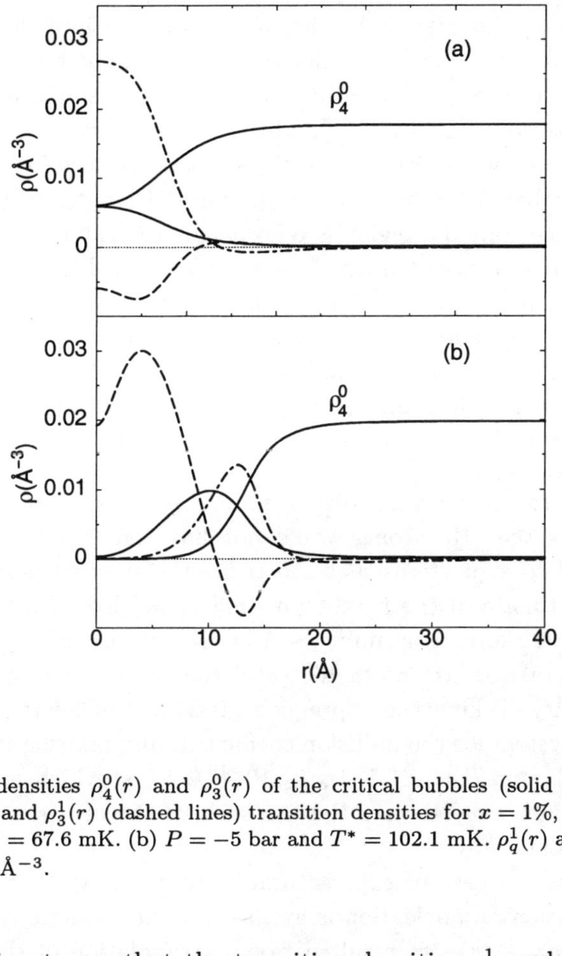

Fig. 15. Particle densities $\rho_4^0(r)$ and $\rho_3^0(r)$ of the critical bubbles (solid lines), and the $\rho_4^1(r)$ (dash-dotted lines) and $\rho_3^1(r)$ (dashed lines) transition densities for $x = 1\%$, corresponding to: (a) $P = -8$ bar and $T^* = 67.6$ mK. (b) $P = -5$ bar and $T^* = 102.1$ mK. $\rho_q^1(r)$ are drawn in arbitrary units, and $\rho_q^0(r)$ in Å^{-3}.

It is interesting to see that the transition densities ρ_q^1 evolve from those corresponding to 'volume oscillations' (Fig. 15, panel (a)) to 'surface oscillations' for ^{4}He, and a mixed surface-volume type for ^{3}He (panel (b)), to eventually become pure surface oscillations for both isotopes when we go from the spinodal towards the saturation line.

The different surface tensions involved in these processes, together with the existence of a ^{3}He-^{4}He segregation curve at negative pressures down to $x \sim 2.4\%$ (see Fig. 11) are the cause of the structures displayed in Fig. 14. Using the surface or volume character of the transitions densities as a useful guide, the figure can be understood as follows. Below $x \sim 2.4\%$, no pure ^{3}He drop can 'co-exist' with the homogeneous mixture, and the critical configurations look as drawn in Fig. 15. Above $x \sim 2.4\%$, the situation changes, and the existence of a segregation line allows the system to develop critical configurations resembling pure ^{3}He drops 'co-existing' with the mixture. Thus, one first finds the kind of configurations that, as before, correspond to pressures close to the spinodal line and originate a rise in T^* (left-hand side of all curves in Fig. 14). Next, one finds ^{3}He-rich droplets embedded in the mixture, whose interface vibration originates the first decrease of $T^*(P)$, followed by a ^{3}He volume vibration which causes the rise at the second maximum. At pressures closer to zero, the ^{3}He drop is reabsorbed, the critical configuration is that of a ^{4}He bubble covered with ^{3}He, and the second decrease of the $T^*(P)$ curve is eventually associated with surface vibrations of the mixture free-surface. Here we do not give any further detail since, as in the pure ^{4}He case, only the part of the $T^*(P)$ curve near the spinodal region is relevant for the cavitation problem.

$P_h(T^*)$ is shown as circles (squares) on the curves in Fig. 14. The circles correspond to $(V\mathcal{T})_e = 10^4\,\text{Å}^3$ sec, and the squares to $10^{14}\,\text{Å}^3$ sec. Compared to the pure ^{4}He case, and depending on the $(V\mathcal{T})_e$ value, for $x = 1\%$ T^* has been reduced by a factor of 4 or 5, respectively. Fig. 16 shows P_h as a function of T for the above-mentioned x values and $(V\mathcal{T})_e = 10^{14}\,\text{Å}^3$ sec. Thermal and quantum regimes are displayed. The dashed line is the extrapolation of the thermal regime to temperatures close to $T = 0$.

Notice that the smallest x value displayed in Figs. 14 and 16 is 0.1%. For small ^{3}He concentrations, the ^{3}He atoms would not have enough time to diffuse and develop the critical configurations that constitute the starting point of the present calculations. An estimate of the maximum time τ_d needed to develop a pure ^{3}He critical drop of 10 Å radius containing ~ 100 ^{3}He atoms, can be obtained from the time it would take a ^{3}He atom to travel the radius of this critical drop at the diffusion velocity. Taking the expression given by Burmistrov i*et al.* [49] for a degenerate Fermi system for the diffusion coefficient, and relating it to the diffusion velocity, [80] it yields, for $T = 0.1$ K, $\tau_d \sim 10^{-13}/x^{4/3}$ s, which for concentrations above $\sim 0.1\%$, is much smaller than the timescale of the microwave bursts used in cavitation experiments. [38]

We now address the case of supersaturated mixtures at $P \geq 0$ for which experimental information on nucleation is available. A first estimate of the degree of supersaturation $\Delta x_{cr} \equiv x - x_s$ resulted from extrapolation of the measured ^{3}He chemical potential excess $\Delta\mu_3 \equiv \mu_3(x) - \mu_3(1)$ along the demixing line. This extrapolation yielded $\partial\Delta\mu_3/\partial x \geq 0$ up to $x \sim 16$ %, thus giving [81] $\Delta x_{cr} \sim 10\%$. Large critical supersaturation values have also been obtained in microscopic calculations. [76]

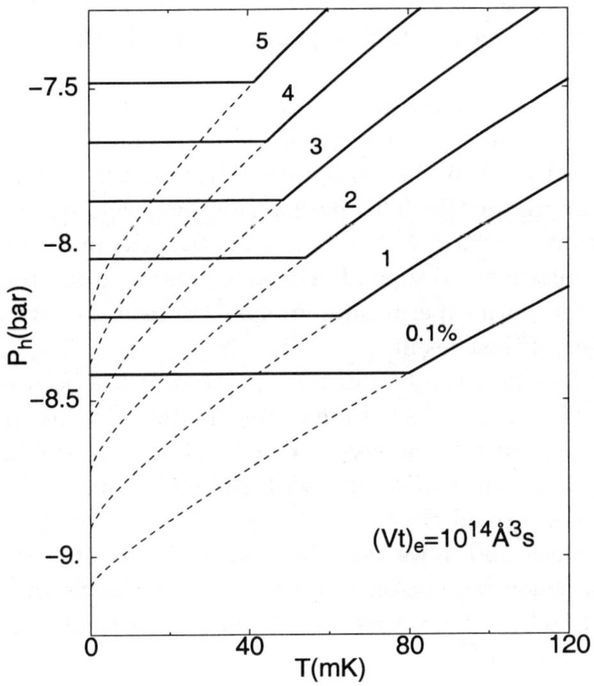

Fig. 16. Homogeneous cavitation pressure P_h as a function of T for the same x values as in Fig. 14. The dashed line is the extrapolation of the thermal regime to temperatures below T^*.

The degree of critical supersaturation reached in recent experiments on supersaturated helium mixtures at $P \geq 0$ is very small, of the order of 1 % in the experiments carried out by Rudavskii and coworkers, [67–69] and below 0.5 % in those of Satoh and coworkers. [65,66] Actually, the measurements in the 60's [63,64] also yielded small Δx_{cr}. This is indeed an intriguing observation. Being so close to the saturation line, the capillarity approximation should work well, and one could use Eqs. (2.2)-(2.3) with the same $\mathcal{S}$ but with [49,70] $\mathcal{V} = 4\pi\rho_3\Delta\mu_3/3$ to evaluate U_{max} and R_c. We have seen that to observe nucleation one must have $(V\tau)_e \cdot J \sim 1$, and thus $U_{max} = kT\ln[(V\tau)_e J_{0T}]$. Typical values of the logarithm [27,49,72] are about 80. Thus, $U_{max} \sim$ 8K if $T \sim$ 100 mK. However, using the experimental surface tension σ of the ^{3}He-^{4}He interface, the saturation ρ_3 value ($\sim$ 0.017 K Å^{-2} and $\sim$ 0.016 Å^{-3}, respectively), and a linear approximation to the experimental chemical potential excess values [81] $\Delta\mu_3 \sim 2.3\,\Delta x$ (K) valid near the saturation line, one has

$$U_{max} \sim \frac{6.1 \times 10^{-2}}{(\Delta x)^2} \text{ (K)}. \tag{4.18}$$

This equation yields $\Delta x_{cr} \sim 10\,\%$ if $U_{max} \sim$ 8 K, i.e., one order of magnitude larger than experiment. Taking $\Delta x \sim 0.004$, which is in the range of experimental values of Satoh *et al.*, [65] one has $U_{max} \sim 3800$ K, i.e., two orders of magnitude larger than the value at which phase separation triggered by ^{3}He drops nucleation would be possible.

An explanation of the discrepancy between theory and experiment is that phase separation may be caused by heterogeneous nucleation on vortices. Indeed, at the low temperatures involved in the experiments ^{4}He is still superfluid due to the limited solubility of ^{3}He in ^{4}He. This mechanism was proposed by Jezek *et al.*, [82] and it has been further refined. [83,84] It turns out that the presence of vortex lines in the mixture decreases the degree of critical supersaturation from $\Delta x_{cr} \sim 10\%$ to ~ 1 %. At present, both experimental groups [85,86] seem to have adhered to this possible explanation of their results.

The presence of vortices can be taken for granted. It has been recognized for a long time that any container of superfluid ^{4}He, treated in a conventional fashion, will be permeated *ab initio* by numerous quantized vortices stabilized by surface pinning. [87] One should not confuse this with the interesting problem of vorticity nucleation [88–90] in superfluid ^{4}He.

To see how it comes about, we again use the capillarity approximation applied to a vortex with a hollow core instead of a droplet. The energy per unit length of a singly quantized hollow core vortex line of radius a in a ^{3}He-^{4}He mixture with $x > x_{cr}$ may be written as [82]

$$E_v = 2\pi\sigma a - \pi a^2 \rho_3 \Delta\mu_3 + \pi \frac{\hbar^2}{m_4} \rho_4 \ln\left(\frac{a_\infty}{a}\right) , \tag{4.19}$$

where σ is the surface tension of the ^{3}He-^{4}He interface and a_∞ is a large enough radius. Minimizing E_v with respect to a we get the radius of the vortex line:

$$a = 2a_0 \frac{\mu_c}{\Delta\mu_3}\left[1 \pm \sqrt{1 - \frac{\Delta\mu_3}{\mu_c}}\right] , \tag{4.20}$$

with $a_0 = \hbar^2\rho_4/(2\sigma m_4)$ and $\mu_c = \sigma^2 m_4/(2\hbar^2\rho_3\rho_4)$. The plus sign in front of the root corresponds to a maximum of E_v (unstable vortex with radius $a = a_>$) and the minus sign corresponds to a minimum (metastable vortex with radius $a = a_<$). a_0 is the radius of the stable vortex in the case of a mixture at saturation, i.e., $\Delta\mu_3 = 0$. At $P = 0$, taking $\rho_4 = 0.020\,\text{Å}^{-3}$ and $\hbar^2/m_4 \sim 12$ K Å^2, one gets $a_0 = 7.1$ Å and $\mu_c = 0.038$ K. When $\Delta\mu_3 < 0$ the homogeneous mixture is stable, and the only sensible sign in front of the root is minus. Therefore, the corresponding vortex line is stable, and if $|\Delta\mu_3| \ll \mu_c$ the vortex radius is $\sim a_0$. When $\Delta\mu_3 > 0$ the homogeneous mixture is metastable, and provided $\Delta\mu_3 < \mu_c$, a metastable vortex line with radius $a_<$ as well as an unstable (critical) vortex with radius $a_>$ are obtained. However, when $\Delta\mu_3 = \mu_c$ both vortex lines collapse onto a single line (the situation is quite similar to what we have discussed for metastable electron bubbles), and for $\Delta\mu_3 > \mu_c = 0.038$ K the vortex is no longer stable and the mixture will undergo phase separation. Using the experimental data, [81] this implies that $\Delta x_{cr} \leq 1.6\%$, which is considerably smaller than the value obtained if nucleation seeds were ^{3}He droplets. It is worth stressing that this value is an upper limit of the actual critical supersaturation, as we have not yet taken into account the fact that the stable vortex may destabilize by quantum or thermal fluctuations.

These simple estimates have been confirmed by DFT calculations of vortex lines in helium mixtures. [83] To this end, the energy density for the mixture $f(\rho_3, \rho_4)$, Eq. (4.3), has been supplemented with a centrifugal energy term associated with the vortex velocity field in the Feynman-Onsager approximation, namely $\hbar^2 \rho_4/(2m_4 r^2)$, where r is now the radial distance to the vortex line placed along the z axis (cylindrical symmetry). The Euler-Lagrange equations are solved again for given x and P, and the metastable and unstable configurations, which are both solutions of the EL for the *same* P and x conditions, are obtained. The barrier height per unit vortex length is

$$\Delta\Omega_{max} = 2\pi \int r dr \left[\omega(\rho_3^c, \rho_4^c) - \omega(\rho_3^m, \rho_4^m)\right] , \tag{4.21}$$

where ρ_q^c and ρ_q^m are the particle densities of the critical and metastable vortices, respectively. It is worth noting that $\Delta\Omega_{max}$ is a *finite* quantity: there is no need to introduce any r cutoff which would have been unavoidable if we had described either configuration separately.

In Fig. 17 we have plotted the critical and metastable density profiles corresponding to a configuration with $P = -1.66$ bar, $x = 1\%$ (top panel), and to a configuration with $P = 0.91$ bar, $x = 8\%$ (bottom panel). This figure illustrates that metastable and critical vortex line configurations can indeed be found in both metastability regions, namely the region of negative pressures, and the region of supersaturated mixtures. The detailed DFT calculations have confirmed the results obtained from the hollow core vortex model. [83]

These pinned vortices, which permeate the mixture, generate self-sustaining vortex tangles with a typical radius of curvature. The effect that the curvature of vortices may have on the instability of a supersaturated mixture has been addressed by Jezek *et al.* [84] using ring vortices as nucleation seeds instead of vortex lines.

Fig. 18 shows the experimental data of Chagovets *et al.* [69] compared with calculations in which the nucleation seed is a vortex ring of 500 Å radius. [84] The agreement between theory and experiment is fair, and the P-dependence of Δx_{cr} is correctly reproduced.

Besides critical supersaturation, T^* has also been determined for positive pressures measuring the temperature below which Δx_{cr} becomes almost T independent; [65] it is a few tens of mK. A rather detailed theoretical discussion of the nucleation process in liquid helium mixtures near the demixing line has been presented by Barranco *et al.* [41] In this work ^{3}He droplets and ^{4}He vortex lines filled with ^{3}He were considered as nucleation seeds. Neither nucleation configuration was able to simultaneously reproduce the current experimental values of Δx_{cr} and T^*: At $P \sim 0$ vortex lines yield $\Delta x_{cr} \sim 1.3\,\%$ and $T^* \sim 1$ mK, whereas ^{3}He droplets yield $\Delta x_{cr} \sim 19\,\%$ and $T^* \sim 28$ mK. The origin of this disagreement is still unknown.

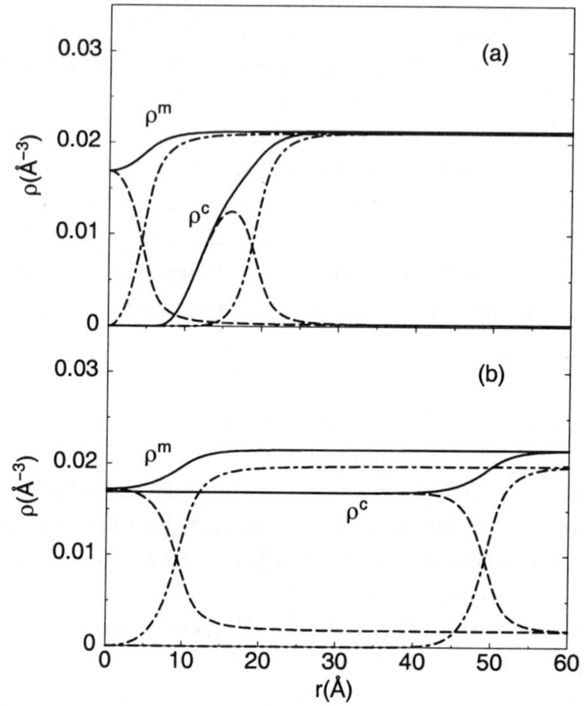

Fig. 17. Panel (a): vortex profiles for $x = 1\%$ and $P = -1.66$ bar. Panel (b): vortex profiles for $x = 8\%$ and $P = 0.91$ bar. The solid lines represent the total particle density, and the dash-dotted (dashed) lines, the ρ_4 (ρ_3) densities. Critical (metastable) configurations are denoted as ρ^c (ρ^m).

Acknowledgments

We would like to thank J. Navarro and R. J. Lombard for their contribution to the development of our nucleation project, and S. Balibar, J. Boronat, F. Caupin, V. K. Chagovets, E. Chudnovsky, L. B. Dubovskii, and H. J. Maris for the many useful discussions we have had on different aspects of the problem. This work has been performed under grant 2000SGR00024 from Generalitat de Catalunya. D. M. J. acknowledges the CONICET (Argentina) and the Generalitat de Catalunya ACI program for financial support.

References

1. D. W. Oxtoby, *J. Phys.: Condens. Matter* **4**, 7627 (1992).
2. D. W. Oxtoby and R. Evans, *J. Chem. Phys.* **89**, 7521 (1988).
3. X. C. Zeng and D. W. Oxtoby, *J. Chem. Phys.* **95**, 5940 (1991).
4. Q. Xiong and H. J. Maris, *J. Low Temp. Phys.* **77**, 347 (1989).

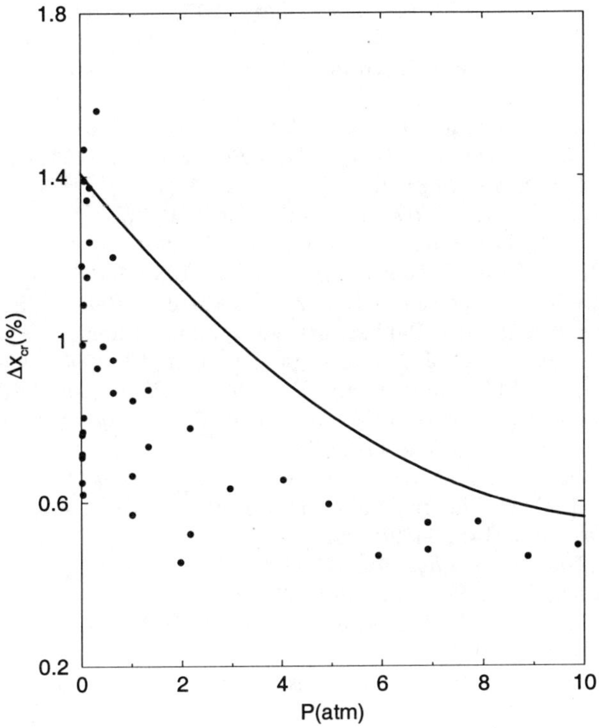

Fig. 18. Critical supersaturation Δx_{cr} as a function of P (atm). The dots are the experimental data, [69] and the line is the calculation of Jezek *et al.* [84]

5. M. Guilleumas, M. Pi, M. Barranco, J. Navarro, and M. A. Solís, *Phys. Rev.* **B47**, 9116 (1993).
6. D. M. Jezek, M. Guilleumas, M. Pi, M. Barranco, and J. Navarro, *Phys. Rev.* **B48**, 16582 (1993).
7. H.J. Maris and S. Balibar, *Phys. Today* pag. 29, February 2000.
8. D. N. Sinha, J. S. Semura, and L. C. Brodie, *Phys. Rev.* **A26**, 1048 (1982).
9. D. Lezak, L. C. Brodie, J. S. Semura, and E. Bodegom, *Phys. Rev.* **B37**, 150 (1988).
10. J. A. Nissen, E. Bodegom, L. C. Brodie, and J. Semura, *Phys. Rev.* **B40**, 617 (1989).
11. M. S. Pettersen, S. Balibar, and H. J. Maris, *Phys. Rev.* **B49**, 12062 (1994).
12. H. Lambaré, P. Roche, S. Balibar, H. J. Maris, O. A. Andreeva, C. Guthmann, K. O. Keshishev, and E. Rolley, *Eur. Phys. J.* **B2**, 381 (1998).
13. F. Caupin, P. Roche, S. Marchand, and S. Balibar, *J. Low Temp. Phys.* **113**, 473 (1998).
14. I. M. Lifshitz and Yu. Kagan, *Sov. Phys. JETP* **35** 206 (1972) [*Sov. Phys. JETP* **35**, 206 (1972)].
15. V. A. Akulichev, *Ultrasonics*, January 1986, pag. 8.
16. M. Barranco, M. Pi, A. Polls, and X. Viñas, *J. Low Temp. Phys.* **80**, 77 (1990).
17. A. Guirao, M. Centelles, M. Barranco, M. Pi, A. Polls, and X. Viñas, *J. Phys. Condens.*

Matter **4**, 667 (1992).
18. S. C. Hall and H. J. Maris, *J. Low Temp. Phys.* **107**, 263 (1997).
19. J. Boronat, J. Casulleras, and J. Navarro, *Phys. Rev.* **B0**, 3427 (1994).
20. C. E. Campbell, R. Folk, and E. Krotscheck, *w Temp. Phys.***105**, 13 (1996).
21. G. H. Bauer, D. M. Ceperley, and N. Goldenfeld, *Phys. Rev.* **B61**, 9055 (2000).
22. J. Casulleras and J. Boronat, *Phys. Rev. Lett.* **84**, 3121 (2000).
23. F. Caupin, S. Balibar, and H. J. Maris, *Phys. Rev. Lett.* **87** 145302 (2001).
24. Q. Xiong and H. J. Maris, *Phys. Rev. Lett.* **63**, 1078 (1989).
25. M. A. Solís and J. Navarro, *Phys. Rev.* **B45**, 13080 (1992).
26. F. Caupin and S. Balibar, *Phys. Rev.* **B64**, 064507 (2001).
27. Q. Xiong and H. J. Maris, *J. Low Temp. Phys.* **82**, 105 (1991).
28. The actual experimental points of Ref. 27 and these of Ref. 10 lie along the solid theoretical curve in Fig. 5 (S. Balibar, private communication).
29. J. Classen, C.-K. Su, and H. J. Maris, *Phys. Rev. Lett.* **77**, 2006 (1996).
30. J. Classen, C.-K. Su, M. Mohazzab, and H. J. Maris, *Phys. Rev.* **B57**, 3000 (1998).
31. C.-K. Su, C. E. Cramer, and H. J. Maris, *J. Low Temp. Phys.* **113**, 479 (1998).
32. H. J. Maris, *J. Low Temp. Phys.* **120**, 173 (2000).
33. R. Jackiw, C. Rebbi and J. R. Schrieffer, *J. Low Temp. Phys.* **122**, 587 (2001).
34. P. M. McConnell, M. L. Chu Jr., and R. D. Finch, *Phys. Rev.* **A1**, 411 (1970).
35. F. Dalfovo, *Phys. Rev.* **B46**, 5482 (1992).
36. H. J. Maris, *J. Low Temp. Phys.* **94**, 125 (1994).
37. E. M. Chudnovsky, *Phys. Rev.* **A46**, 8011 (1992).
38. S. Balibar, C. Guthmann, H. Lambaré, P. Roche, E. Rolley, and H. J. Maris, *J. Low Temp. Phys.* **101**, 271 (1995).
39. M. Guilleumas, M. Barranco, D. M. Jezek, R. J. Lombard, and M. Pi, *Phys. Rev.* **B54**, 16135 (1996).
40. H. J. Maris, *J. Low Temp. Phys.* **98**, 403 (1995).
41. M. Barranco, M. Guilleumas, D. M. Jezek, R. J. Lombard, J. Navarro, and M. Pi, *J. Low Temp. Phys.* **117**, 81 (1999).
42. R. P. Feynman, *Statistical Mechanics* (Benjamin, New York, 1972).
43. J. S. Langer, *Ann. Phys. (N.Y.)* **41**, 108 (1967).
44. S. Coleman, *Phys. Rev.* **D15**, 2929 (1977).
45. C. G. Callan and S. Coleman, *Phys. Rev.* **D16**, 1762 (1977).
46. T. Nakamura, Y. Kanno, and S. Takagi, *Phys. Rev.* **B51**, 8446 (1995).
47. A. Galindo and P. Pascual, *Quantum Mechanics*, Springer-Verlag, Berlin (1990).
48. V. I. Goldanskii, *Sov. Phys. Dokl.* **4**, 74 (1959).
49. S. N. Burmistrov, L. B. Dubovskii, and V. L. Tsymbalenko, *J. Low Temp. Phys.* **90**, 363 (1993).
50. V. A. Akulichev and V. A. Bulanov, *Sov. Phys. Acoust.* **20**, 501 (1975).
51. C.-K. Su and H. J. Maris, *J. Low Temp. Phys.* **110**, 485 (1998).
52. S. Balibar and F. Caupin, private communication.
53. S. Balibar, F. Caupin, P. Roche, and H. J. Maris, *J. Low Temp. Phys.* **113**, 459 (1998).
54. F. Caupin, S. Balibar, and H. J. Maris, *J. Low Temp. Phys.* in print (2002).
55. D. M. Jezek, M. Barranco, and M. Pi, *Phys. Rev.* **B60**, 3048 (1999).
56. S. N. Burmistrov and L. B. Dubovskii, *Sov. Phys. JETP* **73**, 1020 (1991).
57. D. M. Jezek, M. Guilleumas, M. Pi, and M. Barranco, *Physica* **B284-288**, 214 (2000).
58. H. Goldstein, *Classical Mechanics*, 2nd edition Addison-Wesley, Reading, MA. (1980).
59. F. Gantmacher, *Lectures in Analytical Mechanics*, Mir Publishers, Moscow (1970)
60. L. Landau and E. Lifchitz, *Mécanique des fluids*, Editions Mir, Moscou (1971).
61. G. Baym and C. Pethick, *Landau Fermi-liquid Theory*, J. Wiley and Sons, Inc. New

York (1991).
62. J. M. Parpia, D. J. Sandiford, J. E. Berthold, and J. D. Reppy, *Phys. Rev. Lett.* **40**, 565 (1978).
63. J. Landau, J. T. Tough, N. R. Brubaker and D. O. Edwards, *Phys. Rev. Lett.* **23**, 283 (1969).
64. G. E. Watson, J. D. Reppy, and R. C. Richardson, *Phys. Rev.* **188**, 384 (1969).
65. T. Satoh, M. Morishita, M. Ogata and S. Katoh, *Phys. Rev. Lett.* **69**, 335 (1992).
66. T. Satoh, M. Morishita, S. Katoh, K. Hatakeyama, and M. Takashima, *Physica* **B197**, 397 (1994).
67. V. A. Maidanov, V. A. Mikheev, N. P. Mikhin, N. F. Omelaenko, E. Ya. Rudavskii, V. K. Chagovets, and G. A. Sheshin, *Sov. J. Low Temp. Phys.* **18**, 663 (1992).
68. V. A. Mikheev, E. Ya. Rudavskii, V. K. Chagovets, and G. A. Sheshin, *Low Temp. Phys.* **20**, 485 (1994).
69. V. Chagovets, I. Usherov-Marshak, G. Sheshin, and E. Ya. Rudavskii, *J. Low Temp. Phys.* **110**, 479 (1998).
70. I. M. Lifshitz, V. N. Polesskii and V. A. Khokhlov, *Sov. Phys. JETP* **47**, 137 (1978).
71. L. D. Landau and I. M. Lifshitz, *Statistical Physics Part 1*, Pergamon Press, Oxford (1981).
72. M. Guilleumas, D. M. Jezek, M. Pi, M. Barranco and J. Navarro, *Phys. Rev.* **B51**, 1140 (1995).
73. F. Dalfovo, Ph. D. thesis, University of Trento (1989).
74. D. O. Edwards and M. S. Pettersen, *J. Low Temp. Phys.* **87**, 473, (1992).
75. M. Guilleumas, M. Pi, M. Barranco, D. M. Jezek, and J. Navarro, *Phys. Rev.* **B52**, 1210 (1995).
76. E. Krotscheck and M. Saarela, *Phys. Rep.* **232**, 1 (1993).
77. J. K. Hoffer and D. N. Sinha, *Phys. Rev.* **A33**, 1918 (1986).
78. M. Barranco, M. Pi, S. M. Gatica, E. S. Hernández, and J. Navarro, *Phys. Rev.* **B56**, 8997 (1997).
79. D. M. Jezek, M. Guilleumas, M. Pi, and M. Barranco, *Europhys. Lett.* **38**, 601 (1997).
80. F. Reif, *Fundamentals of Statistical and Thermal Physics*, McGraw-Hill Book Company, Singapore (1985).
81. P. Seligmann, D. O. Edwards, R. E. Sarwinski, and J. T. Tough, *Phys. Rev.* **181**, 415 (1969).
82. D. M. Jezek, M. Guilleumas, M. Pi, and M. Barranco, *Phys. Rev.* **B51**, 11981 (1995).
83. D. M. Jezek, M. Guilleumas, M. Pi, and M. Barranco, *Phys. Rev.* **B55**, 11092 (1997).
84. D. M. Jezek, M. Pi, M. Barranco, R. J. Lombard, and M. Guilleumas, *J. Low Temp. Phys.* **112**, 303 (1998).
85. S. N. Burmistrov, L. B. Dubovskii, and T. Satoh, *J. Low Temp. Phys.* **110**, 479 (1998).
86. V. Chagovets, E. Ya. Rudavskii, G. A. Sheshin, and I. Usherov-Marshak, *J. Low Temp. Phys.* **113**, 1005 (1998).
87. D. D. Awschalom and K. W. Schwarz, *Phys. Rev. Lett.* **52**, 49 (1984).
88. G. Williams and R. Packard, *J. Low Temp. Phys.* **33**, 459 (1978).
89. R. M. Bowley, P. V. E. McClintock, F. E. Moss, and P. C. E. Stamp, *Phys. Rev. Lett.* **44**, 161 (1980).
90. C. M. Muirhead, W. F. Vinen, and R. J. Donnelly, *Proc. R. Soc. Lond.* **A402**, 225 (1985).

CHAPTER 8

EXCITATIONS OF SUPERFLUID ^{4}He IN CONFINEMENT

B. Fåk
ISIS Facility, Rutherford Appleton Laboratory,
Chilton, Didcot, OX11 0QX, England
and *Commissariat à l'Energie Atomique, DRFMC/SPSMS/MDN,*
38054 Grenoble Cedex 9, France
E-mail:B.Fak@rl.ac.uk

H. R. Glyde
Department of Physics and Astronomy, University of Delaware,
Newark, Delaware 19716
E-mail: glyde@udel.edu

Neutron inelastic scattering studies of the elementary excitations in confined superfluid ^{4}He are reviewed. Both recent work on helium in porous silica glass (aerogel, Vycor, etc.) and earlier work on helium films on graphite surfaces are discussed. The global picture emerging from these studies is that the three-dimensional excitations are essentially the same as in bulk helium. The characteristic feature of confined helium is the existence of additional layer modes that propagate in the first few liquid layers near the solid-liquid interface. The dispersion and gap energy of these layers modes depend on the substrate. The layer modes are believed to be at the origin of the differences in macroscopic properties compared to bulk helium. Experiments suggest the existence of a localized condensate in Vycor.

1. Introduction

The impact of confinement and disorder on the superfluid and thermodynamic properties of liquid ^{4}He has been a topic of great interest since the 1960s. [1–4] The superfluid transition temperature T_c of liquid ^{4}He in porous media and confined to surfaces is depressed below the transition temperature in bulk liquid ^{4}He, $T_\lambda = 2.172$ K. In fully filled aerogel, Vycor, and Geltech silica, for example, $T_c = 2.167$ K, [5] 1.95–2.01 K, [5,6] and 0.725 K, [7] respectively. Also, the temperature dependence of the superfluid density $\rho_S(T)$ below T_c can be significantly modified. At low enough coverage, ^{4}He in Vycor behaves like an ideal gas. [8]

The characteristic phonon-roton excitations in bulk superfluid ^{4}He have been actively studied by neutron scattering since 1957. Landau initially proposed the

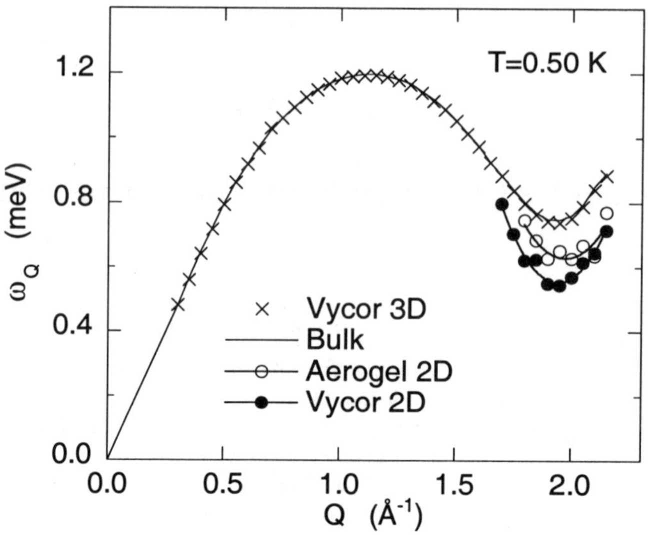

Fig. 1. Dispersion curve of the elementary excitations in ^{4}He in confinement. The three-dimensional (3D) phonon-roton curve is the same in bulk liquid ^{4}He (line) as for helium in Vycor (crosses) and in aerogel (not shown). The open and solid circles show the dispersion for the two-dimensional (2D) layer modes in aerogel and Vycor, respectively, which propagate in the liquid helium layers adjacent to the media walls.

existence of excitations in superfluid ^{4}He having energies of the phonon-roton form (see Fig. 1) as a basis for his remarkable theory of superfluidity. [9] In contrast, London proposed that Bose-Einstein condensation (BEC) was the origin of superfluidity. [10] The existence of a condensate is also sufficient to produce a dispersion curve of the phonon-roton form in a Bose fluid. [11,12] In bulk liquid ^{4}He, superfluidity, BEC, and well defined phonon-roton excitations all disappear at the same temperature, T_λ.

The nature of the excitations in porous media is similarly of great interest. Confinement and disorder offer a new arena in which T_c and $\rho_S(T)$ are modified to explore the relationship between the condensate, the excitations, and superfluidity. In two dimensions (2D), the loss of superfluidity is attributed to the unbinding of vortex pairs. [13] Below T_c, $\rho_S(T)$ is related to the density of vortices with no clear connection to BEC or to the density of (phonon-roton) excitations. As 2D films thicken, there will be a crossover from 2D to 3D behavior. The study of phonon-roton excitations in films on surfaces and in porous media as a function of film thickness is therefore most interesting.

Equally, liquid ^{4}He in porous media is a model example of "bosons in disorder", which can be related to other dirty Bose systems. [14] Predictions of the impact of disorder on the excitations can be tested. [15] Similarly, calculations of the structure and the excitations of liquid ^{4}He on surfaces have been made, [16,17] as well as predictions of the roton gap energy in two-dimensional ^{4}He. [18,19] These can be tested directly against neutron scattering data.

It is only relatively recently that the elementary excitations of superfluid ^{4}He in confinement and disorder have been successfully studied by neutron scattering. Different types of media have been investigated. Examples are randomly or partly oriented graphite surfaces and disordered porous media with a large variety of porosity, pore sizes, and pore-size distributions. The porous media include different types of aerogel, xerogel, Vycor, and Geltech silica. The excitations of helium in these media show very strong similarities, even between media with flat surfaces such as graphite and media with irregular surfaces such as aerogel. This observation suggests that a global picture of the excitations in confinement can be presented. Unfortunately, reading the original literature without some guidance can be misleading, since the neutron scattering measurements have been difficult and full of pit falls. For these reasons, we believe it is timely to review the experimental results of the excitations in confined helium.

The aim of this review is to present a unified picture of the most significant neutron scattering results of the excitations of superfluid ^4He in confinement and disorder. We will not discuss the dynamic properties of ^3He in confinement, [20] nor measurements at high wave vectors Q, which probe single-particle properties (atomic momentum distributions) rather than collective dynamics. The review is organized as follows. The global picture emerging from neutron scattering studies of superfluid ^4He confined in different media is presented in Sec. 2. Section 3 identifies the pit falls and difficulties in the analysis of the data and discusses general experimental aspects. Many of the original results presented in the literature are in fact artefacts of the data analysis. The understanding of these effects has been a crucial ingredient in arriving at the global picture presented in this review. The following sections 4–6 treat the results from the different media: graphite, aerogel, and Vycor, respectively. Finally, Sec. 7 discusses some interpretations and open questions and identifies some areas where further experimental and theoretical work is needed.

2. Global picture

In this section, we present a global picture of the excitations of superfluid ^4He in confinement. References to the literature can be found in the more detailed account of the experimental results in Secs. 4–6. Superfluid ^4He in all media studied to date supports well defined three-dimensional (3D) phonon-roton-like excitations. The energies and widths (inverse lifetimes) of these excitations, when carefully analyzed, are the *same* as in bulk helium within current available precision. In particular, the temperature dependence of the excitation energies and widths and the wave-vector dependence of the intensity in the modes are the same as in bulk helium. No well-defined 3D excitations are observed until the equivalent of 1.5 liquid layers of ^4He coat the surfaces (substrate plus solid layers). The only significant differences from the bulk occur when the coverage (filling) is reduced: it seems that the maxon energy in liquid ^4He on graphite decreases and that the roton energy in aerogel increases. Both these observations suggest the presence of low-density helium layers for thin films.

A particularly interesting observation concerning the 3D excitations is that the temperature dependence of the dynamic structure factor $S(Q, \omega)$ of ^{4}He in Vycor is similar to that in bulk ^{4}He while the superfluid transition temperature T_c is lower. This means that well-defined excitations persist above T_c, suggesting the existence of a localized condensate for temperatures $T_c < T < T_\lambda$.

All systems studied show the existence of layer modes, or indirect signs thereof. Layer modes are two-dimensional (2D) excitations propagating in the first liquid layers close to the media walls, i.e. close to the solid-liquid interface. In all media, the layer modes give rise to a relatively broad peak in $S(Q, \omega)$. There is either one broad or several sharp modes. The layer modes have a roton-like dispersion in dense aerogels and in Vycor, where they are observed only near the roton wave vector ($Q_R = 1.925$ Å^{-1}). On graphite, it seems that both dispersive and dispersionless (flat) layer modes are observed. Typical 2D roton (gap) energies are 0.55 meV on graphite and in Vycor, and 0.63 meV or higher in aerogel. The intensity of the layer modes starts to grow when there are approximately 1.5 liquid layers present, and saturates at a filling of about 4–5 liquid layers.

Another type of 2D excitation is also seen on flat surfaces such as graphite. Here, a strongly dispersive excitation originating from the free surface (the liquid-gas interface) is observed. It corresponds to a ripplon, which is a quantized capillary wave.

A general observation is that the first 1–1.5 liquid layers do not support any well-defined excitations, neither 2D nor 3D, in any of the media studied. These layers appear "inert", despite being liquid. They sit on top of the solid layer(s). On graphite, there are two crystalline layers with different densities, while it is believed that there is only one amorphous layer in aerogel and Vycor, if one can talk of layers in these cases with irregular surfaces.

3. Experimental aspects

The effects of confinement on the 3D phonon-roton excitations are very small or perhaps absent. Special care is thus needed in analyzing these excitations. Additional intensity in the tails of the phonon-roton peak arising from layer modes or from multiple scattering can easily be mistaken for shifts and broadenings of the peak, if not identified and accounted for in the analysis. Such problems have hampered many of the neutron scattering experiments of superfluid ^{4}He in almost all of the confining media studied. These experimental aspects will be briefly discussed in this section.

The presence of hydrogen, probably in form of OH^- groups bound to the surfaces of aerogel and Vycor, results in strong isotropic incoherent elastic scattering and leads also to multiple scattering: the inelastic scattering of a neutron by ^{4}He is followed or preceded by elastic scattering from the hydrogen. This type of multiple scattering destroys the Q information of $S(Q, \omega)$ from the ^{4}He, but does not change the energy transfer. Its signature is therefore a phonon-roton density-of-states like

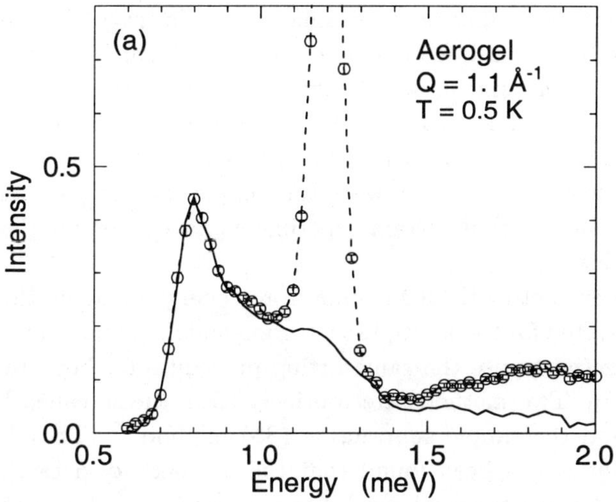

Fig. 2. The multiple scattering (solid line) from helium in aerogel (extracted at low wave vectors) resembles the phonon-roton density of states and is essentially Q independent in aerogel and Vycor. The circles and dashed line are the raw data at the maxon wave vector before the multiple scattering is subtracted. As seen, all scattering at energies below the peak is due to multiple scattering. The broad feature at higher energies is the multiphonon contribution.

feature (see Fig. 2). Fortunately, this contribution can be accurately identified and subtracted from the data by comparing with the scattering from bulk ^{4}He at selected wave vectors, in particular at low wave vectors. A problem is that the multiple scattering depends on temperature, since the density-of-states of the phonon-roton excitations depends on temperature. Measurements of ^{4}He on graphite suffer from similar problems, with additional complications coming from the Bragg peaks of the substrate, which introduce a stronger Q dependence of the multiple scattering.

The amount of multiple scattering can be reduced by deuterating the samples in the case of aerogel. The most common method is to flush the aerogel with deuterium gas at relatively high temperatures, which partly replaces the hydrogen by deuterium. Even better results are obtained by making the aerogel from deuterated chemicals and ensuring that it is never exposed to air. In standard Vycor, multiple scattering is not a problem, since the absorption from B_2O_3 impurities is high. However, this absorption also severely reduces the scattered intensity from the ^{4}He, and the key to the most successful experiments on Vycor was to use a non-absorbing boron isotope in the fabrication process.

It is also important to avoid having too much bulk liquid around the sample. One method is to use aerogels grown in-situ in the neutron scattering cells, thereby reducing the amount of helium between the aerogel and the cell walls. Another method is underfilling, where the sample is filled only to about 95% of the full capacity, which requires precisely known adsorption isotherms for the actual sample used. This method has the advantage of reducing the amount of helium in cracks in the sample, but increased care is needed to ensure thermal equilibrium.

Since the effects of confinement are small, it is necessary to have accurate measurements of the bulk liquid, made under conditions identical to the measurements of the confined system, and preferentially at the same time. An accurate calibration of the thermometers is also needed. The best method is probably to calibrate the thermometers in-situ, using the vapor pressure of ^{3}He, as was done by some groups. A good thermal contact between the sample and the thermometer(s) is also required, and it appears that certain experiments have given unreliable results due to problems of this type.

When finally extracting the 3D phonon-roton energies and widths from the data, after having corrected for the multiple scattering and including a model for the layer modes, it is essential to use the same fitting procedure for both the confined ^{4}He and the bulk data. Two methods are routinely used, the so-called Wood-Svensson (WS) method and the simple subtraction (SS) method. [21] They have both their shortcomings, but it has been found that the SS method gives the same results independent of the instrumental resolution, while the results of the WS method depend on the resolution. A final remark: it is easier to analyze 3D phonon-roton excitations in very high-resolution measurements, at least at low temperatures, as the height of the main peak is then so high that the multiple scattering and the layer modes will have a much smaller effect.

4. Films on graphite

Historically, the first neutron scattering studies of the excitations in confined helium considered films on graphite substrates. Measurements began with the pioneering studies of Lambert *et al.* [22] at the Institut Laue-Langevin (ILL) and the early work of Carneiro *et al.* [23] and Thomlinson *et al.* [24] at Brookhaven National Laboratory. Since the 1980s, there has been a major ongoing program at the ILL. [25–32]

Different graphite substrates with large surface to volume ratio have been used over the years. They are all characterized by having flat graphite surfaces with the hexagonal c axis perpendicular to the surface. These surfaces are more or less aligned depending on the material. The most common substrates are graphitized carbon powder (Graphon) and exfoliated and recompressed graphite (Grafoil or Papyex), the latter being more homogeneous and uniform. [20] The results for the ^{4}He excitations are slightly different for different materials, but these differences are not significant compared to the difficulties of analyzing and interpreting the neutron scattering data. In films of at least 4 layers of ^{4}He on exfoliated graphite, the first two layers are solid, forming 2D triangular lattices with densities of 0.115 atoms/Å^2 and 0.095 atoms/Å^2, respectively. [27,30] Subsequent layers are liquid with densities of approximately 0.078 atoms/Å^2, [30] which is similar to the density of bulk liquid ^{4}He at saturated vapor pressure (SVP). It is possible that the third layer has a slightly higher density. [24] Graphite preplated with two solid layers of Ne or H_2 has also been investigated. [26,27,29] The third layer is then liquid or possibly partly solid ^{4}He. There appears to be no significant difference in the excitation spectrum

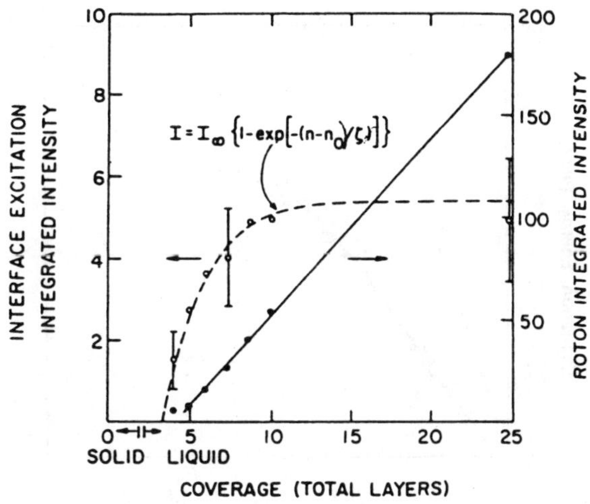

Fig. 3. Integrated intensity of the 3D bulk-like roton (solid circles and line) and the 2D layer mode (open circles and dashed line) as a function of helium film thickness on graphite. From Thomlinson *et al.* [24]

between preplated graphite and ^{4}He-only systems.

The first measurements focused on the excitations at wave vectors in the roton region. [22,23] No well-defined peaks in $S(Q,\omega)$ were seen until four layers (two liquid layers) were deposited. For two or more liquid layers, $S(Q,\omega)$ showed a well-defined peak interpreted as arising from the excitation of a bulk-like 3D roton. Additional intensity below the 3D roton energy was also observed, [22] which was interpreted as the 2D roton predicted by Padmore. [18] Although the poor statistics prevented further interpretation, the measurements by Lambert *et al.* showed all the basic features.

Measurements with much improved statistics were performed by Thomlinson *et al.* [24] They showed that the additional intensity originated from excitations in the first 2–5 liquid layers and was therefore indeed due to a 2D "layer" mode, propagating in the first few liquid layers. The integrated intensity in the layer mode increased with increasing coverage and saturated after 3–4 liquid layers were deposited (see Fig. 3). They also showed that the main peak in $S(Q,\omega)$ was a bulk mode, the 3D roton. At SVP the energy of the 2D roton was 0.54 meV compared to the 3D roton energy of 0.742 meV. No well-defined excitations were observed in films of 1.5 liquid layers or less.

Lauter, Godfrin, and collaborators have reported extensive measurements of excitations in liquid ^{4}He films on graphite. [25–32] Their data at the roton wave vector for thin films (2–8 liquid layers) confirm previous results: a 3D bulk-like roton plus additional intensity attributed to layer modes. Most importantly, their measurements were extended to the phonon and maxon regions, covering wave vectors between 0.25 and 2.0 Å^{-1}. In addition to the 3D rotons and the 2D layer modes,

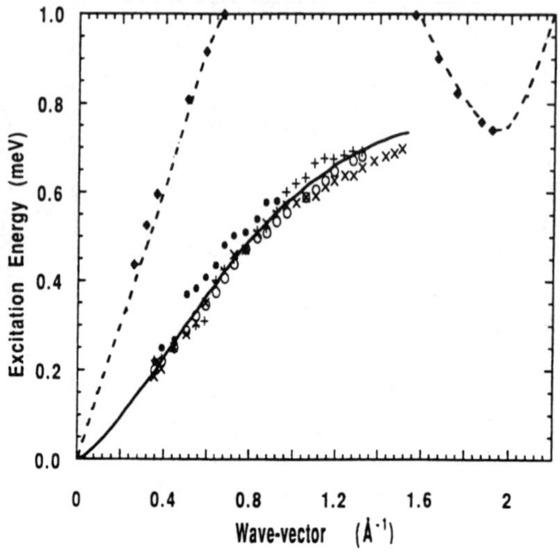

Fig. 4. The ripplon dispersion (symbols and solid line) for different film thicknesses of ^{4}He on graphite. The dashed line with symbols is the 3D bulk roton in the cell completely filled with helium. From Lauter *et al.* [31]

a new mode was discovered, the ripplon. [29,30,31] The ripplon, which can be viewed as a quantized capillary wave, propagates along the surface of the free liquid. The atomic displacements in the mode are perpendicular to the liquid-vapor interface. At low wave vectors, the ripplon dispersion relation is $\omega^2 = (\alpha/\rho)Q^3$, where α is the surface tension and ρ is the liquid density. [33] At wave vectors observable by neutrons the dispersion is approximately linear in Q, up to $Q \approx 1$ Å^{-1} where the ripplon energy flattens and reaches a maximum value of approximately 0.7 meV at $Q = 1.5$ Å^{-1} (see Fig. 4). The ripplon is not clearly separated from the layer modes beyond $Q = 1.5$ Å^{-1}. Lauter *et al.* demonstrated elegantly that the layer modes originate from the solid-liquid interface [26] and that the ripplons originate from the liquid-gas interface (free surface). [31] They found that the layer modes persisted while the ripplons disappeared when the free surface was suppressed by filling the sample completely with liquid ^{4}He. Thus, liquid ^{4}He films support three excitations: a 2D ripplon on the liquid surface, a 3D phonon-roton mode within the liquid film, and 2D layer modes propagating in the liquid layers adjacent to the solid-liquid interface near the substrate.

In what follows, we summarize the main experimental findings concerning the 3D bulk-like excitations and the 2D layer modes of ^{4}He films on graphite, based on the available literature. [22–32] The energy and width (inverse lifetime) of the 3D excitations are very similar to those in bulk helium. In some work, small changes in the energy or the width have been reported, but in most cases these certainly arise from the additional intensity in the tails of the main peak. These tails originate from layer modes or from multiple scattering. In work where corrections for the multiple scattering have been made and the layer modes have been incorporated in the fitting, the energy and width of the 3D excitations are identical to bulk helium. The only exception is possibly the softening of the maxon observed as the coverage is reduced, as shown in Fig. 5. [29,30] This softening suggests the existence of low-density layers in thin films of helium on graphite, while the opposite might be expected. The integrated intensity of the 3D excitations scales linearly with the amount of condensed helium, and extrapolates to zero at approximately 1.5 liquid layers (see Fig. 3). [24] There are indications that the phonons disappear somewhat quicker than maxons and rotons as the coverage is reduced. [29,30] This could be because long wavelength phonons need longer defect-free regions to propagate than rotons, which have a wavelength of the order of atomic distances.

Two-dimensional layer modes propagate in the first few liquid layers close to the substrate. Their integrated intensity scales initially with the amount of condensed liquid, saturates at about 3–4 liquid layers, and extrapolates to zero at approximately 1–1.5 liquid layers (see Fig.r 3). [24] The intensity of the layer modes is very weak, making a quantitative analysis difficult. Also, above the roton energy, multiple scattering contributions makes any extraction of layer modes highly uncertain. Several layer modes seem to be observed. Lauter *et al.* reported dispersionless modes that would correspond to excitations perpendicular to the substrate. [26–30] Such flat modes might explain the small Kapitza resistance observed in ^4He. The work of Clements *et al.* shows several modes below the roton energy. [32,34] A slightly dispersive branch is seen between the ripplon and the 3D phonon mode for wave vectors below 0.8 $Å^{-1}$. Near the roton wave vector, one broad or two sharp, slightly

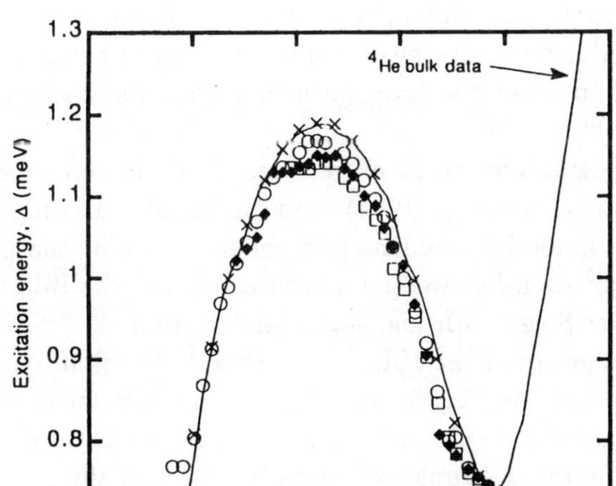

Fig. 5. Dispersion relation in the maxon region for different film thicknesses of ^4He on graphite. The maxon softens with reduced film thickness. The crosses are data from the cell completely filled with ^4He and the line is bulk data. From Lauter *et al.* [30]

dispersive modes are observed, similar to those in aerogel and Vycor. The gap energies of these modes are 0.5 and 0.6 meV, respectively, if we follow the assumption of Clements *et al.* that there are two sharp modes. Thomlinson *et al.* [24] resolved only one layer mode at the roton wave vector with an energy of 0.54 meV.

Calculations suggest that helium layers on surfaces support a two-dimensional layer mode with a phonon-roton-like dispersion. [17,32] Krotscheck *et al.* also report dispersionless modes that are essentially standing modes propagating between the film surface and the solid layer. [17] The energy of these modes would presumably depend on the film thickness. However, it is difficult to determine with precision the energies of the dispersionless modes reported by Lauter *et al.* [29,30] Thus, while the additional intensity observed in the roton region is entirely consistent with a layer mode, the situation is less clear concerning the energy. It would be interesting with experimental and theoretical studies of the temperature dependence of the excitations of helium on graphite.

5. Aerogel

Aerogels are porous solids formed by a sol-gel process. They have a highly tenuous structure of irregularly connected silica (SiO_2) globules and strands with a large distribution of pore sizes, from a few Å to a few hundred Å, [35] and a mean free path of typically 1000 Å. Porosities range from 87 to 99.5%. Small-angle x-ray and neutron diffraction measurements show fractal-like correlations on length scales up to 650 Å. Macroscopic measurements show that the superfluid transition temperature is decreased below T_λ by only 5 mK while the critical exponent for the superfluid flow density increases from 0.67 in bulk helium to 0.75 for helium in aerogel. [5] There has been no detailed published work on the structure of the helium layers in aerogel, but it appears as if only the first layer forms a solid, an amorphous solid, all other layers being liquid. [36]

Neutron scattering studies of the excitations of ^{4}He in aerogel and other porous media began much more recently (1994) than studies of films on graphite. The first measurements were made by Coddens and collaborators at the Laboratoire Léon Brillouin (LLB), [37,38] soon followed by measurements at the ISIS spallation source (and at the ILL) by Sokol, Stirling, and collaborators. [39–45] Experiments were also performed by a group at the ILL. [46–53] Already the first experiments showed that multiple scattering (see Sec. 3) was a major problem for the interpretation of the data. Much improved results are obtained if it is identified as in Fig. 2 and subtracted, or if deuterated samples are used to substantially reduce it. [46,51,52]

Coddens and collaborators studied both base-catalyzed and neutral-reaction aerogels of 96–96.5% porosity, partially or fully filled with superfluid helium at temperatures of 1.6–1.8 K. [37,38] They elegantly demonstrated that the broad Q-independent scattering centered at the roton energy was due to multiple scattering, as it disappeared when the incident neutron wave vector was made too low to excite the roton. The multiple scattering was relatively strong despite the deuterium-gas treatment of the samples. Due to the limited counting statistics and the coarse energy resolution of $\Delta E = 160\ \mu$eV, no difference was found between the excitation spectrum of helium in aerogel and bulk ^{4}He.

Sokol, Stirling, and collaborators made most of their measurements [39–44] on the backscattering spectrometer IRIS at ISIS, which has a high energy resolution of 15–20 μeV and a rather coarse Q resolution. They used non-deuterated aerogels with 90 and 95% porosity, for which the multiple scattering was an order of magnitude higher than the best samples available. [46] The measurements were performed at temperatures between 1.3 and 2.3 K on slightly underfilled samples, to ensure that no bulk liquid was present. They found initially an increased broadening with temperature of the 3D phonon-roton excitations compared to bulk ^{4}He. [39] This effect is probably due to thermalization problems and/or that the bulk ^{4}He reference run was not made at the same time, and hence under slightly different conditions. Their finding that the temperature dependence of the roton energy was not the same for helium in aerogel and bulk ^{4}He, [40,42] has not been confirmed. [51] It appears as if thermalization problems of the cryostat might be at the origin. [54] No layer modes were observed, which is not surprising in view of the limited signal-to-noise ratio on IRIS at that time. The roton in 95% porous aerogel was measured for temperatures between 0.077 and 1.2 K with very high energy resolution, $\Delta E \sim 1\ \mu$eV, using the IN10 spectrometer at the ILL. [45] Measurements in a related system, 72% porous xerogel, have also been made on IN6 at the ILL as a function of coverage at $T = 1.25$ K. [55]

The group at the ILL made most of their measurements on the IN6 time-of-flight spectrometer ($\Delta E \approx 110\ \mu$eV) and on the IN12 triple-axis spectrometer ($\Delta E = 46$ or 110 μeV) at temperatures between 0.5 and 2.25 K. [46–52] They used aerogel samples of 87 and 95% porosity that were either fully deuterated or deuterium-gas treated to reduce multiple scattering. The aerogels were grown directly in the sample cells used for the neutron scattering measurements to reduce the amount of bulk liquid, and the thermometers were calibrated against the ^{3}He vapor pressure. Bulk ^{4}He reference measurements were made simultaneously. Small shifts and broadenings of the phonon-roton excitations in aerogel were initially found with respect to bulk helium. [46] However, these effects arised from additional scattering by layer modes, which were not identified at that time. [51] Measurements beyond the roton wave vector ($Q \gtrsim 2.4$ Å^{-1}) were made on IN12 [46] and on IRIS, [53] using the same fully deuterated sample.

We will now discuss the main results of the above mentioned neutron scattering experiments on helium in aerogel. These measurements show that superfluid ^{4}He in aerogel (fully filled) supports well-defined 3D phonon-roton excitations at low temperatures ($T \lesssim 1.4$ K). These excitations are identical to those in bulk helium, within the precision of present neutron scattering techniques. [45,51] The small modifications of the excitation energy and width reported in early work most likely arise because additional intensity in the tails of the main phonon-roton peak was not recognized and subtracted. If this additional intensity is accounted for, the 3D

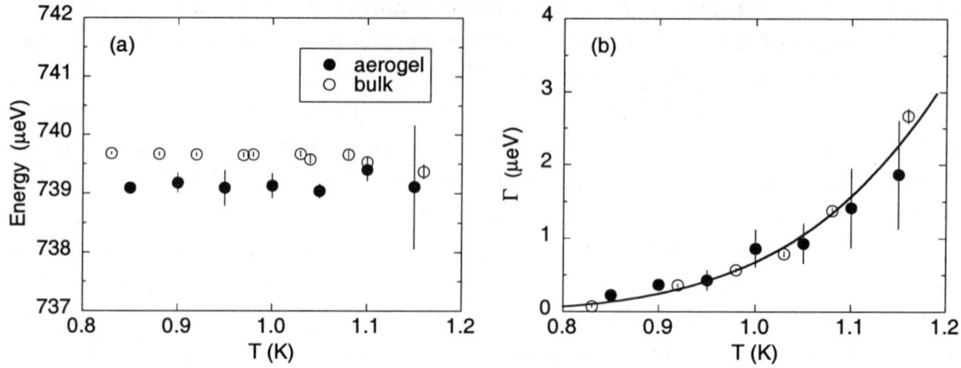

Fig. 6. Temperature dependence of the 3D roton energy (a) and line width [HWHM] (b) for superfluid helium in aerogel (solid circles) and in bulk ^{4}He (open circles) from the high-resolution measurements of Anderson *et al.* [45]

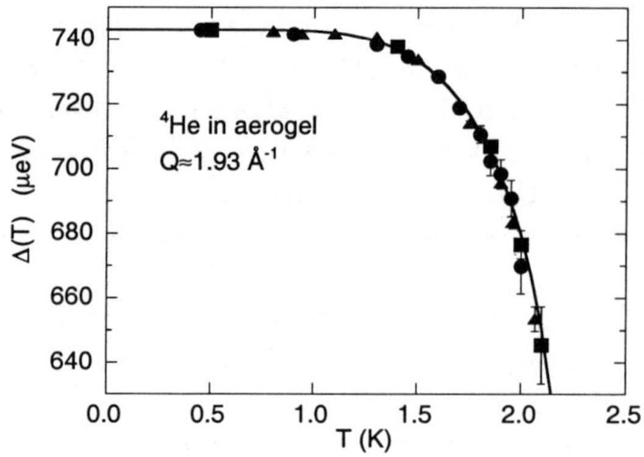

Fig. 7. Temperature dependence of the 3D roton energy $\Delta(T)$ in different aerogel samples (symbols) compared to bulk ^{4}He (line). From Fåk *et al.* [51]

excitations are the same as in bulk helium. The additional scattering is due to layer modes or multiple scattering. Measurements with sufficiently high energy resolution allows the separation of the main peak from the tails. This was very convincingly shown in the high-resolution measurements of Anderson *et al.* (see Fig. 6). [45] They found that the intrinsic width of the roton was below 0.1 μeV at low temperatures ($T = 0.08$ K).

The temperature dependence of the 3D phonon-roton excitations of ^{4}He in aerogel is also the same as in bulk helium. This was shown for the intrinsic width by Anderson *et al.* [45] for temperatures up to 1.2 K (see Fig. 6) and for temperatures up to 2 K by Plantevin *et al.* [52] The temperature dependence of the 3D roton energy in aerogel is also the same as in bulk, as clearly shown in Fig. 7. [49–52]

The filling dependence of the 3D excitations has also been studied in denser aerogels of 87% porosity. [51] The integrated intensity of the 3D phonon-roton excitation scales with the filling fraction (see Fig. 8), and extrapolates to zero at a total of three layers (1–2 liquid layers). The roton energy appears to be higher than the bulk value at low coverages (see Fig. 8), suggesting the existence of low-density layers.

In addition to the 3D phonon-roton excitations, clear evidence for layer modes is seen in denser aerogel, but only near the roton wave vector (see Fig. 9). [51] This is in contrast to helium films on graphite, where layer modes are observed over a large range of Q values. As for graphite, the additional scattering can be attributed to one broad or several sharp layer modes. The layer modes appear in $S(Q,\omega)$ as additional intensity immediately below the 3D roton peak. The integrated intensity of the layer modes increases with the filling fraction (see Fig. 8). It extrapolates to zero at 3 total layers (1–2 liquid layers), i.e. the same as for the 3D excitations, and saturates at a total of 6 layers. The layer modes have a roton-like dispersion (see Fig. 1) with an energy gap of 0.63–0.72 meV, which is higher than for graphite or Vycor. Different experiments give different values for the gap energy of the layer mode, [51] and it is not clear at present whether different aerogels have different gap energies or if there are several layer modes with different energies.

There are few theoretical calculations that deal with the excitations of superfluid ^{4}He in aerogel. It is not clear how calculations of excitations on flat surfaces carry over to aerogel, which is characterized by irregular surfaces. Aerogel is probably also a rather inhomogenous media, due to the wide distribution of open volume sizes.

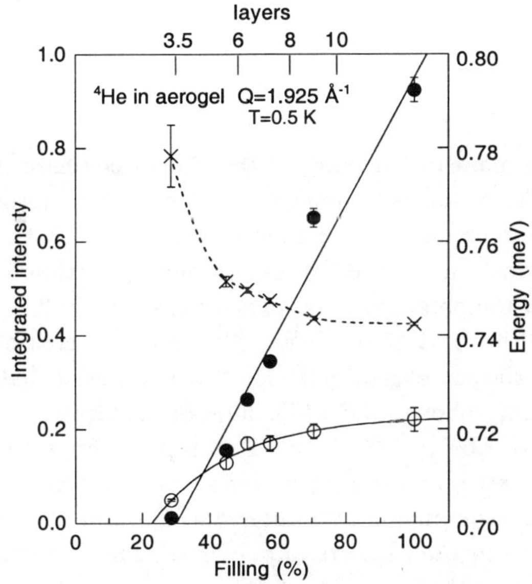

Fig. 8. Integrated intensity of the 3D bulk-like roton (closed circles) and the 2D layer mode (open circles) in superfluid ^{4}He in 87% porous aerogel at low temperatures as a function of filling. The solid lines show a linear fit to the 3D mode and a fit of $I(n) = I_\infty\{1 - \exp[-(n - n_0)/\zeta]\}$ to the 2D mode. The crosses show the energy of the 3D bulk-like roton. The dashed line is a guide to the eye. From Fåk *et al.* [51]

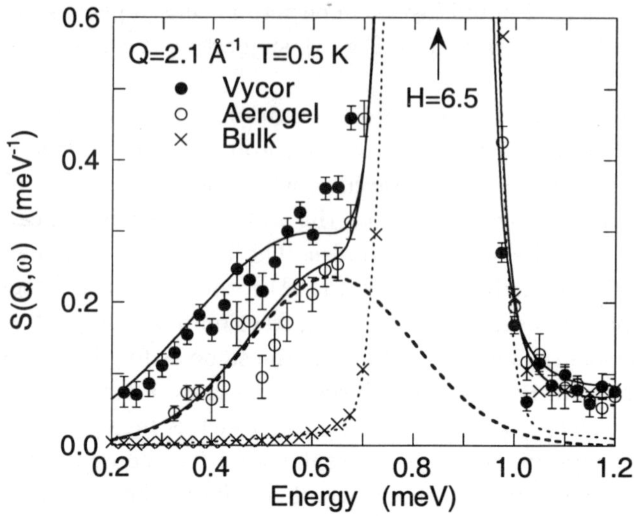

Fig. 9. Dynamic structure factor near the roton wave vector for helium in aerogel (open circles) and Vycor (solid circles) compared to bulk ^{4}He (crosses). The 2D layer modes are clearly seen in the low-energy tail ($\sim$ 0.5 meV) of the main 3D phonon-roton peak.

For partly filled samples, small open regions are likely to fill up and have more layers than larger cavities, due to capillary condensation. It is not clear whether there are one or several layer modes, and theoretical calculations would assist data interpretation.

6. Vycor

Porous Vycor glass is made by leaching out the B_2O_3-rich phase of a phase-separated borosilicate glass. The result is a sponge-like silica-rich material, consisting of an open network of approximately 70 Å diameter channels and with a porosity of about 30%. Due to the fabrication method, Vycor contains approximately 3.5% B_2O_3. The superfluid transition temperature is suppressed from $T_\lambda = 2.172$ K in the bulk to T_c in the range 1.95 to 2.03 K in Vycor, [5,6] while the critical exponent for the superfluid density is the same as in bulk ^{4}He. [5] It is believed that the first ^{4}He layer on Vycor forms an amorphous solid while subsequent layers are liquid.

The strong neutron absorption of the ^{10}B isotope in the remaining B_2O_3 impurities in Vycor makes neutron scattering experiments very difficult. The first inelastic neutron scattering measurements by Lauter and Godfrin at the ILL and by Coddens and collaborators at the LLB are unpublished. The first successful experiment used fully filled standard Vycor on IN6. [56] This was soon followed by more accurate work (also on IN6) using isotopic Vycor, where the ^{10}B isotope was replaced by non-absorbing ^{11}B. [52,57] The negligible absorption of this particular sample greatly improved the quality of the data and made a quantitative analysis possible, after correction for multiple scattering (which becomes important with reduced absorption).

The results from these measurements are that the 3D phonon-roton excitations are the same in fully filled Vycor as in bulk helium at all temperatures. The indications of an increased width and a modified energy (compared to bulk) as a function of temperature in the absorbing sample are most likely due to poor statistics and unresolved layer modes. [56] No differences in the phonon-roton excitations from the bulk were observed in the non-absorbing sample. [57]

In addition to the 3D excitations, additional intensity is observed near the roton wave vector, interpreted as due to layer modes. [56,57] There are no published results on the filling dependence of the excitations in Vycor, and hence no proof for that this additional intensity comes from two-dimensional excitations. However, the scattering is very similar to the layer modes in aerogel and on graphite, leaving little doubt about its origin (see Fig. 9). The dispersion of the layer modes is roton-like, with a roton gap energy of 0.55 meV (see Fig. 1). [57] This energy is the same as the roton energy of the layer modes of ^{4}He on graphite (0.54 meV), but lower than that in aerogel. Layer modes with an energy of 0.53 meV explain the observed specific heat attributed to helium layers in Vycor. [58]

The biggest surprise was perhaps that the dynamic structure factor $S(Q, \omega)$ for the 3D excitations has a temperature dependence in Vycor similar to that in bulk helium (see Fig. 10), even though the superfluid transition temperature T_c is lower. In particular, well-defined maxons and rotons were observed above the superfluid transition temperature $T_c \approx 1.95$ K in Vycor. [57] Estimates of the dead volume between the Vycor sample and the sample cell and in cracks in the Vycor sample show that the amount of bulk liquid collecting in these volumes cannot account for the observed effect.

If well-defined maxons and rotons is the signature of the existence of a Bose condensate, as proposed in the Glyde-Griffin interpretation, [59,60,61] the observation of such excitations above T_c suggests the existence of a localized condensate above T_c. In other words, between the superfluid transition temperature T_c in Vycor and that of the bulk liquid, T_λ, there is a condensate localized in the larger pores in the sample, and it is only below T_c that different localized condensates connect and allows the superfluid to flow through the sample. We note that T_c is determined by torsional oscillator measurements, which measure the percolating superfluid in the sample. The existence of a localized condensate in helium due to disorder has been predicted theoretically, [62,63,64] and it would be very interesting if its observation could be confirmed. Direct measurements of the condensate fraction in Vycor using high-momentum-transfer neutron scattering are under way.

7. Discussion

The energies and lifetimes of the phonon-roton excitations of superfluid ^{4}He in fully filled aerogel and Vycor are the same as in bulk liquid ^{4}He within current precision (5 μeV). Particularly, the width of the roton excitation is unobservably

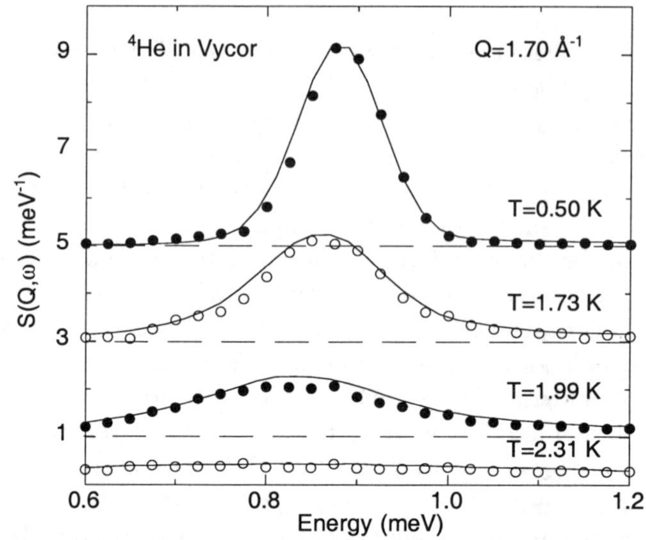

Fig. 10. Temperature dependence of the dynamic structure factor of liquid helium in Vycor (symbols) compared to bulk ^{4}He (lines) at a wave vector between the maxon and roton regions. There is clearly a relatively well-defined excitation at $T = 1.99$ K $\gtrsim T_c$ in Vycor. From Glyde *et al.* [57]

small at low temperatures ($\Gamma \leq 0.1$ μeV at $T = 0.08$ K) [45] and its temperature dependence is the same as in bulk ^{4}He up to T_λ. [52] Whatever differences have been observed tend to disappear when measured with increased precision and compared with simultaneously measured bulk values.

This result is somewhat surprising since the excitation energies are a sensitive function of the liquid density. [60] It suggests that the liquid density in which the excitations propagate in aerogel and Vycor is predominantly at the bulk SVP density. The excitation energies in superfluid films at low temperatures on graphite are also much the same as in the bulk within observed precision. No 3D phonon-roton excitations propagate until at least 3.5 total (1.5 liquid) layers are deposited on graphite, [24] and until the equivalent of 3.5 total layers in aerogel. [51] On graphite, the fourth layer is believed to be at the bulk density. Thus the layers in which the 3D excitations propagate are probably at bulk density, which is consistent with the equality of the excitation energies in films on graphite and the bulk. Any variation in density would also introduce a width to the excitations. A finite width for phonons, which propagate with a sound velocity $c \simeq 200$ m/s, is expected in confinement. This has not been observed.

The exceptions to the above findings are: (i) in partially filled aerogel, the roton energy lies above the bulk value; [51] (ii) in films on graphite, the maxon energy may lie below the bulk value for low coverages; [29] and (iii) in Geltech silica, which has 25 Å diameter pores, the excitations appear to differ from bulk values at partial fillings. [65]

There have been several predictions for the change in excitation energies in Bose fluids arising from disorder. In a dilute Bose gas, Zhang predicted that white noise random disorder would reduce the sound velocity of phonons and introduce a width. [66] New excitations at low energy (low $\hbar\omega$), especially at long wavelength, have been predicted when disorder is introduced. [67] Theoretical calculations have suggested that a gap could open up in the phonon dispersion curve in the long wavelength limit in disorder. [63] Plantevin *et al.* searched for a gap and new excitations at low wave vectors ($Q \simeq 0.2$ Å^{-1} corresponding to $\hbar\omega \approx 0.325$ meV) in 95% porous aerogel but found no departure from the bulk. [46] A major problem in this search is that the elastic scattering ($\omega = 0$) even from fully deuterated aerogel samples is still large enough to mask small inelastic contributions at low ω. Boninsegni and Glyde calculated $S(Q, \omega)$ of superfluid ^{4}He containing small (2 Å) hard spheres randomly placed. [68] $S(Q, \omega)$ was both broadened and displaced in energy with new weight at low ω. Apparently, disorder on short length scales is needed to modify $S(Q, \omega)$ significantly.

The new excitations observed in porous media not seen in the bulk are the layer modes. They have been observed in aerogel, [51] xerogel, [55] Vycor, [56,57] and Geltech silica. [65] The layer modes are pictured as 2D "phonons" propagating in the liquid layers adjacent to the media walls. At the roton wave vector at least, the layer modes are the same as those observed in liquid ^{4}He films on graphite. [22,24,32] In porous media, the layer modes have been observed in the wave-vector range $1.7 \leq Q \leq 2.15$ Å^{-1}. In this Q range, they have a "roton-like" dispersion (see Fig. 1). Below $Q = 1.7$ Å^{-1}, the intensity in the mode becomes too small to be observed or the mode overlaps with the phonon-roton peak or the multiple scattering. [52] The mode intensity increases with wave vector up to $Q = 2.15$ Å^{-1}, the maximum value investigated. It would be most interesting to determine whether layer modes exist up to higher wave vectors and what their intensity and dispersion might be. In films on graphite, it seems that both dispersive [32] and dispersionless [29] layer modes are observed. Calculations of films on substrates show propagating dispersive 2D modes. [32]

In Vycor, [57] the 2D roton energy $\Delta_{2D} = 0.55 \pm 0.01$ meV agrees well with the energy of 0.53 meV extracted by Brewer *et al.* [4,58] from the layer contribution to the specific heat. It is also consistent with the gap energy of 0.50 meV obtained by Kiewiet *et al.* [69] from the superfluid density in Vycor for $T \leq 1.4$ K. Further connections between excitation energies observed by neutron scattering and thermodynamic or transport properties would be most interesting. These, however, must be restricted to low temperatures ($T \leq 1.5$ K), where the excitation energies are sharply defined.

In bulk liquid ^{4}He, the integrated intensity in the characteristic phonon-roton excitation for $Q \gtrsim 0.7$ Å^{-1} scales approximately with the superfluid density $\rho_S(T)$. There is no well-defined excitation above T_λ in the normal phase (for $Q \gtrsim 0.7$ Å^{-1}). In Vycor, a well-defined excitation was observed above $T_c \approx 1.95$ K. At $T = 1.99$ K, where $\rho_S(T) = 0$ in Vycor, approximately one half of the intensity is still in the elementary excitation. Glyde and Griffin proposed that the intensity in the elementary excitation should scale as $n_0(T)$. [59] The result in Vycor suggests that there is still some condensate above T_c, perhaps a localized condensate. [64] That is, there could be superfluid Bose-condensed regions on dimensions of the pore size in larger open regions. However, these regions do not extend across the whole media. These locally condensed regions would not be observed in a torsional oscillator measurement. [2,5] Bose-Einstein condensation on short length scales in confinement has been suggested to explain thermal expansion data [70] and discussed theoretically. [64,71] Further theoretical exploration of this interesting concept is needed.

Helium in porous media is subject to both confinement and disorder. The confinement comes from the limited pore size and the quasi-two dimensional character of the media walls/substrate, while the disorder comes from the irregular structures of the pores. A legitimate question is which of these two effects is dominating the changes in the microscopic and macroscopic properties. By comparing the results from ^{4}He in aerogel and Vycor with ^{4}He on graphite, which to a rather high degree presents atomically well ordered planar substrates, a preliminary answer would be that confinement is more important. The reasons are that the 3D excitations are identical in these two types of materials and that the 2D layer modes show many similarities. Theoretical work would be very welcome here, particularly on the length scales of disorder needed to have an impact on the excitations.

Experimentally, work is in progress to study porous media that present long-range order, such as zeolites. Another trend is to go to smaller pore sizes, in order to increase the effects of confinement. The first studies of high-porosity aerogels ($\approx$ 95%) did not permit direct observation of layer modes, although indirect signs were seen such as apparent broadenings and shifts of the 3D excitations. However, in higher density aerogels [51] as well as in xerogel, [55] both with a porosity of 87%, layer modes were observed. Another system which has just started to be investigated is 50% porous Geltech, [65] where the pore size is very small, of the order of 25 Å, and where the superfluid transition temperature is suppressed to 0.725 K.

Acknowledgments

We would like to thank O. Plantevin, J. Bossy, G. Coddens, H. Schober, and R.T. Azuah for very fruitful collaboration on the neutron scattering experiments. These measurements would not have been possible without the efforts of J.R. Beamish, N. Mulders, and D.S. Danielson, who specifically prepared our samples. We have also benefitted from helpful discussions with C.R. Anderson, K.H. Andersen, E. Krotscheck, H. Godfrin, H.J. Lauter, W.G. Stirling, P.E. Sokol, R.M. Dimeo, L. Puech, and E. Wolf. HRG was supported in part by the National Science Foundation through research grant DMR-9972011.

References

1. D. F. Brewer, in *The Physics of Liquid and Solid Helium*, Part II, eds. K. H. Benneman and J. B. Ketterson (Wiley, New York, 1978), p. 573.
2. J. D. Reppy, *J. Low Temp. Phys.* **87**, 205 (1992).
3. M. Chan, N. Mulders, and J. Reppy, *Physics Today*, August 1996, p. 30.
4. D. F. Brewer, *Physica* B **280**, 4 (2000).
5. M. H. W. Chan, K. I. Blum, S. Q. Murphy, G. K. S. Wong, and J. D. Reppy, *Phys. Rev. Lett.* **61**, 1950 (1988).
6. G. M. Zassenhaus and J. D. Reppy, *Phys. Rev. Lett.* **83**, 4800 (1999).
7. S. Miyamoto and Y. Takano, *Czech. J. Phys.* **46**, 137 (1996).
8. J. D. Reppy, B. C. Crooker, B. Hebral, A. D. Corwin, J. He, and G. M. Zassenhaus, *Phys. Rev. Lett.* **84**, 2060 (2000).
9. L. D. Landau, *J. Phys. U.S.S.R.* **11**, 91 (1947).
10. F. London, *Nature* **141**, 643 (1938).
11. N. N. Bogoliubov, *J. Phys. U.S.S.R.* **11**, 23 (1947).
12. J. Gavoret and P. Nozières, *Ann. Phys.* **28**, 349 (1964).
13. J. M. Kosterlitz and D. J. Thouless, *Prog. Low Temp. Phys.* VIIB, 371 (1978).
14. U. C. Täuber and D. R. Nelson, *Phys. Repts.* **289**, 157 (1997).
15. M. Ma, P. Nisamaneephong, and L. Zhang, *J. Low Temp. Phys.* **93**, 957 (1993).
16. B. E. Clements, E. Krotscheck, and C. J. Tymczak, *Phys. Rev.* B **53**, 12 253 (1996).
17. E. Krotscheck, M. D. Miller, and R. Zillich, *Physica* B **280**, 59 (2000).
18. T. C. Padmore, *Phys. Rev. Lett.* **32**, 826 (1974).
19. W. Götze and M. Lücke, *J. Low Temp. Phys.* **25**, 671 (1976).
20. H. Godfrin and H. J. Lauter, *Prog. in Low Temp. Phys.* Vol. XIV, ed. W. P. Halperin, (Elsevier, Amsterdam, 1995), p. 213; H. Godfrin and R. E. Rapp, *Adv. Phys.* **44**, 113 (1995).
21. K. H. Andersen, W. G. Stirling, R. Scherm, A. Stunault, B. Fåk, H. Godfrin, and A. J. Dianoux, *J. Phys. Condens. Matter* **6**, 821 (1994); K. H. Andersen and W. G. Stirling, *ibid.* 5805.
22. B. Lambert, D. Salin, J. Joffrin, and R. Scherm, *J. Phys. (Paris), Lett.* **38**, L377 (1977).
23. K. Carneiro, W. D. Ellenson, L. Passell, J. P. McTague, and H. Taub, *Phys. Rev. Lett.* **37**, 1695 (1976).
24. W. Thomlinson, J. A. Tarvin, and L. Passell, *Phys. Rev. Lett.* **44**, 266 (1980).
25. H. J. Lauter, H. Wiechert, and C. Tiby, *Physica* **107**B, 239 (1981).
26. H. J. Lauter, H. Godfrin, C. Tiby, H. Wiechert, and P. E. Obermayer, *Surface Science* **125**, 265 (1983).
27. H. J. Lauter, H. Godfrin, and H. Wiechert, in *Phonon Physics*, eds. J. Kollar *et al.* (World Scientific, Singapore, 1985), p. 842.
28. H. J. Lauter, V. L. P. Frank, H. Godfrin, and H. Wiechert, in *Elementary Excitations in Quantum Fluids*, eds. K. Ohbayashi and M. Watabe, Springer Series in Solid State Sciences Vol. 79 (Springer-Verlag, Berlin, 1989), p. 99.
29. H. J. Lauter, H. Godfrin, V. L. P. Frank, and P. Leiderer, in *Excitations in Two-Dimensional and Three-Dimensional Quantum Fluids*, eds. A.F.G. Wyatt and H.J. Lauter (Plenum, New York, 1991), p. 419.
30. H. J. Lauter, H. Godfrin, and P. Leiderer, *J. Low Temp. Phys.* **87**, 425 (1992).
31. H. J. Lauter, H. Godfrin, V. L. P. Frank, and P. Leiderer, *Phys. Rev. Lett.* **68**, 2484 (1992).
32. B. E. Clements, H. Godfrin, E. Krotscheck, H. J. Lauter, P. Leiderer, V. Passiouk, and C.J. Tymczak, *Phys. Rev.* B **53**, 12 242 (1996).

33. D. O. Edwards and W. F. Saam, *Prog. in Low Temp. Phys.* Vol. VII A, ed. D. F. Brewer, (North Holland, Amsterdam, 1978), p. 283.
34. Note that Clements *et al.* [32] present the data at constant scattering angle rather than constant Q as is customary. The wave vectors k given in the paper correspond therefore to elastic scattering, and need to be modified at finite energy transfers.
35. J. V. Porto and J. M. Parpia, *Phys. Rev.* B **59**, 14 583 (1999).
36. H. Godfrin, J. Klier, V. Lauter-Pasyuk, H. Lauter, P. Leiderer, *ILL Experimental Report* 6-01-177 (1998), unpublished.
37. J. de Kinder, G. Coddens, and R. Millet, *Z. Phys.* B, *Condensed Matter* **95**, 511 (1994).
38. G. Coddens, J. de Kinder, and R. Millet, *J. Non-Cryst. Solids* **188**, 41 (1995).
39. M. R. Gibbs, P. E. Sokol, R. T. Azuah, W. G. Stirling, and M. A. Adams, *Physica* B **213-214**, 462 (1995).
40. P. E. Sokol, M. R. Gibbs, W. G. Stirling, R. T. Azuah, and M. A. Adams, *Nature* **379**, 616 (1996).
41. M. R. Gibbs, P. E. Sokol, W. G. Stirling, R. T. Azuah, and M. A. Adams, *J. Low Temp. Phys.* **107**, 33 (1997).
42. R. M. Dimeo, P. E. Sokol, D. W. Brown, C. R. Anderson, W. G. Stirling, M. A. Adams, S. H. Lee, C. Rutiser, and S. Komarneni, *Phys. Rev. Lett.* **79**, 5274 (1997).
43. P. E. Sokol. R. M. Dimeo, D. W. Brown, C. R. Anderson, W. G. Stirling, M. A. Adams, S. H. Lee, C. Rutiser, and S. Komarneni, *Physica* B **241-243**, 929 (1998).
44. R. M. Dimeo, P. E. Sokol, C. R. Anderson, W. G. Stirling, and M. A. Adams, *J. Low Temp. Phys.* **113**, 369 (1998).
45. C. R. Anderson, K. H. Andersen, J. Bossy, W. G. Stirling, R. M. Dimeo, P. E. Sokol, J. C. Cook, and D. W. Brown, *Phys. Rev.* B **59**, 13 588 (1999).
46. O. Plantevin, B. Fåk, H. R. Glyde, J. Bossy, and J. R. Beamish, *Phys. Rev.* B **57**, 10 775 (1998).
47. H. R. Glyde, B. Fåk, and O. Plantevin, *J. Low Temp. Phys.* **113**, 537 (1998).
48. O. Plantevin, Ph. D. Thesis, Université Joseph Fourier, Grenoble, France (1999).
49. B. Fåk, O. Plantevin, and H. R. Glyde, *Physica* B **276-278**, 806 (2000).
50. B. Fåk, O. Plantevin, and H. R. Glyde, *J. Phys. IV (France) Colloque* **10**, Pr7: 163 (2000).
51. B. Fåk, O. Plantevin, H. R. Glyde, and N. Mulders, *Phys. Rev. Lett.* **85**, 3886 (2000).
52. O. Plantevin, B. Fåk, H. R. Glyde, N. Mulders, J. Bossy, G. Coddens, and H. Schober, *Phys. Rev.* **B 63**, 224508 (2001).
53. R. T. Azuah, H. R. Glyde, J. R. Beamish, and M. A. Adams, *J. Low Temp. Phys.* **117**, 113 (1999).
54. P. E. Sokol, private communication.
55. C. R. Anderson, W. G. Stirling, K. H. Andersen, P. E. Sokol, and R. M. Dimeo, *Physica* B **276-278**, 820 (2000).
56. R. M. Dimeo, P. E. Sokol, C. R. Anderson, W. G. Stirling, K. H. Andersen, and M. A. Adams, *Phys. Rev. Lett.* **81**, 5860 (1998).
57. H. R. Glyde, O. Plantevin, B. Fåk, G. Coddens, P. S. Danielson, and H. Schober, *Phys. Rev. Lett.* **84**, 2646 (2000).
58. D. F. Brewer, A. J. Symonds, and A. L. Thomson, *Phys. Rev. Lett.* **15**, 182 (1965).
59. H. R. Glyde and A. Griffin, *Phys. Rev. Lett.* **65**, 1454 (1990).
60. H. R. Glyde, *Excitations in Liquid and Solid Helium* (Oxford University Press, Oxford, 1994); *J. Low Temp. Phys.* **93**, 349 (1993).
61. A. Griffin, *Excitations in a Bose-Condensed Liquid* (Cambridge University Press, Cambridge, 1993).

62. D. J. Thouless, *Phys. Rep., Phys. Lett.* **13C**, 93 (1974).
63. W. Krauth and N. Trivedi, *Europhys. Lett.* **14**, 627 (1991); W. Krauth, N. Trivedi, and D. Ceperley, *Phys. Rev. Lett.* **67**, 2307 (1991); T. Giamarchi and P. Le Doussal, *Phys. Rev. Lett.* **76**, 3408 (1996).
64. K. Huang, in *Bose Einstein Condensation*, eds. A. Griffin, D. Snoke, and S. Stringari (Cambridge University Press, Cambridge, 1995), p. 31.
65. O. Plantevin, B. Fåk, and H. R. Glyde, *J. Phys. IV (France) Colloque* **10**, Pr7: 177 (2000).
66. L. Zhang, *Phys. Rev. B* **47**, 14 364 (1993).
67. M. Makivić, N. Trivedi, and S. Ullah, *Phys. Rev. Lett.* **71**, 2307 (1993).
68. M. Boninsegni and H. R. Glyde, *J. Low Temp. Phys.* **112**, 251 (1998).
69. C. W. Kiewiet, H. E. Hall, and J. D. Reppy, *Phys. Rev. Lett.* **35**, 1286 (1975).
70. P. Thibault, J. J. Préjean, and L. Puech, *Czech. J. Phys.* **46**, 149 (1996).
71. K. Huang and H.-F. Meng, *Phys. Rev. Lett.* **69**, 644 (1992).

CHAPTER 9

MICROSCOPIC SUPERFLUIDITY OF SMALL ^{4}He AND PARA-H_2 CLUSTERS INSIDE HELIUM DROPLETS

J. P. Toennies
Max-Planck-Institut für Strömungsforschung
37073 Göttingen, Germany
E-mail: jtoenni@gwdg.de

The present review describes recent molecular beam experiments in which large ^{4}He or ^{3}He liquid droplets consisting typically of 10^3 to 10^4 atoms are produced and doped by pick-up of single atomic or molecular chromophores. The spectroscopy of these single particles has led to new detailed insight into the elementary microscopic interactions of the probe particles with their environment. In the visible the spectral features are unusually sharp with line widths comparable to those of the free molecules. The phonon wings of vibronic transitions give direct evidence that the droplets are superfluid. In the infra-red well defined rotational lines appear that indicate that the molecules rotate freely inside the liquid. From the intensities of the sharp lines temperatures of 0.37 K and about 0.14 K are determined for ^{4}He and ^{3}He droplets, respectively. These experiments demonstrate that superfluid ^{4}He droplets provide a new ultra cold uniquely gentle matrix for high resolution spectroscopy. At the same time the molecular spectra contribute new microscopic insight into the intriguing phenomenon of superfluidity. This last aspect will be emphasized in this review. Several reviews which emphasize more the new opportunities for high resolution spectroscopy, [1–4] an introductory overview [5] and a special issue of the Journal of Chemical Physics have recently been published. [6–8]

1. Introduction

The helium isotopes ^{4}He and ^{3}He are unusual among the elements in that they do not exhibit a triple point and consequently remain in the liquid state at atmospheric pressures down to the lowest temperatures $T = 0$. They are also the only liquids exhibiting superfluidity below $T_c = 2.18$ K (^{4}He) and $T_c = 2.4 \cdot 10^{-3}$ K (^{3}He). Superfluidity manifests itself through many different apparently unrelated phenomena, such as a vanishingly small viscosity, the fountain effect, the ability to creep out of a container thereby defying gravity or the extraordinary ability to conduct heat much more efficiently than even the purest metals. These many strange macroscopic properties, some of which were discovered in 1938 [9,10] when superfluidity was first recognized as a new phenomenon, inspired Fritz London in 1954 to proclaim that "superfluid helium, also called liquid helium II, is the only representative of a particular fourth state of aggregation beside the solid and gaseous states". [11] Since London's time it is now realized that superfluidity and superconductivity are closely related and very ubiquitous phenomena which occur in solids, in the nuclear matter in stars [12] as well as in nuclei [13] and also in elementary particle physics, e.g. Higgs boson. [14]

Landau was the first to formulate a microscopic theory of superfluidity by postulating that the elementary excitations in a superfluid are dominated by highly coherent phonons at low energies and at energies above about 8.5 K by another type of excitation called rotons. [14] Compared to the excitations in ordinary classical liquids the dispersion curves in the superfluid are sharply defined in energy as in a solid. The sharp dispersion curves coupled with the two fluid model provide a unifying framework within which many of the above macroscopic observations [15,16] can be described. Nevertheless on a microscopic level the physical mechanisms behind many of the unusual properties of both the quantum liquids ^{3}He and ^{4}He are not fully understood even now over more than 60 years after the discovery of superfluidity. Today the statement made in 1984 that the "connection between Bose condensation and superfluidity (in ^{4}He) remains a deep and complex problem" [17] is still valid despite considerable progress in explaining many related phenomena in Bose-Einstein condensed gases. [18]

Several different avenues have been pursued in recent years to acquire a deeper more microscopic insight into these fascinating phenomena. The earlier predominantly macroscopic investigations have in recent years been extended to exploring superfluidity in diverse porous substances with confined geometries (e.g. in aerogels [19]) or in quasi-one-dimensional systems (anapore and nuclepore filters [20]) or in thin films on the surfaces of solids. [21] Attempts at implanting molecules [22] or atoms [23] inside the bulk liquid to serve as a microscopic probe have met with only limited success. Because of their high mobility the impurities quickly clump together and either rise to the surface, fall to the bottom or precipitate on the walls. [22] Sophisticated pulsed laser experiments have made it possible to circumvent these difficulties by ablating metal atoms inside bulk liquid helium and interrogating their spectra by a second laser pulse. [24] So far, however, these and related investigations have been restricted to metastable excited dimers and atoms of helium, the alkali atoms, the alkaline earth atoms as well as the corresponding ions and a few other metals. In most cases the spectral features are broad, essentially because of the strong repulsive exchange interactions of the outer electrons of these open shell species with the helium environment.

About ten years ago it was found that finite sized droplets of helium could pick up closed shell molecules singly [25] and that these molecules were located in the interior of the droplets. [26] It was also discovered that dimers and larger numbers of atoms or molecules and even specified mixtures can be prepared inside helium droplets. [25,27,28] Both from experiment and from theory [29,30] it is now clear that virtually all closed shell atoms (e.g. rare gases) and closed shell molecules (e.g. HF, H_2O, OCS, and SF_6) exhibit a *heliophilic* behavior and are localized inside the droplets. Open shell atoms (e.g. alkali metals Na, K, etc. or the alkaline earth metals Ba and Ca) and presumably most open shell molecules, such as free radicals, are *heliophobic.* Since the outer electrons of these systems interact with the droplets largely via strongly repulsive potentials they cannot be drawn into the interior and remain on the surface of the droplets.

The remainder of this review is organized in the following order. The next Section describes the experimental techniques and results on the droplet sizes, pick-up of foreign molecules and the spectroscopic method. Section 3 reviews theoretical and experimental evidence for the superfluidity of helium droplets. The spectroscopic evidence for free rotations of molecules is presented in Section 4. The explanation for their increased moments of inertia is the topic of Section 5. Finally in Section 6 the evidence for superfluidity of small para-H_2 clusters is presented. This Chapter then closes with a brief summary and outlook.

2. Experimental aspects

2.1. *Production of droplets in free jet expansions*

Beams containing helium clusters and droplets are readily produced by passing the high purity gas at a high pressure P_o (10-200 bar) and at a low temperature T_o (20-3 K) either through a nozzle [31] or through a narrow thin walled orifice with diameter d (5-10 microns) into a vacuum. [31-33] In the ensuing radial expansion the particle density n falls off with the inverse square of the distance z measured from the orifice ($n(z) \propto (d/z)^2$). To a very good approximation the changes in the thermodynamic state of the gas in the course of the expansion can be approximated by assuming an adiabatic (isentropic) process. Thus, for example, the temperature at some point downstream from the nozzle z is given by

$$T(z) = T_o \left(\frac{n(z)}{n_o} \right)^{\gamma - 1} , \tag{2.1}$$

where γ is the ratio of the specific heats ($\gamma = 1.67$ for atoms). The sharp drop in the density $n(z)$ beyond the orifice leads to a rapid decrease in the temperature. The decrease in temperature has been estimated to be extremely fast approaching values of about 10^{11} K/sec. [34] The temperature drops continuously as long as the density is large enough so that collisions still occur in significant numbers to assure local equilibrium.

In the case of helium a unique quantum effect leads to a dramatic increase in the atom-atom scattering cross section and the related collision rate as the temperature drops below about 1 K. At these low temperatures the cross section is dominated by s-wave scattering and approaches a value of $\sigma_{T\to 0} = 8\pi a^2$, where a is the scattering length. The latter is related to the binding energy by the relationship $a \propto |\epsilon_b|^{-1}$. [35] Since the binding energy of the helium dimer is only $\epsilon_b = 1.1 \cdot 10^{-3}$ K ($\cong 10^{-7}$ eV) [36] the scattering length is very large, $a \cong 100$ Å. Thus the asymptotic value of the cross section is 259,000 Å^2, [37] which is one of the largest known cross sections involving atoms and molecules. This huge cross section leads to a significant enhancement in the collision induced cooling in the final stages of the expansion. The final temperatures have been found to be as low as 10^{-2} K or even colder. [38,39]

Provided that the particle density is large enough so that there are still sufficient numbers of three-body collisions at these low temperatures the atoms will have the tendency to condense to form clusters. Thermodynamically the transition from the gas to liquid phase is determined by the minimization of the Gibbs free energy $G = H - TS$, where H is the enthalpy and S is the entropy of the medium. Thus as T approaches zero the term which accounts for the accompanying decrease in the entropy $-T\Delta S$ is suppressed and the aggregation of the atoms to liquid droplets is then favored by the negative heat released, $-\Delta H$.

The production of clusters and droplets can be conveniently described in terms of a trajectory in the P, T phase diagram of ^{4}He [33] (see Fig. 1). Such a thermodynamical description is, of course, only approximate because of the finite size of the droplets. Initially the system evolves from the source stagnation conditions P_o, T_o along the adiabatic curve obeying

$$T^{\gamma} P^{1-\gamma} = T_o^{\gamma} P_o^{1-\gamma} \quad . \tag{2.2}$$

Immediately after crossing over into the superfluid He II region the system will, before it can respond, continue on for some distance into a metastable supercooled gaseous state. Then as condensation begins the heat released will raise the temperature and bring the system trajectory back to the equilibrium vapor phase curve as the droplets continue to grow in size.

After a reduced distance of about $z/d \approx 10^3$ the droplets leave the region of extensive collisions and continue their forward motion in vacuum without further encounters. There the droplets undergo very rapid evaporation with a time constant of about 10^{-7} seconds [40,41] and their internal temperatures drop until the rate of evaporation becomes negligible. After travelling typical apparatus distances of about one meter, corresponding to a flight time of about 10^{-3} seconds, the final temperatures have been calculated to be 0.38 K (^{4}He) [40] and 0.15 K (^{3}He) [41]. The rapid loss in temperature explains the downward continuation of the state trajectory along the phase line in the lower part of Fig. 1. Fig. 2 illustrates the evaporation induced decrease in droplet sizes and temperatures as a function of time after the droplets have left the region of extensive collisions.

Droplets have also been produced by expanding mixtures of ^{4}He and ^{3}He. These droplets have been found experimentally [42,43] and theoretically [44] to consist of an inner core of nearly pure ^{4}He atoms and an outer shell of ^{3}He atoms. The latter, being on the outside, determine the temperature and bestow these mixed droplets with a temperature of only 0.15 K. [42,43] As will be discussed later many of the remarkable properties of He droplets derive from the low-temperatures resulting from the rapid efficient evaporative cooling. This also endows them with a huge effective heat capacity making them robust isothermal *nano-cryostats* with a well

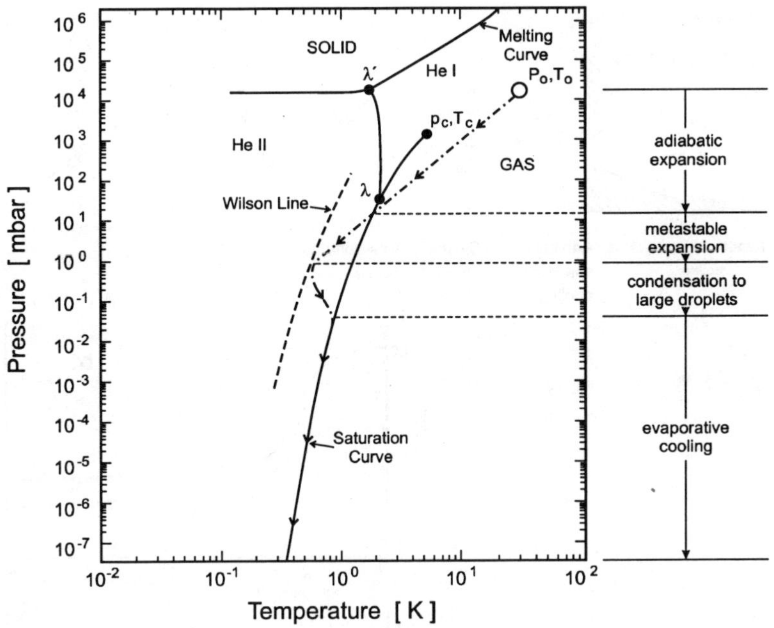

Fig. 1. Pressure-temperature phase diagram of ^{4}He showing the transitions between the gaseous, liquid non-superfluid HeI and superfluid HeII and the solid states. The dashed line shows the state trajectory of the gas in a free jet expansion with stagnation conditions of P_o, T_o which lead to the formation of large droplets. [33]

defined temperature. The accompanying drawback is the difficulty to change their temperatures so that the experiments are presently constrained to either 0.37 K or 0.15 K.

2.2. *Sizes of He droplets*

The sizes of the droplets and their size distributions were first measured in scattering experiments in which a secondary beam of a heavy rare gas, usually krypton, crossed the droplet beam. [45] From a careful analysis of the deflected droplets it was found that the entire momentum of the Kr atoms was imparted to the droplets, which are consequently deflected. [45] Thus from the angle of deflection the momentum of the droplets could be determined. [51] Since the velocity of the droplet beam is sharply peaked the distribution of deflection angles can be directly related to the droplet mass and number size distributions. With this technique deflection angles down to 10^{-3} radians have been resolved and droplets with sizes up to 10^4 helium atoms could be determined reliably. The inset in Fig. 3 shows some typical size distributions which follow a log-normal distribution, [46]

$$f(N) = \frac{1}{N\sigma\sqrt{2\pi}} \exp\left[\frac{-(ln\ N\ -\ \mu)^2}{2\sigma^2}\right] \quad , \tag{2.3}$$

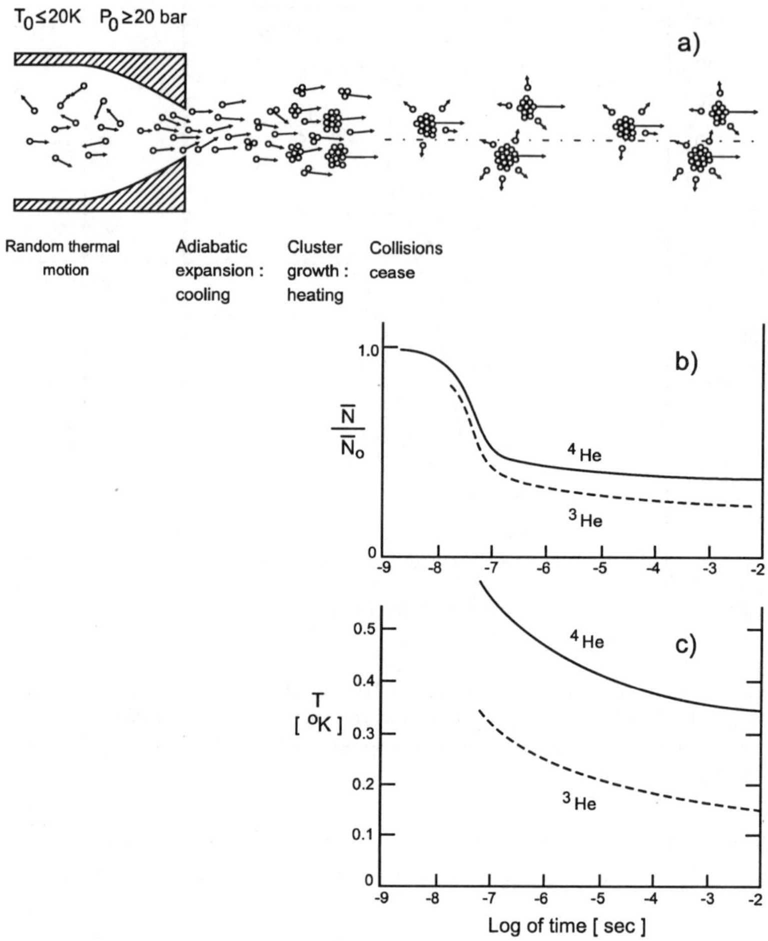

Fig. 2. (a) Schematic diagram of the collision processes leading to the formation of large droplets. (b) After collisions cease the droplets evaporate atoms loosing about half of their initial numbers. [40,41] (c) As a result of evaporation the temperature falls rapidly approaching values of about 0.37 K (^{4}He droplets) [40] and 0.15 K (^{3}He droplets) [41] after about 10^{-3} seconds.

where μ and σ are fit parameters. The half-widths of the measured distributions (2.3) are approximately given by $\Delta N_{1/2} \cong 0.8\overline{N}$, where $\overline{N}$ is the average number of atoms in the droplet.

Fig. 3 also displays the dependence of the mean number sizes $\overline{N}$ and liquid drop diameters on the source stagnation conditions P_o and T_o, measured for a 5 micron diameter orifice. These mean sizes have since been confirmed by several other experiments including one which relies on the measured relative depletion of the mass spectrometer signal following laser absorption, which is inversely proportional to the average number of He atoms in the droplet $\overline{N}$. [47] At the present time there is no satisfactory theory for predicting droplets sizes for a given orifice and source

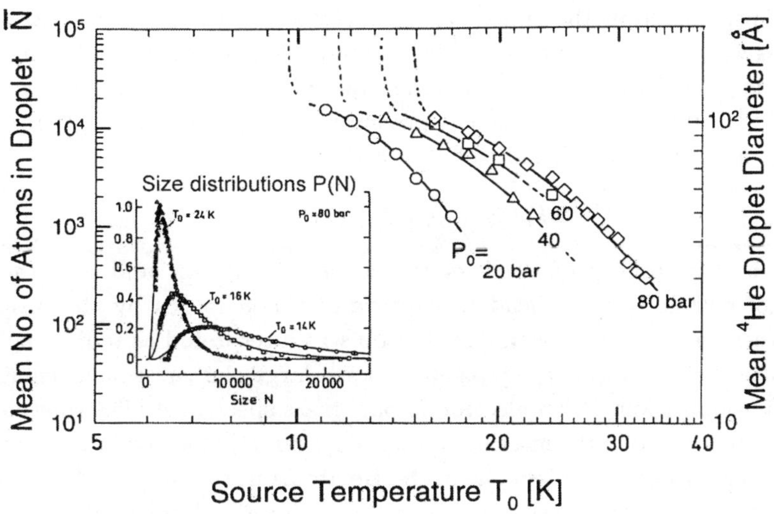

Fig. 3. The mean number sizes and diameters of ^{4}He droplets are plotted as a function of the source stagnation temperature T_o for four different source pressures P_o. The insert shows some typical size distributions. [45] The diameters were calculated assuming a liquid drop with the bulk density and a sharp edge: $d = 4.44 \cdot \overline{N}^{1/3}$ Å. The actual diameters at 10% of the bulk density are larger by about 6 Å. [51]

stagnation conditions although several semiempirical rules for estimating droplet sizes have been reported. [48–50]

The measurements of droplet sizes have been combined with measurements of the total integral cross sections of the droplets σ to obtain information on the densities of the droplets. Since $\sigma \propto (v_{drop})^{2/3}$, where v_{drop} is the total volume of the droplets, which are assumed and confirmed to be nearly spherical, [51] the volume of the droplets can be measured. Knowing both $\overline{N}$ and v_{drop} the average particle density of the droplets $\overline{\rho} = \overline{N} \; / \; v_{drop}$ has been determined. [51] Moreover, with an appropriate model for the radial density distribution, a more sophisticated analysis of the data has made it possible to estimate the radial distance over which the density distribution falls off at the droplet edges for both ^{4}He [51] and ^{3}He [52] droplets. The 90%-10% fall-off distance is 6.4 ± 1.3 Å and 6.7 ± 1.3 Å, respectively, independent of droplet sizes for $\overline{N}$=10^3–10^4 in reasonable agreement with the fall-off length of the free bulk-liquid helium surface.

2.3. *Pick-up of foreign molecules*

The first mass spectroscopy studies of He droplets beams revealed some persistent unidentified peaks which were subsequently attributed to pump oil molecules. [33] Oil molecules with very small partial pressures of only 10^{-9} mbar are present in the residual vacuum of high vacuum systems evacuated by conventional diffusion pumps. In these beam experiments they collide with and stick to the liquid droplets on their passage from the source to the detector. This facile pick-up of foreign molecules was then confirmed by dedicated mass spectroscopy studies in which the pressure in one of the vacuum chambers was increased by backfilling with a rare gas. [25,26,53] With increasing rare gas pressure ion fragments of larger rare clusters RG_n^+ appeared and were found to follow a Poisson distribution,

$$P_k(\overline{z}) = \frac{\overline{z}}{k!} \exp(-\overline{z}) \quad , \tag{2.4}$$

where k is the number of particles in the neutral cluster and $\overline{z}$ is the average number of collisions which lead to capture of the particles by the droplet. The good agreement with a Poisson distribution was interpreted as indicating that the atoms, after being captured, aggregate to form large clusters which remain largely intact after ionization. Other electron impact mass spectral studies showed that the clusters were located in the interior and were not directly ionized but rather that the impacting electron first ionizes one of the He atoms in the droplet. [26] After a series of rapid delocalizing charge transfer processes within the droplet the positive charge eventually becomes localized either on the foreign particle cluster [26,54] or on another He atom. The large amount of heat released in the ensuing electrostriction around the nascent ion is rapidly dissipated by evaporation of the droplet leaving behind the intact ionized cluster. [54] More recent refined studies have revealed however some fragmentation in the case of embedded Ne [55], Ar [56] and Xe [57] clusters.

The deflection scattering technique has also been used to investigate the transfer of energy to the droplet by the captured particle. [53] The evaporative decrease in the droplet size after the droplets have captured increasing number of atoms was found to be proportional to the sum of the kinetic and internal energies of the particles entering the droplet plus the energy released in binding with the He atoms in the droplet. If other atoms or molecules were present inside the droplet then also the energy released in the binding to them leads to the evaporation of additional He atoms. The total energy transferred divided by the heat needed to evaporate an atom, 7.2 K (^{4}He) or 2.7 K (^{3}He), then yields the total number of atoms evaporated. For example, the capture of a single Xe atom in a ^{4}He droplet leads to a loss of about 560 ^{4}He atoms. [53] These pick-up experiments then paved the way for carrying out spectroscopic studies of embedded particles either singly or as clusters of selected sizes and composition.

2.4. *Apparatus used in spectroscopic studies*

The apparatus developed in Göttingen for spectroscopic studies is shown schematically in Fig. 4. [58] After the droplets have picked up one or more atoms or molecules they are exposed to a laser beam, which enters the apparatus coaxially but antiparallel to the droplet beam. Since both beams have rather narrow angular distributions, they have a large overlap over nearly the entire length of the apparatus thereby enhancing the absorption over that expected in a crossed beam arrangement. The energy deposited in the droplets following absorption of a laser photon leads to the

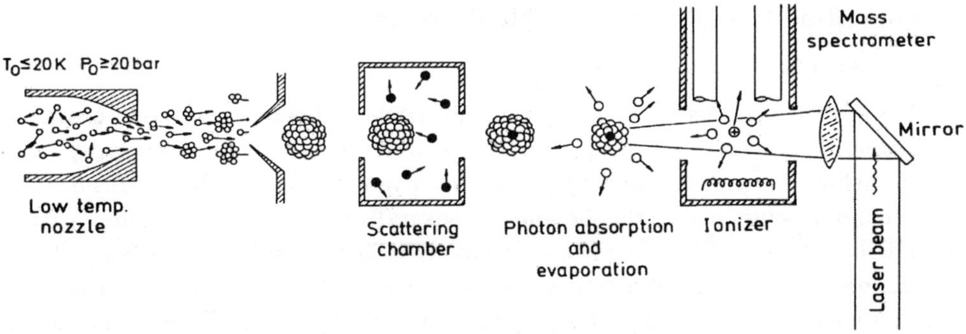

Fig. 4. Schematic diagram of the laser depletion apparatus used for the pick-up and spectroscopy of chromophore molecules inside helium droplets. [58] The spectral features are detected via the decrease in the mass spectrometer signal. The signal decrease is due to the reduction in droplet size resulting from the extensive evaporative loss of He atoms as the photon energy deposited in the chromophore relaxes and temporarily heats up the droplet.

evaporative loss of several hundred He atoms. The decrease in the overall size of the droplet reduces the electron-impact cross section for ionization and the mass spectrometer signal shows a distinct dip at resonance. In the case of helium this widely used technique of *depletion spectroscopy* has its highest sensitivity because of its very small heat of evaporation. For example, one IR photon at 2000 cm^{-1} corresponds to an energy of 2800 K and leads to the loss of about 400 atoms. For a typical droplet with $\overline{N} = 6 \cdot 10^3$ this leads to a 7% depletion dip in the mass spectrometer signal.

There are presently (summer 2001) about 22 apparatus of this type in operation worldwide. In some of these the beam is detected with a cryogenically cooled bolometer. [59] Different laser systems are in use in the infra-red. In addition to the tunable diode lasers used in Göttingen, tunable color-center lasers are in operation in North Carolina [60] or cavity amplified F-center lasers are employed in Princeton. [61] Several other groups are using line tunable CO_2 lasers. [62] In the visible spectral region dye lasers and parametrically tunable laser systems are also in operation in several laboratories.

This simple apparatus has proven to be very universal and with only minor modifications has been employed for spectroscopic investigations of a wide range of embedded particles including small closed shell molecules such as HF [63], OCS [64,65] and SF_6 [58,66], a number of metal atoms such as Ag [67], Al [68], Eu [69], Mg [70], large molecules such as tetracene ($C_{18}H_{12}$) [71,72], pentacene ($C_{24}H_{14}$) [71], C_{60} [73] and some large biological active molecules such as porphin ($C_{20}H_{14}N_4$) [74] and pthalocyanine ($C_{32}H_8N_{18}$) [74,75], as well as several amino acids [76,77] and even a very large molecule used in photonic devices called PTCDA (3, 4, 9, 10-perylenetetra-carboxylicdianhydride) [78]. Open shell alkali atoms [79], dimers [80] and trimers [81], which remain on the droplet surface, have also been extensively investigated. [3]

3. Superfluidity in finite sized ^{4}He droplets

3.1. *Theoretical predictions*

Well before these spectroscopic experiments nuclear theoreticians had recognized that superfluid droplets are ideal model finite-sized many-particle quantum systems. Liquid helium has the advantage that the weak van der Waals interactions are well known and depend only on the distance between the atoms. Thus the complications in understanding nuclei arising from the largely unknown and complex tensor forces encountered between their constituents are avoided. Moreover the helium isotopes offer an ideal opportunity for studying the differences between Fermi statistics, in the case of ^{3}He, and Bose statistics in the case of ^{4}He. The first definite evidence for a Bose condensate fraction inside ^{4}He droplets came from the 1988 variational calculations of Lewart, Pandharipande and Pieper. [82] These authors were also the first to calculate the radial particle density distributions in small helium clusters with up to 70 atoms for $T = 0$ K. Interestingly they found that the relative Bose *condensate fraction* increases from about 10% at the cluster center, as in the bulk superfluid, to about 100% at the surface, an effect which was also predicted earlier for thin films. [83]

The first direct theoretical evidence for *superfluid* behavior came a year later from the path integral Monte Carlo calculations of the normal fraction as a function of temperature for clusters with 64 and 128 atoms by Sindzingre, Klein and Ceperley. [84] To test for superfluidity these authors simulated the effect of a rotating field on the response of the small liquid cluster in a way somewhat related to the famous Andronikashvili rotating bucket experiment. [85] For clusters with 64 and 128 atoms a substantial decrease in the normal fraction and a corresponding increase in the superfluid fraction was found at temperatures somewhat below the bulk superfluid transition temperature of 2.18 K. Superfluidity in droplets was later confirmed by Rama Krishna and Whaley. [86] The extent of off-diagonal long range order, which is directly related to the Bose condensate has been studied theoretically [87] and confirmed in calculations of droplets with 10^3 ^{4}He atoms. [88] There is now some theoretical evidence that only a few atoms are needed to support a Bose condensate [87] or even superfluidity. [90]

Radial particle density distributions have also been calculated using mostly density functional theory which, however, cannot distinguish between the normal and superfluid components and, moreover, is restricted to $T = 0$ K. Results are available from several groups both for large pure ^{4}He droplets [30,51] (see Figs. 5a and b) and mixed droplets [91] also with various foreign molecules in the center (see Figs. 5c and d) [30,91,92]. These calculations predict that closed shell atoms and molecules are located in the interior and that the local density in the first helium shell next to the molecules is increased by up to a factor of four compared to the density in the central region of pure droplets. To a large extent this increase in density is due to an enhanced radial localization and substantial disorder is still expected within the nearest neighbor shell. [93] Some guidance on the effect of impurities on the condensate fraction comes from Krotscheck's 1985 pioneering calculations of the density distributions and condensation fractions of thin films of ^{4}He on a flat substrate. [83] The most recent droplet calculations take account of the anisotropy of the van der Waals potential between the He atom and the molecule. They reveal a large degree of localization also with regard to different polar positions. [92,94,96] An example, shown in Fig. 6, for the linear molecule OCS in superfluid ^{4}He indicates that the molecule is surrounded by four fairly distinct rings of ^{4}He atoms with varying densities. These new simulations also provide detailed information on the relative superfluid and non-superfluid fractions and their distribution in the immediate vicinity of the chromophore molecule as discussed at greater length in Subsection 5.1.3.

Considerable effort has also gone into calculating the dispersion curves for the elementary excitations in He droplets. The simplest model assumes a classical liquid drop with a sharp boundary as developed to explain the vibrations of nuclei. [97] This model was first applied to He droplets by Stringari and coworkers [98] and later further developed by Tamura *et al.* [99] These models predict the existence of low frequency surface modes, also called Rayleigh modes, [98,100] which are similar to ripplons at the surface of a liquid, as well as higher frequency compressional modes which have their greatest amplitudes in the interior. As first pointed out by Toennies and Vilesov [101] only the surface excitations are excited at the typical temperatures of ^{4}He droplets with $N \approx 10^3$-10^4 and the internal compressional modes are quiescent. Thus the impurity molecules in the interior of the droplets are expected to be in a medium devoid of excitations. Variational Monte Carlo calculations [102] and more recently optimized variational calculations [103,104] have also been used to calculate the low-lying collective excitations of ^{4}He droplets with up to 10^3 atoms. These calculations predict for the elementary excitations a bulk-like, smeared-out maxon-roton dispersive behavior, which depends on the cluster/droplet size, in addition to a surface ripplon branch, which is linear with respect to an effective radial wave number. [103]

3.2. *Experimental evidence for superfluidity*

As mentioned in the Introduction the absorption and the laser stimulated emission spectra in the visible region of open shell metal atoms and ions inside the *bulk* liquid generally exhibit rather broad features. [24] The broad lines and large shift between the absorption and emission lines have been explained by a bubble model which takes account of the strong electron induced exchange repulsion between the outer valence electrons and the surrounding He atoms. Similar effects are known from earlier studies of single electrons and neutral excited states of metastable He atoms and He_2 excimers. [105] In the case of closed shell organic molecules the coupling of the electronically excited states with the helium environment is expected to be greatly reduced since a significant part of the change in the electron density distribution is contained within the skeleton defined by the atoms of the molecule.

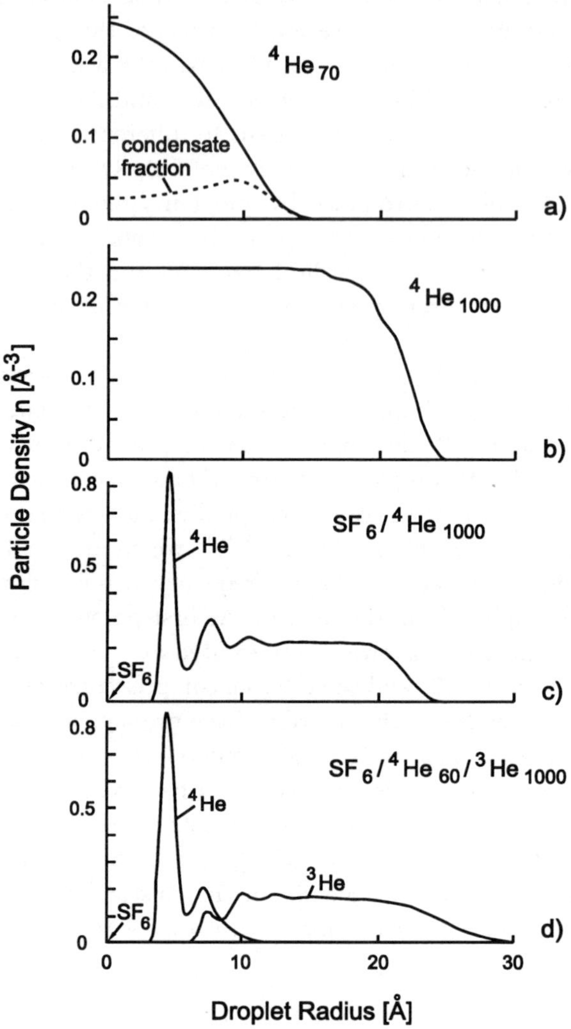

Fig. 5. Calculated particle radial density distributions. (a) Total and Bose condensate fraction density distributions calculated for a ^{4}He cluster consisting of 70 atoms; [82] (b) for ^{4}He atoms in a droplet with 10^3 atoms; [51] (c) for ^{4}He atoms surrounding a central SF_6 molecule in a droplet with 10^3 atoms. [30] (d) The radial density distributions of 60 ^{4}He atoms surrounding a SF_6 molecule in the center of a ^{3}He droplet consisting of 10^3 ^{3}He atoms. [91] In (c) and (d) the SF_6 molecules have been assumed to be spherically symmetric.

This phenomenon was first revealed in the case of the $S_1 \leftarrow S_0$ electronic transition of the glyoxal molecule ($C_2H_2O_2$) measured by the depletion technique inside ^{4}He droplets [106] in the near UV region of about 22,000 cm^{-1} ($\lambda \approx 450$ nm). The spectrum is dominated by an intense 0^0_0 line due to the transition from the electronic and vibrational ground state to the ground vibrational (v=0) state of the

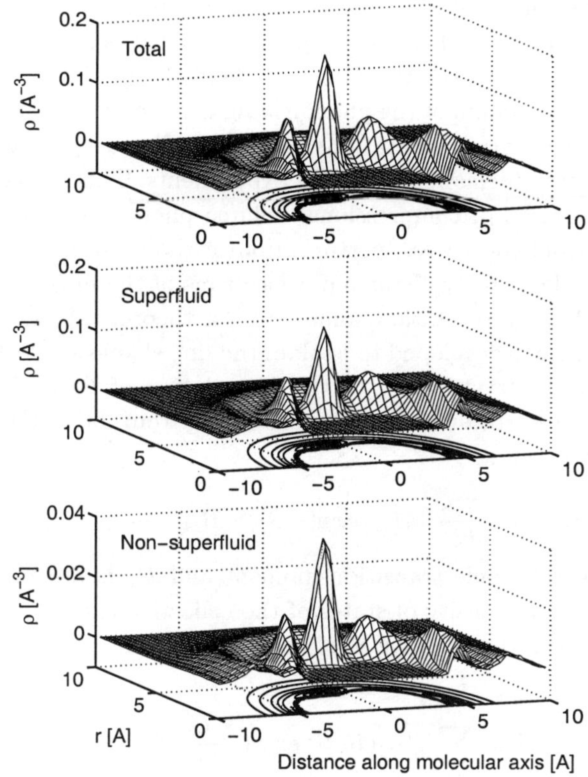

Fig. 6. Path integral Monte Carlo calculation of the anisotropic particle density distributions of ^{4}He atoms surrounding an OCS molecule in a cluster consisting of 64 ^{4}He atoms. Top panel shows the total density; middle panel, the local superfluid density; bottom panel, the local non-superfluid density, in Å^{-3}. The OCS molecule is oriented O-C-S from negative to positive along the molecular axis. Distances are in Å. [92]

electronically excited S_1 manifold and a few additional sharp lines at higher frequencies involving the vibrational excitation of the ν_7, ν_5 and ν_8 normal modes. All these lines are red-shifted to lower frequencies by 30.6 cm^{-1}, which is an order of magnitude smaller than, for example, found in a cold solid Ar matrix where the lines are also much broader with widths in the range, 100–200 cm^{-1}. [107] A careful examination of the line profiles reveals that they are all accompanied by small tails which extend to higher frequencies. The 0_0^0 region of the spectrum obtained with greater laser intensities is shown at higher resolution in Fig. 7a. By increasing the laser intensity the central feature becomes saturated, whereas the weaker tail region is greatly enhanced. The central intense features are assigned to a zero phonon line (ZPL) and the feature at higher frequencies to a phonon wing (PW). Similar features are well known from low temperature matrix spectroscopy where the

ZPL is attributed to pure excitation of the molecule and the PW to a simultaneous excitation of both the molecule and the collective excitations of the surrounding medium. [108] Compared to other low temperature matrices the PW in glyoxal is unusual since it is separated from the ZPL by a distinct gap of about 6 cm^{-1} (8.6 K). The energy at the maximum of the phonon wing is close to that of the roton (8.5 K) minimum in the elementary excitations as first predicted by Landau [15] and since confirmed by neutron inelastic diffraction experiments. This suggests that the PW is due to the excitation of compressional volume phonons of the entire superfluid He droplet as a result of a weak perturbation of the surrounding medium by the sudden change in the density of the outer electrons of the molecule.

Support for this interpretation comes from a theoretical calculation based on the Huang-Rhys theory developed to explain the line shapes in electronic excitation of impurities in solid crystals. [109] In the formulation of Pryce [110] the frequency dependence of the absorption cross section resulting from the coupling only to single phonons can be written as

$$\sigma_1(\omega) = \frac{4\pi^2\omega}{\hbar c}\,|M_{01}|\,exp(-S)\;S\;B_1(\omega-\omega_{01}) \quad , \tag{3.1}$$

where M_{01} is the electronic transition moment and B_1 is a complicated function which accounts for the density of states of the collective excitations of the droplet. All the information on the coupling to phonons is contained in the quantity S, which is given by

$$S \propto \sum_Q |f_{1Q}(R_0)|^2\, exp\left(\frac{-Q^2}{2\beta^2}\right) \quad , \tag{3.2}$$

where $f_{1Q}(R_0)$ is the amplitude of the wavefunction of the excited collective excitations of the droplet, R_0 is the radius of the molecule, and β is the range parameter of the chromophore molecule-liquid helium interaction. The final result obtained with a number of simplifying approximations is that the PW spectrum is proportional to the density states of the surrounding medium:

$$\sigma_1(\omega) \propto \left|\frac{d\,\hbar\omega(Q)}{dQ}\right|^{-1} . \tag{3.3}$$

As shown in Fig. 7b the density of collective states of superfluid helium is characterized by a gap leading up to a sharp peak at 8.65 K (5.9 cm^{-1}). This corresponds to the roton in the dispersion curve for elementary excitations proposed by Landau [15] and the other weaker peak at somewhat higher energies (13.7 K) is related to the maximum (maxon) in the dispersion curve. The distinct gap and sharp rise at the roton maximum are important features arising from the energetic sharpness of the dispersion curve, which has been established by neutron inelastic scattering to set in abruptly at temperatures below the superfluid transition temperature. [111,112]

Confirmation for this interpretation in terms of superfluid behaviour comes from experiments inside ^3He droplets (Fig. 7c), where no gap but only a broad continuous band extending to higher frequencies is observed. [2,4,113] This difference in the PW

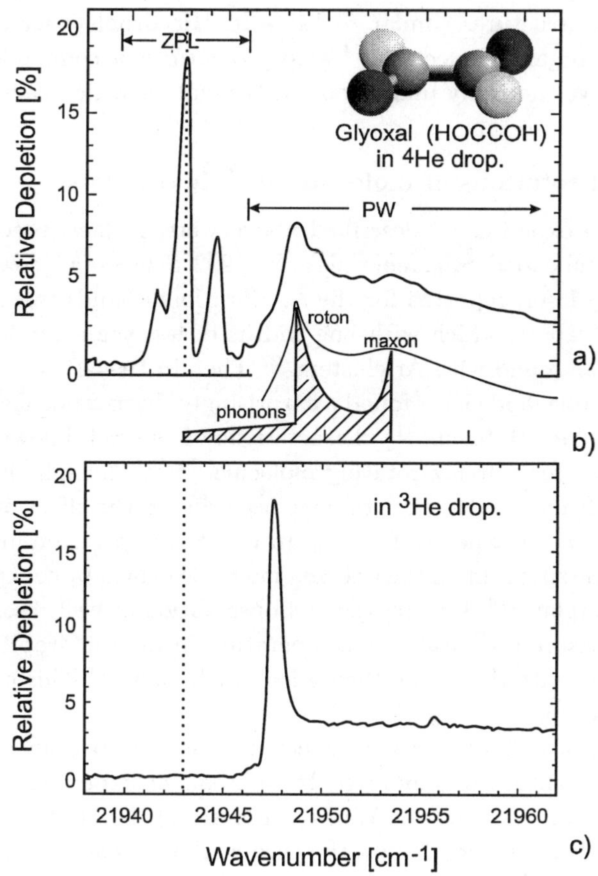

Fig. 7. In a) the mass spectrometer depletion signal is plotted as a function of the frequency in the vicinity of the zero phonon line of the $S_1 \leftarrow S_0$ electronic transition of a single glyoxal ($C_2O_2H_2$) molecule (see inset) inside a ^{4}He droplet consisting of 5500 atoms. [106] b) The phonon wing (PW) is explained by the density of states of superfluid ^{4}He (dashed area), determined from the dispersion curve of collective excitations of superfluid helium plotted as a function of energy. The overlaying curve includes other terms in the theory as well as multiphonon contributions (adapted from Ref. 106). c) The same region of the spectrum but measured for glyoxal inside a non-superfluid ^{3}He droplet consisting of $5 \cdot 10^4$ atoms. [2,3,113] The disappearance of the phonon wing confirms the interpretation of the phonon wing in the ^{4}He droplets as evidence for their superfluidity.

is expected on the basis of the more complicated collective excitations in liquid ^{3}He which involve in addition to a *phonon-maxon* dispersion curve a particle-hole pair branch at low frequencies. [112] It is gratifying to find that the shift of the ZPL in ^{3}He is less by about 25% than in ^{4}He which is just the difference in the densities. Thus an extrapolation of the frequency to zero density is expected to agree with the frequency in the free molecule!

Phonon wing structures similar to those in ^4He droplets have been observed with many other organic molecules, [74] albeit sometimes accompanied by split ZPL features [77] or as yet not fully understood additional inelastic peaks. [74]

4. Unhindered rotations of molecules in ^4He droplets

The 1996 glyoxal experiments described above were, in fact, preceeded by infrared spectra of embedded SF_6 molecules. In 1992 Scoles and coworkers, using a line tunable CO_2 laser, reported for the ν_3-vibrational bands two absorption lines about 0.67 cm^{-1} apart, which with line widths of less than $\Delta\nu = 0.5$ cm^{-1} were much sharper than found with Ar clusters. [27] Then in 1994-1995 our group using a continuously tunable diode laser found a surprisingly sharp sequence of well resolved rotational lines ($\Delta\nu = 0.01$ cm^{-1}) characteristic of the well known P-, Q- and R-branches expected for a freely rotating molecule. [58,66] The split lines seen earlier could not be confirmed and appear to have been due to the effect of some impurity molecules which was attached to the SF_6. In Fig. 8 the spectrum in ^4He droplets is compared with the spectrum of the free SF_6 molecule cooled by coexpansion in a free jet seeded with argon. [114] The unexpected observation of well resolved lines made it possible to measure the rotational temperature inside the droplets to be 0.37 K in good agreement with the earlier theoretical predictions. [41] This phenomenon was dubbed *molecular superfluidity.* [64]

The discovery of a sharp rotational spectrum has lead to a number of important conclusions all of which are supported by subsequent experiments on more than 15 molecules and several van der Waals clusters: (1) The lack of any anomalous lines confirms earlier evidence [26] that the chromophore molecules are located in the interior of the droplets and not on the surface as first reported on the basis of the absorption lines observed in 1992. [27] (2) The Boltzmann distribution of the rotational lines and the excellent agreement with the predicted droplet temperatures confirms that the internal degrees of freedom are in equilibrium with the helium bath. (3) The excellent fit of the spectra by the same Hamiltonian as for the free molecule indicates that the helium bath has no effect on the symmetry of the rotating molecule. [42] (4) From the rotational frequencies it is possible to determine both the rotational energy constant B and the centrifugal distortion constant D. B is found to be a factor of 2.8 smaller than for the free molecule and, since $B = \hbar^2/(2I)$, where I is the moment of inertia, the latter is increased by the same factor. Surprisingly D was found to be greater than for the free SF_6 molecule by about four orders of magnitude which is also found for OCS. [65] Finally, (5) The lines are red shifted by only a small amount, which for SF_6 is 1.42 cm^{-1}. This is much less than expected from the line shifts in other rare gas cryomatrices.

The results for the vibrational line shifts and the average line widths as well as for the rotational energy constants and their ratios with respect to the free molecule values, A_0, B_0 and C_0 are tabulated for 15 molecules in Table 1 in order of decreasing magnitude of B_0. An examination of the ratios B_0/B_{He} reveals a distinct trend to

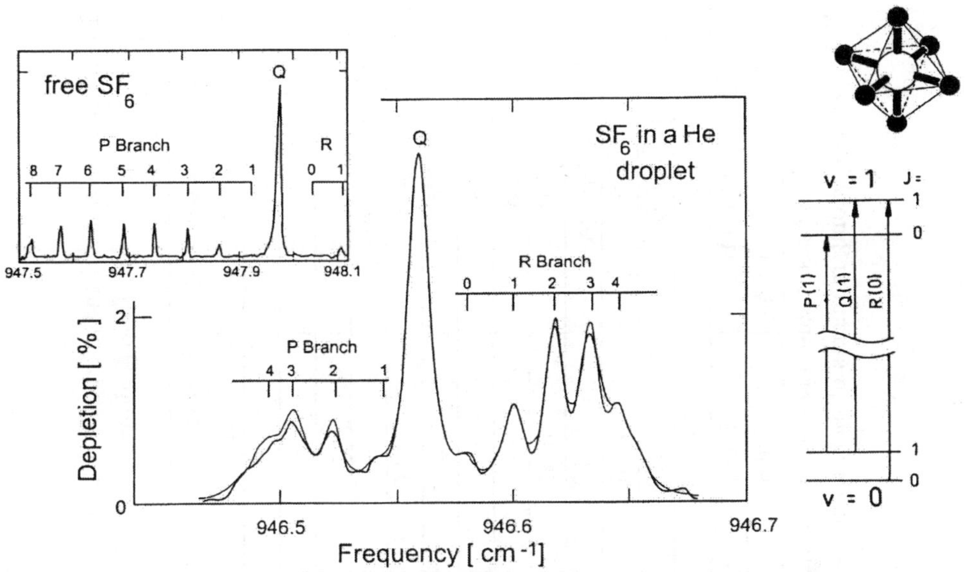

Fig. 8. Infra-red depletion spectrum showing resolved rotational transitions with $\Delta J = -1$ (P-branch), $\Delta J = 0$ (Q-branch) and $\Delta J = +1$ (R-branch) accompanying the $v = 0 \rightarrow 1$ vibrational transition of the ν_3 normal mode of a single SF_6 in a ^4He droplet consisting of 2700 atoms. The thin line curve is a best fit of the spectrum based on the Hamiltonian of a free molecule. [66] The inset shows the analogous spectrum of free SF_6 molecules in a seeded beam. [114] Note the reduced spacing of the lines in the He droplet experiment indicating a factor three decrease in the rotational constant.

larger values of the ratio with decreasing value of B_0. However there are significant deviations from this trend as seen, for example, by comparing HCCCN ($B_0 = 1.53$ cm^{-1}, and $B_0/B_{\rm He} \geq 3.0$) with HCN which has nearly the same B_0 value ($B_0 = 1.47$) but a much smaller ratio $B_0/B_{\rm He} = 1.23$. This would seem to suggest that the greater anisotropy of the HCCCN molecule leads to a compensating effect. But then the ratio $B_0/B_{He} = 2.8$ found for SF_6, which is, at least in first approximation, a spherical molecule, [115] would appear to be anomalous. Also it is interesting to find that molecules with $B_0 > 6$ cm^{-1}, which happens to be about the energy of the roton in superfluid helium, all have $B_0/B_{\rm He} \cong 1.0$. This suggests that the rotations of these molecules because of their greater rotational energy are largely decoupled from the helium bath. On the other hand, the small ratios found for HCN, DCN, HCCH, and DCCH despite their much smaller values of B_0 raise doubts as to the generality of this explanation. Thus although the overall general trends in the observed behavior are rather clear there are a number of apparent anomalies, which await future clarification by either models or theories.

The large increase in the moments of inertia found in ^{4}He droplets is also of considerable interest in connection with measurements of the breakdown of superfluidity at the interfaces of thin He films with the supporting solid substrates. [21]

Table 1: The ratios of the average[a] rotational energy constants A_{He}, B_{He} and C_{He} of molecules in He droplets relative to the free molecule values A_0, B_0, C_0 in order of decreasing of rotational constants B_0[b].

Molecule	Vibrat. Line Shift [cm^{-1}]	Average Line Width [cm^{-1}]	A_0, B_0, C_0 [cm^{-1}]	A_0/A_{He}, B_0/B_{He}, C_0/C_{He}[c] (=I_{He}/I_0)	Ref.
H_2O	—	—	27.8, 14.5, 9.3	1.0	116
HF	—	—	20.9	1.0	63
NH_3	—	—	9.94	1.3	117
$C_2O_2H_2$(glyoxal)	—	—	1.84,0.160,0.147	2.87,2.22,2.09	118
HCCCN	-0.176	$3.23\cdot10^{-2}$	1.53	3.0	61
HCN	-0.470	$3.10\cdot10^{-2}$	1.47	1.23	61
DCN	—	—	1.20	1.20	119
HCCH	—	—	1.19	1.13	95
DCCH	-0.06	$3.06\cdot10^{-2}$	0.99	1.15	61
CH_3CCH(propyne)	-0.207	$3.39\cdot10^{-2}$	0.28	3.9	61
OCS	-0.557	$0.5\text{-}3.0\cdot10^{-2}$	0.20	2.7	65
CF_3CCH	-0.369	$5.27\cdot10^{-2}$	0.105	2.75	61
SF_6	-1.415	$1.0\cdot10^{-2}$	0.091	2.8	66,123
$(CH_3)_3CCCH$(TBA)	1.171	$9.3\cdot10^{-2}$	0.09	3.0	61
$(CH_3)_3SiC\equiv CH$(TMSA)	0.865	$2.25\cdot10^{-2}$	0.0655	4.5	61

[a] The listed values are averages of the B-values of the ground (v=0) and fundamental vibrational states.
[b] The corresponding moments of inertia can be obtained using the relationships $I\cdot B = 16.851\ cm^{-1}\cdot amu\cdot Å^2 = 24.25 K\cdot amu\cdot Å^2$.
[c] If only one value is listed it is for B_0.

Depending on the substrate the superfluid fraction is greatly suppressed in the first layer or a fraction thereof next to the solid.[120] Similar effects are also found in studies of porous media.[121] A detailed microscopic understanding of these thin film phenomena has been hampered by the difficulties in producing atomically clean and perfect single crystal surfaces for these experiments. In contrast to solids the surfaces of single molecules are atomically *clean* and *smooth* and perfectly well defined. Moreover, their interaction potentials with He atoms are in many cases much better known than for the interactions with solid surfaces. Furthermore the sharp spectroscopic features provide very precise quantitative information which is only available in thin film and porous media by using neutron inelastic diffraction experiments requiring very large laboratory experimental facilities.[122]

Another important droplet experiment indicates that the free rotations are directly related to the superfluidity of the ^{4}He environment.[64] Since the linear OCS molecule has a simpler spectrum than for SF_6, and also has a large transition dipole moment it has been used in the more recent experiments to be described in this and the subsequent sections. Inside ^{4}He droplets OCS exhibits a particularly well separated sequence of rotational lines with a width of only about 150 MHz (Fig. 9a).[65] As expected for a linear molecule, and found in the infrared spectrum of the free molecule, the Q-branch is also missing in the droplet spectrum. This again confirms that the molecular symmetry is not affected by the helium environment even though the B and D constants are changed by nearly the same factors of 2.8 and 10^{-4} as found for SF_6.[66,42]

To test for the effect of the superfluid environment on the rotations of OCS, it would have been desirable to heat the ^{4}He droplets to transform them into the normal non-superfluid state. Despite attempts using collisional heating of the droplets[123] this has not been possible. Instead the OCS molecules were embedded inside ultrapure ^{3}He droplets. Their lower temperatures of 0.15 K are still greater than the superfluid transition at about $3{\cdot}10^{-3}$ K so that the ^{3}He droplets are not expected to be superfluid. Furthermore, because of its more complex collective excitations,[112] liquid ^{3}He is expected to behave in much the same manner as a classical liquid and rather like normal liquid ^{4}He. Inside the ^{3}He droplets (see Fig. 9b) the IR spectrum consists of only a single broad feature with a factor 40 larger line width of about 6 GHz than in the ^{4}He droplets. This increase in line width is attributed to the high rate of collisions with the surrounding liquid which prevents the molecule from rotating freely.[124] Similar broad features are also found, for example, in the infra-red spectra of molecules such as HCl in liquid HCl.[125] Thus, it was concluded on the basis of this difference in the spectra that the sharp rotational features are related in some way to the superfluidity of the droplets. If the sharp spectra were merely a consequence of the very weak van der Waals energies and large zero point energies characteristic of He atoms then the lines should have been narrower in the colder ^{3}He droplets. An analysis of the overlapping lines in the pure ^{3}He droplets indicates that there the moments of inertia are about 5 times larger than for the free molecules.[129]

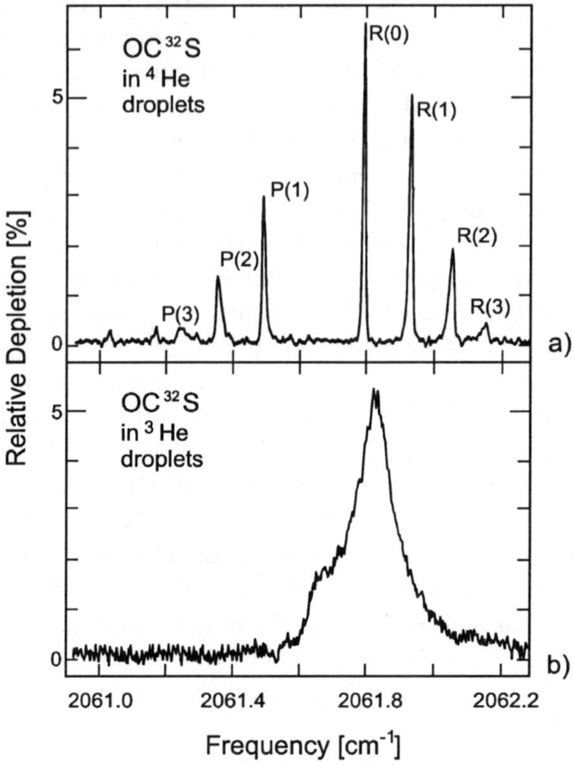

Fig. 9. Comparison of the depletion spectrum of single OCS molecules embedded in (a) a pure ^{4}He droplet ($\overline{N}_4 = 6{\cdot}10^3$ atoms) and (b) in a pure ^{3}He droplet ($\overline{N}_3 = 1.2 \cdot 10^4$ atoms). [64] The disappearance of the sharp rotational lines in the non-superfluid ^{3}He droplets in (b) is interpreted as indicating that the sharp rotational lines are a manifestation of the superfluidity of the ^{4}He droplets.

This simple experiment lead to another experiment which provides further insight into the phenomenon of free rotations. After embedding a single OCS molecule inside a large ^{3}He droplet ($N_3 \cong 10^4$), ^{4}He atoms were added one-by-one [64] by passing the droplet beam through a second scattering chamber filled with ^{4}He gas at several different pressures. Because of their high mobility in liquid ^{3}He [126] and smaller zero point energy the ^{4}He atoms replace the ^{3}He atoms next to the OCS molecule, thereby gradually surrounding and coating the molecule within the large ^{3}He droplets. Fig. 10 shows a series of infra-red spectra measured with increasing numbers of ^{4}He atoms. When only a few ^{4}He atoms are added the spectrum is not appreciably changed. Presumably the attached ^{4}He atoms form asymmetric complexes with the OCS molecule. With 60 added ^{4}He atoms, however, narrow rotational lines again begin to emerge. With additional ^{4}He atoms the spectra sharpen further and with 1000 ^{4}He atoms have a line width of only 50 MHz, which is even narrower than found in pure ^{4}He droplets. [128] But instead of the seven lines seen

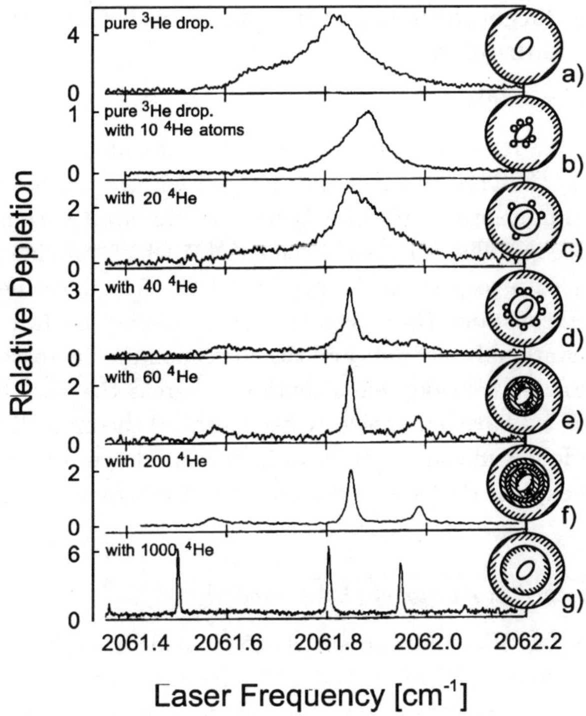

Fig. 10. A series of OCS IR spectra in (a) pure ^{3}He droplets, ($N_3 = 10^4$) and (b)-(g) in ^{3}He droplets with increasing average numbers of added ^{4}He atoms.[64,128] The reappearance of distinct peaks with 60 ^{4}He atoms is interpreted as indicating that about 60 ^{4}He atoms are needed to restore nearly free rotations, which is a microscopic manifestation of superfluidity. The spectra become even sharper (g) when 1000 ^{4}He atoms are added.

in pure ^{4}He droplets (Fig. 10a) only altogether three lines are found (Fig. 10f-h). The fewer lines and the reduced line width are explained by the lower temperature of 0.15 K of the outer ^{3}He layer. The reappearance of a sharp rotational spectrum is interpreted as indicating that the superfluidity effect sets in gradually, and is almost complete, with the addition of about 60 atoms, which correspond to about two shells of ^{4}He atoms. Density functional calculations [91] reveal that the second shell is heavily infiltrated by ^{3}He atoms to about 30%. According to recent thin film studies ^{3}He significantly reduces the superfluid fraction and superfluid transition temperature. [127] Thus it is likely that only a single shell of ^{4}He atoms (17-20) atoms may, in fact, be sufficient for the molecules to rotate freely. Finally it is noted that despite the lower temperatures in the interior of the mixed droplets the rotational constants are the same as in the pure ^{4}He droplets within about 1%, which is also about the size of the errors.

These experiments with mixed droplets have lead to many new avenues of research some of which are discussed next.

5. Anomalously large moments of inertia of molecules in superfluid ^{4}He droplets

5.1. *Theoretical models*

The infra-red spectra reveal that in many respects the molecules behave as if they were in a vacuum. [130] Thus most of the spectral features such as the Boltzmann progression of line intensities, lack of a Q-branch, etc. are the same as in ordinary molecular spectroscopy. The features of the spectra which provide new information are the line shifts with respect to the spectra of the free molecules and the much different rotational constants B and centrifugal stretching constants. Especially the reduction in the rotational constants and the corresponding increase in the moments of inertia has attracted considerable attention, whereas the line shifts [132] and the large decrease in D [133] have only recently been studied theoretically. Essentially the following three different models in order of increasing degree of sophistication have been proposed to explain the observed increase in the moments of inertia.

5.1.1. *Rigidly attached atom model (RAA model)*

The initial results for SF_6 could be simply explained by noting that the van der Waals potential between a He atom and a SF_6 molecule has its deepest minima at the 8 octahedral sites each formed by three of the F-atoms. [66] By assuming 8 He atoms to be nested in these sites and rigidly attached at a distance equal to the minimum of the He-SF_6 van der Waals potential [115] it was possible to explain the observed increase in the moments of inertia. The same model has also been adapted to explain the factor 2.8 increase in the moments of inertia found for OCS. [66] In this case the cylindrical symmetry and the deep van der Waals well at the waist position nearest the middle C-atom ($R_m = 3.4$ Å) [131] suggest that 6 He atoms are attached. This number of He atoms just fills an entire ring, called a donut ring, around the molecule. The spacing of the atoms in the donut ring is then equal to 3.6 Å which is the interatomic distance in bulk helium. The existence of a donut is also consistent with the calculated density distributions of He atoms in the surrounding first shell [92,94,132] which show the largest density at the waist position (see Fig. 6). Very recent molecular dynamics calculations show that on adding He atoms one by one the first added atoms fill these same sites at the waist position. [134] Moreover the first 7-8 added ^{4}He atoms can explain the observed reduced rotational constant. These new calculations are in excellent agreement with the donut ring concept.

Another version of this model has recently been introduced to explain the further increase in the moments of inertia on adding one and greater numbers of H_2 molecules to OCS. [135] These experiments are discussed in more detail in the next section. In the modified so-called *shell model* a single complete layer of He atoms is assumed to surround the chromophore. By reducing the effective masses of the He atoms from 4 a.m.u. to 0.55 a.m.u. the increase in the moment of inertia could also be explained. A uniform layer is, in fact, also quite consistent with the anisotropic density distributions shown in Fig. 6, which indicate that in addition to the 6 atom ring with the highest density the helium atoms form 3 additional lower density rings with between about 2 and 6 atoms yielding total of about 17 atoms altogether. The shell model has the attractive feature that it also explains the shift in the direction of the transition dipole in the OCS-H_2 complexes, which cannot be explained by the donut model. [136,137] This effect may however be related to more subtle effects related to the quantum nature of the surrounding helium bath.

The various RAA models have the merit of extreme simplicity. The donut model correctly emphasizes the importance of the potential anisotropy and thus it implies that, in the case of the nearly spherical molecules NH_3, H_2O and HF, the increase in the moments of inertia should be much less as is indeed observed (see Table 1). From the practical point of view the major flaw of this model is that it is not easily applied to explain the moments of inertia of larger molecules especially those with a lower symmetry. In one attempt to explain a similar factor 2.8 increase in the moments of inertia of glyoxal the model was found to fail completely. [138] Conceptually it suffers from completely ignoring the many-body quantum mechanical nature of the helium superfluid. Also it does not explicitely take account of backflow corrections which are known to play an important role in microscopic superfluid hydrodynamics. [139]

5.1.2. *The superfluid hydrodynamical model (SH-model)*

In this model the backflow of the superfluid is fully accounted for by treating the surrounding liquid as an irrotational (inviscid) continuous fluid. [94] This model is known to explain many of the *macroscopic* flow properties of the superfluid. In the calculations in which this model has been applied to molecular rotations in He droplets the increase in the particle density in the immediate vicinity of the molecule (see Fig. 6) resulting from the anisotropic van der Waals potential is also accounted for. This model has been found to slightly overestimate the increased moments of inertia of seven linear molecules including OCS by less than 25%. Despite the good agreement the model has a number of conceptual drawbacks. The major assumption that continuum hydrodynamics is valid even in the highly disturbed and distorted first shell of He atoms at the atomic level seems questionable. Moreover as recently pointed out by Whaley and collaborators it may be important to include the quantization of the local angular momentum in different local regions in the fluid. [96] Nor does the model take account of the effect of the molecule on depleting the superfluid fraction. This is well known from healing length measurements of thin films on solid substrates. [120] The calculations within this model are rather complex so that the model cannot be readily applied to asymmetric top molecules.

5.1.3. *Theoretical simulations*

The most advanced and extensive theoretical calculations of the enhanced moments of inertia come from the group of Birgitta Whaley at Berkeley. [92] Several different quantum mechanical simulation methods including the Diffusion Monte Carlo

(DMC) technique, which is restricted to T=0 K, and the Feynman path integral Monte Carlo (PIMC) method for finite temperatures have been applied to the three model systems SF_6, OCS and HCN. [92] These calculations take full account of the superfluid nature of the liquid within the framework of the two-fluid model of Tisza and Landau. [15,16] In the standard two-fluid model the normal fluid component ρ_n accounts for the thermal excitations, whereas the superfluid component ρ_s is entirely devoid of excitations but includes the depletion of the condensate resulting from the He-He interactions. In the bulk the total density is equal to the sum of these two components $\rho_{tot} = \rho_n + \rho_s$. To explain the effect of the molecule on the surrounding superfluid helium a new concept *the non-superfluid component* is introduced. This component accounts for the partial immobilization of a fraction of the He atoms in the immediate vicinity of the molecule in a similar way as known from healing length measurements [120] referred to in the previous paragraph. Since the dynamical and particle exchange behavior of these atoms is mostly determined by the forces exerted by the chromophore van der Waals potential and are only indirectly affected by the temperature, they cannot be simply ascribed to the normal component. The fraction of the non-superfluid component is established with the aid of the PIMC calculations in the formulation of Ceperley. [140] Accordingly in the non-superfluid component the Feynman permutation exchanges with other nearby He atoms are restricted to chains with less than 6 atoms, while in the bulk superfluid longer chains extending to infinite length are encountered.

It is interesting to recognize some similarities between the non-superfluid component and the rigidly attached atoms of the RAA donut model. An important distinction to the RAA model is that the non-superfluid component is not completely isolated from the rest of the helium bath, but undergoes permutational exchanges albeit of a reduced length. Fig. 6 illustrates the results on the density distributions from PIMC calculations for OCS in large He clusters ($N_4 = 64$) and shows the large anisotropy in the distribution of the He atom density in both the non-superfluid and superfluid components.

Another important concept coming out of these simulations is that of *adiabatic following* which accounts for the forces exerted by the anisotropy of the molecular van der Waals potential potential on the surrounding He atoms. If the anisotropic potential barrier is greater than the rotational energy of the molecule then the non-superfluid fraction is effectively coupled to the rotations. Obviously in this respect there is another close correspondence with the RAA donut model. The effective numbers of non-superfluid [92] atoms in the case of SF_6 is 8 as in the donut model but, presently, for OCS only 3.2 are predicted to be effectively non-superfluid [92] as opposed to 6 in the donut model.

The Whaley group has also analyzed the hydrodynamical contributions to the moments of inertia. In their two fluid model these effects can only be associated with the superfluid fraction. In the case of SF_6 their analysis predicts that the non-superfluid component smoothens the anisotropy and the remaining hydrodynamical effect is only about 3% of the total increase. For OCS the situation is complicated by the fact that when the molecule is excited from j=0 to j=1 the associated superfluid angular momentum amounts to 0.55 h which is a significant fraction of that involved in the molecular transition. Since their simulations confirm the irrotational behavior of the superfluid at the microscopic level this angular momentum cannot, in fact, be accomodated by the superfluid. They therefore conclude that a classical hydrodynamical model may no longer be justified on these grounds alone. For this reason the true size of the actual hydrodynamical correction is probably difficult to assess, at least, at the present time.

The theory confirms the importance of altogether three terms which contribute to the effective moment of inertia of the molecule [64,65,92]

$$I_{eff} = I_0 + I_{ns} + I_s \quad , \tag{5.1}$$

where I_0 is the moment of inertia of the bare molecule, I_{ns} is the contribution from the non-superfluid component which adiabatically follows the rotating molecule. The superfluid back flow contribution I_s seems to be small provided the angular momentum of the superfluid is small. Otherwise its contribution at present appears to be difficult to pin down precisely.

These calculations provide a rather complete picture of the microscopic processes involved in molecular rotations inside the superfluid. It is gratifying that some of the basic mechanisms of the simpler RAA- and SH-models can be found in the more exact theoretical simulations. The numerical simulations are, however, the only ones which take correct account of the exchange permutations which is an essential property of the boson superfluid.

5.2. *Experimental studies*

Recently some of the shortcomings of the RAA-donut model have been brought to light by a number of sophisticated spectroscopy experiments. [135,136] Just as the ^{4}He atoms have been found to replace the less tightly bound ^{3}He atoms next to the OCS molecule, H_2 molecules and their isotopes HD and D_2, which are known to have a factor four greater van der Waals potential depth, [141] are expected to replace the ^{4}He atoms next to the OCS chromophore molecule. In Fig. 11 the infra-red spectra for single bare OCS molecules inside mixed ^{4}He/^{3}He droplets are composed with spectra with about one attached para-hydrogen (pH_2) molecule and for different pH_2 pick-up pressures. From the spectra in Fig. 11c it is seen that for each additional added H_2 molecule the band structures are shifted to the red by about 0.2 cm^{-1} which is just sufficient that in most cases the P-, Q- and R-branches each are rather well separated. In these spectra clusters consisting of a central OCS molecule with up to 8 attached pH_2 molecules could be identified. The resulting identification of each of the complexes could be confirmed by the spectra measured in pure ^{3}He droplets where the collapsed spectra similar to those in Fig. 9 and 10 were still sufficiently sharp to confirm the positions of the band origins.

With the addition of only one pH_2 molecule there is already a dramatic proliferation in the number of lines (Fig. 11a). [136,137] The well-resolved spectrum consisting

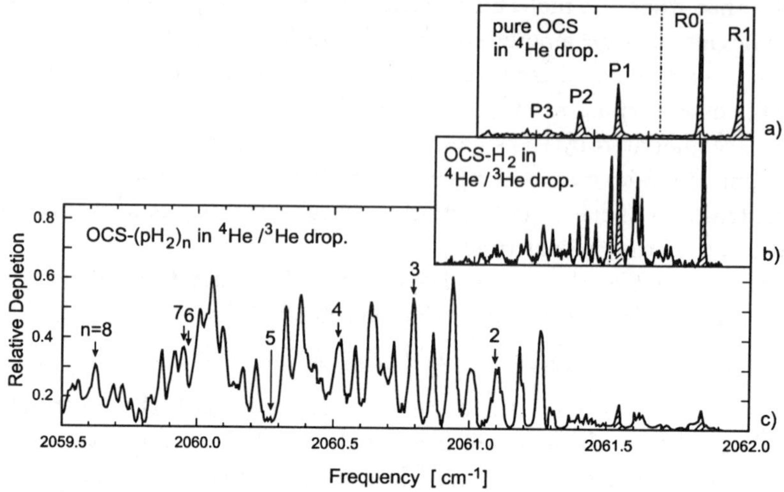

Fig. 11. Comparison of the IR spectra of (a) pure OCS measured in a ^{4}He droplet ($N_4 = 3{\cdot}10^3$) [64] and of (b) OCS with an attached pH_2 molecule in a mixed ^{4}He/^{3}He droplet [136,137] and of (c) OCS-$(pH_2)_n$ clusters with n=2-8 in a mixed ^{4}He/^{3}He droplet ($N_4 = 500$, $N_3 = 10^4$). [135,143] Similar spectra as in (b) have been evaluated for the other hydrogen isotopomers via Kraitchman's equations [142] to predict the structure of the OCS-H_2 complexes in very good agreement with the structures of the free complexes. The arrows in (c) indicate the expected positions of the Q-branches for the corresponding clusters.

of at least 11 lines could be very well fitted by assuming an OCS-H_2 complex with the H_2 molecule attached to a waist site thereby forming an asymmetric top in which the H_2 molecule rotates freely around the OCS molecule. [137] As for the bare OCS molecule the moments of inertia were found to be much larger than expected for the free complex. However by also measuring the spectra for HD and D_2, which were equally well resolved, the dependence of the moments of inertia on the mass of the attached hydrogen molecule could be ascertained. [136,137] By assuming the same structure for all three isotopomers the differences in the moments of inertia were used, via Kraitchman's equations, [142] to determine the structural parameters. Surprisingly the structure obtained agrees remarkably well within the mutual errors with the structure recently determined for the free OCS-H_2 van der Waals complex via high resolution infra-red spectroscopy in a pulsed free jet expansion. [144] Moreover the line shifts for all three complexes are the same for pH_2, oH_2, oD_2, pD_2 and HD and amount to only 0.61 cm^{-1}. This unexpected result suggests a strategy for structural determinations based on the rotational spectra which avoids the complications arising from the enhanced moments of inertia.

The increases in the moments of inertia of the OCS-H_2 (HD, D_2) complexes are found to be even larger and amount to a factor of 4 to 9 compared to the free complexes. This greater increase in comparison with the factor 2.8 found for the bare OCS molecule has been interpreted using the modified RAA shell model discussed in the previous section. Despite the fact that the H_2 molecule is lighter than the displaced He atom the effect could only be explained by a large increase in the effective mass of the H_2 molecule to 10 a.m.u. With this modification the shell model could also explain the observed nearly linear increase in the moments of inertia for end-over-end rotations of the entire complex with increasing numbers of attached H_2 or D_2 molecules. [135] The remarkable effect of these *foreign* species leading to the large effective masses is attributed to their inability to participate in exchange permutations with the surrounding ^{4}He bath atoms. Conversely, it can be argued that it appears that permutation exchanges are instrumental in keeping the effective molecular moments of inertia as small as possible.

Another important finding coming from the $H_2(D_2)$ experiments is the disappearance of the Q-branch for the OCS-$(pH_2)_n$ clusters for n=5 and 6. For n=5 this is clearly apparent from an examination of the spectrum in Fig. 11c. In the corresponding OCS-$(oD_2)_n$ clusters [143] a distinct clearly resolved intense Q-branch was found for all the clusters. This difference in Q-branches is surprising since the van der Waals potentials of H_2 and D_2 both with the OCS molecule and with He atoms are identical and the difference in masses is expected to have only a minor influence. The pH_2 molecules do however differ in an important respect in that they are composite bosons with a total nuclear spin of I=0 and are, therefore, *indistinguishable* from each other. The oD_2 molecules, although also bosons, have however both I=0 and I=2 nuclear spin components with relative abundances of 1/6 and 5/6, respectively. Because of the $2I+1$ projections of the dominant I=2 component these particles are essentially all *distinguishable* both from each other and, of course, with respect to the He atoms. [145] Since the volume occupied by the $H_2(D_2)$ molecules is not much different from that occupied by the He atoms [146] their insertion in sites otherwise occupied by He atoms should provide a sensitive probe of the role of symmetry in the dynamical behavior of the He atoms which they replace.

The vanishing of the Q-branch can now be understood as due to the high energy needed to thermally excite the ring rotations involving identical particles, which for a ring of symmetry C_i is given by

$$E_i(ring) = A K^2 \quad . \tag{5.2}$$

In equation (5.2) K is an angular momentum quantum number and A is the rotational constant of the new donut ring of H_2-molecules which has replaced the ^{4}He donut ring. For symmetry reasons a ring of i equally spaced identical particles can only take on values of K=0, i, $2i$, $3i$, ... For a symmetric top, in this case OCS-$(pH_2)_n$ with n=5 or 6, with its transition dipole moment oriented along the symmetry axis, Q-branch ($\Delta J = 0$) transitions are only allowed if the ring rotations are excited implying that $K > 0$. For n=5 and 6 equally spaced molecules the lowest possible K values are K=5 and 6, respectively. But since the corresponding ring rotational energies according to (5.2) are 2.2 K (n=5) or 3.2 K (n=6) they are not appreciably excited at the low ambient temperatures of 0.15 K. Thus for a ring of identical particles the Q-branch is expected to be absent as has been confirmed by calculations of the spectral intensities. [143] The distinguishable oD_2 molecules behave, on the other hand, as if they were classical particles for which all states K=0, 1, 2, ... are allowed. Since the K=1 level is appreciably excited, even at the low temperatures, a Q-branch is possible as indeed found in the spectra. This experiment also has been shown to provide indirect confirmation for the likely existence of 5 or 6 ^{4}He atoms in a donut ring surrounding the waist of the OCS-molecule as assumed in the simple RAA-model. [143]

As discussed above these experiments could be explained using the two different RAA models, the donut and the shell models. Each is able to account for one or the other different facets of the observations. But the fact that one model cannot account for all the phenomena underlines their basic inadequacy and that they do not fully encompass the true underlying physical phenomena. They have nevertheless been useful constructs for understanding various aspects of the complex many-particle quantum mechanical superfluid behavior at the atomistic level. They have also been useful in designing the experiments described in the next section.

6. Evidence for superfluidity in para-hydrogen clusters inside superfluid ^{4}He droplets

As discussed above pH_2 molecules at the temperatures of helium droplets are expected to be all in their lowest rotational state j=0. Since, moreover, their total nuclear spin is I=0 they are, just as the ^{4}He atoms, indistinguishable spinless bosons and should also exhibit superfluidity. In 1972 Ginzburg and Sobyanin estimated, using the London formula, that the transition temperature to the superfluid state in pH_2 should occur at about to 6 K. [147] Since hydrogen solidifies at its triple point at 13.95 K (P_{tp}= 0.072 bar) the realization of superfluidity requires very extensive supercooling. Therefore it is not surprising that all attempts to create superfluid pH_2 have so far been unsuccessful. [148] More recently in 1991 Ceperley and coworkers used Feynman path integral Monte Carlo simulations to explore superfluidity in small pure pH_2 clusters. [149] They used the same rotating field response as a criterium for testing for superfluidity as in their previous calculations of pure ^{4}He clusters.[84] In these calculations the reduction in the calculated moments of inertia of the clusters is compared to their classical values following along ideas first proposed by Landau. Whereas clusters with 33 molecules did not exhibit a superfluid transition they found that smaller clusters with only 13 and 18 molecules underwent a superfluid transition at about 2 K. They attributed these unexpected findings to the 30% decreased density in the small clusters resulting from their large surface to volume ratio and the resulting reduced coordination of the molecules. Thus they suggested that by artificially increasing the spacing between the molecules by, for example, using substrates with adsorbate spacer atoms superfluidity might be achievable in two dimensional films. [150]

Our observations of free rotations of small clusters assembled inside ^{4}He droplets and the possibilities to choose between two temperatures of 0.37 K (pure ^{4}He

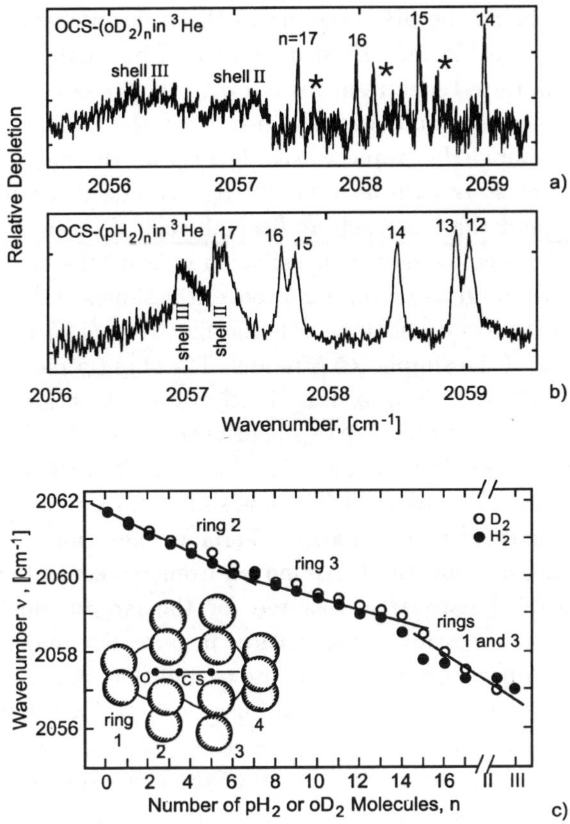

Fig. 12. Spectra of OCS-$(oD_2)_n$ clusters (a) and OCS-$(pH_2)_n$ clusters (b) inside pure ^3He droplets, consisting of about $5 \cdot 10^3$ atoms, in the frequency region where the first shell is filled. The asterisks denote peaks resulting from the inclusion of pD_2 impurities. (c) Measured frequencies of the collapsed spectra of OCS-$(pH_2)_n$ and OCS-$(oD_2)_n$ clusters inside ^3He droplets are plotted as a function of n. [151]

droplets) or 0.15 K (mixed $^4He/^3He$ droplets) has opened up another strategy in the search for superfluidity of hydrogen. This involves attaching small pH_2 clusters to a central OCS chromophore [151] inside helium droplets. The previous experiments with small numbers of attached $H_2(D_2)$ molecules and the good agreement with the RAA shell model demonstrated that the OCS molecule can serve as a template for attaching possibly up to about 17-19 H_2 molecules in four coaxial rings around the axis of the molecule (see Fig. 6). As in the previous experiments the infra-red spectra of the central chromophore molecule can be expected to provide information on the angular momentum and on the moments of inertia along all three axis of the cluster formed. Especially the axial moment of inertia might then serve as an indicator for the transition to a superfluid state in the same way as in the calculations by Ceperley *et al.* [149]

By further increasing the pH_2 pressure in the pick-up chamber the sizes of the OCS-$(oH_2)_n$ clusters inside the ^{4}He droplets could be further increased. Fig. 12a shows some of the infra-red spectra measured with the largest clusters. The numbers of attached H_2 molecules could be identified from the positions of the collapsed peaks measured in pure ^{3}He droplets. Fig. 12b shows the measured line shifts for the entire series of clusters with n=1-17. [151] It was indeed totally unexpected but very gratifying to find that for each additional peak there is a well defined shift which is sufficiently large to make it possible to identify the spectra of each of the clusters even for these large values of n. Moreover the slopes of the different segments of the line shift curve are consistent with the first 6 $H_2(D_2)$ molecules filling the donut ring positions of the simple RAA model. The slightly reduced slope for $n > 6$ is also consistent with the filling of less tightly bound positions in adjacent rings and finally the abrupt fall-off for $n \geq 14$ suggests that now the H_2 molecules occupy the pole cap positions where their effect on the axial vibrations is expected to be especially large as indeed is observed. The line shifts corresponding to the formation of the next outer shells are much smaller. Therefore they can no longer be resolved since these molecules are now further removed from the central chromophore.

Fig. 13 compares the spectra measured for the largest clusters for both isotopes and for the two accessible temperatures in pure ^{4}He and in mixed ^{4}He/^{3}He droplets. [151] Indeed, once more it is surprising to find that the resolution is sufficient to clearly resolve the individual rotational lines in the P- and R-branches even for clusters with n= 15, 16 and 17 oD_2 molecules. In part, of course, the remarkable resolution in these spectra can be attributed to the lower temperatures available in the mixed droplets (see for example Fig. 10g). By assuming, as before, the same Hamiltonian as for a free cluster the individual rotational lines in these spectra could be very well simulated as shown by the best fit in the bottom part of Fig. 13b for the oD_2 containing clusters. The surprisingly good fit confirms that the clusters have the expected nearly perfect symmetric top structure and the expected temperatures of 0.15 K imposed by the outer ^{3}He shell. In both sets of the OCS-$(oD_2)_n$ spectra the distinct Q-branches indicate extensive excitation of the axial rotations at 0.37 and 0.15 K. This is not unexpected in view of the discussion in Subsection 5.2. These large clusters have much larger moments of inertia than those with only a single donut ring. Moreover since the oD_2 molecules are essentially distinguishable there are no symmetry restrictions on the occupied K-states (see (5.2)) so that they are easily excited.

In the case of the indistinguishable pH_2 clusters shown in Fig. 13c and d, such symmetry restrictions might play a similar role as in the case of 5- and 6-membered donut rings. However, in view of the fact that their overall axial symmetry depends on the product of the structural symmetries of the individual rings, which are expected to exhibit both odd and even symmetries, the overall total symmetries can be estimated to be about the same as for the oD_2 molecules despite the fact that they are indistinguishable bosons. This is borne out by the spectra in the warmer pure ^{4}He droplets displayed in Fig. 13c where the spectra are dominated by distinct

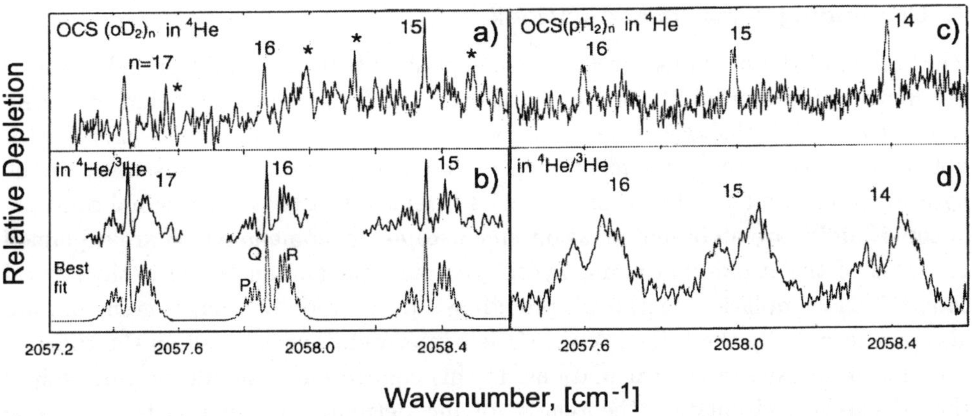

Fig. 13. Comparison of the IR spectra measured for three different OCS-$(oD_2)_n$ clusters in helium droplets (a and b) with $n = 15$, 16, and 17 in pure ^{4}He droplets with about 8000 atoms ($T = 0.38$ K) and in mixed ^{4}He/^{3}He droplets with 10^4 ^{3}He atoms and 500 ^{4}He atoms ($T = 0.15$ K). Comparison of IR spectra of OCS-$(pH_2)_n$ clusters with $n = 14$, 15, and 16 in the same droplet environments (c and d). The asterisks in (a) denote peaks resulting from the inclusion of pD_2 impurities. The bottom part of (b) shows best-fit simulations based on a free symmetric top Hamiltonian with spectral resolution of $\delta\nu = 0.01$ cm^{-1}. [151]

Q-branches.

Thus the effect of the different symmetries, which is apparent from the side-by-side comparison of the oD_2 and pH_2 clusters afforded by Fig. 13, cannot explain why the Q-branches disappear completely in the three much better resolved pH_2 cluster spectra measured in the colder mixed ^{4}He/^{3}He droplets (see Fig. 13d). The only way in which it has been possible to explain the sharp drop-off of the Q-branch intensity in the pH_2 clusters has been to assume a large decrease in the axial moment of inertia of the complex. A greatly reduced moment of inertia raises the axial rotational energy to such an extent that the rotations can no longer be excited at the low ambient temperatures just as discussed in Subsection 5.2 in connection with donut ring rotations. Additional simulations of the spectra could rule out that the disappearance of the Q-branch is not merely a consequence of the reduction in the Boltzmann factor due to the decrease in temperature from 0.37 to 0.15 K. [151] Therefore it is argued that the effect must be due to an inherent change in the angular momentum response of the cluster as its temperature is reduced. This behavior is, as illustrated by the path integral Monte Carlo calculations of the Ceperley group, [149] just that which is expected for the transition to a superfluid state. Thus it can be regarded as a fortunate circumstance that the superfluid transition temperature happens to fall somewhere between the two available temperatures of 0.37 and 0.15 K. It is very likely that the observed much lower transition temperature compared to the simulations for pure clusters [149] can be explained in terms of the detrimental effect of the cylindrical holding potential provided by the OCS molecule.

7. Concluding remarks and outlook

The results of the previous spectroscopic experiments were all interpreted in terms of *microscopic* superfluidity or simply as evidence for superfluidity. Here it is necessary to caution that the true meaning of these concepts on the atomic level is not at all clear as was also mentioned in connection with our assessment of the hydrodynamical model. The problems arise from the fact that the original concepts of superfluidity are all based per se on *macroscopic* phenomena which are explained in terms of the usual continuum hydrodynamical phenomena. The path integral Monte Carlo simulations, although providing a description of what is going on, are also not necessarily the full story since their interpretation is based on the theoretical algorithm used in the calculations. In this connection it should be emphasized that Feynman's circulation theorem involving permutation rings is the only one which we possess for interpreting microscopic atomic many-particle coherent quantum mechanics. [152] This problem of interpretation is nicely illustrated by a recent alternative explanation of the differences in the OCS spectra measured in pure ^{4}He and in pure ^{3}He droplets presented by Babichenko and Kagan. [153] According to their theory these differences are directly related to the different wavelengths of the elementary excitations in the droplets. In ^{4}He they are dominated by phonons with wavelengths much greater than the size of the molecule and thus their effect on the rotations of the molecules is very small. In ^{3}He where the low frequency excitations are dominated by the creation of particle-hole pairs, which have much shorter wavelengths, their interaction with the molecule is much stronger. [154] Thus the differences in the Bose and Fermi statistics as reflected in the elementary excitations can explain the observations without invoking any of the hydrodynamic concepts. Kagan has also adapted this model to explain the return of free rotations when 60 ^{4}He atoms are added to the ^{3}He droplets. [155] This explanation is completely different than provided by the PIMC calculations discussed in Subsection 5.1.3. Thus there are still many open issues in connection with these droplet experiments, as, in fact, is also the case for the many *macroscopic* superfluid phenomena.

In this connection it is of some interest to point out that there are additional experiments in finite-sized droplets which reveal superfluid phenomena which are more closely related to our conventional macroscopic understanding. In one recent experiment the transmission of ^{3}He atoms through large ^{4}He droplets was observed. [156] In this experiment the free particle behavior of ^{3}He atoms inside superfluid ^{4}He, which is the basis of helium dilution refrigerators, has been confirmed for free droplets. These observations are indirectly related to the frictionless motion of microscopic particles moving at effective velocities below the Landau velocity of 58 m/sec. The only previous experiments which demonstrated frictionless motion of microscopic particles were carried out with ions. [157] Similar experiments are not possible with droplets because all ions, closed shell neutral molecules and atoms have much great binding energies to the droplets. If these particles would have sufficiently low kinetic energies to enable them to move inside the droplets without friction then they would not be able to leave the droplets once they reach the droplet surface. Since ^{3}He atoms have a binding energy of only 2.5 K to bulk liquid ^{4}He there is only a narrow window of incident energies below the roton minimum of 8.5 K and greater than their binding energy in which the ^{3}He atoms can pass through the liquid ^{4}He without creating excitations and also escape the droplet. These were just the conditions under which the frictionless transmission of ^{3}He atoms was recently found experimentally. [156]

Another experiment involves resonantly injecting [158] single electrons into large He droplets where they form large metastable bubbles. Motivated by the observation that these bubble lifetimes were of the order of only 10^{-3} sec compared to predicted lifetimes of 10^{12} sec this behavior was carefully studied in a dedicated beam experiment on both ^{3}He and ^{4}He droplets; with and without a superimposed electric field. [159,160] The 15 orders of magnitude shorter lifetimes found in the ^{4}He droplets could then be interpreted in terms of a frictionless bouncing back and forth of the electron bubbles inside the superfluid ^{4}He droplets, while in the ^{3}He droplets their motion was viscously impeded. [159,160] These microscopic experiments thus reveal an atomic behavior consistent with Landaus original idea of superfluidity involving frictionless motion at velocities below a critical velocity. They do suggest, as do the earlier experiments on ion transmission through the bulk superfluid, [157] that some of the macroscopic concepts also carry over to the microscopic world.

Finally, it is tempting to speculate about future developments in the field of He droplet spectroscopy. One very recent heuristic experimental accomplishment has been the improvement of pulsed nozzles to achieve high densities and sub-degree Kelvin temperatures in free jet seeded beams. [162] With this technique the group of Even and Jortner have been able to produce clusters of chromophore molecules with up to 20 attached ^{4}He atoms. [163] This development, which has great potential for further development, will hopefully soon enable the gap in the spectroscopy of molecules in helium from small free He clusters to the clusters and droplets described in this review to be bridged.

The question of whether droplets can support vortices is still not resolved. [164] Recent calculations have shown that embedded molecules are expected to pin down and stabilize vortices in the droplet interior. [161] However since all the high resolution spectra could be explained without invoking the presence of vortices there is no completely convincing evidence for their presence in droplets at the present time. Perhaps, however, fogs containing droplets, which have recently been produced by piezo-driven transducers just under the liquid helium surface in cryostats, [165] may provide a way to manipulate droplets using external fields in such a way that vortices can be created and their decay can be followed?

Another avenue for creating large superfluid drops has recently been discovered in our laboratory. There instead of expanding the gas to produce droplets (see Subsection 2.1) superfluid liquid ^{4}He at source temperatures T_o less than the superfluid transition temperature of $T_\lambda = 2.2$ K is squirted at high pressure into vacuum. [166] The preliminary results suggest that an intense beam of large micron-sized droplets are produced. Because of the large size of these droplets it may be possible to equilibrate them to an external gas pressure and thereby influence their equilibrium temperatures. In this way perhaps in the future it may be possible to bridge the temperature gap for spectroscopy between the superfluid and normal fluid phases.

Thus there are many new horizons opening up. Judging by past experience we can be confident that the enigmatic liquid, called superfluid helium, still has many wonderful surprises in store.

Acknowledgments

The research reported in this review was carried out in close collaboration with Andrey Vilesov and Boris Sartakov. We have profitted greatly from many discussions with Franco Gianturco, Yuri Kagan and Birgitta Whaley. Eckhard Krotscheck and Moses Chan provided several important references. The editors and also F. Müller-Hoissen have facilitated and helped in the final editing of the electronic version of the manuscript. My thanks go to all of the above as well as to the many students who carried out the experiments.

References

1. J.P. Toennies and A.F. Vilesov, *Ann. Rev. Phys. Chem.* **49**, 1 (1998).
2. S. Grebenev, M. Hartmann, A. Lindinger, N. Pörtner, B. Sartakov, J.P. Toennies and A.F. Vilesov, *Physica* **B280**, 65 (2000).
3. J.P. Higgins, J. Reho, F. Stienkemeier, W.E. Ernst, K.K. Lehmann and G. Scoles in: Atomic and Molecular Beams (R. Campange, editor), Springer, Berlin, 2001, p. 755.
4. E. Lugovoi, J.P. Toennies, S. Grebenev, N. Pörtner, A.F. Vilesov and B. Sartakov in: Atomic and Molecular Beams (R. Campargue, editor) Springer, Berlin, 2001, p. 31
5. J.P. Toennies, K.B. Whaley, and A.F. Vilesov, *Physics Today*, Feb. 2001.
6. F. Stienkemeier and A.F. Vilesov, *J. Chem. Phys.*, submitted.
7. C. Callegari, I.K. Lehmann, R. Schmied and G. Scoles, *J. Chem. Phys.*, submitted.
8. J. Northby *J. Chem. Phys.*, submitted.
9. P. Kapitsa, *Nature* **141**, 74 (1938).
10. J.F. Allen and A.D. Misener, *Nature* **141**, 75 (1938).
11. F. London, Superfluids, Vol. II, John Wiley, New York, 1954, page xi.
12. C.J. Pethick and D.G. Ravenhall, *Annu. Rev. Nucl. Part. Sci.* **45**, 429 (1995).
13. A.B. Migdal, *Soviet. Phys. JETP* **37**, 176 (1960).
14. For a discussion of bosonic phenomena see Bose-Einstein Condensation (A. Griffin, B.W. Snoke and S. Stringari, eds.) Cambridge University Press, Cambridge, UK 1995.
15. L. Landau, *J. Phys. USSR* **5**, 71 (1941); L. Landau, *J. Phys. USSR* **11**, 91 (1947).
16. L. Tisza, *Nature* **141**, 913 (1938); *Compt. Rend. Acad. Sci. Paris* **207**, 1035 (1938); *ibid.* **207**, 1186 (1938).
17. H.B. Ghassib and G.V. Chester, *J. Chem. Phys.* **81**, 585 (1984).
18. F. Dalfovo, S. Giorgini, L.P. Pitaevskii, and S. Stringari, *Rev. Mod. Phys.*, **71**, 463 (1999).
19. M. Chan, N. Mulders and J. Reppy, *Physics Today*, August 1996.
20. see for example A.H. Wootters, M.P. Lilly and R.B. Hallock, *J. Low Temp. Phys.*, **110**, 561 (1998).
21. R.B. Hallock, *Physics Today*, June 1998.

22. I. Silvera, *Phys. Rev.* **B29**, 3899 (1984).
23. R.E. Boltnev, E.B. Gordon, V.V. Khmelenko, I.N. Krushinskaya, M.V. Martynenko, A.A. Pelmenev, E.A. Popov, A.F. Shestakov, *Chem. Phys.* **189**, 367 (1994); *J. Chim. Phys.* **92**, 362 (1995).
24. B. Tabbert, H. Günther, G. zu Putlitz, *J. Low Temp. Phys.* **109**, 653 (1997).
25. A. Scheidemann, J.P. Toennies and J.A. Northby, *Phys. Rev. Lett.* **64**, 1899 (1990).
26. A. Scheidemann, B. Schilling and J.P. Toennies, *J. Phys. Chem.* **97**, 2128 (1993).
27. S. Goyal, D.L. Schutt and G. Scoles, *Phys. Rev. Lett.* **69**, 933 (1992); *J. Phys. Chem.* **97**, 2236 (1993).
28. M. Hartmann, R.E. Miller, J.P Toennies and A.F. Vilesov, *Science* **272**, 1631 (1996).
29. K.E. Kürten and M.L. Ristig, *Phys. Rev.* **B31**, 1346 (1985).
30. F. Dalfovo, *Z. Phys.* **D29**, 61 (1994).
31. J. Gspann in Physics of Electronic and Atomic Collisions (S. Datz, editor) North Holland, Amsterdam 1982, pp. 79-96.
32. A.P.J. van Deursen and J. Reuss, *J. Chem. Phys.* **63**, 4559 (1975); P.W. Stephens and J.G. King, *Phys. Rev. Lett.* **51**, 1538 (1983).
33. H. Buchenau, E.L. Knuth, J. Northby, J.P. Toennies and C. Winkler, *J. Chem. Phys.* **92**, 6875 (1990).
34. J. Farges, M.F. de Feraudy, B. Raoult, and G. Torchet, *Surf. Sci.* **106**, 95 (1981).
35. J. Lekner, *Mol. Physics* **23**, 619 (1972).
36. R.E. Grisenti, W. Schöllkopf, J.P. Toennies, G.C. Hegerfeldt, T. Köhler and M. Stoll, *Phys. Rev.Lett.* **85**, 2294 (2000).
37. F. Luo, G. Kim, G.C. Mc Bane, C.F. Giese and W.R. Gentry, *J. Chem. Phys.* **98**, 9687 (1993).
38. J.P. Toennies and K. Winkelmann, *J. Chem. Phys.* **66**, 3965 (1977).
39. J. Wang, V.A. Shamamian, B.R. Thomas, J.M. Wilkinson, J. Riley, C.F. Giese and W.R. Gentry, *Phys. Rev. Lett.* **60**, 969 (1988).
40. D.M. Brink and S. Stringari, *Z. Phys.* **D15**, 257 (1996).
41. A. Guirao, M. Pi and M. Barranco, *Z. Phys.* **D21**, 185 (1991).
42. J. Harms, M. Hartmann, J.P. Toennies, A.F. Vilesov and B. Sartakov, *J. Mol. Spectr.* **185**, 204 (1997).
43. J. Harms, M. Hartmann, B. Sartakov, J.P. Toennies and A.F. Vilesov, *J. Chem. Phys.* **110**, 5124 (1999).
44. M. Barranco, M. Pi, S.M. Gatica, E.S. Hernández and J. Navarro, *Phys. Rev.* **B56**, 8997 (1997).
45. M. Lewerenz, B. Schilling and J.P. Toennies, *Chem. Phys. Lett.* **206**, 381 (1993).
46. For a review on log-normal distributions see A.L. Koch, *J. Theoret. Biol.* **12**, 276 (1966).
47. M. Hartmann, N. Pörtner, B. Sartakov, J.P. Toennies and A.F. Vilesov, *J. Chem. Phys.* **110**, 5109 (1999).
48. J. Wörmer, R. Karnbach, M. Joppien and T. Möller, *J. Chem. Phys.* **104**, 8269 (1996).
49. O.F. Hagena, *Z. Phys.* **D4**, 291 (1987).
50. E.L. Knuth, *J. Chem. Phys.* **107**, 9125 (1997).
51. J. Harms, J.P. Toennies and F. Dalfovo, *Phys. Rev.* **B58**, 3341 (1998).
52. J. Harms, J.P. Toennies, M. Barranco and M. Pi, *Phys. Rev.* **B63**, 184513-1 (2001).
53. M. Lewerenz, B. Schilling and J.P. Toennies, *J. Chem. Phys.* **102**, 8191 (1995).
54. B.E. Callicoatt, K. Förde, L.F. Jung, T. Ruchti, and K.C. Janda, *J. Chem. Phys.* **109**, 10195 (1998); M. Ovchinnikov, B.L. Grigorenko, K.C. Janda, and V.A. Apkarian, *J. Chem. Phys.* **108**, 9351 (1998).
55. T. Ruchti, K. Förde, B.E. Callicoatt, H. Ludwigs and K.C. Janda, *J. Chem. Phys.*

109, 10679 (1998).
56. B.E. Callicoatt, K. Förde, T. Ruchti, L. Jung, K.C. Janda, *J. Chem. Phys.* **108**, 9371 (1998).
57. T. Ruchti, B.E. Callicoatt and K.C. Janda, *Phys. Chem. Chem. Phys.* **2**, 2075 (2000).
58. The apparatus was first described in R. Fröchtenicht, J.P. Toennies and A.F. Vilesov, *Chem. Phys. Lett.* **229**, 1 (1994).
59. S. Goyal, D.L. Schutt and G. Scoles, *Acc. Chem. Res.* **26**, 123 (1993).
60. K. Nauta, D.T. Moore and R.E. Miller, *Faraday Discuss.* **113**, 261 (1999).
61. C. Callegari, A. Conjusteau, I. Reinhard, K.K. Lehmann and G. Scoles, *J. Chem. Phys.* **113**, 10535 (2000).
62. F. Huisken, *Adv. Chem. Phys.* **81**, 63 (1992).
63. D. Blume, M. Lewerenz, F. Huisken and M. Kaloudis, *J. Chem. Phys.* **105**, 8666 (1996).
64. S. Grebenev, J.P. Toennies and A.F. Vilesov, *Science* **279**, 2083 (1998).
65. S. Grebenev, M. Hartmann, M. Havenith, B. Sartakov, J.P. Toennies and A.F. Vilesov, *J. Chem. Phys.* **112**, 4485 (2000).
66. M. Hartmann, R.E. Miller, J.P. Toennies and A. Vilesov, *Phys. Rev. Lett.* **75**, 1566 (1995).
67. A. Bartelt, J.D. Close, F. Federmann, N. Quaas and J.P. Toennies, *Phys. Rev. Lett.* **77**, 3525 (1996); F. Federmann, K. Hoffmann, N. Quaas and J.D. Close, *Phys. Rev. Lett.* **83**, 2548 (1999).
68. J. Reho, U. Merker, M.R. Radcliff, K. Lehmann and G. Scoles, *J. Phys. Chem.* **A104**, 3620 (2000).
69. A. Bartelt, J.D. Close, F. Federmann, K. Hoffmann, N. Quaas and J.P. Toennies, *Z. Phys.* **D39**, 1 (1997).
70. J. Reho, U. Merker, M.R. Radcliff, K. Lehmann and G. Scoles, *J. Chem. Phys.* **112**, 8409 (2000).
71. M. Hartmann, A. Lindinger, J.P. Toennies and A.F. Vilesov, *J. Phys. Chem.* **A105**, 6369 (2001).
72. A. Lindinger, J.P. Toennies and A.F. Vilesov, *Phys. Chem. Chem. Phys.* **3**, 2581 (2001).
73. J.D Close, F. Federmann, K. Hoffmann and N. Quaas, *Chem. Phys. Lett.* **276**, 393 (1997).
74. M. Hartmann, A. Lindinger, J.P. Toennies and A.F. Vilesov, to be published.
75. A. Slenczka, B. Dick, M. Hartmann, and J.P. Toennies, *J. Chem. Phys.* submitted.
76. A. Lindinger, J.P. Toennies and A.F. Vilesov, *J. Chem. Phys.* **110**, 1429 (1999).
77. A. Lindinger, E. Lugovoi, J.P. Toennies and A.F. Vilesov, *Z. Phys. Chem.* **215**, 401 (2001).
78. F. Stienkemeier and M. Wewer, private communication; see also Ref. 6.
79. F. Stienkemeier, J. Higgins, C. Callegari, S.I. Kanorsky, W.E. Ernst and G. Scoles, *Z. Phys.* **D38**, 253 (1996).
80. J. Higgins, C. Callegari, J. Reho, F. Stienkemeier, W.E. Ernst, M. Gutowsky and G. Scoles, *J. Phys. Chem.* **A102**, 4952 (1998).
81. J. Higgins, W.E. Ernst, C. Callegari, J. Reho, K.K. Lehmann, G. Scoles and M. Gutowski, *Phys. Rev. Lett.* **77**, 4532 (1996); *Science* **273**, 629 (1996).
82. L.S. Lewart, V.R. Pandharipande and S.C. Pieper, *Phys. Rev.* **B37**, 4950 (1988).
83. E. Krotscheck, *Phys. Rev.* **B32**, 5713 (1985).
84. P. Sindzingre, M.L. Klein and D.M. Ceperley, *Phys. Rev. Lett.* **63**, 1601 (1989).
85. E.L. Andronikashvili, *J. Phys. USSR* **10**, 201 (1946).
86. M.V Rama Krishna and K.B. Whaley, *Phys. Rev. Lett.* **64**, 1126 (1990).

87. C.E. Campbell, *J. Low Temp. Phys.* **93**, 907 (1993).
88. S.A. Chin, *J. Low Temp. Phys.* **33**, 921 (1993).
89. E. Cheng, M.A. McMahon, and K.B. Whaley, *J. Chem. Phys.* **104**, 2669 (1996).
90. J. Lekner, *J. Phys.: Cond. Matt.* **12**, 4327 (2000).
91. M. Pi, R. Mayol, and M. Barranco, *Phys. Rev. Lett.* **82**, 3093 (1999).
92. Y. Kwon, P. Huang, M.V. Patel, D. Blume and K.B. Whaley, *J. Chem. Phys.* **113**, 6469 (2000).
93. E.B. Gordon and A.F. Shestakov, *Low Temp. Phys.* **26**, 1 (2000).
94. C. Callegari, A. Conjusteau, I. Reinhard, K.K. Lehmann, G. Scoles and F. Dalfovo, *Phys. Rev. Lett.* **83**, 5058 (1999). Erratum: *Phys. Rev. Lett.* **84**, 1848 (2000).
95. Results from K. Nauta and R.E. Miller, reported in Ref. 94.
96. F. Paesani, F.A. Gianturco, K.B. Whaley, *J. Chem. Phys.*, submitted June 2001.
97. A. Bohr and B.R. Mottelson, Nuclear Structure, Vol. II Nuclear Deformations, W.A. Benjamin, Reading USA 1975, Appendix 6A.
98. M. Casas and S. Stringari, *J. Low Temp. Phys.* **79**, 135 (1990).
99. A. Tamura, *Phys. Rev.* **B53**, 14475 (1996); *Z. Phys.* **D40**, 225 (1997).
100. D. Eichenauer, A. Scheidemann and J.P. Toennies, *Z. Phys.* **D8**, 295 (1988).
101. J.P. Toennies and A.F. Vilesov, *Chem. Phys. Lett.* **235**, 596 (1995).
102. M.V. Rama Krishna and K.B. Whaley, *J. Chem. Phys.* **93**, 746 (1990).
103. S.A. Chin and E. Krotschek, *Phys. Rev.* **B52**, 10405 (1995).
104. R.Guardiola, J. Navarro and M. Portesi, *Phys. Rev.* **B63**, 224519 (2001).
105. C.M. Surko and F. Reif, *Phys. Rev.* **175**, 229 (1968).
106. M. Hartmann, F. Mielke, J.P. Toennies, A.F. Vilesov and G. Benedek, *Phys. Rev. Lett.* **76**, 4560 (1996).
107. L.J. van Ijzendoorn *et al.*, *J. Chem. Phys.* **85**, 1812 (1986).
108. R.I. Personov, in Spectroscopy and Excitation Dynamics of Condensed Molecular Systems (V.M. Agranovich and R.M. Hochstrasser, eds.), North Holland, 1983, Chapt. 10.
109. D.L. Dexter, *Solid State Phys.* **6**, 353 (1958).
110. M.H.L. Pryce, in Phonons in Perfect Lattices and in Lattices with Point Imperfections, (R.W.H. Stevenson, editor), Oliver and Boyd, Edingburgh and London, 1966, p. 403.
111. D.G. Henshaw and A.D.B. Woods, Proc. 7th Int. Conf. Low Temp. Phys. (G.M. Graham and A.C. Hollis Hallett, eds.), Univ. Toronto Press, Toronto, 1961, p. 539.
112. H.R. Glyde and E.C. Svensson, *Methods of Expt. Phys.* **23B**, 303 (1987).
113. N. Pörtner, Dissertation, Göttingen University, 2000.
114. G. Schweizer, Dissertation, Bonn University, 1983.
115. R.T. Pack, E. Piper, G.A. Pfeiffer and J.P. Toennies, *J. Chem. Phys.* **80**, 4940 (1984).
116. R. Fröchtenicht, M. Kaloudis, M. Koch, and F. Huisken, *J. Chem. Phys.* **105**, 6128 (1996).
117. M. Behrens, U. Buck, R. Fröchtenicht, M. Hartmann, F. Huisken and F. Rohmund, *J. Chem. Phys.* **109**, 5914 (1998).
118. N. Pörtner, J.P. Toennies and A. Vilesov, to be published.
119. A. Conjusteau, C. Callegari, I. Reinhard, K.K. Lehmann and G. Scoles, *J. Chem. Phys.* **113**, 4840 (2000).
120. G.A. Csáthy and M.H.W. Chan, *J. Low. Temp. Phys.* **121**, 451 (2000).
121. D.A. Crowell, F.W. van Keuls, and J.D. Reppy, *Phys. Rev.* **B55**, 12620 (1997).
122. H.J. Lauter, H. Godfrin and P. Leiderer, *J. Low. Temp. Phys.* **87**, 425 (1992); B.E. Clements, H. Godfrin, E. Krotscheck, H.J. Lauter, P. Leiderer, V. Passiouk and C.J. Tymczak, *Phys. Rev.* **B53**, 12242 (1996).

123. M. Hartmann, N. Pörtner, B. Sartakov, J.P. Toennies and A.F. Vilesov, *J. Chem. Phys.* **110**, 5109 (1998).
124. R.G. Gordon, *J. Chem. Phys.* **44**, 1830 (1966).
125. G. Herzberg, Molecular Spectra and Molecular Structure, II. Infrared and Raman spectra of Polyatomic molecules, D. Van Nostrand Co., Princeton, 1945, p. 532.
126. D.O. Edwards and M.S. Pettersen, *J. Low Temp. Phys.* **87**, 473 (1992); J. He, T.G. Culman, H.H. Hjort, D.O. Edwards, *Phys. Rev. Lett.* **80**, 3292 (1998).
127. G.A. Csáthy and M.H.W. Chan, *Phys. Rev. Lett.* **87**, 045301-1 (2001).
128. S. Grebenev, Dissertation, Göttingen University, 2000.
129. S. Grebenev, B. Sartakov, J.P. Toennies and A.F. Vilesov, in preparation.
130. D. Ceperley, *Physics World*, p. 19, June 1998.
131. K. Higgins and W. Klemperer, *J. Chem. Phys.* **110**, 1383 (1999).
132. F. Paesani, F.A. Gianturco and K.B. Whaley, *J. Chem. Phys.*, submitted.
133. K.K. Lehmann, *J. Chem. Phys.* **114**, 4643 (2001).
134. F. Paesani and F.A. Gianturco, *Europhys. Lett.*, submitted.
135. S. Grebenev, E. Lugovoi, B.G. Sartakov, J.P. Toennies and A.F. Vilesov, *Faraday Discuss.* **118**, 19 (2001).
136. S. Grebenev, B. Sartakov, J.P. Toennies and A.F. Vilesov, *J. Chem. Phys.* **114**, 617 (2001).
137. S. Grebenev, B. Sartakov, J.P. Toennies and A.F. Vilesov, in preparation.
138. N. Pörtner, J.P. Toennies and A.F. Vilesov, in preparation.
139. For an excellent discussion on the role of backflow in hydrodynamics see: G. Birkhoff, Hydrodynamics, Princeton University Press, Princeton, 1960 p. 148 ff.
140. D.M. Ceperley, *Rev. Mod. Phys.* **67**, 270 (1995).
141. K. Higgins and W. Klemperer, private communication.
142. W. Gordy and R.L. Cook, Microwave Molecular spectra, Wiley, New York, 1984, Chapter XIII.
143. S. Grebenev, B. Sartakov, J.P. Toennies and A.F. Vilesov, to be submitted.
144. J. Tang and A.R.W. McKellar, to be published.
145. L. Landau and E.M. Lifshitz, Quantum Mechanics, Second edition Pergamon Press, Oxford, 1965, paragraph 105 p. 394ff.
146. The van der Waals potential parameters for He-He are $\epsilon = 10.8$ K and $R_m = 2.94$ Å [K.T. Tang, J.P. Toennies and C.L. Yiu, *Phys. Rev. Lett.* **74**, 1546 (1995)], while those for He-H_2 are 13.2 K and 3.07 Å, respectively [F.M. Tao, *J. Chem. Phys.* **100**, 4946 (1994)].
147. V.L. Ginzburg and A.A. Sobyanin, *JETP Lett.* **15**, 242 (1972).
148. H.J. Maris, G.M. Seidel, and T.E. Huber, *J. Low Temp. Phys.* **51**, 471 (1983).
149. P. Sindzingre, D.M. Ceperley, M.L. Klein, *Phys. Rev. Lett.* **67**, 1871 (1991).
150. M.C. Gordillo and D.M. Ceperley, *Phys. Rev. Lett.* **79**, 3010 (1997).
151. S. Grebenev, B. Sartakov, J.P. Toennies and A.F. Vilesov, *Science* **289**, 1532 (2000).
152. R.P. Feynman, *Phys. Rev.* **94**, 262 (1954).
153. V.S. Babichenko and Yu. Kagan, *Phys. Rev. Lett.* **83**, 3488 (1999).
154. M. Morshita, T. Kurada, A. Sawada and T. Saton, *J. Low Temp. Phys.* **76**, 387 (1989).
155. Y. Kagan, private communication.
156. J. Harms and J.P. Toennies, *Phys. Rev. Lett.* **83**, 344 (1999).
157. P.V.E. McClintock and R.M. Bowley, *Progr. Low Temp. Phys.*, Vol. XIV (W.P. Halperin, ed.) Elsevier, 1995.
158. U. Henne and J.P. Toennies, *J. Chem. Phys.* **108**, 9327 (1998).
159. M. Fárnik, U. Henne, B. Samelin and J.P. Toennies, *Phys. Rev. Lett.* **81**, 3892 (1998).

160. M. Fárnik, B. Samelin and J.P. Toennies, *J. Chem. Phys.* **110**, 9195 (1999).
161. F. Dalfovo, R. Mayol, M. Pi and M. Barranco, *Phys. Rev. Lett.* **85**, 1028 (2000).
162. U. Even, I. Al-Hroub and J. Jortner, *J. Chem. Phys.* **115**, 2069 (2001).
163. U. Even, J. Jortner, D. Noy, C. Cossart-Magos and N. Lavi, *J. Chem. Phys.* **112**, 8068 (2000).
164. J.D. Close, F. Federmann, K. Hoffmann and N. Quaas, *J. Low Temp. Phys.* **111**, 661 (1998).
165. H. Kim, K. Seo, B. Tabbert, and G.A. Williams, *J. Low Temp. Phys. 121*, 621 (2000).
166. R. Grisenti and J.P. Toennies, in preparation.

Index